Lecture Notes in Computer Science   9612

Commenced Publication in 1973
Founding and Former Series Editors:
Gerhard Goos, Juris Hartmanis, and Jan van Leeuwen

More information about this series at http://www.springer.com/series/7409

Elisabeth Métais · Farid Meziane
Mohamad Saraee · Vijayan Sugumaran
Sunil Vadera (Eds.)

Natural Language Processing and Information Systems

21st International Conference on Applications
of Natural Language to Information Systems, NLDB 2016
Salford, UK, June 22–24, 2016
Proceedings

 Springer

Editors
Elisabeth Métais
Conservatoire National des Arts et Métiers
Paris
France

Farid Meziane
University of Salford
Salford
UK

Mohamad Saraee
University of Salford
Salford
UK

Vijayan Sugumaran
Oakland University
Rochester, MI
USA

Sunil Vadera
University of Salford
Salford
UK

ISSN 0302-9743   ISSN 1611-3349   (electronic)
Lecture Notes in Computer Science
ISBN 978-3-319-41753-0   ISBN 978-3-319-41754-7   (eBook)
DOI 10.1007/978-3-319-41754-7

Library of Congress Control Number: 2016942797

LNCS Sublibrary: SL3 – Information Systems and Applications, incl. Internet/Web, and HCI

Printed on acid-free paper

This Springer imprint is published by Springer Nature
The registered company is Springer International Publishing AG Switzerland

Preface

This volume of *Lecture Notes in Computer Science* (LNCS) contains the papers presented at the 21st International Conference on Application of Natural Language to Information Systems, held at MediacityUK, University of Salford, during June 22–24, 2016 (NLDB 2016). Since its foundation in 1995, the NLDB conference has attracted state-of-the-art research and followed closely the developments of the application of natural language to databases and information systems in the wider meaning of the term.

The NLDB conference is now a well-established conference that is attracting participants from all over the world. The conference evolved from the early years when most of the submitted papers where in the areas of natural language, databases and information systems to encompass more recent developments in the data and language engineering fields. The content of the current proceedings reflects these advancements. The conference also supports submissions on studies related to languages that were not well supported in the early years such as Arabic, Tamil, and Farsi.

We received 83 papers and each paper was reviewed by at least three reviewers with the majority having four or five reviews. The conference co-chairs and Program Committee co-chairs had a final consultation meeting to look at all the reviews and make the final decisions on the papers to be accepted. We accepted 17 papers (20.5 %) as long/regular papers, 22 short papers, and 13 poster presentations.

We would like to thank all the reviewers for their time, their effort, and for completing their assignments on time despite tight deadlines. Many thanks to the authors for their contributions.

May 2016

Elisabeth Métais
Farid Meziane
Mohamad Saraee
Vijay Sugumaran
Sunil Vadera

Organization

Conference Chairs

Elisabeth Métais   Conservatoire National des Arts et Metiers, Paris, France
Farid Meziane   University of Salford, UK
Sunil Vadera   University of Salford, UK

Program Committee Chairs

Mohamad Saraee   University of Salford, UK
Vijay Sugumaran   Oakland University Rochester, USA

Program Committee

Hidir Aras   FIZ Karlsruhe, Germany
Imran Sarwar Bajwa   The Islamia University of Bahawalpur, Pakistan
Pierpaolo Baslie   University of Bali, Italy
Nicolas Béchet   IRISA, France
Chris Biemann   TU Darmstadt, Germany
Sandra Bringay   LIRMM, France
Johan Bos   Groningen University, The Netherlands
Goss Bouma   Groningen University, The Netherlands
Mihaela Bornea   IBM Research, USA
Cornelia Caragea   University of North Texas, USA
Christian Chiarcos   University of Frankfurt, Germany
Raja Chiky   ISEP, France
Kostadin Cholakov   HU Berlin, Germany
Philipp Cimiano   Universität Bielefeld, Germany
Isabelle Comyn-Wattiau   CNAM, France
Ernesto William De Luca   Potsdam University of Applied Sciences, Germany
Bart Desmet   Ghent University, Belgium
Zhou Erqiang   University of Electronic Science and Technology, China
Antske Fokkens   VU Amsterdam, The Netherlands
Vladimir Fomichov   National Research University Higher School of Economics, Russia
Thierry Fontenelle   CDT, Luxemburg
André Freitas   University of Passau, Germany/Insight, Ireland
Debasis Ganguly   Dublin City University, Ireland
Ahmed Guessoum   USTHB, Algiers, Algeria

Contents

Poster Papers

Full Papers

A Methodology for Biomedical Ontology Reuse

Nur Zareen Zulkarnain[1](✉), Farid Meziane[1], and Gillian Crofts[2]

[1] School of Computing, Science and Engineering, Informatics Research Centre,
University of Salford, Salford M5 4WT, UK
`N.Z.Zulkarnain@edu.salford.ac.uk`, `F.Meziane@salford.ac.uk`
[2] School of Health Science, University of Salford, Salford M6 6PU, UK
`G.Crofts@salford.ac.uk`

Abstract. The abundance of biomedical ontologies is beneficial to the
development of biomedical related systems. However, existing biomedical ontologies such as the National Cancer Institute Thesaurus (NCIT),
Foundational Model of Anatomy (FMA) and Systematized Nomenclature of Medicine-Clinical Terms (SNOMED CT) are often too large to be
implemented in a particular system and cause unnecessary high usage of
memory and slow down the system's processing time. Developing a new
ontology from scratch just for the use of a particular system is deemed as
inefficient since it requires additional time and causes redundancy. Thus,
a potentially better method is by reusing existing ontologies. However,
currently there are no specific methods or tools for reusing ontologies.
This paper aims to provide readers with a step by step method in reusing
ontologies together with the tools that can be used to ease the process.

Keywords: Ontology · Ontology reuse · Biomedical ontology ·
BioPortal

1 Introduction

Biomedical systems are an integral part of today's medical world. Systems such
as electronic patient records and clinical decision support systems (CDSS) have
played an important role in assisting the works of medical personnel. One area
that could benefit from the development of biomedical systems is ultrasound
reporting. In ultrasound, reports generated have more value compared to the
image captured during the examination [2]. Variations in ultrasound reporting
impacts the way a report is interpreted as well as in decision making. Thus,
the standardization of these reports is important. In order to achieve this goal,
ontologies are used to understand the reports and structure them according to
a certain format [16] as well as recognizing the relationships between the parts
of the text composing the report.

General and established domains such as medicine have existing ontologies
that cover the general concepts in the domain. Examples of these ontologies
include the National Cancer Institute Thesaurus (NCIT), Foundational Model

© Springer International Publishing Switzerland 2016
E. Métais et al. (Eds.): NLDB 2016, LNCS 9612, pp. 3–14, 2016.
DOI: 10.1007/978-3-319-41754-7_1

of Anatomy (FMA) and Systematized Nomenclature of Medicine-Clinical Terms (SNOMED CT). These ontologies however are often too large to be manipulated or processed in a specific application. Thus, a domain specific ontology is needed to solve this problem. Building a new domain specific ontology from scratch would not be efficient since this will cause redundancy and takes a lot of time. Thus, ontology reuse has been potentially seen as a better alternative. This paper discusses how ontology reuse has been done before and proposes a methodology to reuse ontologies together with the existing tools that can be used to ease the reuse process. The development of the Abdominal Ultrasound Ontology (AUO) as the knowledge base for an ultrasound reporting system developed by Zulkarnain et al. [16] is used in this paper to explain the proposed ontology reuse methodology.

2   Related Work

Ontology reuse can be defined as a process where a small portion of existing ontologies is taken as an input to build a new one [3]. The process of reusing large existing ontologies allows their use without slowing down the process of an application. Ontology reuse also increases interoperability [14]. Indeed, when an ontology is reused by several other new ontologies, interoperability between these ontologies can be achieved much easier since they share several features such as classes naming method and concept modelling.

Even though ontology reuse brings a lot of benefits, there are currently no tools that provide adequate support for the ontology reuse process [8,14] which hinders the effort of ontology reuse. There is also no one specific method agreed in reusing ontologies. Even so, most ontology reuse methodologies that have been used in previous works [1,4,5,11,12,14,15] falls along the line of these four steps: (i) Ontology selection for reuse, (ii) Concept selection, (iii) Concept customization and (iv) Ontology integration.

The first step for ontology reuse is to select the ontology to be reused. Ontology selection is done according to several criteria according to the needs of the new ontology, for example the language of the ontology, its comprehensiveness and its reasoning capabilities. Once the ontology for reuse is chosen, the next step would be to select the concepts that would be reused. One or several ontologies can be selected for reuse depending on the needs of the new ontology. Russ et al. [11] in their work merged two aircraft ontologies where most of its concepts were selected to develop a broader aircraft ontology. Shah et al. [12] on the other hand reused just one ontology; SNOMED CT where he selected the concepts needed then adds other relevant concepts not included in SNOMED CT.

Concepts selected are then translated into the same semantic language and then merged. In Caldarola et al.'s work [4] this includes manually translating metadata to better understand concepts. Alani [1] in developing his ontology has merged several ontologies that contain different properties for one same concept which resulted in additional knowledge representation. Several different concepts have also been selected from different ontologies which are then compared and merged. Finally, the ontology will be integrated into the system or application.

In this research, these four steps serve as a guideline in developing an ontology reuse methodology for the biomedical domain. The methodology proposed in this research will allow for the ontology to be reused from multiple existing ontologies and suggest tools that would help in each step of the methodology. The ontology developed, Abdominal Ultrasound Ontology (AUO), will serve two purposes in this research: (i) it will be used to standardize the development of ultrasound reports and enforce the use of standard terminology and (ii) to analyse the reports written in Natural Language (English free-text) with the aim of automatically transforming them into a structured format.

3   The Proposed Methodology

In developing a new ontology by reusing existing biomedical ones, proper planning and execution are important in order to ensure the modularity of the concepts reused. Thus, the ontology reuse methodology developed in this paper, adopted the general four steps mentioned in Sect. 2 and summarised in Fig. 1.

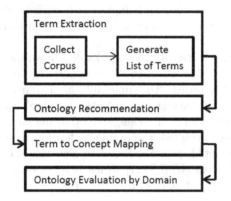

Fig. 1. Ontology reuse methodology

3.1   Term Extraction

The first step in ontology reuse or even in developing one from scratch is to decide on its scope and domain. In this case of developing the Abdominal Ultrasound Ontology (AUO), the scope and domain of the ontology is abdominal ultrasound. 49 sample ultrasound reports have been collected and used as the basis of our ontology corpus. These sample reports were obtained from the Radiology Departments in a large NHS Trust incorporating 4 regionally based hospitals in Manchester and Salford. Once we have our corpus, the next step is to extract relevant terms from the corpus to generate a list of terms for reuse. Two biomedical term extraction applications; (i) TerMine[1] and (ii) BioTex[2] have been used

---

[1] http://www.nactem.ac.uk/software/termine.
[2] http://tubo.lirmm.fr/biotex/.

Table 1. Comparison of biomedical term extraction using TerMine and BioTex

	TerMine	BioTex
Language	English	English
License	Open	Open
POS Tagger	GENIA Tagger/Tree Tagger	Tree Tagger
Terms found	241 (GENIA Tagger) 232 (Tree Tagger)	761
Extraction type	Multi-word extraction	Multi-word and single-word extraction

for the extraction. All 49 sample reports were submitted to both applications and the results are shown in Table 1.

From this comparison, BioTex was chosen as the better biomedical term extractor in this research because of its ability to extract more terms compared to TerMine. BioTex is an automatic term recognition and extraction application that allows for both multi-word and single-word extraction [7]. It is important that the term extractor is able to extract not only multi-word but also single-word terms.

For example, if the sentence "Unremarkable appearances of the liver with no intrahepatic lesions" was submitted to both applications, TerMine will only extract two multi-word terms "Unremarkable appearance" and "intrahepatic lesion" while BioTex will extract not only the two multi-word terms but also "liver" which is a single word term. If single-word terms such as "liver", "kidney" and "spleen" were not extracted, the ontology developed would be incomplete. Terms which are extracted from BioTex were also validated using the Unified Medical Language System (UMLS) [7] which is a set of documents containing health and biomedical vocabularies and standards.

3.2   Ontology Recommendation

The next step after obtaining a list of terms for ontology reuse would be to select the suitable ontology to be reused. Three important criteria were used for selecting the ontology in this research: (i) Ontology coverage - To which extend does the ontology covers the terms extracted from the corpus? (ii) Ontology acceptance - Is the ontology being accepted in the medical field and how often is it used? and (iii) Ontology language - Is the ontology written in OWL, OBO or other semantic languages? Initial review resulted in choosing FMA, SNOMED-CT and RadLex as suitable candidates because of their domain coverage, acceptance in the biomedical community and language which is OWL. In order to verify this, an ontology recommender was developed using BioPortal's ontology recommender API[3] which is an open ontology library that contains ontologies with domains that range from anatomy, phenotype and chemistry to experimental conditions [10].

---

[3] http://bioportal.bioontology.org.

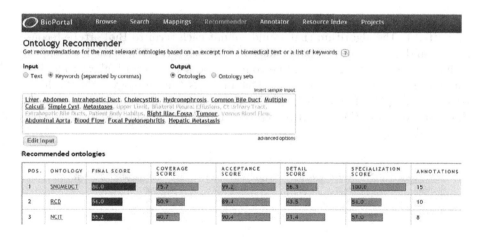

Fig. 2. BioPortal's ontology recommender

BioPortal has an ontology recommender available on its portal that can be used to obtain suggestions on suitable ontology to be reused for certain corpus. The ontology recommender makes a decision according to the following three criteria: (i) Coverage - Which ontology provides most coverage to the input text?, (ii) Connectivity - How often the ontology is mapped by other ontologies? and (iii) Size - Number of concepts in the ontology [6]. When a list of terms is submitted to the recommender, it will give a recommendation of 25 ontologies which are ranked from the highest to lowest scores (see Fig. 2). The final score is calculated based on the following formula:

$$FinalScore = (CoverageScore * 0.55) + (AcceptanceScore * 0.15)$$
$$+ (KnowledgeDetailScore * 0.15) + (SpecializationScore * 0.15) \quad (1)$$

The coverage score is given based on the number of terms in the input that are covered by the ontology. The acceptance score indicates how well-known and trusted the ontology is in the biomedical field. Knowledge detail score on the other hand indicates the level of details in the ontology; i.e. does the ontology have definitions, synonym or other details. Specialization score is given based on how well the ontology covers the domain of the input. An example is given in Fig. 2 where 21 terms where submitted. There is however a limitation in using it on the portal whereby it only allows for 500 words to be submitted. This limitation has prompted us to develop our own recommender by manipulating the data from BioPortal's ontology recommender API. We first develop the recommender that would give 25 ontology recommendations just like how it would be in the BioPortal's recommender. However, it seems that 761 terms were too big for the recommender's server to handle. Because of this, a recommender that would suggest ontology for each term was developed.

A list of terms, in this case the 761 terms that have been extracted, are submitted to the algorithm that would submit each term to BioPortal's recommender and get ontology recommendations for each term. Then, the frequency of each ontology recommended will be counted and sorted from highest to lowest. The recommender has ranked NCI Thesaurus as the ontology with the highest frequency (341) followed by SNOMED CT (140) and RadLex (37). Figure 3 shows an excerpt of the result from processing 761 terms using the recommender we have developed.

(a)

1	Keywords	Ontology Name	Score
2	liver	NCIT	0.91
3	abdomen	NCIT	0.882
4	gallbladder	NCIT	0.938
5	us	0	0
6	pancreas	NCIT	0.847
7	spleen	NCIT	0.911
8	kidneys	OMIM	0.779
9	portal vein	NCIT	0.868
10	cbd	CCO	0.89
11	duct	NCIT	0.883
12	dilatation	SNOMEDCT	0.85
13	lesion	NCIT	0.879
14	yr old male	NCIT	0.906
15	duct dilatation	MP	0.78
16	portal	NIFSTD	0.855
17	size	NCIT	0.851

(b)

1	Ontology Name	Frequency
2	NCIT	341
3	SNOMEDCT	140
4	RADLEX	37
5	MESH	30
6	LOINC	21
7	RCD	19
8	MEDDRA	12
9	OMIM	8
10	CRISP	8
11	SWEET	8
12	CCO	7
13	DCM	7
14	SOPHARM	6
15	NIFSTD	4
16	NMR	4
17	HUPSON	4

Fig. 3. (a) Ontology recommendation for each term, (b) Ranking of ontology recommended

3.3   Term to Concept Mapping

Once the ontology for reuse has been selected, the next step in building the abdominal ultrasound ontology is to map the terms extracted to concepts in the ontology which is done by referring to the result from BioPortal's Search API. The API allows us to insert several parameters to perform concept search which in this case, the parameters used are "q" to specify the term that we would like to search for, and "ontologies" which specifies the ontology where we would like to look for the term. Once these parameters have been submitted, the API will return a concept if there is a match with the term submitted. The concept will be returned with several other properties such as the preferred label, definition, synonym, match type and the terms relationship with its children, descendant, parents and ancestors.

In previous works by Mejino et al. [9] and Shah et al. [13], term to concept mapping was done by referring to the existing ontology and mapping it into the new one by deleting and adding concepts in the ontology to make it complete. Using BioPortal's API consumes less time and work as the terms are queried according to the provided parameters. This will also ensure the accuracy of the relationship between concepts and its children, descendant, parents and ancestors since there are links that can be clearly seen in the API result.

There was an intention to auto populate these data into Protègè (the OWL editor that was used in this research) by taking advantage of the option of saving the results in XML compared to JSON. However, there are two reasons why this is not possible at the moment. The first reason was that data from the API does not give the complete properties of a concept. For example, parents and ancestors were provided as links which makes it hard for the data to be manipulated since the properties of the parents and ancestors can only be obtained after the link is visited. The second reason is there are terms which matched several concepts in the ontology. For example, the term "calculus" could mean "branch of mathematics concerned with calculation" or "an abnormal concretion occurring mostly in the urinary and biliary tracts, usually composed of mineral salts". Thus, it is important to know in which context it is being used in order to adopt the correct meaning.

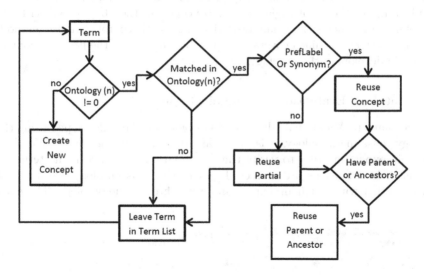

Fig. 4. Term to concept mapping guide

In deciding whether a term should be reused or not, the term to concept mapping guide (see Fig. 4) was used. Firstly, a term from the term list will be queried using the Search API in the ontology with the highest frequency which in this case is NCIT. If there is a match, we will see whether the match is a preferred label, synonym or partial match. A preferred label (PrefLabel) match means that the API found a concept that has an exact match to the term while synonym match means that the term is found as a synonym to the concept. Partial match on the other hand means that there is no exact match for the term but there are at least two concepts that match the term. For example, for the term "intrahepatic biliary", there are no concepts that match the term exactly.

However, there is the concept "intrahepatic" which is an anatomy qualifier in NCIT and the concept "duct" in NCIT which is an organ that matches.

If the match is a PrefLabel or synonym match, the concept will be reused. If the match is partial, the concepts that make up the term will also be reused. However, the term would still remain in the term list so that it could be compared to concepts in other ontologies. After the concept has been reused, we will find out if the term has a parent or ancestors. If there is, the parent or ancestors will also be reused. Once all terms have been searched, this process will then be repeated for the remaining recommended ontologies.

In this research, all terms are first searched in NCIT followed by SNOMED CT and RadLex. The Abdominal Ultrasound Ontology modelling follows the modelling of NCIT since it is the main ontology being reused. When merging ontologies from SNOMED CT and RadLex into the ontologies reused from NCIT, we would first find a parent that would be suitable for the concept. If no such parent exists, the parent and ancestors of the concept will then be reused. This is done to ensure the modularity of the ontology developed. If no match is found in any of these ontologies, a new concept will then be created with the help of domain experts. The main objective of using this ontology reuse methodology is to achieve as much coverage as possible and reduce the need for domain experts in developing the ontology.

3.4   Ontology Evaluation by Domain Expert

Once a complete Abdominal Ultrasound Ontology has been developed using the ontology reuse methodology, it is important that the ontology be evaluated by a domain expert in order to verify that the relationship between the terms as well as their definitions are correct. In evaluating this ontology, we have sat down together with a domain expert and went through the ontology. There are

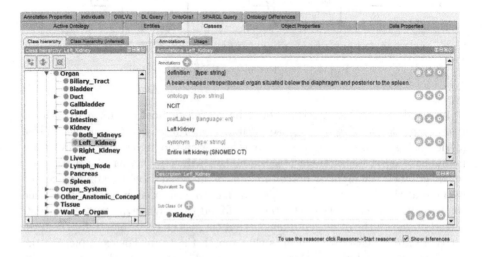

Fig. 5. Snapshot of the Abdominal Ultrasound Ontology

some corrections that need to be done but overall, the domain expert believes that the 92.6 % ontology coverage is enough to cover all the important concepts that an abdominal ultrasound report would need. For the other 7.4 % terms that have no match in the ontology, some of it were caused by human error whereby spelling mistakes were made by the reporter. As for the rest of it, the domain expert will help in giving definitions and suggestions on where it would fit in the ontology. Out of the 7.4 % terms that have no match in the ontology, there are also several terms that the domain expert believes we can omit since these words should not be in an ultrasound report for good practice. Examples of such words are "comet tail", "NAD", and "hepatopetal". Figure 5 shows a snapshot of the complete Abdominal Ultrasound Ontology.

4 Result and Discussion

The ontology reuse methodology used to develop the Abdominal Ultrasound Ontology (AUO) has given the highest number of concept match compared to using only one ontology. This can be proved by performing a term to concept matching using the 761 terms extracted from the sample ultrasound report corpus. Figure 6 shows the comparison of total matches according to type (PrefLabel match, synonym match, partial match and no match) between NCIT, SNOMED CT and AUO. Between NCIT and SNOMED CT, NCIT has the higher concept match total with 151 PrefLabel matches, 79 synonyms matches and 438 partial matches. SNOMED CT on the other hand has only 98 PrefLabel matches,

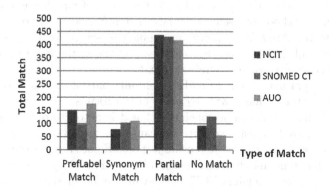

	NCIT	SNOMED CT	AUO
PrefLabel Match	151	98	176
Synonym Match	79	104	111
Partial Match	438	431	418
No Match	93	128	56

Fig. 6. Breakdown of total match according to type against NCIT, SNOMED CT and Abdominal Ultrasound Ontology (AUO) (Color figure online)

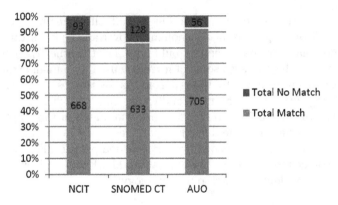

Fig. 7. Percentage of total match and no match in NCIT, SNOMED CT and AUO (Color figure online)

104 synonyms matches and 431 partial matches. The reason SNOMED CT has lower PrefLabel matches compared to synonyms is because of its naming convention. For example, the preferred label for "kidney" is "kidney structure" and "entire gallbladder" for "gallbladder". When writing report, radiologist often used simpler words like "kidney" and "gallbladder" instead of "kidney structure" and "entire gallbladder" thus, when term to concept matching was performed, SNOMED CT returned more synonym matches compared to PrefLabel.

Compared to NCIT and SNOMED CT, AUO returns the highest total match where it has 176 PrefLabel matches, 111 synonym matches and 418 partial matches. The reason AUO returns the most number of matches is because the ontology reuse methodology selects the best match from different ontologies and merge it into the AUO. Its exhaustive mapping in several ontologies based on the ontology rank has ensured that almost all terms in the corpus are covered by AUO. Whenever possible, a PrefLabel match will be inserted in the ontology. If not, a synonym match will be added then only partial matches are included to ensure the ontology has a wide coverage of the corpus.

From the analysis, it can be concluded that it is better to reuse from several ontologies compared to just one. This is because reusing several ontologies offers better term coverage compared to reusing just one. Figure 7 shows the percentage of total match and no match in all three ontologies. If ontology reuse was done by mapping the 761 terms against NCIT, there will only be an 87.8 % of coverage. If the mapping were done against SNOMED CT, the percentage of coverage would be only 83.2 % which is lower than NCIT. However, the percentage of coverage increases to 92.6 % when several ontologies were reused; which in this case are NCIT, SNOMED CT, and RadLex.

The percentage of no match is also very small (7.4 %) which means that the AUO covers almost all the terms in the corpus. After ontology evaluation with domain expert, the percentage of no match has been reduced to only 5 % after the domain expert included new concepts which before this have no match in any of the other ontologies being reused. The reason there is still 5 % of no match is because there are several term in the corpus that the domain experts believe are poor usage of terms to describe findings in an ultrasound report. The domain expert believes that this is bad practice and the medical ultrasound experts are now slowly cutting down the usage of such words thus making it irrelevant to be in the AUO. Another reason for the 5 % of no match is spelling errors made by ultrasound reporters. This is not a concern for now but for future work, we could consider using the ontology to also correct and understand these errors.

NCIT has a total of 113,794 classes while SNOMED CT has 316,031 classes. However, there are only 668 and 633 matches respectively for each NCIT and SNOMED CT regarding abdominal ultrasound terminology. On the other hand, AUO has only 509 classes which is less than 0.5 % of either NCIT or SNOMED CT but still managed to have 705 matches which is more than the matches NCIT and SNOMED CT each gets. This is because of the specialization of the ontology. Since the ontology has an intended purpose in an application, it is much better and more efficient to build a domain specific ontology through reuse. It definitely would not be efficient to store a large ontology such as NCIT and SNOMED CT and use only less than 0.3 % of it. This is because it would take a lot of storage space and it will also slow down the application since the application will need to go through the whole ontology to find a match. Thus the better way to develop an ontology based application is to build a new domain specific ontology through ontology reuse methodology.

5 Conclusion

Ontology reuse can be beneficial in developing domain specific ontologies for application system whereby it reduces development time and redundancy. The lack of proper methodology and tools in reusing ontology has hindered this effort. Thus, this paper proposed a methodology to reuse ontology together with supporting tools that would make the ontology reuse process much easier. The development of AUO using this methodology has proven that ontology reuse is beneficial in developing a small domain specific ontology which has wide coverage of the terminology used in the application system compared to using a large general domain ontology. It is hoped that the proposed ontology reuse methodology would encourage more usage of ontology in medical system without the development of similar domain ontologies that would cause redundancy.

References

1. Alani, H.: Position paper: ontology construction from online ontologies. In: Proceedings of the 15th International Conference on World Wide Web, WWW 2006, pp. 491–495. ACM, New York (2006)
2. Boland, G.: Enhancing the radiology product: the value of voice-recognition technology. Clin. Radiol. **62**(11), 1127 (2007)
3. Bontas, E.P., Mochol, M., Tolksdorf, R.: Case studies on ontology reuse. In: Proceedings of the 5th International Conference on Knowledge Management IKNOW 2005 (2005)

4. Caldarola, E.G., Picariello, A., Rinaldi, A.M.: An approach to ontology integration for ontology reuse in knowledge based digital ecosystems. In: Proceedings of the 7th International Conference on Management of Computational and Collective IntElligence in Digital EcoSystems, pp. 1–8. ACM (2015)
5. Capellades, M.A.: Assessment of reusability of ontologies: a practical example. In: Proceedings of AAAI 1999 Workshop on Ontology Management, pp. 74–79. AAAI Press (1999)
6. Jonquet, C., Musen, M.A., Shah, N.H.: Building a biomedical ontology recommender web service. J. Biomed. Semant. 1(S1), 1 (2010)
7. Lossio-Ventura, J.A., Jonquet, C., Roche, M., Teisseire, M.: Biotex: a system for biomedical terminology extraction, ranking, and validation. In: 13th International Semantic Web Conference (2014)
8. Maedche, A., Motik, B., Stojanovic, L., Studer, R., Volz, R.: An infrastructure for searching, reusing and evolving distributed ontologies. In: Proceedings of the 12th International Conference on World Wide Web, WWW 2003, pp. 439–448. ACM, New York (2003)
9. Mejino Jr., J.L., Rubin, D.L., Brinkley, J.F.: FMA-RadLex: an application ontology of radiological anatomy derived from the foundational model of anatomy reference ontology. In: AMIA Annual Symposium Proceedings, vol. 2008, p. 465. American Medical Informatics Association (2008)
10. Noy, N.F., Shah, N.H., Whetzel, P.L., Dai, B., Dorf, M., Griffith, N., Jonquet, C., Rubin, D.L., Storey, M.-A., Chute, C.G., et al.: Bioportal: ontologies and integrated data resources at the click of a mouse. Nucleic acids research, pp. W170–W173 (2009)
11. Russ, T., Valente, A., MacGregor, R., Swartout, W.: Practical experiences in trading off ontology usability and reusability. In: Proceedings of the 12th Workshop on Knowledge Acquisition, Modeling and Management (KAW 1999), pp. 16–21 (1999)
12. Shah, T., Rabhi, F., Ray, P.: OSHCO: a cross-domain ontology for semantic interoperability across medical and oral health domains. In: 2013 IEEE 15th International Conference on e-Health Networking, Applications Services (Healthcom), pp. 460–464, October 2013
13. Shah, T., Rabhi, F., Ray, P., Taylor, K.: A guiding framework for ontology reuse in the biomedical domain. In: 2014 47th Hawaii International Conference on System Sciences (HICSS), pp. 2878–2887. IEEE (2014)
14. Simperl, E.: Reusing ontologies on the semantic web: a feasibility study. Data knowl. Eng. 68(10), 905–925 (2009)
15. Uschold, M., Clark, P., Healy, M., Williamson, K., Woods, S.: An experiment in ontology reuse. In: Proceedings of the 11th Knowledge Acquisition Workshop (1998)
16. Zulkarnain, N.Z., Crofts, G., Meziane, F.: An architecture to support ultrasound report generation and standardisation. In: Proceedings of the International Conference on Health Informatics, pp. 508–513 (2015)

All links were last followed on January 25, 2016.

Tamil Morphological Analyzer Using Support Vector Machines

T. Mokanarangan[✉], T. Pranavan, U. Megala, N. Nilusija, G. Dias,
S. Jayasena, and S. Ranathunga

Department of Computer Science Engineering, University of Moratuwa,
Moratuwa, Sri Lanka
{mokanarangan.11,pranavan.11,megala.11,nilu.11,gihan,
sanath,surangika}@cse.mrt.ac.lk

Abstract. Morphology is the process of analyzing the internal structure of words. Grammatical features and properties are used for this analysis. Like other Dravidian languages, Tamil is a highly agglutinative language with a rich morphology. Most of the current morphological analyzers for Tamil mainly use segmentation to deconstruct the word to generate all possible candidates and then either grammar rules or tagging mismatch is used during post processing to get the best candidate. This paper presents a morphological engine for Tamil that uses grammar rules and an annotated corpus to get all possible candidates. A support vector machines classifier is employed to determine the most probable morphological deconstruction for a given word. Lexical labels, respective frequency scores, average length and suffixes are used as features. The accuracy of our system is 98.73 % and a F-measure of .943, which is more than the same reported by other similar research.

Keywords: Tamil · Morphological analyzer · Support vector machine · Natural language processing · Dravidian languages

1 Introduction

Morphological analysis is the process of segmentation of words into their component morphemes, and the assignment of grammatical morphemes to grammatical categories and lexical morphemes to lexemes [1]. Tamil language is morphologically rich and agglutinative. Each word is pinned with morphemes and during morphological construction, the original form of the word changes, hence making the morphological deconstruction tough.

Morphological analysis is the basis for many natural language processing tasks such as Named Entity Recognition, Part of Speech Tagging and Machine translation. Morphological analysis can provide a wealth of information. For Tamil in particular, like many other Dravidian languages, a good morphological analyzer can extract many information about a word ranging from verb or noun to tense and gender due to its rich morphology.

Previous attempts on morphological analysis for Tamil have been made using three approaches: rule based, machine learning based, and hybrid approaches that combine

© Springer International Publishing Switzerland 2016
E. Métais et al. (Eds.): NLDB 2016, LNCS 9612, pp. 15–23, 2016.
DOI: 10.1007/978-3-319-41754-7_2

both the rule based and the machine learning approaches. This paper outlines an approach that uses a morphological engine encompassing all grammar rules in Tamil that generates all possible candidates for a word along with the Part of Speech (PoS) tags for each morpheme. These PoS tags, respective frequency scores, average length and suffixes are used as features in a Support Vector Machines (SVM) classifier to select the best candidate out of the candidate list.

Rest of the paper is organized as follows. The next section discusses the previous attempts on building morphological analyzers for Tamil. Third section describes our approach and the fourth section gives the evaluation results. Final section discusses future work and concludes the paper.

2 Related Work

First ever Tamil morphological analyzer was built by AU-KBC Research Centre in 2003 [2]. Since then research on Tamil morphological analysis was continued in two directions, using machine learning and using rule based approaches. Selvam and Natarajan [3] carried out research on morphological analysis and PoS tagging for Tamil using a rule based approach via projection and induction techniques. Another morphological analyzer for Tamil was implemented using the sequence labelling based machine learning approach [4]. It was a supervised machine learning approach and a corpus with morphological information was used for training. Another approach used the open source platform apertium [5]. Apertium tool uses the computational algorithm called Finite State Transducers for one-pass analysis and generation, and the database is based on the morphological model called Word and Paradigm. In a very recent research, a rule-based morphological analyzer was presented [6]. Researchers have used a set of rules, a postposition suffix word list and a root word dictionary developed from classical Tamil text. Not considering all the grammar rules coupled with high ambiguity has been the problem for this approach.

Our approach drew inspiration from morphological analyzers designed for two different languages: first from an Arabic morphological analyzer [6]. In this approach, text is broken down into each of the hundreds and thousands of possible lexical labels, which represent their constituent elements including lemma ID and part-of-speech. Features are computed for each lexical token based on their local and document-level context. Based on these features the support vector machines classifier is implemented to do the classification. The second method was from a compound word splitting approach for German [7]. This approach introduced methods to learn splitting rules from monolingual and parallel corpora. These rules were then evaluated against a gold standard [7].

3 Our Approach

3.1 Outline

As show in the Fig. 1 the first step is to get all possible lexical units of a single word and annotate each lexical unit with part of speech tags. In some cases one lexical unit

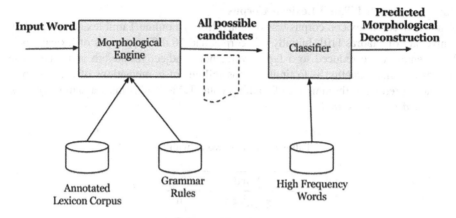

Fig. 1. Outline of the morphological analyzer

can have more than one part of speech tag. For example ஓடு (ōṭu) can mean 'Roof' and 'Run'. Morphological rules of Tamil can also affect the spelling of the root.

Consider the following example for the word ஓடினான் (Transliteration - ōṭinān, Translation – ran) (Table 1).

In the next step, the annotated lexical labels along with other features were fed into the SVM classifier. The SVM then predicts the best candidate for a certain word. The reason to choose SVM over other available options such as multilayer perceptron and boosted is the best trade-off SVM provided between accuracy and training time. This is explained in detail below in Sect. 4.

Table 1. All possible combinations for the word ஓடினான் (transliteration - ōṭinān, translation – ran)

Morphological Deconstruction	Lexical Labels
ஓடு + இன் + ஆன்	<v><Idainilai><part>
ஓடி + ன்+ ஆன்	<n><Idainilai><part>
ஓடு + இன் + ஆ + ன்	<v><Idainilai><part><part>
ஓடி + ன் + ஆ + ன்	<n><Idainilai><part><part>

3.2 Data Sources

To generate all possible candidates, annotate with PoS labels for each lexical label, and to get the total frequency of each word, we used two sources: a lexicon corpus along with PoS annotations, and a list of high frequency words along with the frequency score for each word.

3.2.1 Annotated Tamil Lexicon Corpus

Annotated Tamil lexicon corpus was obtained from an online Tamil lexicon created by University of Madras [10]. Initially this corpus had 16 different types of lexical labels but eventually we reduced to 5 types: verb, noun, adjective, adverb and other. The purpose of this reduction is to limit the possibilities of combinations of lexical labels and hence reducing the amount of training data. Table 2 illustrates a sample of how words and tags are stored.

Table 2. Lexicon words with tag

Word	Tag
அஃகல்	n
அஃகான்	n
அஃகு	v

3.2.2 High Frequency Words List

The high frequency word list was built using the usage data obtained by crawling Tamil Wikipedia and other Tamil news websites. Each entry in this list has the word and the word count. Here the word count was used to calculate the frequency score.

3.3 Morphological Engine

Morphological Engine is the vital part in the system. Encompassing all the grammar rules regarding morphological construction, this engine generates all possible candidates along with their lexical labels. Some of the rules in morphological engine are shown in Fig. 2.

As illustrated in the diagram. The word கிளியை (Transliteration: Kiliyai, Translation: 'the parrot') can be deconstructed under two grammar rules:

Grammar rule 1: உயிர் முன் உயிர் புணர்தல் (Transliteration – "Uyir munn uyir punarthal", Translation: "Vowel on Vowel morphological construction").

கிளியை - கிளி> + Yakaram> + ஐ

Grammar rule 2: இயல்பு புணர்ச்சி (Transliteration – "Iyalpu punarchi", Translation: "Natural morphological construction").

கிளியை - கிளி + யை
கிளியை - கிளி + ய் + ஐ

Based on the last letter of the first word and the first letter of the second word the grammar rules define the morphological construction. To ensure that all grammar rules and all types of morphological deconstruction is covered, two Tamil grammar books [9, 10] were followed to obtain 14 rules. Using these rules all the candidate are

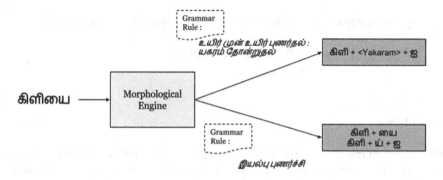

Fig. 2. Grammar rules and morphological deconstruction

generated and then lexical labels. To get all the approaches a finite state machine that uses brute force approach to get all possible combinations was implemented.

The word given in the diagram is relatively easy to deconstruct, now consider a complicated word: ஓடிச்சென்றான் (Transliteration – Oodichchendran, Meaning – 'He ran').

All candidates possible for this is:

- ஓடி+<EkaramVallinam>+சென்று+ஆன்,
- ஓடு+<Idainilai>+ச்+சென்றான்
- ஓடி+<EkaramVallinam>+சென்றான்
- ஓடு+<Idainilai>+ச்+செல்+<Idainilai>+ஆன்
- ஓடு+இ+<MellinamVallinamUyirMun>+சென்றான்
- ஓடு+இ+<MellinamVallinamUyirMun>+சென்று+ஆன

3.4   Classifier

3.4.1   Features Set

Frequency Based Scores

This frequency based approach was proposed by Koehn and Knight [8] to split compound words in German. The more frequent a word occurring in a training corpus, the bigger the statistical basis to estimate translation probabilities, and the more likely the correct translation probability distribution is learned. This insight leads to define a splitting metric based on word frequency [8].

Given the count of words in the corpus, the split S with the highest geometric mean of word frequencies of its parts pi (n being the number of parts) is selected. Here $count(p_i)$ is frequency count the word p_i obtained from the high frequency words list.

$$argmax_s(\prod_{p_i \in S} count(p_i))^{\frac{1}{n}}$$

Consider the following example: ஓடி+<EkaramVallinam>+சென்று+ஆன்

- Frequency scores $= \left(\text{count}(\text{ஓடி}) + \text{count}(\text{சென்று}) + \text{count}(\text{ஆன்}) \right)^{\frac{1}{3}}$

$$= (550 + 2524 + 120)^{\frac{1}{3}}$$
$$= 14.33$$

Lexical Labels

Other important feature set is the lexical labels generated by the morphological engine. The morphology engine has been developed for the particular case of Tamil and the particular set of lexical labels. This tagging order gives more priority to more commonly occurring patterns and indirectly covers more subtle grammar patterns in Tamil.

Suffix (விகுதி)

Tamil is a morphologically rich language with many morphemes pinned to each word. But in many cases, certain morphemes do not appear as suffixes for certain type of words. For example 'ஐ' suffix is not present in a verb. In retrospect, the model was fed with the final suffix of a word as a feature to the system to differentiate verbs, adjective, adverbs and noun stem based words.

Average Length

This is a new feature that has not been tried in any previous approaches. When the model was tried on with only the above mentioned features, it was found that for some compound noun cases, the morphological deconstruction was going a step further.

For example:-
மாநகரசபை −மா+நகரம்+சபை (Expected)
-    மா+நக+ர+ச+பை (Output without average length feature)

Therefore, to eliminate this issue, we introduced a threshold feature called average length. It is obtained by calculating the average length of the lexical parts in the candidate. This feature was found out based on the factor analysis carried out on the training data.

3.5 Training Data

Over 70,000 words were manually labelled and used as training data. Correct morphological disambiguation candidate was labelled as 'Yes' while mismatches were labelled as 'No'.

3.6 Prediction

Using the training data, a probabilistic model was built using the SVM classifier. The candidates are then classified using the classifier and the one with highest probability of classified as 'Yes' is selected. This probabilistic model not only provides us with the best candidate but also if there is ambiguity the top candidates are displayed. This feature can come in handy while implementing a Part of Speech tagger for Tamil.

4 Evaluation Results

Upon generating all the candidates, the next step is to feed the data into the classifier to select the best suitable candidate. We selected SVM because of the best trade-off between accuracy and time taken to build the model. Table 3 illustrates the comparison of accuracies between various classifiers.

Table 4 illustrates the difference in accuracy by using average length and not using average length as features.

Table 5 illustrates the difference in accuracy by using average length and not using frequency scores as features.

Table 3. Comparisons of accuracy between various classifiers

	Multilayer perceptron	Boosted decision tree	Support vector machine
Correctly classified instances	91.236 %	85 %	98.73 %

Table 4. Accuracy difference between with and without using average length

Accuracy without using average length	92.83 %
Accuracy using average length	98.73 %

Table 5. Accuracy difference between with and without using frequency scores

Accuracy without using frequency scores	47.36 %
Accuracy using frequency scores	98.73 %

Tables 6 and 7 show the results obtained from 10-Fold cross validation test for over 30,000 words. Table 6 illustrates the overall accuracy of the system while Table 7 illustrates the detailed accuracy by class.

Table 6. Stratified cross validation

Correctly classified instances	98.7376 %
Kappa statistic	0.8869
Mean absolute error	0.0265
Root mean squared error	0.1033
Relative absolute error	23.9067 %
Root relative squared error	43.8834 %

Table 7. Detailed accuracy by class

TP rate	FP rate	Precision	Recall	F-measure	Range of coverage area	Class
0.89	0.01	0.888	0.899	0.89	0.99	Yes
0.99	0.10	0.994	0.993	0.99	0.99	No

5   Conclusion and Future Work

We presented a morphological engine for Tamil that uses grammar rules and an annotated corpus to get all possible candidates. A support vector machines classifier was employed to determine the most probable morphological deconstruction for a given word. Lexical labels, respective frequency scores, average length and suffixes are used as features. The accuracy of our system is 98.73 %, which is more than the same reported by other similar research.

Tamil is a morphologically rich language. Computationally, each root word of can take a few thousand inflected word-forms, out of which only a few hundred will exist in a typical corpus. This morphological analyzer which uses a different approach from previous approaches have proved to be effective.

Though the main intention of this approach is to tackle the ambiguity sometimes this approach fails when encountering name entities. It tends to break into meaningless morphological disambiguation. This is a pitfall that should be taken care of in the further researches.

Since most Dravidian language share the same characteristics, hoping that this approach can be used in other languages to get a highly accurate morphological analyzer. The analyzer not only outputs the construct the deconstructed morphology but also the lexical labels.

As future work we intend to build on this approach and along with it build a PoS tagger and Name Entity recognizer that uses the features extracted from morphological analyzer. Once these goals have been achieved we eventually hope to build a successful Tamil machine translator and eventually preserve an ancient endangered language.

References

1. Jayan, J.P., Rajeev, R., Rajendran, S.: Morphological analyzer and morphological generator for Malayalam - Tamil machine translation. Int. J. Comput. Appl. (0975 – 8887) **13**(8), 15–18 (2011)
2. Au-kbc.org. Tamil Morphological Analyzer (2015)
3. Selvam, M., Natarajan, A.M.: Improvement of rule based morphological analysis and POS tagging in Tamil language via projection and induction techniques. Int. J. Comput. **3**(4), 357–367 (2009)
4. Anand Kumar, M., Dhanalakshmi, V., Soman, K.P., Rajendran, S.: A sequence labeling approach to morphological analyzer for Tamil language. Int. J. Comput. Sci. Eng. **2**(6), 1944–1951 (2010)
5. Parameshwari, K.: An implementation of APERTIUM morphological analyzer and generator for Tamil. Probl. Parsing Indian Lang. **11**, 41–44 (2011)
6. Akilan, R., Naganathan, E.R.: Morphological analyzer for classical Tamil texts: a rule-based approach. Int. J. Innov. Sci. Eng. Technol. **1**(5), 563–568 (2014)
7. Shah, R., Dhillon, P.S., Liberman, M., Foster, D., Maamouri, M., Ungar, L.: A new approach to lexical disambiguation of Arabic text. In: Proceedings of the 2010 Conference on Empirical Methods in Natural Language Processing, Cambridge, Massachusetts, pp. 725–735, 09–11 October 2010
8. Koehn, P., Knight, K.: Empirical methods for compound splitting. In: Proceedings of the Tenth Conference on European Chapter of the Association for Computational Linguistics, Budapest, Hungary, 12–17 April 2003
9. Nuhman, M.A.: அடிப்படைத்தமிழ்இலக்கணம், Revised edn, pp. 93–260. Poobalasingam Publications, Sri Lanka (2010)
10. Naavalar, A.: தமிழ்இலக்கணம், 10th edn, pp. 88–180. Poobalasingam Publications, Sri Lanka (2008)

The Role of Community Acceptance in Assessing Ontology Quality

Melinda McDaniel[1], Veda C. Storey[2], and Vijayan Sugumaran[3(✉)]

[1] Department of Computer Science, Georgia State University,
Atlanta, GA 30302, USA
mmcdaniel16@student.gsu.edu
[2] Department of Computer Information Systems,
J. Mack Robinson College of Business, Georgia State University,
Box 4015, Atlanta, GA 30302, USA
vstorey@cis.gsu.edu
[3] School of Business Administration, Oakland University,
Rochester, MI 48309, USA
sugumara@oakland.edu

Abstract. Ontologies are crucial for the Semantic Web to flourish. Several communities are beginning to develop and maintain ontology repositories in different domains. Although a developer can often find multiple ontologies in the library that fit a particular domain, he or she then must select which of the potential ontologies would be most suitable for a specific purpose. Users, therefore, need a way to assess the quality of the ontologies stored in the library based upon a broad set of criteria; for example, the level of acceptance by the community of which it is a part. The history of an ontology's development and the authority an ontology receives via links from other ontologies can be used to assess the level of endorsement within the group that shares its domain. This research defines metrics for history and authority within a community and shows how they can be weighted for a particular task. A case study demonstrates the usefulness of these metrics and discusses why they should be incorporated in any broad metrics suite that is used to rank ontologies in a library.

1 Introduction

The Semantic Web is "a set of standards for knowledge representation and exchange that is aimed at providing interoperability across applications and organizations" [1]. The degree of this interoperability between human and software agents depends upon how many communities they have in common and how many ontologies they share [1]. An ontology, which has been called the third component of the Semantic Web, is defined simply as a group of consistent and related terms [1] and more formally as "a formalization of a shared conceptualization" [2]. The latter definition, and the idea that the conceptualization is "shared" is expanded further by Hepp et al. (2006) who asserted that "ontologies are not just formal representations of a domain, but much more community contracts about such formal representations" [3].

© Springer International Publishing Switzerland 2016
E. Métais et al. (Eds.): NLDB 2016, LNCS 9612, pp. 24–36, 2016.
DOI: 10.1007/978-3-319-41754-7_3

A community consists of a set of relationships between people sharing a common interest [4]. An online community can then be considered as a community that employs the Internet for communication among its members [4]. Berners-Lee and Kagal described the Semantic Web as composed of overlapping online communities of varying sizes and fractal in nature, as membership in these communities changes frequently [1]. Many online communities allow members to participate fully in the site through contributing and accessing information, as well as by commenting on the information added by other members. The BioPortal ontology repository [5], for example, considers anyone who uses this portal to be a member and allows them to actively contribute to the content in the library — a fact that its designers claim should increase the quality of that content [7].

This feeling of shared responsibility within a community for the overall improvement of the ontological content is consistent with what Shadbolt and Berners-Lee have asserted will greatly reduce the effort involved in developing an ontology as the size of the community grows [6]. Noy et al. contend that the Wisdom of the Crowd could even replace knowledge experts when a consensus is able to be reached within a community [7]. Reaching this consensus, however, is not always easy, requiring time and effort, and a large number of dedicated participants. Therefore, the degree of participation in the process of revising, adopting, expanding and reviewing of any ontology is a factor in the assessment of that ontology's value.

The selection of an ontology from among the options available in an ontology repository should be made based upon a broad set of attributes that may be weighted depending upon the requirements of each application [24]. One of the attributes to include in such a list of criteria should be the acceptance of the ontology within its community. Metrics to assess this acceptance should include measures of how many community members endorse the ontology, how long the ontology has been available, how much active participation has been done by community members in the ontology's development. This community acceptance attribute is difficult to assess, with metrics to measure it not applied successfully in the past [18]. While much work has been carried out developing metrics related to syntactic, semantic and pragmatic aspects of ontologies, the social quality of ontologies has not been thoroughly investigated. The objective of this research, therefore, is to do so.

This research introduces new metrics for social quality assessment, defines them formally, applies them to existing ontologies, and analyzes the challenges involved in using them. The result is to show how these attributes provide valuable insight into ontology quality and should, therefore, be included in any rigorous ontology evaluation. The results of this assessment could promote interoperability between systems and help progress the use of ontologies in the Semantic Web. Terms related to social quality assessment used in this paper are defined in Table 1.

The next section provides an overview of prior work on assessing ontology quality based on its social valuation. Sections 3 and 4 present history and authority metrics for assessing ontology social quality, and outlines the implementation of these metrics. Section 5 describes a case study validating the results of the social quality metrics. Section 6 summarizes the work and suggests future research directions.

Table 1. Definitions of terms related to social quality assessment

Term	Definition	References
Authority	"The degree of reputation of an ontology in a given community or culture"	Stvilia et al. [8]
Community	"A set of relationships where people interact socially for mutual benefit"	Andrews [4]
History	"The way that a particular subject or object has developed or changed throughout its existence"	History [9]
Online Community	"A social network that uses computer support as the basis of communication among members instead of face-to-face interaction"	Andrews [4]
Revising	"The act of thinking, comparing, deciding, choosing then taking action"	Sudol [10]
Revision	"The act of making changes to a written document to make it better"	Horning and Becker [11]
Social Network	"A set of people (or organizations or other social entities) connected by a set of socially-meaningful relationships"	Wellman [12]
Social Quality	"The level of agreement among participants' interpretations"	Su and Ilebrekke [13]

2   Related Research

In the decade and a half since the introduction of the Semantic Web [14], much work has been carried out on ontology evaluation. Many researchers have addressed the complexity of choosing a high-quality ontology for a particular task or domain. Attributes considered to be valid measures of ontology quality include adaptability, clarity, comprehensiveness, conciseness, correctness, craftsmanship, relevance, reusability, richness and stability as well as many others [14]. Numerous metrics have been developed to assess these and other aspects of ontology quality. Specific metrics which assess one particular attribute and broad suites of metrics that attempt to provide an overall picture of an ontology's quality have been developed [15–25].

D'Aquin and Noy (2012) defined an ontology library as "a Web-based system that provides access to an extensible collection of ontologies with the primary purpose of enabling users to find and use one or several ontologies from this collection" [26]. Although ontologies should reside in libraries and be developed and endorsed by communities that share a common interest [6], little work has been conducted to develop a means for assessing the amount of recognition received by each ontology within a library. To provide a comprehensive picture of an ontology's quality, factors such as how much the ontology is being used, how many other ontologies refer to this one as an authority, and how long the ontology has been in existence, should all be taken into consideration [18].

2.1 Ontology Role in Communities

A community can no longer be considered as a physical place, but, rather, as a set of relationships between people who interact socially for their mutual benefit [4]. An online community is a social network that uses the Internet to facilitate the communication among its members rather than face-to-face meetings [4]. These virtual social networks are frequently used for information sharing and problem solving among members who share common interests [12].

Ontologies have been defined as formal representations of a domain, but in order for those representations to be meaningful, they must be agreed upon by the members of a community [6]. This type of meaningful discourse between members of a group is a dynamic social process consisting of shared topics being added, expanded, revised or even discarded. Therefore, an ontology representing the shared communication between members should not be static, but should be able to reflect the community consensus of meaning at any particular time [1]. When a community shows its approval of an ontology by actively participating in its ongoing evolution, the quality of the ontology is more likely to be high within that community [26]. A way of measuring this type of active participation would be helpful in assessing community endorsement of a particular ontology.

2.2 Metrics Suites

The usefulness of metrics to provide a quantified measurement of ontological quality has long been recognized [19] with many metric suites being created that attempt to provide a broad picture of many aspects of an ontology's quality. OntoQA [19], OQuaRE [25], OntoMetric [17], and AKTiveRank [20] are a few of the most comprehensive suites of metrics. Table 2 summarizes these, and other, metric suites currently available for broad ontology assessment, identifies the number of metrics, and specifies how many of them measure an ontology's social importance within a particular library.

Table 2. Examples of broad metrics suites

Assessment approach	Total metrics	Social metrics	Description of social assessment
Protégé-2000 (Noy et al.) [15]	Varies	0	none
OntoClean (Guarino and Welty) [16]	Varies	0	none
OntoMetric (Lozano-Tello and Gómez-Pérez) [17]	160	3	Assesses whether an ontology fits a system's requirements
Semiotic Metrics Suite (Burton-Jones et al.) [18]	10	2	Assesses History and Authority of an ontology by counting ontology links

(Continued)

Table 2. (*Continued*)

Assessment approach	Total metrics	Social metrics	Description of social assessment
OntoQA (Tartir et al.) [19]	12	0	none
AKTiveRank (Alani et al.) [20]	Varies	0	Ranks ontologies based on user criteria
OQual (Gangemi et al.) [21]	Varies	0	none
Biomedical Ontology Recommender web service (Jonquet et al.) [22]	8	2	Assesses ontologies based on page rankings
ROMEO (Yu et al.) [23]	varies	0	none
(Vrandečić) [24]	8	0	none
OQuaRE (Duque-Ramos et al.) [25]	14	0	none

2.3 Social Assessment Within Metrics Suites

Although communities should support the development, maintenance and endorsement of ontologies [6], very few assessment systems have a means by which to measure an ontology's value within its community. OntoMetric [17], the BioPortal Recommender [22], and the Semiotic Metrics Suite [18] are among the few suites that attempt to assess an ontology's acceptance within a community as one of the factors to measure its quality. Unfortunately, none of these assessment suites are able to fully evaluate the level of acceptance an ontology receives within its community.

OntoMetric [17] contains approximately 160 metrics for assessing ontology quality, which focus primarily on the fitness of an ontology for a particular software project for which it will be used. However, only three of its metrics relate to its relationship with other ontologies. The large number of metrics makes the OntoMetric system difficult to employ [19]. The OntoMetric system reflects the fact that part of the suitability of an ontology for a given project is the methodology used to create it. It, therefore, assesses the social acceptance of that methodology by counting the number of other ontologies that were created with it, the number of domains that have been expressed with its developed ontologies, and how important the ontologies developed with this methodology have become. Unfortunately, in most situations, a user must attempt to answer these questions (perhaps by conducting additional research) as well as to provide an answer expressed on a scale between "very low" and "very high," reducing the accuracy of the results in this factor's assessment.

The BioPortal recommender system includes Acceptance metrics as part of the ranking system that it provides as a tool for choosing an ontology for a particular purpose [22]. Users enter desired keywords and the recommender system presents a list of ontologies from the BioPortal repository containing the keywords. The list of applicable ontologies is ranked in order of each ontology's score on four individually weighted attributes, one of which is the Acceptance of the ontology within the BioPortal community. The other three attributes that are included in the Recommender system are Coverage, Detail of knowledge and Specialization. Unfortunately, the metrics used by the BioPortal recommender system to assess Acceptance are based on factors such as the number of site visits to the BioPortal website, membership in the UMLS database and mentions in the BioPortal journal, so those metrics cannot be used on ontologies in other libraries without access to this information.

The Semiotic Metrics Suite developed by Burton-Jones et al. [18] is based upon the theory of semiotics, the study of signs and their meanings, and builds upon Stamper et al.'s [27] framework for assessing the quality of signs. One of the layers of the framework is the Social layer, which evaluates a sign's usefulness on a social level by evaluating its "potential and actual social consequences" and asks the question "Can it be trusted?" [27]. The Semiotic Metrics Suite includes the Social layer, which measures an ontology's recognition within a community by two metrics: (1) *Authority* which measures the link from an ontology to other ontologies in the same library; and (2) *History* which measures the frequency with which these links are employed. Unfortunately the calculations for these measurements require information that is not available for most ontologies. The number of links from other ontologies to a particular one, and the number of times the linking ontologies have been used for other applications are usually not provided by ontology libraries, making these metrics difficult to use for ontology assessment. This research introduces new Authority and History metrics using information available for most ontologies and includes a case study demonstrating their effectiveness.

3   Metrics for Assessing Social Quality

Social Quality is "the level of agreement among participants' interpretations" [27] and reflects the fact that, because agents and ontologies exist in communities, agreement in meaning is essential within the community. This research proposes two new metrics to measure the level of an ontology's recognition within its community by measuring its authority within the library and the history of its participation and use in the library. These metrics can be combined to determine the overall assessment for Social Quality within the library.

Stvilia defines Authority as the "degree of reputation of an ontology in a given community or culture" [8]. One way to measure Authority is by the number of other ontologies that link to it as well as how many shared terms there are within those linked ontologies. More authoritative ontologies signal that the knowledge they provide is accurate or useful [18].

Another social metric is the History of an ontology. The history of a conceptualization is a valuable part of its definition [3]. The History metric measures the number of years an ontology has existed in a library, as well as the number of revisions made to it during the course of its residence there. Ontologies with longer histories are expected to be more dependable because each new revision should improve upon the previous version showing a pattern of active participation by community members resulting in additions and modifications.

3.1 Social Quality Metric

The Social Quality metric is computed by the combined weighted scores on these two measurements defined as SQ. The weights of the History and Authority metrics could be equivalent, but it is possible for a user to adjust the significance of each for a particular task by varying the values of the weights.

Definition 1: The Social Quality (SQ) of an ontology is defined as the weighted average of Authority (SQa) and History (SQh) where w_a represents the percentage assigned by the user to the authority attribute and w_h represents the weight assigned to the history attribute.

$$SQ = w_a * SQa + w_h * SQh$$

3.2 Social Authority Metric

The Authority of a particular ontology is determined by the number of other ontologies that link to it. By scanning all of the other ontologies in the library looking for links to this ontology, two counts are determined: the number of total links to the ontology; and the number of ontologies which include 1 or more references to it. The two counts are weighted depending on the user's task and the result is normalized between 0, meaning no links at all, to a score of 100, indicating that this ontology is the one in the library with the most links to it. The equation for computing this metric is defined as SQA. External links can also be considered in the determination of SQA if available. Many ontologies, such as the Gene Ontology [28], are in multiple libraries. SQA should then take into consideration all of the links to the Gene Ontology from all of the libraries for which it is a part.

Definition 2: The Social Quality Authority (SQA) of an ontology is defined as the weighted average of the number of linking ontologies (LO) and the number of total linkages (LT) where w_o represents the percentage assigned by the user to the number of linking ontologies and w_t represents the weight assigned to the total number of links.

$$SQA = w_o * LO + w_t * LT$$

3.3 Social History Metric

History is determined by calculating the number of years that an ontology has been a member of a community as well as the number of revisions to the ontology that have been made during those years. The two counts are weighted depending on the user's task and the result normalized between 1, indicating only one submission that was never updated, to a score of 100, indicating this ontology is the one in the library with the most total revisions over the longest number of years.

Definition 3: The Social Quality History (*SQH*) of an ontology is defined as the weighted average of the number of years it has been in the library (Y) and the number of submissions (including revisions) that have been uploaded (S) where w_y represents the percentage assigned by the user to the number of years and w_s represents the weight assigned to the total number of submissions.

$$SQH = w_y * Y + w_s * S$$

4 Implementation

A system has been developed to assess community recognition of an ontology by applying the revised Social Quality metrics. This system can be employed by any community containing an ontology repository, and aids in the selection of an ontology when multiple options are available. By entering relevant keywords and desired metric weights into the system, a user retrieves a set of potential ontologies containing the keywords. The system then assesses the Authority of each of those ontologies by searching all the other ontologies in the repository counting the number of ontologies that link to each of the potential ontologies as well as the total number of links. Each ontology in the list of potential ontologies then has its History assessment computed by counting the number of years each ontology has been stored in the library and the number of revisions made to the ontology during that time. The Authority and History metrics are then weighted according to the metric weights entered by the user and the list of potential ontologies is sorted in decreasing order of the overall Social Quality score. At this time the user receives a list of recommended ontologies that contain the desired keywords and that rank high in social recognition from the community. The specific steps carried out for Social Quality metric assessment and ontology ranking are shown in Fig. 1.

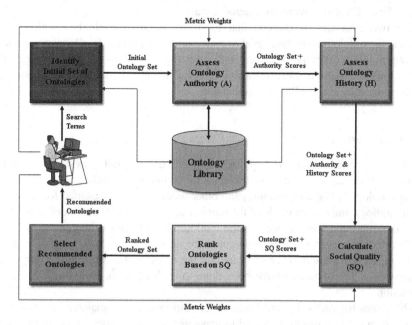

Fig. 1. Social quality assessment and ranking of ontologies

5 Case Studies

To obtain an understanding of how information about an ontology's acceptance within its community could help a user choose an appropriate ontology from a list of options, two case studies were carried out using the social quality metrics. The BioPortal ontology library was chosen for both studies as an example of a large, well maintained ontology repository that has been deemed useful to the biomedical community [5]. The BioPortal website was also selected because of the availability of additional information included in the library that could be used to examine the results of the case studies with other information on its ontology profile pages. The BioPortal website allows members of the community to contribute reviews to its ontologies, list projects, and make suggestions. BioPortal also keeps track of the number of site visits for each of the ontologies, and provides annotation and term mappings services for its ontologies [7].

The first case study applied the social quality metrics to all 383 of the ontologies currently in the library, ranking them from highest Social Quality score to lowest. This case study was carried out to assess whether the highest-ranking ontologies in the library were in actuality the ones that were most endorsed by the biomedical community. The second case study searched the BioPortal library for ontologies matching key terms and determining a list of recommended ontologies ranked by their Social Quality assessments as well as using our SQ metric. The ontology list for each term was then examined to ascertain whether the highest-ranking ontologies on each list was actually more likely to be frequently accessed than the ontologies that showed up later on the list. In both case studies, all metrics were weighted equally in the overall determination of Social Quality. It is possible to weight the individual metrics differently, depending on the particular task requirements. However, for the purposes of the case studies, all metrics were considered equally.

The two case studies showed that useful information could be obtained from assessing the ontologies on their level of endorsement within the BioPortal community. By examining the difference between ontologies high on the list to the ontologies that ranked lower, a pattern can be easily observed about whether the ontologies are well-supported by the BioPortal membership.

5.1 Case Study 1

The Authority metrics were first applied to all 383 of the ontologies currently part of the BioPortal library assigning equal weights to the number of links and the number of linking ontologies. For each ontology, all other ontologies were scanned for references to that ontology and counts made of the number of ontologies that included at least one reference to the ontology, as well as the total number of links to the ontology.

The History metric was then computed for all of the BioPortal ontologies using equal weighting for the number of years that the ontology has been in the library and the number of revisions that have been done to each of them, including the original submission.

The scores for Authority and History were individually normalized between 1 and 100 and then the two scores averaged to generate the overall Social Quality metric for each of them. The list of ontologies was ordered from 100 to 1 in order to identify the most highly ranked ontologies at the top of the list.

Table 3 shows the highest ranked ontologies from the library on the combined social metrics. Examining other information available on the BioPortal website, it is clear that the ontologies that scored high on the Social Quality metric were the ones involved in many biomedical projects and with good reviews from members who have used them. On the other hand, 70 of the ontologies tested scored only 1 out of 100 on the combined social quality metrics. Exploring the BioPortal website revealed that these 70 had no other ontologies linking to them and no revisions after the initial submission, which was often several years prior, and were not currently involved in any listed projects.

Table 3. Highest ranked ontologies from BioPortal using History and Authority metrics

Name of ontology	Authority	History	Combined metrics
Gene Ontology (GO)	69	100	85
Human Phenotype Ontology (HP)	51	100	75
Mosquito Insecticide Resistance Ontology (MIRO)	17	100	58
Mass Spectrometry Ontology (MS)	85	27	56
Systems Biology Ontology (SB)	10	100	55
Minimal Anatomical Ontology (MAT)	100	2	51
Sequence Types and Features Ontology (SO)	60	42	51
Human Disease Ontology (HD)	1	100	51
Mammalian Phenotype Ontology (MP)	38	61	50
Plant Trait Ontology (PTO)	7	79	43

5.2 Case Study 2

The BioPortal repository was searched for ontologies containing each of ten preselected keywords. A list of applicable ontologies was generated for each of the keywords, and each ontology's Social Quality score determined by applying the method outlined in Case Study 1. Each potential ontology list was sorted in descending order to identify the highest results for each of the terms based upon social quality. The keyword searches each retrieved at least 30 potential ontologies. The results of ranking these ontology lists in reverse order of Social Quality was used to identify the best candidates for a possible task requiring each of the keywords. The top three ontologies recommended for each of the keywords are shown in Table 4.

Additional information provided by the BioPortal website showed that these listed ontologies are favorably reviewed and frequently accessed. In comparison, ontologies retrieved by the keyword search but scoring low on the Social Quality metric, were accessed infrequently, indicating little use within the community. For example, the Current Procedural Terminology ontology (CPT), which ranked highest for two of the keywords and obtained an SQ score of 53, received over 35,000 site visits in the last two years. In contrast, the Bone Dysplasia Ontology (BDO) containing the same two keywords, received an SQ score of 2 and only received 894 site visits.

Table 4. Top three ontologies recommended for each term with corresponding SQ scores.

Keyword	Number of potential ontologies	Highest scoring	Second highest	Third highest
Cell	47	CPT (53)	AURA (35)	FLU (29)
Temperature	39	MAT (52)	ACGT-MO (51)	MA (40)
Disease	46	ACGT-MO (51)	NMR (43)	AURA (35)
Blood	37	MAT (52)	ACGT-MO (51)	AURA (35)
Cortisol	30	CPT (53)	MESH (9)	PMA (21)
Sucrose	33	AURA (35)	DDO (34)	PR (26)
Water	45	AURA (35)	MESH (5)	CCONT (25)
Patient	45	AERO (51)	ACGT-MO (51)	CTX (27)
Oxygen	31	DCM (37)	AURA (35)	MESH (33)
Cerebrum	30	MA (40)	CTX (37)	MESH (33)

6   Conclusions and Future Work

This research has introduced two metrics for assessing the authority and history of an ontology within a community and illustrated their effectiveness by applying them to approximately four hundred of the ontologies in the BioPortal library. Results from that case study showed that application of the metrics was feasible and provided useful information regarding ontology recognition within its community.

Future work will consider other factors such as the number of times an ontology has been viewed or downloaded; user comments/ranking of ontologies; and the usability of ontologies to gain a more comprehensive view of the social quality metric. In addition, the social quality metrics need to be incorporated into broad metric suites that assess various attributes of ontologies. When users select an ontology from a number of options, a broad overview is required that considers syntax, semantics, pragmatics, as well as social acceptance, to make an appropriate recommendation to a user. Furthermore, it is necessary for any recommendation system to consider the task for which an ontology will be needed. Merely matching keywords is not enough to select an appropriate ontology; the specific characteristics of the actual task to be completed must also be taken into account.

Acknowledgement. This research is supported by the departments of Computer Information Systems and Computer Science, Georgia State University and a 2016 School of Business Administration Spring/Summer Research Fellowship from Oakland University.

References

1. Berners-Lee, T., Kagal, L.: The fractal nature of the semantic web. AI Mag. **29**(3), 29 (2008)
2. Gruber, T.R.: Toward principles for the design of ontologies used for knowledge sharing. Int. J. Hum. Comput. Stud. **43**(5), 907–928 (1995)

3. Hepp, M., Bachlechner, D., Siorpaes, K.: OntoWiki: community-driven ontology engineering and ontology usage based on Wikism. In: Proceedings of 2006 International Symposium on Wikis. ACM (2006)
4. Andrews, D.C.: Audience-specific online community design. Commun. ACM **45**(4), 64–68 (2002)
5. Noy, N.F., Shah, N.H., Whetzel, P.L., Dai, B., Dorf, M., Griffith, N., Jonquet, C., Rubin, D.L., Storey, M.A., Chute, C.G., Musen, M.A.: BioPortal: ontologies and integrated data resources at the click of a mouse. Nucleic Acids Res. **37**, 197 (2009)
6. Shadbolt, N., Hall, W., Berners-Lee, T.: The semantic web revisited. IEEE Intell. Syst. **21**(3), 96–101 (2006)
7. Noy, N.F., Griffith, N., Musen, M.A.: Collecting community-based mappings in an ontology repository. In: Sheth, A.P., Staab, S., Dean, M., Paolucci, M., Maynard, D., Finin, T., Thirunarayan, K. (eds.) ISWC 2008. LNCS, vol. 5318, pp. 371–386. Springer, Heidelberg (2008)
8. Stvilia, B., et al.: A framework for information quality assessment. J. Am. Soc. Inf. Sci. Technol. **58**, 1720–1733 (2007)
9. History. In: Macmillan Online Dictionary. http://www.macmillandictionary.com
10. Sudol, R.: Revising: new essays for teachers of writing. National Council of Teachers of English, 1111 Kenyon Rd., Urbana, IL (1982)
11. Horning, A., Becker, A.: Revision: History, Theory, and Practice. Parlor Press, Anderson (2006)
12. Wellman, B.: An electronic group is virtually a social network. Cult. Internet **4**, 179–205 (1997)
13. Su, X., Ilebrekke, L.: A comparative study of ontology languages and tools. In: Pidduck, A., Mylopoulos, J., Woo, C.C., Ozsu, M. (eds.) CAiSE 2002. LNCS, vol. 2348, pp. 761–765. Springer, Heidelberg (2002)
14. Berners-Lee, T., Hendler, J., Lassila, O.: The semantic web. Sci. Am. **284**(5), 28–37 (2001)
15. Noy, N.F., Fergerson, R.W., Musen, M.A.: The knowledge model of Protege-2000. In: Knowledge Engineering and Knowledge Management Methods, pp. 17–32 (2000)
16. Guarino, N., Welty, C.: Evaluating ontological decisions with OntoClean. Commun. ACM **45**, 61–65 (2002)
17. Lozano-Tello, A., Gómez-Pérez, A.: Ontometric: a method to choose the appropriate ontology. J. Database Manag. **15**, 1–18 (2004)
18. Burton-Jones, A., Storey, V.C., Sugumaran, V., Ahluwalia, P.: A semiotic metrics suite for assessing the quality of ontologies. Data Knowl. Eng. **55**(1), 84–102 (2005)
19. Tartir, S., Arpinar, I.B., Moore, M., Sheth, A.P., Aleman-Meza, B.: OntoQA: metric-based ontology quality analysis (2005)
20. Alani, H., Brewster, C., Shadbolt, N.R.: Ranking ontologies with AKTiveRank. In: Cruz, I., Decker, S., Allemang, D., Preist, C., Schwabe, D., Mika, P., Uschold, M., Aroyo, L.M. (eds.) ISWC 2006. LNCS, vol. 4273, pp. 1–15. Springer, Heidelberg (2006)
21. Gangemi, A., Catenacci, C., Ciaramita, M., Lehmann, J.: Modelling ontology evaluation and validation. In: Sure, Y., Domingue, J. (eds.) ESWC 2006. LNCS, vol. 4011, pp. 140–154. Springer, Heidelberg (2006)
22. Jonquet, C., Musen, M.A., Shah, N.H.: Building a biomedical ontology recommender web service. Biomed. Seman. **1**, S1 (2010)
23. Yu, J., Thom, J.A., Tam, A.: Requirements-oriented methodology for evaluating ontologies. Inf. Syst. **34**, 766–767 (2009)
24. Vrandečić, D.: Ontology evaluation. In: Handbook on Ontologies, pp. 293–313 (2009)

25. Duque-Ramos, A., Fernández-Breis, J.T., Stevens, R., Aussenac-Gilles, N.: OQuaRE: a SQuaRE-based approach for evaluating the quality of ontologies. J. Res. Pract. Inf. Technol. **43**, 159 (2011)

26. d'Aquin, M., Noy, N.F.: Where to publish and find ontologies? A survey of ontology libraries. Web Seman. Sci. Serv. Agents World Wide Web **11**, 96–111 (2012)

27. Stamper, R., Liu, K., Hafkamp, M., Ades, Y.: Understanding the roles of signs and norms in organizations-a semiotic approach to information systems design. Behav. Inf. Technol. **19**(1), 15–27 (2000)

28. Ashburner, M., et al.: Gene ontology: tool for the unification of biology, the gene ontology consortium. Nat. Genet. **25**, 25–29 (2000)

How to Complete Customer Requirements
Using Concept Expansion for Requirement Refinement

Michaela Geierhos[(✉)] and Frederik Simon Bäumer

Heinz Nixdorf Institute, University of Paderborn,
Fürstenallee 11, 33102 Paderborn, Germany
{geierhos,fbaeumer}@hni.upb.de
http://wiwi.upb.de/seminfo

Abstract. One purpose of requirement refinement is that higher-level requirements have to be translated to something usable by developers. Since customer requirements are often written in natural language by end users, they lack precision, completeness and consistency. Although user stories are often used in the requirement elicitation process in order to describe the possibilities how to interact with the software, there is always something unspoken. Here, we present techniques how to automatically refine vague software descriptions. Thus, we can bridge the gap by first revising natural language utterances from higher-level to more detailed customer requirements, before functionality matters. We therefore focus on the resolution of semantically incomplete user-generated sentences (i.e. non-instantiated arguments of predicates) and provide ontology-based gap-filling suggestions how to complete unverbalized information in the user's demand.

Keywords: Requirement refinement · Concept expansion · Ontology-based instantiation of predicate-argument structure

1 Introduction

In the Collaborative Research Center "On-The-Fly Computing", we develop techniques and processes for the automatic ad-hoc configuration of individual service compositions that fulfill customer requirements[1]. Upon request, suitable basic software and hardware services available on world-wide markets have to be automatically discovered and composed. For that purpose, customers have to provide software descriptions. These descriptions are subject to the same quality standards as software requirements collected and revised by experts: They should be complete, unambiguous and consistent [11]. To achieve these quality goals, techniques and approaches are often proposed, which are also used in classical requirement engineering. Popular examples are controlled languages and formal methods, which indeed can achieve the goal of clarity and completeness but are hardly to use for non-experts and therefore miss the target group.

---

[1] Refer to http://sfb901.uni-paderborn.de for more information.

© Springer International Publishing Switzerland 2016
E. Métais et al. (Eds.): NLDB 2016, LNCS 9612, pp. 37–47, 2016.
DOI: 10.1007/978-3-319-41754-7_4

From the customer's point of view, natural language (NL) cannot be formalized without partial loss in expressiveness [24]. Even if some details are missing, native speakers are able to understand what was meant in a concrete situation. Since the complexity and unrestrictedness of NL makes the automated requirement extraction process more difficult, we want to support the requirement refinement process by filling individual knowledge gaps. We therefore define incompleteness as missing information in user-generated requirements necessary for a feasible implementation or selection of the wanted software application.

Besides consistency, clarity and verifiability [9,12], completeness is one of the fundamental quality characteristics of software requirements. Although the notion of completeness is widely discussed, there is broad consensus on the negative impact of incomplete requirements for a software product [8], the product safety [10] or an entire project [15,25]. However, users are encouraged to express their individual requirements for a wanted software application in NL to improve user acceptance and satisfaction [6]. Therefore, our approach analyzes unrestricted NL requirement descriptions in order to provide suggestions how to elaborate the user's incomplete software specifications where possible.

This paper is structured as follows: In Sect. 2, we provide a brief overview of related work before we describe by means of a concrete example how an ontology can help to refine NL requirement descriptions (see Sect. 3). Finally, we present our next development steps and conclude in Sect. 4.

2   Related Work

The term of *incomplete requirements* often refers to the complete absence of requirements within a requirements documentation. Here, we focus on existing but incomplete requirements, which we call *incomplete individual requirements* [6]. In general, the preparation of checklists for request types as well as the application of "project-specific requirement completeness guidelines and/or standards" is recommended for hand-crafted requirement gathering [6]. Other approaches for the identification (and compensation) of incompleteness are often based on third-party reviews and are therefore affected by subjectivity, limited views, and even again inconsistency [17,28]. Especially, the perception of completeness can widely vary due to explicit and implicit assumptions [1].

Especially NL requirement descriptions suffer from incompleteness if not all required arguments of a predicate are given. For example, "send" is a three-place predicate because it requires the agent ("sender"), the theme ("sent") and the beneficiary argument ("sent-to"). If the beneficiary is not specified here, it is unknown whether one or more recipients are possible. But how many arguments does any predicate require? This information is provided by linguistic resources like FrameNet [2] to enable to automatic recognition of incompleteness [13,14] and its compensation [16]. For instance, RAT (Requirements Analysis Tool) can deal with incomplete (i.e. missing arguments of predicates) and missing NL requirements [27]. It therefore uses glossaries, controlled syntax and domain-specific ontologies. Another application is RESI (Requirements

Engineering Specification Improver), which can point out linguistic errors in requirements by using WordNet [18] and asks the user to give suggestions for improvement.

Since it remains unclear when some state of completeness will be reached [6], Ferrari et al. [5] developed the Completeness Assistant for Requirements (CAR), which is a tool to measure the completeness of requirement specifications by aligning mentioned requirements to descriptions in transcripts. Here, completeness is reached when all concepts and dependencies mentioned in the additional input documents are covered by the user's statement. This approach extracts terms and their dependencies from the existing document sources in order to create a kind of benchmark but without any use of external linguistic resources (e.g. ontologies).

Incompleteness occurs not only on the requester side (user requirements) but on the service provider side (service specifications) in software projects. The reasons therefor range from knowledge gaps on the requester side to conscious omission of information due to non-disclosure agreements on the provider side [21,22]. Even the perception of (in)completeness can vary due to explicit and implicit assumptions [1]. Previous work in the OTF context considered incomplete requirements as far as only (semi-)formal specifications – not NL – were supported as input format [19,22]. Geierhos et al. (2015) consider NL requirements and discuss an approach based on domain-based similarity search to compensate missing information. Their goal is to allow unrestricted requirement descriptions but to support the user by a "how to complete" feature [7]. Here, NL requirements descriptions that are unique in form and content are reduced to their main semantic cues and stored in templates for matching purposes (between requests and the provider's offers). Another way of reducing incompleteness is the transformation of NL requirements into formal specifications [4,23], which goes along with information loss due to the restrictions of the formal language [7].

3 Ontology-Based NL Requirement Refinement

3.1 Customer Requirements as Input Data

Given the following sample NL requirement description by a user:

"I *want* to *send* e-mails with large attachments."

We assume our input to be at least one sentence but do not limit it to a maximum number of words or sentences. Moreover, we probably have to deal with spelling or grammatical errors. Furthermore, we expect epistemic modality expressed by auxiliaries like "want" which is important for the ranking requirements according to user's priority. In order to detect the wanted functionality, we have to recognize the relevant parts of a sentence. We therefore start with the predicate argument analysis because especially full verbs (i.e. predicates) can be put on a level with function names ("send") that developers would choose. Furthermore, the noun phrases surrounding the predicate bear additional information specifying the input ("e-mails") or output parameters of the same function or hint at another one (e.g. "large attachments").

But how can we distinguish between parameters and further specifications of objects associated with the expressed functionality? For this purpose, we pursue two directions: predicate argument analysis to identify unspoken but semantically missing information (see Sect. 3.2) and ontology look-up for concept clarification and expansion (see Sect. 3.3).

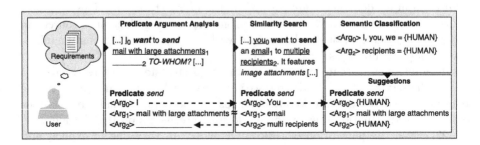

Fig. 1. Processing pipeline for knowledge gap filling during requirements analysis

3.2 Predicate Argument Analysis

In Fig. 1, we pick up the idea of semantic dependency [5] and combine it with external linguistic resources such as PropBank [20] to analyze the predicate-argument structure of NL requirement descriptions. Based on the results of this step, we are able to identify missing information and therefor provide similar instantiated arguments from our messaging ontology (see Fig. 2) in order to generate requirement-specific predicate-argument templates.

Preprocessing. Since NL requirement descriptions vary widely in quality and scope, further preprocessing is necessary: (1) sentence splitting, (2) filtering, (3) lemmatization, and (4) POS tagging. First we limit the recognition of a predicate and its argument(s) to the scope of a sentence, then we separate the off-topic sentences from the on-topic ones which we lemmatize and annotate their syntactic structure. We need the morpho-syntactic information to look-up in the PropBank [20] how many arguments (noun phrases) can be instantiated for the identified predicate (here: "send").

Semantic Role Labeling. Per each sentence, a semantic role labeler (SRL)[2] assigns iteratively semantic roles such as agent, theme, beneficiary represented by $Arg_0, ..., Arg_n$ to the arguments of a recognized predicate. In order to identify the different arguments, we use PropBank [20]. PropBank contains verbal propositions and their arguments. For the predicate *send* ("send.01"), we can obtain

---

[2] Curator's SRL [3] is used because of its convincing results on user-generated text.

three possible arguments: "sender" (Arg_0), "sent" (Arg_1), "sent-to" (Arg_2). By matching this information to our sample input sentence given in Sect. 3.1, the third argument position Arg_2 ("sent-to") remains non-instantiated.

In this case, it is irrelevant to whom (in person) the e-mail should be sent. However, the number of recipients matters for the functionality of the software application: addressing a single person or a group of people. Since we cannot expect the user to specify a concrete number of addressees (e.g. "to a group of 4 persons"), it is sufficient to distinguish the number (i.e., between plural, e.g. "to my friends" and singular, e.g. "to Peter"). Of course, it is not helpful to only point out the user's mistake without providing concrete suggestions. Based on the shared user stories, it would be possible to fill the gap with argument candidates such as "to my friends" or "to Peter". But if we suggested the user to elaborate his NL requirement description by replacing the undefined sender by e.g. Peter, this would be quite confusing. For this reason, we have to learn different variants for recipients, which will later be classified according to their semantic type (e.g. human being).

Grouping Argument Candidates. Semantically is the key to an intuitive user guidance. We therefore gathered 77,649 unique software descriptions and their corresponding reviews from download.com. Each record contains information about the rating, platform, publisher, version and the (sub-)category the software application belongs to (e.g. "messaging → e-mail"). After preprocessing, we applied SRL on the texts and received a very long frequency list of possible arguments for the predicate *send* per category.

Table 1. Possible instances for Arg_2 of *send*

Arg_2	Text	Category
sent-to	"colleagues"	e-mail
sent-to	"all costumers"	e-mail
sent-to	"the (...) mailing list"	e-mail
sent-to	"a specified address"	e-mail

According to the predicate argument analysis, *send* only occurred in 5.7 % of all texts, but it appeared in 36.3 % of all e-mail app descriptions where Arg_2 was non-instantiated in 60.1 % of these texts. To prove these results, we took a random sample of 200 texts and manually searched for utterances of *send* where arguments were unspecified.

As shown in Table 2, Arg_2 is missing in most of the 200 texts. This shows that e-mail writers premise that there has to be at least one recipient and he/she is human. For this reason, it does not seem to be necessary to specify the argument *send-to* in the category "e-mail". Even Arg_0 is often skipped. But the object Arg_1 is specified in most cases. At this point, we can identify a missing argument and provide the user with a list of suggestions sorted by frequency.

Table 2. Absence of Arg_2 in our test set

Argument	Absence
sender (Arg_0)	21 %
sent (Arg_1)	8 %
sent-to (Arg_2)	59 %

Ranking of Argument Suggestions. Table 3 shows the top three suggestions for our sample input in Sect. 3.1 that are provided to the user based on the above described approach. But why is "multiple recipients" on first place?

Table 3. Suggestions for Arg_2 of "send.01"

#	Argument	Semantic type
1	"multiple recipients"	Human
2	"all your customers"	Human
3	"multiple addresses"	Abstract

When we have a look on the most similar customer requirement to this input sentence, we retrieve the following annotated text as result of the predicate argument analysis:

"[...] $you_{\mathrm{Arg_0}}$ want to $send_{\mathrm{S_{01}}}$ an $e-mail_{\mathrm{Arg_1}}$ to $multiple\ recipients_{\mathrm{Arg_2}}$. It features multi [image]$_{\mathrm{IR_0}}$ [attachments]$_{\mathrm{IR_1}}$ [...]"[3].

Here, three possible arguments for *send* in the sense S_{01} (according to Prop-Bank) were identified because of keywords such as "image" and "attachments" which specify the meaning of *send* in our ontology (see next section).

As already mentioned, our approach is also able to semantically group argument candidates (e.g. suggestion no. 1 and no. 2 in Table 3 are typed as *human*). Thus, the user gets one more helpful hint that he or she only has to specify some kind of human being as Arg_2.

3.3   Ontology Look-Up

Disambiguation. In our sample sentence, the predicate *send* can also be used in the sense of faxing (Arg_1 = fax). This may change the possible instantiations for Arg_2 because a standard fax is not sent to "a mailing list" or to "multiple addresses" like an e-mail. This is extremely important because only context-dependent analysis can lead to precise suggestions. Thus, we created our own ontology (see Fig. 2) representing messaging functionality with its entities, relations and corresponding cardinality in order to disambiguate between the

---

[3] See http://download.cnet.com/PS/3000-2369_4-10970917.html for more details.

concepts or word senses (here: "send") respectively. When *send* co-occurs with "e-mail" or "attachment" in the customer requirement, we can limit the search to this domain, even if related terms were initially used.

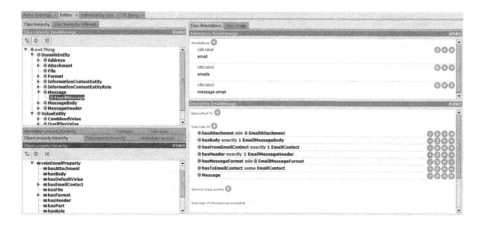

Fig. 2. Snapshot of the ontological representation of an e-mail

Clarification. However, it is still unclear what was meant by "large" in the context of e-mail attachment. We therefore determined the size limits by automatically extracting the maximum sizes of an e-mail from the help websites provided by Gmail, Yahoo! and others. In general, 10 MB is considered as the maximum size of an e-mail but Gmail increased its limit to 25 MB. Thus, we defined the interval for "large" as greater than 10 MB and smaller than 25 MB. Figure 3 shows that we foresee small, medium and large attachments in our ontology representing each different intervals and the derived unit is MB in this domain. So if a customer uses NL to specify requirements for a file upload in the context of messaging, we can clarify for the developer what was probably meant.

Semantic Typing. We therefore use the OpenCyc Ontology[4]. That way, we provide a better ranking of requirement refinement candidates by sorting the different argument instances grouped by semantic type.

When we have a look at Table 3, we can see that "multiple recipients" is Arg_2 in this specific context. In order to determine the semantic type of this noun phrase, we apply the Stanford CoreNLP dependency parser on this word sequence to identify the phrase head. In this case, the adjective "multiple" is tagged as modifier of the head "recipients". We need the phrase head to look up its semantic class, subtypes and its attributed term in the OpenCyc Ontology.

---

[4] http://sw.opencyc.org/.

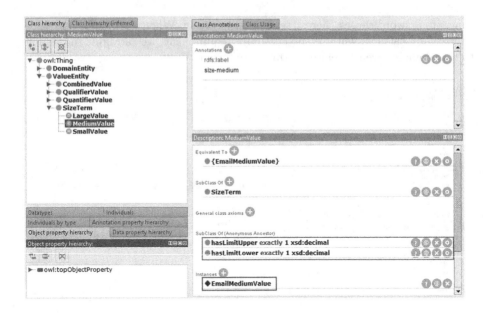

Fig. 3. Mapping vague indications of measurements to sizes for e-mail attachments

Concept Expansion. In order to provide the user with additional suggestions for requirement refinement, we are currently working on the implementation of the Concept Expansion Service offered by the IBM Developer Cloud[5]. It operates on the basis of predetermined concepts (initial seeds) and searches for similar concepts (contextual similarity to the input). The public available demo of the service uses unstructured content extracted from thousands of random websites. This is a good addition to our domain-specific download.com. The expansion process starts with "friends" and "family" together with two predefined concepts (i.e., customers and colleagues) taken from Table 1, which are the heads of Arg_2.

As Table 4 shows, the results are quite close to the input concepts and provide the user with concrete suggestions how to continue a fragmentary requirement description. Since the semantic type of the used input concepts is human, the precision is very high getting results of the same semantic type. But Table 1 also contains "address" and "mailing list" as more abstract term suggestions.

When expanding these input concepts and others, several domain-unspecific suggestions are retrieved. As Table 5 shows, concepts such as (1) "those", (2) "do not involve" and (3) "chapter officers" are part of the expansion. While we are able to identify (1) and (2) as bad results because they do not contain any noun, we cannot yet exclude (3) although it is semantically not close enough to the original input concepts.

---

[5] http://concept-expansion-demo.mybluemix.net.

Table 4. Other possible human instances for Arg_2 of *send*

Heads	Expanded concepts	Prevalence
friends	co-workers	0.73
family	family members	0.70
customers	coworkers	0.69
colleagues	acquaintances	0.68
	extended family	0.66
	close friends	0.66
	loved ones	0.66
	neighbors	0.65

Table 5. Abstract input concepts can lead to poor results

Heads	Expanded concepts	Prevalence
addresses	e-mail service provider	0.69
lists	distribution lists	0.68
individuals	chapter officers	0.67
anyone	those	0.66
	do not involve	0.65
	information seekers	0.65

We are still working on this issue how to decide whether to suggest an expanded concept for requirement refinement, especially when the results may be more confusing than helpful for the user.

4 Conclusion and Future Work

When running out of words, users provide incomplete NL requirement descriptions containing lots of implicit knowledge. In order to assist them with words, we developed a gap-filling matching approach based on a messaging ontology. With regard to the requested software domain, context-sensitive solutions for the instantiation of argument positions of the user input are suggested.

Because no suitable gold standard exists [26], we are building a gold standard for colloquial requirement descriptions, which allows us to evaluate our method on a large data base. However, this is a very time-consuming process, because every sentence must be manually annotated. Unfortunately, our gold standard for the domain "e-mail" only consists of 645 sentences, which cover the nine predicates "to answer", "to attach", "to create", "to delete", "to encrypt", "to reply", "to send", "to share" and "to zip". These are the most frequently used predicates in the e-mail domain from the Download.com data.

As soon as we have enough data to ensure a good coverage, we will perform such a gold standard based evaluation. Moreover, the approach of the concept expansion must be evaluated. For this purpose, first we have to solve the issue how to decide if an expanded concept for requirement refinement should be suggested or not, especially when the results may be more confusing than helpful for the user.

Acknowledgments. This work was partially supported by the German Research Foundation (DFG) within the Collaborative Research Centre On-The-Fly Computing (SFB 901).

References

1. Albayrak, Ö., Kurtoglu, H., Biaki, M.: Incomplete software requirements and assumptions made by software engineers. In: Proceedings of the 9th Asia-Pacific Software Engineering Conference, pp. 333–339, December 2009
2. Baker, C.F., Fillmore, C.J., Lowe, J.B.: The Berkeley FrameNet project. In: COLING-ACL 1998: Proceedings of the Conference, Montreal, pp. 86–90 (1998)
3. Clarke, J., Srikumar, V., Sammons, M., Roth, D.: An NLP curator (or: How I Learned to Stop Worrying and Love NLP Pipelines). In: Proceedings of the 8th International Conference on Language Resources and Evaluation (LREC'12), Istanbul, Turkey, pp. 3276–3283, 23–25 May 2012
4. Fatwanto, A.: Software requirements specification analysis using natural language processing technique. In: Proceedings of the International Conference on Quality in Research QiR 2013, Yogyakarta, pp. 105–110, June 2013
5. Ferrari, A., dell'Orletta, F., Spagnolo, G.O., Gnesi, S.: Measuring and improving the completeness of natural language requirements. In: Salinesi, C., van de Weerd, I. (eds.) REFSQ 2014. LNCS, vol. 8396, pp. 23–38. Springer, Heidelberg (2014)
6. Firesmith, D.G.: Are your requirements complete? J. Object Technol. **4**(2), 27–43 (2005)
7. Geierhos, M., Schulze, S., Bäumer, F.S.: What did you mean? Facing the challenges of user-generated software requirements. In: Loiseau, S., Filipe, J., Duval, B., van den Herik, J. (eds.) Proceedings of the 7th International Conference on Agents and Artificial Intelligence. Special Session on Partiality, Underspecification, and Natural Language Processing (PUaNLP 2015), pp. 277–283. SCITEPRESS - Science and Technology Publications, Lissabon (2015)
8. Ghazarian, A.: A case study of defect introduction mechanisms. In: van Eck, P., Gordijn, J., Wieringa, R. (eds.) CAiSE 2009. LNCS, vol. 5565, pp. 156–170. Springer, Heidelberg (2009)
9. Grande, M.: 100 Minuten für Anforderungsmanagement - Kompaktes Wissen nicht nur für Projektleiter und Entwickler. Springer, Wiesbaden (2011)
10. HSE. Out of control: why control systems go wrong and how to prevent failure. http://automatie-pma.com/wp-content/uploads/2015/02/hsg238.pdf (2003). Accessed 14 Feb 2016
11. Hsia, P., Davis, A., Kung, D.: Status report: requirements engineering. IEEE Softw. **10**(6), 75–79 (1993)
12. IEEE. IEEE Std 830-1998 - Recommended practice for software requirements specifications. Institute of Electrical and Electronics Engineers, New York (1998)
13. Kaiya, H., Saeki, M.: Ontology based requirements analysis: lightweight semantic processing approach. In: Proceedings of the 5th International Conference on Quality Software, pp. 223–230, September 2005

14. Kaiya, H., Saeki, M.: Using domain ontology as domain knowledge for requirements elicitation. In: 14th IEEE International Requirements Engineering Conference, pp. 189–198, September 2006
15. Kamata, M.I., Tamai, T.: How does requirements quality relate to project success or failure? In: Proceedings of the 15th IEEE International Requirements Engineering Conference, pp. 69–78, October 2007
16. Körner, S.J.: RECAA - Werkzeugunterstützung in der Anforderungserhebung. PhD thesis, Karlsruher Institut für Technologie (KIT), Karlsruhe, February 2014
17. Menzel, I., Mueller, M., Gross, A., Doerr, J.: An experimental comparison regarding the completeness of functional requirements specifications. In: Proceedings of the 18th IEEE International Requirements Engineering Conference, pp. 15–24, September 2010
18. Miller, G.A.: WordNet: a lexical database for English. Commun. ACM **38**(11), 39–41 (1995)
19. Naeem, M., Heckel, R., Orejas, F., Hermann, F.: Incremental service composition based on partial matching of visual contracts. In: Rosenblum, D.S., Taentzer, G. (eds.) FASE 2010. LNCS, vol. 6013, pp. 123–138. Springer, Heidelberg (2010)
20. Palmer, M., Gildea, D., Kingsbury, P.: The proposition bank: an annotated corpus of semantic roles. Comput. Linguist. **31**(1), 71–106 (2005)
21. Platenius, M.C.: Fuzzy service matching in on-the-fly computing. In: Proceedings of the 2013 9th Joint Meeting on Foundations of Software Engineering, ESEC/FSE 2013, pp. 715–718. ACM, New York (2013)
22. Platenius, M.C., Arifulina, S., Petrlic, R., Schäfer, W.: Matching of incomplete service specifications exemplified by privacy policy matching. In: Ortiz, G., Tran, C. (eds.) ESOCC 2014. CCIS, vol. 508, pp. 6–17. Springer, Heidelberg (2015)
23. Saeki, M., Horai, H., Enomoto, H.: Software development process from natural language specification. In: Proceedings of the 11th International Conference on Software Engineering, ICSE 1989, pp. 64–73. ACM, New York (1989)
24. Sommerville, I.: Web Chapter 27: formal specification. http://www.SoftwareEngineering-9.com/Web/ExtraChaps/FormalSpec.pdf (2009). Zuletzt abgerufen am 19 Aug 2015
25. Standish Group International. The CHAOS report (1994). https://www.standishgroup.com/sample_research_files/chaos_report_1994.pdf (1995). Accessed 14 Feb 2016
26. Tichy, W.F., Landhäußer, M., Körner, S.J.: nlrpBENCH: a benchmark for natural language requirements processing. In: Multikonferenz Software Engineering & Management 2015, March 2015
27. Verma, K., Kass, A.: Requirements analysis tool: a tool for automatically analyzing software requirements documents. In: Sheth, A.P., Staab, S., Dean, M., Paolucci, M., Maynard, D., Finin, T., Thirunarayan, K. (eds.) ISWC 2008. LNCS, vol. 5318, pp. 751–763. Springer, Heidelberg (2008)
28. Yadav, S.B., Bravoco, R.R., Chatfield, A.T., Rajkumar, T.M.: Comparison of analysis techniques for information requirement determination. Commun. ACM **31**(9), 1090–1097 (1988)

An Information Gain-Driven Feature Study
for Aspect-Based Sentiment Analysis

Kim Schouten[(✉)], Flavius Frasincar, and Rommert Dekker

Erasmus University Rotterdam,
Burgemeester Oudlaan 50, 3062 PA Rotterdam, The Netherlands
{schouten,frasincar,rdekker}@ese.eur.nl

Abstract. Nowadays, opinions are a ubiquitous part of the Web and sharing experiences has never been more popular. Information regarding consumer opinions is valuable for consumers and producers alike, aiding in their respective decision processes. Due to the size and heterogeneity of this type of information, computer algorithms are employed to gain the required insight. Current research, however, tends to forgo a rigorous analysis of the used features, only going so far as to analyze complete feature sets. In this paper we analyze which features are good predictors for aspect-level sentiment using Information Gain and why this is the case. We also present an extensive set of features and show that it is possible to use only a small fraction of the features at just a minor cost to accuracy.

Keywords: Sentiment analysis · Aspect-level sentiment analysis · Data mining · Feature analysis · Feature selection · Information gain

1 Introduction

Nowadays, opinions are a ubiquitous part of the Web and sharing experiences has never been more popular [4]. Information regarding consumer opinions is valuable for consumers and producers alike, aiding in their respective decision processes. Due to the size and heterogeneity of this type of information, computer algorithms are employed to gain insight into the sentiment expressed by consumers and on what particular aspects that sentiment is expressed, and research into this type of algorithms has enjoyed increasingly high popularity over the last decade [8].

Research has led to a number of different approaches to aspect-level sentiment analysis [13], that can be divided into three categories. The first group consists of methods that predominantly use a sentiment dictionary (e.g., [6]). Sentiment values are then assigned to certain words or phrases that appear in the dictionary and using a few simple rules (e.g., for negation and aggregation), the sentiment values are combined into one score for each aspect. The second type is categorized by the use of supervised machine learning methods (e.g., [3]). Using a significant amount of annotated data, where the sentiment is given for

© Springer International Publishing Switzerland 2016
E. Métais et al. (Eds.): NLDB 2016, LNCS 9612, pp. 48–59, 2016.
DOI: 10.1007/978-3-319-41754-7_5

each aspect, a classifier can be trained that can predict the sentiment value for yet unseen aspects. Last, some methods based on unsupervised machine learning are also available, but these usually combine aspect detection and sentiment classification into one algorithm (e.g., [14]).

Supervised learning has the advantage of high performance, and given the fact that sentiment is usually annotated as a few distinct classes (i.e., positive, neutral, and negative), traditional statistical classifiers work remarkably well. Unfortunately, most of these methods are somewhat of a black box: once provided with enough input, the method will do its task and will classify aspect sentiment with relatively good accuracy. However, the inner workings are often unknown, and because of that, it is also not known how the various input features relate to the task. Since most classifiers can deal with large dimensionality on the input, one tends to just give all possible features and let the classifier decide which ones to use. While this is perfectly fine when aiming for performance, it does not give much insight into which particular features are good predictors for aspect sentiment. Knowing which features are relevant is important for achieving insight into the performed task, but it also allows to speed up the training process by only employing the relevant features with possibly only a minor decrease in performance.

This focus on performance instead of explanation is most typically found in benchmark venues, such as the Semantic Evaluation workshops [12]. Here, participants get annotated training data and are asked to let their algorithm provide the annotations for a non-annotated data set. The provided annotations are then centrally evaluated and a ranking is given, showing how each of the participating systems fared against each other. Scientifically speaking, this has the big benefit of comparability, since all of the participants use the same data and evaluation is done centrally, as well as reproducibility, since the papers published in the workshop proceedings tend to focus on how the system was built. The one thing missing, however, is an explanation of why certain features or algorithms perform so great or that bad, since there usually is not enough space in the allowed system descriptions to include this.

Hence, this paper aims to provide insight into which features are useful, using a feature selection method based on Information Gain [11], which is one of the most popular feature filtering approaches. Compared to wrapper approaches, such as Forward Feature Selection, it does not depend on the used classification algorithm. Using Information Gain, we can compute a score for each individual feature that represents how well that feature divides the aspects between the various sentiment classes. Thus, we can move beyond the shallow analysis done per feature set, and provide deeper insight, on the individual feature level.

The remainder of this paper is organized as follows. First, the problem of aspect-level sentiment analysis is explained in more detail in Sect. 2, followed by Sect. 3, in which the framework that is responsible for the natural language processing is described, together with the methods for training the classifier and computing the Information Gain score. Then, in Sect. 4, the main feature analysis is performed, after which Sect. 5 closes with a conclusion and some suggestions for future work.

2   Problem Description

Sentiment analysis can be performed on different levels of granularity. For instance, a sentiment value can be assigned to a complete review, much like the star ratings that are used on websites like Amazon. However, to get a more in-depth analysis of the entity that is being reviewed, whether in a traditional review or in a short statement on social media, it is important to know on which aspect of the entity a statement is being made. Since entities, like products or services, have many facets and characteristics, ideally one would want to assign a sentiment value to a single aspect instead of to the whole package. This challenge is known as aspect-level sentiment analysis [13], or aspect-based sentiment analysis [12], and this is the field this research is focused on.

More precisely, we use a data set where each review is already split into sentences and for each sentence it is known what the aspects are. Finding the aspects is a task that is outside the scope of this paper. Given that these aspects are known, one would want to assign the right sentiment value to each of these aspects. Most of the annotated aspects are explicit, meaning that they are literally mentioned in the text. As such, it is known which words in the sentence represent this aspect. Some aspects, however, are implicit, which means that they are only implied by the context of the sentence or the review as a whole. For these aspects, there are no words that directly represent the aspect, even though there will be words or expressions that point to a certain aspect. Both explicit and implicit aspects are assigned to an aspect category which comes from a predefined list of possible aspect categories.

For explicit aspects, since we know the exact words in the sentence that represent this aspect, we can use a context of n words before and after each aspect from which to derive the features. This allows for contrasting aspects within the same sentence. For implicit aspects, this is not possible and hence we extract the features from the whole sentence. Note that each aspect will have a set of extracted features, since it is the sentiment value of each aspect that is the object

```
<sentence id="1032695:1">
  <text>Everything is always cooked to perfection, the
      service is excellent, the decor cool and understated.</
      text>
  <Opinions>
    <Opinion target="NULL" category="FOOD#QUALITY" polarity="
        " from="0" to="0"/>
    <Opinion target="service" category="SERVICE#GENERAL"
        polarity="" from="47" to="54"/>
    <Opinion target="decor" category="AMBIENCE#GENERAL"
        polarity="" from="73" to="78"/>
  </Opinions>
</sentence>
```

Fig. 1. A snippet from the used dataset showing an annotated sentence from a restaurant review.

of classification. An example from the used data set showing both explicit and implicit aspects (i.e., `target="NULL"` for implicit features) is shown in Fig. 1.

3 Framework

In this section we present the steps of our framework. First, all textual data is preprocessed, which is an essential task for sentiment analysis [5], by feeding it through a natural language pipeline based on Stanford's CoreNLP package [10]. This extracts information like the lemma, Part-of-Speech (PoS) tag, and grammatical relations for words in the text. Furthermore, we employ a spell checker called JLanguageTool[1] to correct obvious misspellings and a simple word sense disambiguation algorithm based on Lesk [7] to link words to their meaning, represented by WordNet synsets. Stop words are not removed since some of these words actually carry sentiment (e.g., emoticons are a famous example), and the feature selection will filter out features that are not useful anyway, regardless of whether they are stopwords or not.

The next step is to prepare all the features that will be used by an SVM [2], the employed classifier in this work. For example, if we want to use the lemma of each word as a feature, each unique lemma in the dataset will be collected and assigned a unique feature number, so when this feature is present in the text when training or testing, it can be denoted using that feature number. Note that, unless otherwise mentioned, all features are binary, denoting the presence or absence of that particular feature. For the feature analysis, the following types of features are considered:

- **Word-based features:**
  - Lemma: the dictionary form of a word;
  - Negation: whether or not one or more negation terms from the General Inquirer Lexicon[2] are present;
  - The number of positive words and the number of negative words in the context are also considered as features, again using the General Inquirer Lexicon;
- **Synset-based features:**
  - Synset: the WordNet synset associated with this word, representing its meaning in the current context;
  - Related synsets: synsets that are related in WordNet to one of the synsets in the context (e.g., hypernyms that generalize the synsets in the context);
- **Grammar-based features:**
  - Lemma-grammar: a binary grammatical relation between words represented by their lemma (e.g., "keep-nsubj-we");
  - Synset-grammar: a binary grammatical relation between words represented by their synsets which is only available in certain cases, since not every word has a synset (e.g., "ok#JJ#1-*cop*-be#VB#1");
  - PoS-grammar: a binary grammatical relation between words represented by PoS tags (e.g., "VB-nsubj-PRP"), generalizing the lemma-grammar case with respect to Part-of-Speech;

---

[1] wiki.languagetool.org/java-api.
[2] http://www.wjh.harvard.edu/~inquirer.

- Polarity-grammar: a binary grammatical relation between synsets represented by polarity labels (e.g., "neutral-nsubj-neutral"). The polarity class is retrieved from SentiWordNet [1], with the neutral class being the default when no entry was found in SentiWordNet;
- **Aspect features:**
  - Aspect Category: the category label assigned to each aspect is encoded as a set of binary features (e.g., "FOOD#QUALITY").

Note that with grammar-based features, we experiment with various sorts of triples, where two features are connected by means of a grammatical relation. This kind of feature is not well studied in literature, since n-grams are usually preferred by virtue of their simplicity. Apart from the effect of Information Gain, the benefit of using this kind of feature will be highlighted in the evaluation section.

With all the features known, the Information Gain score can be computed for each individual feature. This is done using only the training data. Information Gain is a statistical property that measures how well a given feature separates the training examples according to their target classification [11]. Information Gain is defined based on the measure entropy. The entropy measure characterizes the (im)purity of a collection of examples. Entropy is defined as:

$$Entropy(S) = - \sum_i p(i|S) \log_2 p(i|S)$$

with S a set of all aspects and $p(i|S)$ the fraction of the aspects in S belonging to class i. These classes are either positive, negative, or neutral. The entropy typically changes when we partition the training instances into smaller subsets, i.e., when analyzing the entropy value per feature. Information Gain represents the expected reduction in entropy caused by partitioning the samples according to the feature in question. The Information Gain of a feature t relative to a collection of aspects S, is defined as:

$$Information\ Gain(S,t) = Entropy(S) - \sum_{v \in Values(t)} \frac{|S_v|}{|S_t|} Entropy(S_v)$$

where $Values(t)$ is the set of all possible values for feature t. These values again are either positive, negative, or neutral. S_v is the subset of S with aspects of class v related to feature t. S_t is the set of all aspects belonging to feature t. $|\cdot|$ denotes the cardinality of a set.

In this paper, we will analyze the optimal number of features with Information Gain, one of the most popular measures used in conjunction with a filtering approach for feature selection. This is executed as follows. First, the Information Gain is computed for each feature. Next, the IG scores of all the features are sorted from high to low and the top $k\%$ features are used in the SVM. This percentage k can either be determined using validation data or it can be manually set.

Afterwards, the training data is used to train the SVM. The validation data is used to optimize for a number of parameters: the cost parameter C for the SVM, the context size n that determines how many words around an explicit aspect are used to extract features from, and the value for k that determines percentage-wise how many of the features are selected for use with the SVM.

After training, new, previously unseen data can be classified and the performance of the algorithm is computed. By employing ten-fold cross-validation, we test both the robustness of the proposed solution and ensure that the test data and training data have similar characteristics.

4   Evaluation

Evaluation is done on the official training data of the SemEval-2016 Aspect-Based Sentiment Analysis task[3]. We have chosen to only use the data set with restaurant reviews in this evaluation because it provides target information, annotating explicit aspects with the exact literal expression in the sentence that represents this aspect. An additional data set containing laptop reviews is also available, but it provides only category information and does not give the target expression. The used data set contains 350 reviews that describe the experiences people had when visiting a certain restaurant. There were no restrictions on what to write and no specific format or template was required. In Table 1 the distribution of sentiment classes over aspects is given and in Table 2, the proportion of explicit and implicit aspects in this dataset are shown (cf. Sect. 2).

To arrive at stable results for our analysis, we run our experiments using 10-fold cross-validation where the data set is divided in ten random parts of equal size. In this setup, seven parts are used for training the SVM, two parts are used

Table 1. The sentiment distribution over aspects in the used data set

Sentiment	Nr. of aspects	% of aspects
positive	1652	66.1 %
neutral	98	3.9 %
negative	749	30.0 %
total	2499	100 %

Table 2. The distribution of explicit and implicit aspects in the used data set

Type	Nr. of aspects	% of aspects
explicit	1879	75.2 %
implicit	620	24.8 %
total	2499	100 %

---

[3] http://alt.qcri.org/semeval2016/task5/.

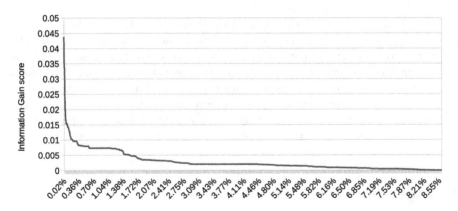

Fig. 2. The Information Gain for all features with a non-zero score, in descending order of Information Gain.

as a validation set to optimize certain parameters (i.e., the C parameter of the SVM, k, the exact percentage of features to be kept, and n, encoding how many words around an aspect should be used to extract features from), and the last part is used for testing. This procedure is repeated ten times, in such a way that the part for testing is different each round. This ensures that each part of the data set has been used for testing exactly once and so a complete evaluation result over the whole data set can be obtained.

From one of the folds, we extracted the list of features and their computed Information Gain. As shown in Fig. 2, the distribution of Information Gain scores over features is highly skewed. Only about 8.6 % of the features actually receives a non-zero score.

Grouping the features per feature type, we can compute the average Information Gain for all features of a certain type. This plot, shown in Fig. 3, shows how important, on average, each of the feature types is. Given the fact that the y-axis is logarithmic, the differences between importance are large. Traditional feature types like 'Negation present' and 'Category' are still crucial to having a good performance, but the new feature type 'Polarity-grammar' also shows good performance. The new 'Related-synsets' and 'POS-grammar' category are in the same league as the traditional 'Lemma' category, having an average Information Gain. Feature types that are less useful are 'Lemma-grammar' and 'Synset-grammar', which are very fine-grained and are thus less likely to generalize well from training to test data.

In Fig. 4, the average in-sample accuracy, as well as the accuracy on the validation and test data are presented for a number of values for k, where k means that the top k percent ranked features were used to train and run the SVM. It shows that when using just the top 1 % of the features, an accuracy of 72.4 % can be obtained, which is only 2.9 % less than the performance obtained when using all features. This point corresponds to a maximum in the performance on the validation data. Other parameters that are optimized using validation data

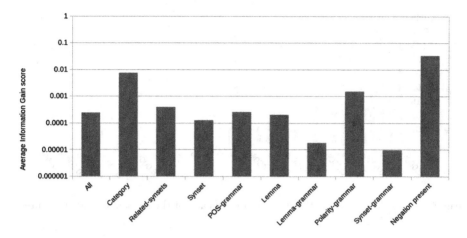

Fig. 3. The average Information Gain for each feature type.

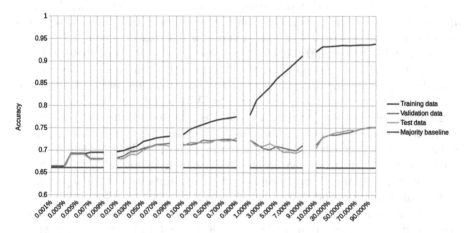

Fig. 4. The average accuracy on training, validation, and test set for each of the subsets of features. (Color figure online)

are C, which is on average set to 1, and n, which is on average set to 4. Note that since all parameters that are optimized with validation data are optimized per fold, their exact value differs per fold and thus an average is given. The picture is split into five different levels of granularity on the x-axis, providing more details for lower values of k. In the first split, the method start at the baseline performance, since at such a low value of k, no features are selected at all. Then, in the second block, one can see that, because these are all features with high Information Gain, the performance on the test data closely tracks the in-sample performance on the training data. However, in the third split, some minor overfitting starts to occur. The best features have already been used, so lesser features make their way into the SVM, and while in-sample performance goes

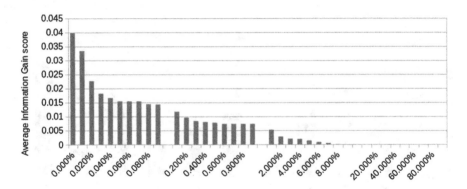

Fig. 5. The average Information Gain score for each of the subsets of added features.

up, out-of-sample performance does not grow as fast. This effect is illustrated even stronger in the fourth block, where performance on the training data goes up spectacularly, while performance on the test data actually goes down. The features that are added here all have a very low Information Gain.

The last block is slightly different because for almost all of these features, roughly 90 % of the total number, the Information Gain is zero. However, as is made evident by the upward slope of the out-of-sample performance, these features are not necessarily useless and the SVM is able to use them to boost performance with a few percent. This is possible due to the fact that, while Information Gain is computed for each feature in isolation, the SVM takes interaction effects between features into account. Hence, while these features may not be useful on their own, given the features already available to the SVM, they can still be of use. This interaction effect accounts for the 2.9 % penalty to performance increase when using feature selection based on Information Gain.

The diminishing Information Gain is also clearly illustrated in Fig. 5. It follows roughly the same setup in levels of detail as Fig. 4, with this exception that the first split is combined into the first bar, since not enough features were selected in the first split to create a meaningful set of bars. Furthermore, while Fig. 4 has a cumulative x-axis, this figure does not, showing the average Information Gain of features that are added in each range (instead of all features added up to that point). Given that the lowest 90 % of the features has an Information Gain of zero, there are no visible bars in the last split.

For each of the feature selections, we can also look at how well each type of feature is represented in that subset. Hence, we plot the percentage of features selected belonging to each of the feature types in Fig. 6. Features whose proportion decrease when adding more features are generally more important according to the Information Gain ranking, while features whose proportion increases when adding more features are generally less important since they generally have a lower Information Gain. This corresponds to the feature types with high bars in Fig. 3. It is interesting to see that 'Negation' and 'Category' are small but strong feature sets, and that 'Related-synsets', while not having many strong features,

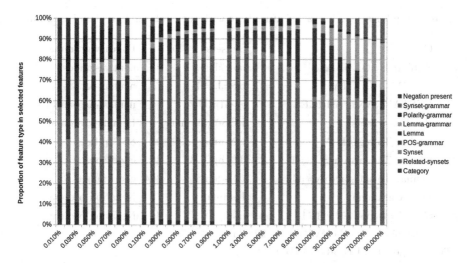

Fig. 6. Proportion of each feature type for each of the cumulative subsets of features. (Color figure online)

has many features that get a non-zero Information Gain, making it still a useful category of features.

Analyzing the top ranked features per feature type in Table 3, some of the features are easily recognized as being helpful to detect sentiment. For instance, the lemma 'not', as well as its corresponding synset in WordNet are good indicators of negative sentiment. Other features, like the 'SERVICE#GENERAL' category feature are not so self-evident. These more generic features, while not directly pointing to a certain sentiment value, are still useful by virtue of their statistics. Again looking at the 'SERVICE#GENERAL' category, we check the dataset and see that about 56 % of all aspects with this category have a negative sentiment, whereas overall, only 30 % of the aspects are negative. This is such a sharp deviation from the norm, that having this category label is a strong sign for an aspect to be negative. It seems that in the used data set, people tend to be dissatisfied with the provided service.

Sometimes features that appear in different categories might still represent (almost) the same information. For instance, the two top 'Lemma-grammar' features are basically the same as the two top 'Synset-grammar' features, corresponding to a phrase like "*some aspect is ok*" or "*some aspect is good*". Another example of this is the lemma 'ok' and its corresponding synset 'ok#JJ#1'.

An interesting type of features is the 'Related-synsets' category. In Table 3, we seen that any synset that is similar to concepts like 'big', 'alarming', and 'satisfactory' are good predictors of sentiment, and this corresponds well with our intuition. Sometimes, a high ranked feature can give insight into how consumers write their reviews. A good example is the 'CD-dep-$' feature in the 'POS-grammar' category, which denotes a concrete price, such as "$100", and is predominantly used in conjunction with a negative sentiment. Apparently, when

Table 3. Top 3 features for each feature type with their Information Gain based rank.

	Category		Synsets		Polarity-grammar
1	SERVICE#GENERAL	3	not#RB#1	6	neutral-amod-positive
40	FOOD#QUALITY	27	ok#JJ#1	7	neutral-amod-neutral
42	RESTAURANT#PRICES	37	good#JJ#1	11	neutral-neg-negative
	Related-synsets		POS-grammar		Lemma-grammar
2	Similar To big#JJ#1	9	NN-amod-JJ	28	ok-cop-be
8	Similar To alarming#JJ#1	25	JJ-cop-VBZ	35	good-cop-be
10	Similar To satisfactory#JJ#1	34	CD-dep-$	374	good-punct-
	Synset-grammar		Lemma		Negation present
29	ok#JJ#1-cop-be#VB#1	5	not	4	Negation present
45	good#JJ#1-cop-be#VB#1	22	do		
705	average#JJ#1-cop-be#VB#1	26	ok		

people are upset about the price of a restaurant, they feel the need to prove their point by mentioning the exact price.

Last, the 'Polarity-grammar' features also score well in terms of Information Gain. The three top features in this category would match phrases such as "good service", "big portions", and "not returning", respectively. Even the 'neutral-amod-neutral' is used in a positive context about 80 % of the time and is therefore a good predictor of positive sentiment. The first and third feature are obvious predictors for positive and negative sentiment, respectively.

In terms of computing time, if we define the training time when using all features to be 100 %, we find that training with 1 % of the features takes about 20 % of the original time, whereas employing only 0.1 % of the features requires just over 1 % of the original time.

5 Conclusion and Future Work

In this paper, filtering individual features using Information Gain is shown to provide good results. With only the 1 % best features in terms of Information Gain, an accuracy is obtained that is only 2.9 % below the accuracy obtained when using all features. Furthermore, training the SVM with 1 % of the features takes only 20 % of the time required to train it using all features. Apart from feature selection, we have shown the effectiveness of a number of relatively unknown types of features, such as 'Related-synsets' and 'Polarity-grammar'. For future work, the set of features can be expanded even further to include a comparison of grammar based features against n-gram based features. Also of interest is the context of an aspect from which we compute the sentiment score. Currently, this is determined using a simple word distance around the aspect words, but this could be done in a more advanced way, for instance using grammatical relations or even Rhetorical Structure Theory [9].

Acknowledgments. The authors are supported by the Dutch national program COMMIT. We would like to thank Nienke Dijkstra, Vivian Hinfelaar, Isabelle Houck, Tim van den IJssel, and Eline van de Ven, for many fruitful discussions during this research.

References

1. Baccianella, S., Esuli, A., Sebastiani, F.: SentiWordNet 3.0: an enhanced lexical resource for sentiment analysis and opinion mining. In: Proceedings of the Seventh International Conference on Language Resources and Evaluation (LREC 2010), vol. 10, pp. 2200–2204 (2010)
2. Chang, C., Lin, C.: LIBSVM: a library for support vector machines. ACM Trans. Intell. Syst. Technol. **2**(3), 27 (2011)
3. Choi, Y., Cardie, C.: Learning with compositional semantics as structural inference for subsentential sentiment analysis. In: Proceedings of the Conference on Empirical Methods in Natural Language Processing 2008 (EMNLP 2008), pp. 793–801 (2008)
4. Feldman, R.: Techniques and applications for sentiment analysis. Commun. ACM **56**(4), 82–89 (2013)
5. Haddi, E., Liu, X., Shi, Y.: The role of text pre-processing in sentiment analysis. Procedia Comput. Sci. **17**, 26–32 (2013)
6. Hu, M., Liu, B.: Mining and summarizing customer reviews. In: Proceedings of 10th ACM SIGKDD International Conference on Knowledge Discovery and Data Mining (KDD 2004), pp. 168–177. ACM (2004)
7. Lesk, M.: Automatic sense disambiguation using machine readable dictionaries: how to tell a pine cone from an ice cream cone. In: Proceedings of the Fifth Annual International Conference on Systems Documentation (SIGDOC 1986), pp. 24–26. ACM (1986)
8. Liu, B.: Sentiment Analysis and Opinion Mining. Synthesis Lectures on Human Language Technologies. Morgan & Claypool Publishers, San Rafael (2012)
9. Mann, W.C., Thompson, S.A.: Rhetorical structure theory: toward a functional theory of text organization. Text Interdiscip. J. Study Discourse **8**(3), 243–281 (1988)
10. Manning, C.D., Surdeanu, M., Bauer, J., Finkel, J., Bethard, S.J., McClosky, D.: The Stanford CoreNLP natural language processing toolkit. In: Proceedings of 52nd Annual Meeting of the Association for Computational Linguistics: System Demonstrations, pp. 55–60. Association for Computational Linguistics (2014)
11. Mitchell, T.M.: Machine Learning, 1st edn. McGraw-Hill Inc., New York (1997)
12. Pontiki, M., Galanis, D., Papageorgiou, H., Manandhar, S., Androutsopoulos, I.: SemEval-2015 task 12: aspect based sentiment analysis. In: Proceedings of the 9th International Workshop on Semantic Evaluation (SemEval 2015), pp. 486–495. Association for Computational Linguistics (2015)
13. Schouten, K., Frasincar, F.: Survey on aspect-level sentiment analysis. IEEE Trans. Knowl. Data Eng. **28**(3), 813–830 (2016)
14. Titov, I., McDonald, R.: A joint model of text and aspect ratings for sentiment summarization. In: Proceedings of the 46th Annual Meeting of the Association for Computational Linguistics: Human Language Technologies (HLT 2008), pp. 308–316. ACL (2008)

ESSOT: An Expert Supporting System for Ontology Translation

Mihael Arcan[1], Mauro Dragoni[2(✉)], and Paul Buitelaar[1]

[1] Insight Centre for Data Analytics, National University of Ireland, Galway, Ireland
{mihael.arcan,paul.buitelaar}@insight-centre.org
[2] FBK-Fondazione Bruno Kessler, via Sommarive 18, 38123 Trento, Italy
dragoni@fbk.eu

Abstract. To enable knowledge access across languages, ontologies, mostly represented only in English, need to be translated into different languages. The main challenge in translating ontologies with machine translation is to disambiguate an ontology label with respect to the domain modelled by the ontology itself; however, a crucial requirement is to have translations validated by experts before the ontologies are deployed. Real-world applications have to implement a support system addressing this task to help experts in validating automatically generated translations. In this paper, we present ESSOT, an Expert Supporting System for Ontology Translation. The peculiarity of this system is to exploit the semantic information of the label's context to improve the quality of label translations. The system has been tested within the Organic.Lingua project by translating the modelled ontology in three languages, whereby the results are compared with translations provided by the Microsoft Translator API. The provided results demonstrate the viability of our proposed approach.

1 Introduction

Nowadays, semantically structured data, i.e. ontologies or taxonomies, typically have labels stored in English only. Although the increasing number of ontologies offers an excellent opportunity to link this knowledge together, non-English users may encounter difficulties when using the ontological knowledge represented in English only [1]. Furthermore, applications in information retrieval or knowledge management, using monolingual ontologies are limited to the language in which the ontology labels are stored. Therefore, to make ontological knowledge accessible beyond language borders, these monolingual resources need to be enhanced with multilingual information [2].

Since manual multilingual enhancement of domain-specific ontologies is very time consuming and expensive, we engage a domain-aware statistical machine translation (SMT) system, called OTTO, embedded within the ESSOT system to automatically translate the ontology labels. As ontologies may change over time, having in place an SMT system adaptable to an ontology can therefore be very beneficial. Nevertheless, the quality of the SMT generated translations relies strongly on the translation model learned from the information stored in parallel corpora.

© Springer International Publishing Switzerland 2016
E. Métais et al. (Eds.): NLDB 2016, LNCS 9612, pp. 60–73, 2016.
DOI: 10.1007/978-3-319-41754-7_6

In most cases, the inference of translation candidates cannot always be learned accurately when domain-specific vocabulary, like ontology labels, appears infrequent in a parallel corpus. Additionally, ambiguous labels built out of only a few words do not express enough semantic information to guide the SMT system in translating a label correctly in the targeted domain. This can be observed in domain-independent systems, e.g. Microsoft Translator,[1] where an ambiguous expression, like *vessel* stored in a medical ontology, is translated as *Schiff*[2] (en. *ship*) in German, but not into the targeted medical domain as *Gefäβ*.

In this paper, we present ESSOT with the domain-aware SMT system, called OTTO, integrated into a collaborative knowledge management platform for supporting language experts in the task of translating ontologies. The benefits of such a platform are *(i)* the possibility of having an all-in-one solution, containing both an environment for modelling ontologies which enables the collaboration between different type of experts and *(ii)* a pluggable domain-adaptable service for supporting ontology translations. The proposed solution has been validated in a real-world context, namely Organic.Lingua,[3] from quantitative and qualitative points of view by demonstrating the effort decrease required by the language experts for completing the translation of an entire ontology.

2   Related Work

In this section, we summarize approaches related to ontology translation and present a brief review of the available ontology management tools with a particular emphasis on their capabilities in supporting language experts for translating ontologies.

The task of ontology translation involves generating an appropriate translation for the lexical layer, i.e. labels stored in the ontology. Most of the previous related work focused on accessing existing multilingual lexical resources, like EuroWordNet or IATE [3,4]. This work focused on the identification of the lexical overlap between the ontology and the multilingual resources, which guarantees a high precision but a low recall. Consequently, external translation services like BabelFish, SDL FreeTranslation tool or Google Translate were used to overcome this issue [5,6]. Additionally, [5,7] performed ontology label disambiguation, where the ontology structure is used to annotate the labels with their semantic senses. Similarly, [8] show positive effect of different domain adaptation techniques, i.e., using web resources as additional bilingual knowledge, re-scoring translations with Explicit Semantic Analysis, language model adaptation) for automatic ontology translation. Differently to the aforementioned approaches, which rely on external knowledge or services, the machinery implemented in ESSOT is supported by a domain-aware SMT system, which provides adequate translations using the ontology hierarchy and the contextual information of labels in domain-relevant background text data.

---

[1] http://www.bing.com/translator/.

[2] Translation performed on 2.3.2016.

[3] http://www.organic-lingua.eu.

Concerning the multilingual ontology management tools, we identified three that may be compared with the capabilities provided by MoKi: *Neon* [9], *VocBench* [10], and *Protégé* [11].

The main features of the *The NeOn toolkit*[4] include the management and the evolution of ontologies in an open, networked environment; the support for collaborative development of networked ontologies; the possibility of using contexts for developing, sharing, adapting and maintaining networked ontologies and an improved human-ontology interaction (i.e. making it easier for users with different levels of expertise and experience to browse and make sense of ontologies).

VocBench[5] is a web-based, multilingual, editing and workflow tool that manages thesauri, authority lists and glossaries using SKOS-XL. Designed to meet the needs of semantic web and linked data environments, VocBench provides tools and functionalities that facilitate both collaborative editing and multilingual terminology. It also includes administration and group management features that permit flexible roles for maintenance, validation and publication.

Protégé[6] is a free, open source visual ontology editor and knowledge-base framework. The Protégé platform supports two main ways of modelling ontologies via the Protégé-Frames and Protégé-OWL editors. Protégé ontologies can be exported into a variety of formats including RDF(S), OWL, and XML Schema.

While the first two, *Neon* and *VocBench*, are the ones more oriented for supporting the management of multilinguality in ontologies by including dedicated mechanisms for modelling the multilingual fashion of each concept; the support for multilinguality provided by *Protégé* is restricted to the sole description of the labels. However, differently from MoKi, none of them implements the capability of connecting the tool to an external machine translation system for suggesting translations automatically.

3   The Organic.Lingua Project

Organic.Lingua is an EU-funded project that aims at providing automated multilingual services and tools facilitating the discovery, retrieval, exploitation and extension of digital educational content related to Organic Agriculture and AgroEcology. More concretely, the project aims at providing, on top of a web portal, cross-lingual facility services enabling users to *(i)* find resources in languages different from the ones in which the query has been formulated and/or the resource described (e.g., providing services for cross-lingual retrieval); *(ii)* manage meta-data information for resources in different languages (e.g., offering automated meta-data translation services); and *(iii)* contribute to evolving content (e.g., providing services supporting the users in content generation).

These objectives are reached in the Organic.Lingua project by means of two components: on the one hand, a web portal offering software components and linguistic resources able to provide multilingual services and, on the other hand,

---

[4] http://neon-toolkit.org/wiki/Main_Page.
[5] http://vocbench.uniroma2.it/.
[6] http://protege.stanford.edu/.

a conceptual model (formalized in the "Organic.Lingua ontology") used for managing information associated with the resources provided to the final users and shared with other components deployed on the Organic.Lingua platform. In a nutshell, the usage of the Organic.Lingua ontology is twofold: *(i)* resource annotation (each time a content provider inserts a resource in the repository, the resource is annotated with one or more concepts extracted from the ontology) and *(ii)* resource retrieval (when web users perform queries on the system, the ontology is used, by the back-end information retrieval system, to perform advanced searches based on semantic techniques). Due to this intensive use of the ontology in the entire Organic.Lingua portal, the accuracy of the linguistic layer, represented by the set of translated labels, is crucial for supporting the annotation and retrieval functionalities.

4 Machine Translation for Ontology Translation

Due to the shortness of ontology labels, there is a lack of contextual information, which can otherwise help disambiguating expressions. Therefore, our goal is to translate the identified ontology labels within the textual context of the targeted domain, rather than in isolation. To identify the most domain-specific source sentences containing the label to be translated we engage the OnTology TranslatiOn System, called OTTO[7] [12]. With this approach, we aim to retain relevant sentences, where the English label *vessel* belongs to the medical domain, but not to the technical domain, which would cause a wrong, out-of-domain translation. This process reduces the semantic noise in the translation process, since we try to avoid contextual information that does not belong to the domain of the targeted ontology.

Statistical Machine Translation. Our approach is based on statistical machine translation, where we wish to find the best translation $\mathbf{e}$, of a string $\mathbf{f}$, given by a log-linear model combining a set of features. The translation that maximizes the score of the log-linear model is obtained by searching all possible translations candidates. The decoder, which is essentially a search procedure, provides the most probable translation based on a statistical translation model learned from the training data.

For a broader domain coverage of an SMT system, we merged several parallel corpora necessary to train an SMT system, e.g. JRC-Acquis [13], Europarl [14], DGT (translation memories generated by the *Directorate-General for Translation*) [15], MultiUN corpus [16] and TED talks [17] among others, into one parallel dataset. For the translation approach, the OTTO System engages the widely used Moses toolkit [18]. Word alignments were built with GIZA++ [19] and a 5-gram language model was built with KenLM [20].

Relevant Sentence Selection. In order to translate an ontology label in the closest domain-specific contextual environment, we identify within the concatenated corpus only those source sentences, which are most relevant to the labels

---

[7] http://server1.nlp.insight-centre.org/otto/.

to be translated. Nevertheless, due to the specificity of the ontology labels, just an *n-gram overlap* approach is not sufficient to select all the useful sentences. For this reason, we follow the idea of [21], where the authors extend the semantic information of ontology labels using Word2Vec [22] for computing distributed representations of words. The technique is based on a neural network that analyses the textual data provided as input and outputs a list of semantically related words. Each input string, in our experiment ontology labels or source sentences, is vectorized using the surrounding context and compared to other vectorized sets of words in a multi-dimensional vector space. Word relatedness is measured through the cosine similarity between two word vectors.

The usage of the ontology hierarchy allows us to further improve the disambiguation of short labels, i.e., the related words of a label are concatenated with the related words of its direct parent. Given a label and a source sentence from the concatenated corpus, related words and their weights are extracted from both of them, and used as entries of the vectors to calculate the cosine similarity. Finally, the most similar source sentence and the label should share the largest number of related words.

OTTO Service in Action. The OTTO service[8] works as a pipeline of tasks. When a user invokes the translation service, all labels contained in the ontology context (where as "context" we mean the set of concepts that are connected directly or with a maximum distance of N arcs with the concept that has to be translated) are extracted from the ontology and stored in the message that is sent to the OTTO service (left part in Fig. 1). When the service receives the translation request, the service looks into the model for the best candidate translations by considering the contextual information accompanying the ontology label to translate. A ranked list based on log probabilities of candidate translations within the JSON output format (Fig. 1) is generated from the OTTO service and sent back to the user that will select, among the proposed translations, the one to save in the ontology.

```
{
    "label2translate":"vessels",
    "concept_context":[
        "blood",
        "medical",
        "disease",
        "biomedical",
    ],
    "translate2":"de"
}
```

```
{
    "possible_translations":{
        "blutgefäßen":-15.8438,
        "gefäßen":-2.4100,
        "halsgefäße":-2.6682
    },
    "time":"24 wallclock secs",
    "source_label":"vessels",
    "best":"gefäßen"
}
```

Fig. 1. JSON representations provided to and from the OTTO system.

---

[8] For more information how to invoke the service, see also: http://server1.nlp.insight-centre.org/otto/rest_service.html.

5 Supporting the Ontology Translation Activity with **MoKi**

The translation component described in the previous section has been integrated in a collaborative knowledge management tool called MoKi [9][23]. It is a collaborative MediaWiki-based[10] tool for modelling ontological and procedural knowledge in an integrated manner[11] and is grounded on three main pillars:

- each basic entity of the ontology (i.e., concepts, object and datatype properties, and individuals) is associated with a wiki page;
- each wiki page describes an entity by means of both unstructured (e.g., free text, images) and structured (e.g. OWL axioms) content;
- multi-mode access to the page content is provided to support easy usage by users with different skills and competencies.

In order to meet the needs of the specific ontology translation task within the Organic.Lingua project, MoKi has been customized with additional facilities: *(i)* connection with the OTTO service that is in charge of providing the translations of labels and descriptions associated with the ontology entities; and *(ii)* user-friendly collaborative features specifically targeting linguistic issues. Translating domain-specific ontologies, in fact, demands that experts discuss and reach an agreement not only with respect to modelling choices, but also to (automated) ontology label translations.

Below, we present the list of the implemented facilities specifically designed for supporting the management of the multilingual layer of the Organic.Lingua ontology.[12]

Domain and Language Experts View. The semi-structured access mode, dedicated to the Domain and Language Experts, has been equipped with functionalities that permits revisions of the linguistic layer. This set of functionalities permits to revise the translation of names and descriptions of each entity (concepts, individuals, and properties).

For browsing and editing of the translations, a quick view box has been inserted into the mask (as shown in Fig. 2); in this way, language experts are able to navigate through the available translations and, eventually, invoke the translation service for retrieving a suggestion or, alternatively, to edit the translation by themselves (Fig. 3).

Approval and Discussion Facilities. Given the complexity of translating domain specific ontologies, translations often need to be checked and agreed upon by a community of experts. This is especially true when ontologies are

---

[9] http://moki.fbk.eu.

[10] Wikimedia Foundation/Mediawiki: http://www.mediawiki.org.

[11] Though MoKi allows to model both ontological and procedural knowledge, here we will limit our description only to the features for building multilingual ontologies.

[12] A read-only version, but with all functionalities available, of the MoKi instance described in this paper is available at https://dkmtools.fbk.eu/moki/3_5/essot/.

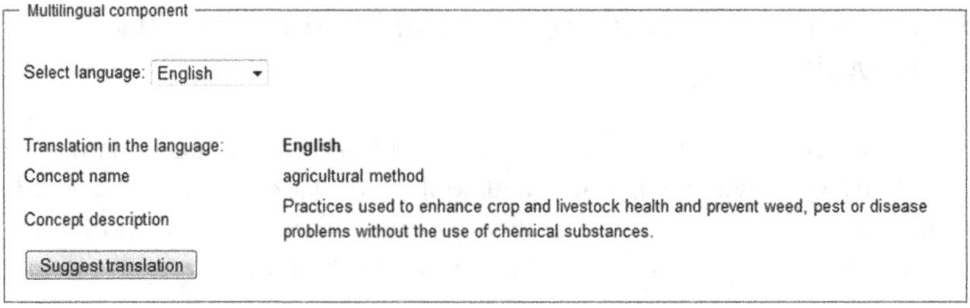

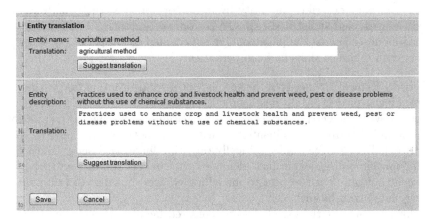

Fig. 2. Multilingual box for facilitating the entity translation

Fig. 3. Quick translation box for editing entities translations

used to represent terminological standards which need to be carefully discussed and evaluated. To support this collaborative activity we foresee the usage of the wiki-style features of MoKi, expanded with the possibility of assigning specific translations of ontology labels to specific experts who need to monitor, check, and approve the suggested translations. This customization promotes the management of the changes carried out on the ontology (in both layers) by providing the facilities necessary to manage the life-cycle of each change.

These facilities may be split in two different sets of features. The first group may be considered as a monitor of the activities performed on each entity page. When changes are committed, approval requests are created. They contain the identification of the expert in charge of approving the change, the date on which the change has been performed, and a natural language description of the change. Moreover, a mechanism for managing the approvals and for maintaining the history of all approval requests for each entity is provided. Instead, the second set contains the facilities for managing the discussions associated with each entity page. A user interface for creating the discussions has been implemented together with a notification procedure that alerts users when new topics/replies, related to the discussions that they are following, have been posted.

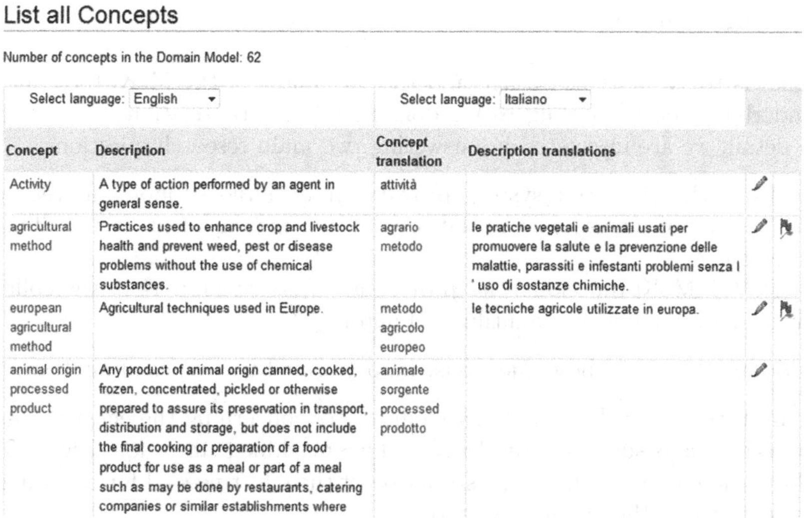

List all Concepts

Number of concepts in the Domain Model: 62

Select language: English ▼   Select language: Italiano ▼

Concept	Description	Concept translation	Description translations	
Activity	A type of action performed by an agent in general sense.	attività		✎
agricultural method	Practices used to enhance crop and livestock health and prevent weed, pest or disease problems without the use of chemical substances.	agrario metodo	le pratiche vegetali e animali usati per promuovere la salute e la prevenzione delle malattie, parassiti e infestanti problemi senza l´ uso di sostanze chimiche.	✎ ▨
european agricultural method	Agricultural techniques used in Europe.	metodo agricolo europeo	le tecniche agricole utilizzate in europa.	✎ ▨
animal origin processed product	Any product of animal origin canned, cooked, frozen, concentrated, pickled or otherwise prepared to assure its preservation in transport, distribution and storage, but does not include the final cooking or preparation of a food product for use as a meal or part of a meal such as may be done by restaurants, catering companies or similar establishments where	animale sorgente processed prodotto		✎

Fig. 4. View for comparing entities translations

"Quick" Translation Feature. For facilitating the work of language experts, we have implemented the possibility of comparing side-by-side two lists of translations. This way, the language expert in charge of revising the translations, avoiding to navigate among the entity pages, is able to speed-up the revision process.

Figure 4 shows such a view, by presenting the list of English concepts with their translations into Italian. At the right of each element of the table, a link is placed allowing to invoke a quick translation box (as shown in Fig. 3) that gives the opportunity to quickly modify information without opening the entity page. Finally, in the last column, a flag is placed indicating that changes have been performed on that concept, and a revision/approval is requested.

Interface and Ontology Multilingual Facilities. In order to complete the set of features available for managing the multilingual aspects of the Organic.Lingua project, MoKi has been equipped with two further components that permit to switch between the languages available for the tool interface: to add a new language to the ontology and to select the language used for showing the ontology in different views.

Through these facilities, it is also possible to add a new language to the MoKi interface and to manage the translation of its labels. This module has been implemented on top of the multilingual features of MediaWiki.

Concerning the ontology, when new labels are added to the ontology, the OTTO service described in Sect. 4 is invoked for retrieving the translations related to its labels and descriptions. Finally, the Ontology Export functionality has been revisited by adding the possibility to choose the export languages, among the available ones.

6   Evaluation

Our goal is to evaluate the usage and the usefulness of the MoKi tools and of the underlying service for suggesting domain-adapted translations.

In detail, we are interested in answering two main research questions:

RQ1. Does the proposed system provide an *effective* support, in terms of the quality of suggested translations, to the management of multilingual ontologies?

RQ2. Do the MoKi functionalities provide an *effective* support to the collaborative management of a multilingual ontology?

In order to answer these questions, we performed two types of analysis:

1. Quantitative: we collected objective measures concerning the effectiveness of the translations suggested by the embedded machine translation service. This information allows to have an estimation of the effort needed for adapting all translations by the language experts.
2. Qualitative: we collected subjective judgements from the language experts involved in the evaluation of the tool on general usability of the components and to provide feedback for future improvements.

Six language experts have been involved in the evaluation of the proposed platform for translating the Organic.Lingua ontology in three different languages: German, Spanish, and Italian. Most of the experts had no previous knowledge of the tool, hence an initial phase of training was necessary.

After the initial training, experts were asked to translate the ontology in the three languages mentioned above. Experts used MoKi facilities for completing the translation task and, at the end, they provided feedback about tool support for accomplishing the task. A summary of these findings and lessons learned are presented in Sect. 6.2.

6.1   Quantitative Evaluation Results

The automatic evaluation on label translations provided by OTTO is based on the correspondence between the SMT output of OTTO and reference translations (gold standard), provided by domain and language experts. For the automatic evaluation we used the BLEU [24], METEOR [25] and TER [26] algorithms.

BLEU is calculated for individual translated segments (n-grams) by comparing them with reference translations. Those scores, between 0 and 1 (perfect translation), are then averaged over the whole *evaluation dataset* to reach an estimate the automatically generated translation's overall quality. METEOR is based on the harmonic mean of precision and recall, whereby recall is weighted higher than precision. Along with standard exact word (or phrase) matching it has additional features, i.e. stemming, paraphrasing and synonymy matching. Differently to BLEU, the metric produces good correlation with human judgement at the sentence or segment level. TER is an error metric (lowers scores

Table 1. Automatic translation evaluation of the Organic.Lingua ontology by the Microsoft Translator API and OTTO System (bold results = best performance)

System	English → German			English → Italian			English → Spanish		
	BLEU	METEOR	TER	BLEU	METEOR	TER	BLEU	METEOR	TER
Microsoft	0.037	0.196	**0.951**	**0.135**	0.286	0.871	0.210	0.369	0.733
OTTO	**0.074**	**0.310**	0.991	0.130	**0.342**	**0.788**	**0.257**	**0.444**	**0.667**
Microsoft n-best	**0.076**	0.279	**0.872**	0.145	0.328	0.829	0.274	0.402	0.657
OTTO n-best	0.074	**0.310**	1.00	**0.150**	**0.408**	**0.719**	**0.333**	**0.523**	**0.566**

are better) for machine translation measuring the number of edits required to change a system output into one of the references.

We evaluate the automatically generated translations into German, Italian and Spanish provided by OTTO and the Microsoft Translator API. Since reference translations are needed to evaluate automatically generated translations, we use the translated labels provided by the domain experts. The Organic.Lingua ontology provides 274 German, 354 Italian and 355 Spanish existing translations out of 404 English labels. As seen in Table 1, with the help of contextual information OTTO significantly outperforms (p-value < 0.05) Microsoft Translator API when translating English labels into German (51.3 % averaged improvement over all metrics) or Spanish (51.7 %) and produces comparable results when translating into Italian (10.5 %).

Since both translation systems can provide additional, less probable translations of an English label, we identify with METEOR the best translation (due the gold standard) out of a set of possible translations. In this setting, the Microsoft Translator API provided on average 1.2 translations per label for German and Italian, and 1.6 for the Spanish, respectively. The OTTO system provided 9.5, 10.4 and 8.9 possible translations for German, Italian and Spanish. This additional information, seen as OTTO n-best and Microsoft n-best in the last part in Table 1 allows us to provide better translation candidates of labels. Compared to the first-best translation evaluation (upper part in Table 1), the translation quality improves for both systems. In the n-best scenario, OTTO demonstrates a 13.7 % averaged improvement in terms of the evaluation metrics over Microsoft and 14.4 % over OTTO first-best scenario when translating into Italian. For Spanish the improvements are 21.8 % and 20.8 %, respectively. Only for German (−2.1 %; −4.9 %) Microsoft Translator performs better.

6.2 Qualitative Evaluation Results

To investigate the subjective perception of the six experts about the support provided for translating ontologies, we analysed the subjective data collected through a questionnaire. For each functionality described in Sect. 5, we provide the information how often each aspect has been raised by the language experts.

Language Experts View
Pros: Easy to use for managing translations (3)
Usable interface for showing concept translations (2)

Approval And Discussion
Pros: Pending approvals give a clear situation about concept status (4)
Cons: Discussion masks are not very useful (5)

Quick Translation Feature
Pros: Best facility for translating concepts (5)
Cons: Interface design improvable (2)

The results show, in general, a good perception of the implemented functionalities, in particular concerning the procedure of translating a concept by exploiting the quick translation feature. Indeed, 5 out of 6 experts reported advantages on using this capability. Similar opinions have been collected about the language expert view, where the users perceived such a facility as a usable reference for having the big picture about the status of concept translations.

Controversial results are reported concerning the approach and discussion facility. On the one hand, the experts perceived positively the solution of listing approval requests on top of each concept page. On the other hand, we received negative opinions by almost all experts (5 out of 6) about the usability of discussion forms. This result shows us to focus future effort in improving this aspect of the tool.

Finally, concerning the "quick" translation facility, 5 out of 6 experts judged this facility as the most usable way for translating a concept. The main characteristic that has been highlighted is the possibility of performing a "mass-translation" activity without opening the page of each concept, with the positive consequence of saving a lot of time.

6.3 Findings and Lessons Learned

The quantitative and qualitative results demonstrate the viability of the proposed platform in real-world scenarios and, in particular, its effectiveness in the proposed use case. Therefore, we can positively answer to both research questions, **RQ1**: the back-end component provides helpful suggestions for performing the ontology translation task, and **RQ2**: the provided interfaces are usable and useful for supporting the language experts in the translation activity. Besides these, there were other insights, either positive and negative, that emerged during the subjective evaluation that we conducted.

The main positive aspect highlighted by the experts was related to the easy and quick way of translating a concept with respect to other available knowledge management tools (see details in Sect. 2), which do not enable specific support for translation. The suggestion-based service allowed effective suggestions and reduced effort required for finalizing the translation of the ontology. However, even if on one hand, the experts perceived such a service very helpful from the point of view of domain experts (i.e. experts that are generally in charge of modelling ontologies, but that might not have enough linguistic expertise for translating label properly with respect to the domain), the facilities supporting a direct interaction with language experts (i.e. discussion form) should be more intuitive, for instance as the approval one.

The criticism concerning the interface design was reported also about the quick translation feature, where some of the experts commented that the comparative view might be improved from the graphical point of view. In particular, they suggested (i) to highlight translations that have to be revised, instead of using a flag, and (ii) to publish only the concept label instead of putting also the full description in order to avoid misalignments in the visualization of information.

Connected to the quick translation facility, experts judged it as the easiest way for executing a first round of translations. Indeed, by using the provided translation box, experts are able to translate concept information without navigating to the concept page and by avoiding a reload of the concepts list after the storing of each change carried out by the concept translation.

Finally, we can judge the proposed platform as a useful service for supporting the ontology translation task, especially in a collaborative environment when the multilingual ontology is created by two different types of experts: domain experts and language experts. Future work in this direction will focus on the usability aspects of the tool and on the improvement of the semantic model used for suggesting translations in order to further reduce the effort of the language experts. We plan also to extend the evaluation on other use cases.

7 Conclusions

This paper presents ESSOT, an Expert Supporting System for Ontology Translation implementing an automatic translation approach based on the enrichment of the text to translate with semantically structured data, i.e. ontologies or taxonomies. The ESSOT system integrates the OTTO domain-adaptable semantic translation service and the MoKi collaborative knowledge management tool for supporting language experts in the ontology translation activity. The platform has been concretely used in the context of the Organic.Lingua EU project by demonstrating the effectiveness in the quality of the suggested translations and in the usefulness from the language experts point of view.

Acknowledgement. This publication has emanated from research conducted with the financial support of Science Foundation Ireland (SFI) under Grant Number SFI/12/RC/2289 and the European Unions Horizon 2020 programme MixedEmotions (Grant Number 644632).

References

1. Gómez-Pérez, A., Vila-Suero, D., Montiel-Ponsoda, E., Gracia, J., Aguado-de Cea, G.: Guidelines for multilingual linked data. In: Proceedings of the 3rd International Conference on Web Intelligence, Mining and Semantics. ACM (2013)
2. Gracia, J., Montiel-Ponsoda, E., Cimiano, P., Gómez-Pérez, A., Buitelaar, P., McCrae, J.: Challenges for the multilingual web of data. Web Semant. Sci. Serv. Agents World Wide Web **11**, 63–71 (2012)
3. Cimiano, P., Montiel-Ponsoda, E., Buitelaar, P., Espinoza, M., Gómez-Pérez, A.: A note on ontology localization. Appl. Ontol. **5**(2), 127–137 (2010)

4. Declerck, T., Pérez, A.G., Vela, O., Gantner, Z., Manzano, D.: Multilingual lexical semantic resources for ontology translation. In: Proceedings of the 5th International Conference on Language Resources and Evaluation, Saarbrücken, Germany (2006)
5. Espinoza, M., Montiel-Ponsoda, E., Gómez-Pérez, A.: Ontology localization. In: Proceedings of the Fifth International Conference on Knowledge Capture, New York (2009)
6. Fu, B., Brennan, R., O'Sullivan, D.: Cross-lingual ontology mapping–an investigation of the impact of machine translation. In: Gómez-Pérez, A., Yu, Y., Ding, Y. (eds.) ASWC 2009. LNCS, vol. 5926, pp. 1–15. Springer, Heidelberg (2009)
7. McCrae, J., Espinoza, M., Montiel-Ponsoda, E., Aguado-de Cea, G., Cimiano, P.: Combining statistical and semantic approaches to the translation of ontologies and taxonomies. In: Fifth workshop on Syntax, Structure and Semantics in Statistical Translation (SSST-5) (2011)
8. McCrae, J.P., Arcan, M., Asooja, K., Gracia, J., Buitelaar, P., Cimiano, P.: Domain adaptation for ontology localization. Web Semant. Sci. Serv. Agents World Wide Web 36, 23–31 (2015)
9. Espinoza, M., Gómez-Pérez, A., Mena, E.: Enriching an ontology with multilingual information. In: Bechhofer, S., Hauswirth, M., Hoffmann, J., Koubarakis, M. (eds.) ESWC 2008. LNCS, vol. 5021, pp. 333–347. Springer, Heidelberg (2008)
10. Stellato, A., Rajbhandari, S., Turbati, A., Fiorelli, M., Caracciolo, C., Lorenzetti, T., Keizer, J., Pazienza, M.T.: VocBench: a web application for collaborative development of multilingual thesauri. In: Gandon, F., Sabou, M., Sack, H., d'Amato, C., Cudré-Mauroux, P., Zimmermann, A. (eds.) ESWC 2015. LNCS, vol. 9088, pp. 38–53. Springer, Heidelberg (2015)
11. Gennari, J., Musen, M., Fergerson, R., Grosso, W., Crubézy, M., Eriksson, H., Noy, N., Tu, S.: The evolution of Protégé: an environment for knowledge-based systems development. Int. J. Hum. Comput. Stud. 58(1), 89–123 (2003)
12. Arcan, M., Asooja, K., Ziad, H., Buitelaar, P.: Otto-ontology translation system. In: ISWC 2015 Posters & Demonstrations Track, Bethlehem, PA, USA, vol. 1486 (2015)
13. Steinberger, R., Pouliquen, B., Widiger, A., Ignat, C., Erjavec, T., Tufis, D., Varga, D.: The JRC-acquis: a multilingual aligned parallel corpus with 20+ languages. In: Proceedings of the 5th International Conference on Language Resources and Evaluation (LREC) (2006)
14. Koehn, P.: Europarl: a parallel corpus for statistical machine translation. In: Conference Proceedings the Tenth Machine Translation Summit, AAMT (2005)
15. Steinberger, R., Ebrahim, M., Poulis, A., Carrasco-Benitez, M., Schlüter, P., Przybyszewski, M., Gilbro, S.: An overview of the european union's highly multilingual parallel corpora. Lang. Resour. Eval. 48(4), 679–707 (2014)
16. Eisele, A., Chen, Y.: Multiun: a multilingual corpus from united nation documents. In: Tapias, D., Rosner, M., Piperidis, S., Odjik, J., Mariani, J., Maegaard, B., Choukri, K., Chair, N.C.C. (eds.) Proceedings of the Seventh Conference on International Language Resources and Evaluation, pp. 2868–2872. European Language Resources Association (ELRA) (2010)
17. Cettolo, M., Girardi, C., Federico, M.: Wit[3]: web inventory of transcribed and translated talks. In: Proceedings of the 16th Conference of the European Association for Machine Translation (EAMT), Trento, Italy, pp. 261–268, May 2012
18. Koehn, P., Hoang, H., Birch, A., Callison-Burch, C., Federico, M., Bertoldi, N., Cowan, B., Shen, W., Moran, C., Zens, R., et al.: Moses: open source toolkit for statistical machine translation. In: Proceedings of the 45th Annual Meeting of the

ACL on Interactive Poster and Demonstration Sessions, pp. 177–180. Association for Computational Linguistics (2007)

19. Och, F.J., Ney, H.: A systematic comparison of various statistical alignment models. Comput. Linguist. **29**, 19 (2003)
20. Heafield, K.: KenLM: faster and smaller language model queries. In: Proceedings of the EMNLP 2011 Sixth Workshop on Statistical Machine Translation, Edinburgh, Scotland, UK, pp. 187–197, July 2011
21. Arcan, M., Turchi, M., Buitelaar, P.: Knowledge portability with semantic expansion of ontology labels. In: Proceedings of the 53rd Annual Meeting of the Association for Computational Linguistics, Beijing, China, July 2015
22. Mikolov, T., Chen, K., Corrado, G., Dean, J.: Efficient estimation of word representations in vector space. In: ICLR Workshop (2013)
23. Dragoni, M., Bosca, A., Casu, M., Rexha, A.: Modeling, managing, exposing, and linking ontologies with a wiki-based tool. In: Calzolari, N., Choukri, K., Declerck, T., Loftsson, H., Maegaard, B., Mariani, J., Moreno, A., Odijk, J., Piperidis, S. (eds.) Proceedings of the Ninth International Conference on Language Resources and Evaluation (LREC-2014), Reykjavik, Iceland, 26–31 May 2014, pp. 1668–1675. European Language Resources Association (ELRA) (2014)
24. Papineni, K., Roukos, S., Ward, T., Zhu, W.J.: BLEU: a method for automatic evaluation of machine translation. In: Proceedings of the 40th Annual Meeting on Association for Computational Linguistics, ACL 2002, pp. 311–318 (2002)
25. Denkowski, M., Lavie, A.: Meteor universal: language specific translation evaluation for any target language. In: Proceedings of the EACL 2014 Workshop on Statistical Machine Translation (2014)
26. Snover, M., Dorr, B., Schwartz, R., Micciulla, L., Makhoul, J.: A study of translation edit rate with targeted human annotation. In: Proceedings of Association for Machine Translation in the Americas (2006)

Adapting Semantic Spreading Activation to Entity Linking in Text

Farhad Nooralahzadeh[1(✉)], Cédric Lopez[2], Elena Cabrio[3], Fabien Gandon[1], and Frédérique Segond[2]

[1] INRIA Sophia Antipolis, Valbonne, France
{farhad.nooralahzadeh,fabien.gandon}@inria.fr
[2] Viseo Research Center, Grenoble, France
{cedric.lopez,frederique.segond}@viseo.com
[3] University of Nice, Nice, France
elena.cabrio@unice.fr

Abstract. The extraction and the disambiguation of knowledge guided by textual resources on the web is a crucial process to advance the Web of Linked Data. The goal of our work is to semantically enrich raw data by linking the mentions of named entities in the text to the corresponding known entities in knowledge bases. In our approach multiple aspects are considered: the prior knowledge of an entity in Wikipedia (i.e. the keyphraseness and commonness features that can be precomputed by crawling the Wikipedia dump), a set of features extracted from the input text and from the knowledge base, along with the correlation/relevancy among the resources in Linked Data. More precisely, this work explores the *collective ranking approach* formalized as a weighted graph model, in which the mentions in the input text and the candidate entities from knowledge bases are linked using the local compatibility and the global relatedness measures. Experiments on the datasets of the Open Knowledge Extraction (OKE) challenge with different configurations of our approach in each phase of the linking pipeline reveal its optimum mode. We investigate the notion of semantic relatedness between two entities represented as sets of neighbours in Linked Open Data that relies on an associative retrieval algorithm, with consideration of common neighbourhood. This measure improves the performance of prior link-based models and outperforms the explicit inter-link relevancy measure among entities (mostly Wikipedia-centric). Thus, our approach is resilient to non-existent or sparse links among related entities.

Keywords: Entity linking · Linked data · Collective entity ranking · Semantic spreading

1 Introduction

The Web publishes information from many heterogeneous sources, and such data can be extremely valuable for several domains and applications as, for instance,

This work has been founded by the French ANR national grant (ANR-13-LAB2-0001).

ⓒ Springer International Publishing Switzerland 2016
E. Métais et al. (Eds.): NLDB 2016, LNCS 9612, pp. 74–90, 2016.
DOI: 10.1007/978-3-319-41754-7_7

business intelligence. A large volume of information on the Web is made of unstructured texts, and contains mentions of named entities (NE) such as people, places and organizations. Names are often ambiguous and could refer to several different named entities (i.e. homonymy) and, vice-versa, entities may have several equivalent names (i.e. synonymy). On the other hand, the recent evolution of publishing and connecting data over the Web under the name of "Linked Data" provides a machine-readable and enriched representation of the world's entities, together with their semantic characteristics. This process results in the creation of large knowledge bases (KB) from different communities, often interlinked to each other as is the case of DBpedia, i.e. the structured version of Wikipedia, Yago and FreeBase[1]. This characteristic of the Web of data empowers both the humans and computer agents to discover more "things" by easily navigating among the datasets, and can be profitably exploited in complex tasks such as information retrieval, question answering, knowledge extraction and reasoning. The *NE recognition and linking* task, i.e. establishing the mapping between the NE mentions in the text to their related resources in KB is of critical importance to achieve this goal. Such mappings (i.e. links) can be used to provide semantic annotations to human readers, as well as a machine-readable representation of the pieces of knowledge in the text, exploitable to improve many Natural Language Processing (NLP) applications. The preliminary step in the entity linking task is identifying the textual boundaries of words, i.e. as a single-word or multiwords in a document such that the recognized boundary corresponds to an individual entity (e.g., person, organization, location, event) in the real world. The detection of entity mentions is currently referred as *named entity recognition*. Then, the main challenge of entity linking is to map each entity mention found in the text to the corresponding entry in a given KB(e.g. DBpedia). More precisely, the task can be described as follows: given a KB '$\mathcal{KB}$' (e.g., Wikipedia, DBpedia) and an unstructured text document $\mathcal{D}$ in which a sequence of NE mentions $\overrightarrow{\mathcal{M}} = \{m_1, m_2, ..., m_i, ...\}$ appear in the text, the aim of entity linking is to output a sequence of entity candidates $\overrightarrow{\mathcal{C}} = \{c^1, c^2, ..., c^i, ...\}$, where $c^i \in \mathcal{KB}$ and corresponds to m_i with a high confidence score. Here m_i refers to a sequence of tokens which is recognized by NER tools as a NE. If an entity mention m_i in text does not have its corresponding entity candidate c^i in the $\mathcal{KB}$, it is labeled as "*NIL*" [16,31].

The state-of-the-art entity linking systems exploit various features extracted from the input text and from the knowledge base. The most important feature is the semantic relatedness among the candidate entities and the effectiveness of this function is a key-point for the accuracy of entity linking system [3]. Most of the entity linking algorithms focus on the extensive link structure of Wikipedia and they consider the overlap of the incoming/outgoing links of entity articles as a notion of their relatedness. However, this approach has a limitation for the entities that do not have explicit links. In order to deal with this limitation, in this paper we answer the following research questions:

---

[1] dbpedia.org; www.mpi-inf.mpg.de/yago/; www.freebase.com.

- How can we configure an accurate entity linking pipeline?
- What is the best approach to deal with non-existent or sparse links among related entities?
- How does the Linked Data-based relatedness measure impact on the performance of the entity linking task?

Taking into account these questions, the main goals of this work are *(i)* to investigate and compare several state-of-art approaches proposed in the literature at each phase of the entity linking pipeline; *(ii)* to introduce and evaluate the Linked Data-based relatedness measure among the entities that we are proposing in our approach, and *(iii)* to compare the effectiveness of this measure against other state-of-art components in entity discovering task. To address these issues, we implemented different components for each step of the entity linking task by employing the open source framework *Dexter* [2]. We designed the inference part as a collective ranking model by constructing the weighted graph among the named entity mentions and their potential candidates. Finally, we evaluated the performance of our approach against publicly available datasets, with TAC (Text Analysis Conference) evaluation tools for the named entity linking task. The rest of the paper is as follows: Sect. 2 presents the related work. Section 3 introduces our approach, whose features are detailed in Sect. 4. In Sect. 5.1 the dataset is described. Section 5 details the experiments and the obtained results. Conclusions end the paper (Sect. 6).

2   Related Work

The entity linking task has been widely studied in the last decade ([6] among many others), and generally consists of 3 steps. The first step is the *mention identification*: the NLP field provides NE recognizers which are particularly efficient for persons, organizations, and locations [13]. To avoid the specification of types of mentions to be recognized, a spotting identification can consider n-grams of words to be compared to entries of a resources (for instance Wikipedia). The second step consists in *candidate generation* which have to be mapped with entities in a KB. Mapping approaches such as exact string matching and partial matching are often used based on string similarity with e.g. Dice score, skip bi-gram score, Hamming distance, Levenshtein, the Jaro similarity. Compared with exact matching, partial matching leads to higher recall, but more noise appears in the candidate entity set. In [10] the authors used also the number of Wikipedia pages linking to an entity's page.

The last step consists in selecting the relevant URI among the candidates. The candidate ranking process relies on the context information provided by a set of features. According to [9] the features can be classified into three groups: *(i)* Extracted from the prior knowledge (Keyphraseness, Commonness, Popularity); *(ii)* Extracted from the text of the document/articles (Local features): Mention-candidate context similarity: use similarity measure to compare the context of the mention with the text associated with a candidate title (the text in the corresponding page); *(iii)* Based on existing links (Global features [3]).

In [3], the authors explored the various type of global features as a relatedness measure among the candidate entities. They showed that the entity linking algorithms benefit from maximizing the relatedness among the relevant entities selected as candidates in ranking process, since this minimizes errors in disambiguating entity-linking. The key role of effective relatedness function in any entity-linking algorithm encouraged us to adapt semantic spreading activation to model another relevancy feature which incorporates multiple types of entity knowledge from the knowledge base.

3   The Entity Linking Task

The task of entity discovery can be divided into a series of steps [9] (Fig. 1): *(i) spotting or mention identification*, i.e. detecting the anchors to link, *(ii) candidate generation and pruning*, i.e. searching for candidate identifiers for each anchor and filtering, *(iii) candidate ranking*, i.e. selecting the best candidate from the results of the searching step. This section, provides a step by step description of our approach. Afterward we explain the employed feature set.

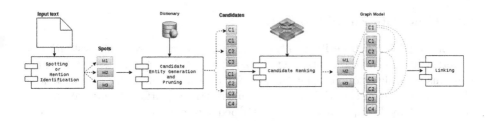

Fig. 1. Entity linking pipeline

3.1   Spotting or Mention Identification

A mention is a phrase used to refer to something in the world like NEs, object, substance, event, philosophy, mental state, rule, etc. (e.g., Stanford University, Berlin, MBA, U.S.Open Tennis). To extract mentions from the text, we implement the following techniques, and then we select the most appropriate to include in our approach, based on their performances (see Sect. 5.3):

– **N-grams linked in lexicon (prior anchor text).** An n-gram is a contiguous sequences of n words from an input text (n represents any integer > 0). Each n-gram is a potential entity mention if it has exact or partial match in the repository. The spotting step uses a dictionary of NEs or a KB, to extract possible mentions of NE in the text. It provides the highest recall, but for large documents it becomes intractable and introduces noise in the mention extraction. We follow a standard approach [16,25] to create an anchor repository to address this issue. The token extractor generates a list of possible spots from a given document. In the extraction phase, the tool produces all the possible n-grams of terms (up to six words in our case), afterward the spotter looks for the matching of each fragment in the spot repository. The filtering process helps us to discard the meaningless mentions, because even stop words may match the anchor text. We use the *keyphraseness* feature [23] to apply this filtering. *Keyphraseness* estimates the probability of a term to be selected as a link to the entities in the knowledge base $\mathcal{KB}$, by counting the number of times where the term was already selected as a anchor, divided by the total number of times where the term appeared. These counts are collected from all Wikipedia articles, permitting to throw out spots that are rarely used as a mention of the relevant entity. In case of overlap among the fragments, we take the union of the fragments boundaries, and if one fragment contains another we consider the longest one as a mention (see Sect. 3.2).

– **Named Entity Tagging.** To avoid the noise and have precise mentions, statistics and symbolic-based filtering could be applied. This will have negative impact on the system recall but will increase the precision. The most prominent named entity tagger for English is Stanford NER [13], which we use to detect the mentions, and segment the text accordingly, in our approach.

3.2   Candidate Entity Generation

At this step, for each entity mention $m_i \in \mathcal{M}$, the entity linking system aims at retrieving a candidate entity set $C(m_i) = (c_1^i, c_2^i, \ldots, c_j^i, \ldots)$ from $\mathcal{KB}$ which contains possible resources that the entity mention m_i may refer to. To achieve this goal, we create a repository which maps strings (surface forms) to entities represented by Wikipedia articles or DBpedia resources [3,19]. The structure of Wikipedia supplies useful features for generating candidate entities such as entity pages (articles), redirect pages, disambiguation pages, bold phrases from the first paragraphs, and hyperlinks inside the articles. Leveraging these features from Wikipedia[2], we build an offline dictionary $\mathcal{D}$ as key-value map $< \mathcal{K}, \mathcal{V} >$ between various names as key ($\mathcal{K}$) and their possible mapping entities as values ($\mathcal{V}$). We exploit $\mathcal{D}$ to generate candidate entities $C(m_i)$. This dictionary compiles a vast amount of information about possible variation of named entities, like abbreviations, alternative names, ambiguous terms, misspellings, initial letters. Based on such dictionary, the simplest way to generate the candidate entity set $C(m_i)$ for mention $m_i \in \mathcal{M}$ is the exact matching between the mention name m_i and the keys ($\mathcal{K}$). If a key is similar to m_i, the set of entities mapped to the key will be added to the candidate entity set $C(m_i)$.

Beside exact matching, we apply partial matching between the mention m_i, and the key ($\mathcal{K}$) in the dictionary. If the mention has a strong string similarity based on *Jaro-Winkler distance* with the name as key in dictionary, then the set of entities assigned to that key will be added to the candidate set. Compared with exact matching, partial matching leads to higher recall, but imposes more noise in the candidate entity set.

---

[2] English Wikipedia dump downloaded on 2015/07/02.

3.3 Candidate Entity Pruning

Since the set of candidates is quite large that contains unrelated candidates, we need to filter some of them to ensure the quality of the entity candidates. To limit the set of candidate entities $C(m_i)$ to the most meaningful ones, the *commonness* property was exploited [3]. The *commonness* of a candidate c_j^i for entity mention m_i is the chance of candidate c_j^i being target of a link by mention m_i as an anchor. It is defined as the fraction between the number of occurrences of mention m_i in $\mathcal{KB}$ actually pointing to candidate c_j^i, and the total number of occurrences of m_i in $\mathcal{KB}$ as an entity mention. Setting a threshold on minimum commonness is a simple and effective strategy to prune the number of candidates, without affecting the recall of the entity linking process [25]. Moreover, we exploit the context-based similarity feature (Sect. 4.3) to prune the candidates that do not have any contextual similarity compared to the context of the target mention in the input text.

3.4 Candidate Ranking

The objective of the ranking module is to create a ranking function $f : \{c_j^i \in C(m_i)\} \to \mathcal{R}$, such that for each entity mention m_i the elements in its corresponding candidates list can be assigned relevance scores using the function, and then be ranked according to the scores. Several approaches have been studied to formulate the ranking function in recent years, and we take inspiration from the successful application of *Graph-based* model in entity linking problem (known under various names like "collective entity linking" [16], "Entity linking via random walk" [15] and "Neighborhood relevance model" [7]). Adopting a graph based model, let $G = (V, E)$ a undirected graph, in which the vertex set $V = \{\mathcal{M} \cup C\}$ contains all entity mentions $m_i \in \mathcal{M}$ in a document and all possible entity candidates of these mentions $c_j^i \in C(m_i)$. The set of edges E includes all the links that represent the mention-entity and entity-entity edges. Each edge between a mention and a candidate entity denotes a "local compatibility" relation; each edge between two candidate entities indicates a "global relatedness/coherence" (Figs. 1 and 2).

The goal is to find the best assignment C that is contextually compatible with mentions in M and has a maximum coherence as the following objective function:

$$C^* = \arg \max_{C} [\Phi(\mathcal{M}, C) + \Psi(C)]$$

where C^* is the best candidate set; $\Phi(\mathcal{M}, C)$ is the local compatibility of mentions and candidates, while $\Psi(C)$ measures the global coherence of selected candidates.

The local compatibility Φ formalizes the intuition that the mention m and a candidate entity c are compatible when their name matches and the context surrounding m has terms similar to the context of the entity candidate c in $\mathcal{KB}$; it has been applied to conventional entity linking approaches [1,6] as discriminative features. Local features work best, when the context is rich enough to uniquely link a mention, which is not always true. Furthermore, the feature

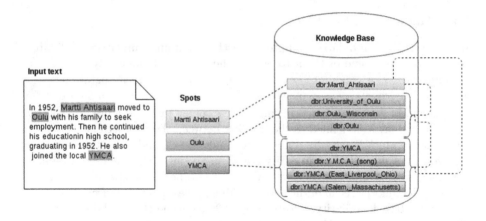

Fig. 2. Graph model for entity linking

sparsity problem arises when there is no similarity between the mention and the related entity to match; therefore no similarity match is applied [15]. With global coherence Ψ, we rely on the idea that the best selected candidates for the neighbor mentions in the text, would be good clues to disambiguate the current mention. It takes into account the relation of the candidate with other candidates within neighbor mentions. In other words, if each neighboring mention of m_i is linked to its correct entity candidate $c_j^k, k \neq i$, the true candidate c_j^i will achieve a high score in Ψ for most of the neighbors c_j^k. To satisfy the objective function, the disambiguation is done collectively on all mentions at the same time. In [16,19,20] authors seek to find a sub-graph embedded in the constructed graph, in which each mention is linked to a single candidate entity. However solving the objective function is a recursive problem [14] and is NP-hard [20], therefore all methods turn to approximate algorithms or heuristics. To solve this disambiguation problem, we design the two following methods:

PageRank-like method: The problem described above can be addressed by a PageRank-like algorithm [11]. In PageRank, the objective is to assign a score of the "importance" to each page. The score (PageRank) of each page depends on the score of pages linking to it. In essence, the PageRank of a page is a weighted sum of pageranks of articles linking to it. A page has a high rank if the sum of the ranks of the pages that link to it is high [27]. In the entity linking problem, we would like to rank the possible candidates associated with a certain mention in the document. So, it can be said that a candidate has a high score if the sum of the scores in the document of the candidates that typically co-occur with it is high. This is formalized as:

$$V^{t+1} = (1 - \lambda) * M * V^t + \lambda * S$$

Where M is a matrix with $(n + m)^2$ dimensions, n as reallocate condition, is the number of mentions and m is the number of candidates. $M[i, j]$ if i denotes the mention and j denotes the candidate are the local feature value

$\phi(m_i, c_j)$, and when i, j represent the candidates, its value will be $\psi(c_i, c_j)$, otherwise it will be 0. The $(n + m) \times 1$ vector S shows the prior importance of each mention $p(m)$, and for the candidates it is 0. V is the $(n + m) \times 1$ preference vector and $\lambda \in (0, 1)$ is the fraction of the reallocation condition at each iteration. To find the final preference vector $V*$, we use the power method with k iteration or until convergence condition. The ith value of vector V^* where i points to the candidate, will be the final rank of each candidate. It should be mentioned that in each iteration, since we have the preference score for each mention in previous iteration, we can update the reallocation vector S with its posterior importance $p_{post}(m_i) = p_{pripr}(m_i) * V[i]$ [16].

Naive method: The expression that we use to select the best assignment C using the maximum aposteriori decision rule can be formalized like:
For each $m_i \in \mathcal{M}$ the best candidate c_*^i is:

$$c_*^i = \arg\max_c \left[\alpha. \sum \phi(m_i, c) + \beta. \sum_{c_j^k \in \mathcal{C}, k \neq i} \psi(c, c_j^k) \right]$$

Here we compute the local compatibility and coherence individually for each candidate and solve the problem linearly by assigning the coefficient α and β, where $\alpha + \beta = 1$. During the annotation process, if there is only one mention to disambiguate and no need to build a graph among the neighbours, we use naive method to find the best candidate entity for mention.

In Sect. 4, the local Φ and global Ψ relatedness features applied in our work are explained.

4   Local (Φ) and Global (Ψ) Features

In this section, we present the features used in this study. Local features rely only on compatibility of the mention and its candidates, while global ones compute the coherence among the candidates.

4.1   Prominence/Prior Feature $\in \Phi$

As a context-independent feature, we choose the *commonness* feature of candidates based on Wikipedia link anchors [23]. It is the popularity of the candidate entity with respect to the mention in the text, and has been found to be very useful in entity linking [19,31]. It tells the prior probability of the appearance of a candidate entity given the mention term. For each surface form that appears as an anchor text in Wikipedia, we compute how often it refers to a particular entity:

$$Commonness(c_j^i) = \frac{x}{y}$$

where x is the number of times a mention m_i links to a candidate entity c_j^i in Wikipedia and y is the number of times a mention m_i appears as anchor text in Wikipedia.

4.2 Feature for Name Variants $\in \Phi$

Variation in entity names is a limitation for information extraction systems. Poor handling of entity name variants results in a low recall. The name string comparison between the mention and the candidate is the most direct feature that can be used. We use approximate string matching based on the *Levenshtein distance* to figure out this feature between mention term and candidate's label.

4.3 Context-Based Feature $\in \Phi$

The key for mapping mentions onto candidates, is the context appearing on both sides of the mapping. For each mention inside the input text, we construct a bag-of-words excluding the mention itself in size of the window w ($w = 6$ in our case) of text around the mention by removing the stop words and punctuation marks. We compute the *TF-IDF* similarity of mentions context regarding to its candidates context. For each candidate, the context is represented as a bag-of-words from the whole Wikipedia entity page (*content*), or the first paragraph of its Wikipedia article (*summary*), or the outgoing to other pages in its Wikipedia page (*links*). We implement each of these representations in *Lucene*[3], as an option than can be selected during the linking process.

4.4 Neighbors Coherence Features $\in \Psi$

We identify a number of previous methods for estimating the coherence between two entities in a knowledge base ($\mathcal{KB}$). In [29], the authors use the category overlap of two entities in Wikipedia. In [24], the authors use the incoming link structure, while [3] introduce various asymmetric and symmetric relatedness features for entity linking. State-of-the-art measures for entity relatedness that perform well in entity linking are based on the extensive link structure of Wikipedia. Most particularly, the *Milne-Witten* and the *Pointwise mutual information* are described in the following.

Milne-Witten (MW). It is a co-citation based similarity, that quantifies the coherence between two entities by the *links* that their Wikipedia articles share [24]. The intuition is that entities occurring often together in Wikipedia articles are related. It takes into account Wikipedia graph structure as its KB ($\mathcal{KB}$) and computes the relatedness between two candidate entities c_i, c_j as:

$$MW(c_i, c_j) = 1 - \frac{log\left(max(|in_{c_i}|, |in_{c_j}|)\right) - log(|in_{c_i} \cap in_{c_j}|)}{log(N) - log\left(min(|in_{c_i}|, |in_{c_j}|)\right)}$$

N is the number of all entities in Wikipedia, $|in_c|$ is the number of incoming links to candidate c.

---

[3] https://lucene.apache.org/.

Pointwise Mutual Information (PMI): Another similarity metric between Wikipedia articles to consider, is based on computing the PMI between the sets of links in the articles to be compared. It is used by [30] for the link-to-Wikipedia task. It is defined between two candidates c_i and c_j as:

$$PMI(c_i, c_j) = \frac{|in_{c_i} \cap in_{c_j}|/N}{(|in_{c_i}|/N)(|in_{c_j}|/N)}$$

Furthermore, as indicated in [3,30], both MW and PMI can be computed using the outgoing links (out_c), and both the incoming and outgoing links ($in_c \cup out_c$).

Wikipedia centric relatedness has no chance to capture the connection between the entities whose corresponding pages do not have links in common. In these cases, previous coherence measures (i.e. the *milne-witten* and *PMI*), do not properly capture the semantic relatedness between entities. In the following, we propose a new measure in order to overcome this limitation.

Semantic Spreading (DiscoveryHub): A relatedness measure for link-poor entities needs to tap into linked open data that is more widely available and can be gathered with reasonable effort. The linked data semantic richness offers plenty of possibilities to retrieve, prioritize, discard, and rank the resources by using the knowledge they contain at the instance and schema levels. To exploit these possibilities, we base our relatedness function on an algorithm [21] that has proven its efficiency for information retrieval: *spreading activation*. We utilize the spreading activation algorithm adaptation (*DiscoveryHub* [21]) which leverages the graph richness thanks to semantic filtering and similarity to devise the proper relevancy measure between candidate entities.

The algorithm presented in *DiscoveryHub* aims to explore the graph obtained from the linked data cloud, namely from *DBpedia*. It identifies a meaningful subset of resources strongly related to the entity. Then these results are presented in a set of relevant entities to the target entity. The semantic graph is exploited thanks to a weight, that is function of the origin entity. It constrains the propagation to certain entity types based on the class-based "semantic pattern" using *rdf:type* and *rdfs:subClassOf* properties. Then, they make use of the *dct:subject* property to compute triple-based similarity measure (commontriple), because the categories constitute a topic taxonomy which is very informative on the resources. It constitutes a valuable basis for computing similarity. The entities reached by the propagation and belonging to the same categories as the origin receive a greater amount of activation. Activation and propagation features of the algorithm expose the relevancy of entities that are not linked explicitly by incoming and outgoing links.

Based on Jaccard index, we define *DiscoveryHub* relatedness measure on the top of this algorithm:

$$DiscoveryHub(c_i, c_j) = \frac{DiscoveryHub(c_i) \cap DiscvoveryHub(c_j)}{DiscoveryHub(c_i) \cup DiscvoveryHub(c_j)}$$

where $DiscoveryHub(c)$ is the set of explicit and implicit neighbours, obtained during the spreading algorithm as relevant resources for the target entity c.

5 Experiments

In the following, we describe the dataset we used in our experiments, the experimental setting itself, and the results obtained by our approach.

5.1 Dataset

To compare our approach with other state-of-art methods, we choose two publicly available datasets published in *NLP Interchange Format* (NIF) [17] and employed in many other systems evaluations.

OKE: It is provided for the Open Knowledge Extraction challenge at ESWC2015. It contains the training and evaluation datasets built by manually annotating 196 sentences. These sentences have been selected from Wikipedia articles reporting the biographies of scholars to cover NEs for people, locations, organizations and roles[4]. It is split into 95 sentences as training set and 101 sentences as evaluation set, by taking into account no overlap between two datasets and equal distribution of entity types within datasets (see Table 1).

KORE50: This small dataset was produced as a subset of the larger AIDA corpus [19], which is based on the dataset of CoNLL 2003 NER task. It aims to capture hard to disambiguate mentions of entities and it contains a large number of first names referring to persons, whose identity needs to deduced from the given context. It consists of 50 sentences from different domains, such as music, celebrities and business and is provided in a TSV (Tab-separated values) format (and converted to NIF format[5]) (see Table 1).

Table 1. OKE and KORE50 datasets overall information

Parameter	OKE-training	OKE-evaluation	KORE50
Nb. of sentences	95	101	50
Nb. of mentions	290	428	148
Nb. of mentions linked to DBpedia	255	321	148

5.2 Evaluation Measures

For the system evaluation, we use the standard metrics of precision, recall and F1 at each phases on the pipeline, over the above described datasets. The TAC

---

[4] https://github.com/anuzzolese/oke-challenge.

[5] http://dashboard.nlp2rdf.aksw.org/.

evaluation tools for NE linking, *neleval*[6], provides a range of linking and clustering evaluation measures. This tool also provides a number of diagnostic measures available to isolate performance of system components and compare to numbers reported elsewhere in the literature. In spotting or mention identification, we consider the *strong-mention-match* measure to evaluate this task. Given that it takes into account all annotations together and gives more importance to documents with more annotations [5], it is called micro-averaged evaluation of entity mentions. In this measure if a span of recognized mention matches a gold mention exactly, then it will be counted as a true positive. For the linking evaluation we choose the *strong-link-match* measure, that is a micro-averaged evaluation of entities. A predicted entity by the system must have the same span and $\mathcal{KB}$ identifier as a gold entity in order to be considered as a true positive. This is equivalent to the strong annotation match presented in [5].

5.3   Experiments and Results over the Training Dataset

In our experiments, we use the English Wikipedia edition (2015-07-02) as the source to mine entities, keyphrases and corresponding context. We prepared the Wikipedia data according to [2] using the core module of the open source framework *Dexter* and implementing extra methods to meet the requirements of our approach. We applied different configurations to investigate the effect of each option in the pipeline of our entity linking approach. The *OKE-training* dataset is employed to tune the best setup model for our study. In the first step, the optimum choice for spotting or mention identification (Sect. 3.1) was explored. As Table 2 shows, Stanford NER provided by yields better result than the wiki-dictionary based N-gram spotting. We evaluate this step applying the *strong-mention-match* measure described above.

Table 2. Evaluation of spotting phase on the OKE-training dataset

Spotter	P %	R%	F1 %
Named entity tagging	69.9	57.2	62.9
N-grams linked in Lexicon	48.7	62.5	54.8

According to these results, we chose the *NE Tagging* method for the spotting phase inside our pipeline. Then, we study the impact of the other parameters as the *ranking method* (Sect. 3.4) and the *global relatedness* (Sect. 4.4) in our approach.

We chose the *summary* field of Wikipedia article representation in *lucene* to compute the context-based similarity feature (Sect. 4.3) and filter the candidate if its commonness (Sect. 4.1) is less than the threshold ($\theta = 0.05$ defined by line-search). In the candidate ranking phase we define empirically $\lambda = 0.1, \epsilon = 1e^{-6}$,

---

[6] https://github.com/wikilinks/neleval/wiki.

$N = 100$ (*max-number-of-iteration*) and $w = 6$ (*window-size*) in PageRank-like method and $\alpha = 0.4, \beta = 0.6$ in naive method (Sect. 3.4). In Table 3, we can see that the *PageRank-like* method including the *DiscoveryHub* as global relatedness has better results than the other configurations. This result indicates that the semantic spreading nature of *DiscoveryHub* relatedness identifies new links among related entities and *PageRank-like* method takes advantage of this favor to assign proper entities to the mentions.

Table 3. Evaluation of candidate ranking on the OKE-training dataset

Candidate ranking	Global relatedness	P%	R%	F1%
PageRank-like	**DiscoveryHub**	**62.6**	**47.2**	**53.8**
Naive	DiscoveryHub	55.8	48.5	52.0
PageRank-like	MW	60.9	46.6	53.0
Naive	MW	56.6	46.0	48.6
PageRank-like	PMI	55.1	47.9	51.2
Naive	PMI	50.9	44.3	47.4

In this phase (see Fig. 2) we are trying to find proper links among the *dbr:Martti_Ahtissan* as a candidate entity for mention *Martti Ahtissar* and the set of candidate for mention *YMCA*. Modeling the graph with MW and PMI relatedness returns graph without any links among these sets. However, using the *DiscoveryHub* coherence features unveil the implicit connection between *dbr:Martti_Ahtissan* and *dbr:YMCA*, contributing in assigning a higher score to *dbr:YMCA*.

It is obvious that the performance on the candidate ranking phase depends on the result of the spotting process in our pipeline. In order to evaluate the performance of the scoring function independently, we assume that the spotting phase extracts the mentions exactly as gold spans (100 % accuracy). Table 4 presents the evaluation of our approach with the most favorable setup.

Table 4. Evaluation of selected ranking approach on the OKE-training dataset

Candidate ranking	P %	R%	F1%
PageRank-like + DiscoveryHub	70.2	69.5	69.9

5.4   Comparison with State-of-the-Art Entity Linking Tools

We conduct an experiment with the final configuration of our approach on the *OKE-evaluation* dataset. We compare our results to the performances of state-of-art systems, i.e. *Adel* (winning system at the OKE challenge for the entity liking task), *AIDA, TagMe* and *DBpedia Spotlight* (Table 5). We use the figures which are reported in [28] for this comparison, since the authors used the public API of these tools with default setting on the *OKE-evaluation* dataset to compare their findings. Our experiment shows that the pipeline which we present in this study outperforms the other systems in the linking task in terms of precision and f1-measure in *OKE-evaluation* dataset.

Table 5. Comparison on the OKE-evaluation dataset

Method	Measure	P%	R%	F1%
Adel-without pruning	Spotting	78.2	**65.4**	**71.2**
	Linking	49.4	46.6	**48**
Adel- with pruning	Spotting	83.8	9.3	16.8
	Linking	**57.9**	6.2	11.1
AIDA	Spotting	55.7	49.1	52.2
	Linking	51.6	43.9	47.4
TagMe	Spotting	39.7	61.7	48.3
	Linking	28.5	**54.9**	37.5
DBpedia spotlight	Spotting	40.3	52.6	45.6
	Linking	28.3	45.7	34.9
Our approach	Spotting	**87.5**	57.1	69.1
	Linking	**73.1**	46.1	**56.5**

We also compare our method by taking into account only the candidate ranking task and assuming full annotation in the mention identification phase, with respect to the same state-of-the-art entity linking systems over *KORE50* dataset. In order to make this comparison, we report the facts from [26], adding our approach's ranking performance. Over this dataset (Table 6) we classify fourth in terms of performances after *Babelfy* and the two *AIDA-KORE* systems. Based

Table 6. State-of-the-art EL systems (Accuracy) on the KORE50 dataset

System	KORE50
Babelfy [26]	71.5
AIDA-KORE-LSH-G [18]	64.6
AIDA-KORE [18]	63.9
Our Ranking Method	**60.6**
MW [24]	57.6
Tagme [12]	56.3
Illinois wikifier [4]	41.7
DBpedia spotlight [22]	35.4

on the error analysis of the results, we found out that this differences comes from the candidate generation phase, and we plan to investigate more this aspect.

6   Conclusion and Future Work

In this work, we implemented and benchmarked different approaches for each step of the entity linking task to build the most accurate pipeline. We also proposed and evaluated a new collective ranking model by constructing the weighted graph among the named entity mentions and their potential candidates. We evaluated the performances of our approach against publicly available datasets, with the TAC evaluation tools. The performance of the graph-based collective entity linking system highly relies on modeling coherence among the candidate entities. Our system incorporates multiple types of entity knowledge from the KB. To deal with non-existent or sparse links among related entities we took advantage of semantic spreading approach to model the entity relatedness measure and it outperformed other Wikipedia-centric measures. Our experiments showed that the Linked-data based measure assists our graph-based model to achieve competitive performance in the target datasets compared to other well-known entity linking systems.

There are several directions for future work: we are interested in the joint inference model [8] to reduce the interdependency nature of our pipeline approach. On the other hand we want to investigate the word embedding impact in computing the contextual similarity between the mentions context and the candidate entities. Finally, we would like to use the *Wikilinks* [32] and *crossWikis* [33] corpora to enrich our repository and improve the candidate generation components.

References

1. Adafre, S.F., de Rijke, M.: Discovering missing links in wikipedia. In: Proceedings of the 3rd International Workshop on Link Discovery, LinkKDD 2005, pp. 90–97. ACM, New York (2005). http://doi.acm.org/10.1145/1134271.1134284
2. Ceccarelli, D., Lucchese, C., Orlando, S., Perego, R., Trani, S.: Dexter: an open source framework for entity linking. In: Proceedings of the Sixth International Workshop on Exploiting Semantic Annotations in Information Retrieval, pp. 17–20 (2013)
3. Ceccarelli, D., Lucchese, C., Orlando, S., Perego, R., Trani, S.: Learning relatedness measures for entity linking. In: Proceedings of the 22nd ACM International Conference on Conference on Information & Knowledge Management, CIKM 2013, pp. 139–148 (2013)
4. Cheng, X., Roth, D.: Relational inference for wikification. In: EMNLP (2013). http://cogcomp.cs.illinois.edu/papers/ChengRo13.pdf
5. Cornolti, M., Ferragina, P., Ciaramita, M.: A framework for benchmarking entity-annotation systems. In: Proceedings of the 22nd International Conference on World Wide Web, WWW 2013, pp. 249–260 (2013)

6. Cucerzan, S.: Large-scale named entity disambiguation based on wikipedia data. In: EMNLP-CoNLL 2007, Proceedings of the 2007 Joint Conference on Empirical Methods in Natural Language Processing and Computational Natural Language Learning, Prague, Czech Republic, 28–30 June 2007, pp. 708–716 (2007)
7. Dalton, J., Dietz, L.: A neighborhood relevance model for entity linking. In: Proceedings of the 10th Conference on Open Research Areas in Information Retrieval, OAIR 2013, pp. 149–156 (2013)
8. Durrett, G., Klein, D.: A joint model for entity analysis: coreference, typing, and linking. In: Proceedings of the Transactions of the Association for Computational Linguistics (2014)
9. Erbs, N., Zesch, T., Gurevych, I.: Link discovery: a comprehensive analysis. In: 2011 Fifth IEEE International Conference on Semantic Computing (ICSC), pp. 83–86, September 2011
10. Fader, A., Soderland, S., Etzioni, O.: Scaling wikipedia-based named entity disambiguation to arbitrary web text. In: Proceedings of WIKIAI (2009)
11. Fernández, N., Arias Fisteus, J., Sánchez, L., López, G.: Identityrank: named entity disambiguation in the news domain. Expert Syst. Appl. **39**(10), 9207–9221 (2012)
12. Ferragina, P., Scaiella, U.: Tagme: on-the-fly annotation of short text fragments (by wikipedia entities). In: Proceedings of the 19th ACM International Conference on Information and Knowledge Management, CIKM 2010, pp. 1625–1628 (2010)
13. Finkel, J.R., Grenager, T., Manning, C.: Incorporating non-local information into information extraction systems by gibbs sampling. In: ACL, pp. 363–370 (2005)
14. García, N.F., Arias-Fisteus, J., Fernández, L.S., Martín, E.: Webtlab: A cooccurrence-based approach to KBP 2010 entity-linking task. In: Proceedings of the Third Text Analysis Conference, TAC 2010, Gaithersburg, Maryland, USA, 15–16 November 2010 (2010)
15. Guo, Z., Barbosa, D.: Robust entity linking via random walks. In: Proceedings of the 23rd ACM International Conference on Conference on Information and Knowledge Management, CIKM 2014, pp. 499–508. ACM (2014). http://doi.acm.org/10.1145/2661829.2661887
16. Han, X., Sun, L., Zhao, J.: Collective entity linking in web text: a graph-based method. In: Proceedings of the 34th international Conference on Research and Development in Information Retrieval, pp. 765–774 (2011)
17. Hellmann, S., Lehmann, J., Auer, S., Brümmer, M.: Integrating NLP using linked data. In: Alani, H., et al. (eds.) ISWC 2013, Part II. LNCS, vol. 8219, pp. 98–113. Springer, Heidelberg (2013)
18. Hoffart, J., Seufert, S., Nguyen, D.B., Theobald, M., Weikum, G.: Kore: keyphrase overlap relatedness for entity disambiguation. In: Proceedings of the 21st ACM International Conference on Information and Knowledge Management, CIKM 2012, pp. 545–554 (2012)
19. Hoffart, J., Yosef, M.A., Bordino, I., Fürstenau, H., Pinkal, M., Spaniol, M., Taneva, B., Thater, S., Weikum, G.: Robust disambiguation of named entities in text. In: Proceedings of the Conference on Empirical Methods in Natural Language Processing, pp. 782–792 (2011)
20. Kulkarni, S., Singh, A., Ramakrishnan, G., Chakrabarti, S.: Collective annotation of wikipedia entities in web text. In: Proceedings of the 15th ACM SIGKDD International Conference on Knowledge Discovery and Data Mining, KDD 2009, pp. 457–466. ACM, New York (2009)

21. Marie, N., Gandon, F.L., Giboin, A., Palagi, É.: Exploratory search on topics through different perspectives with DBpedia. In: Proceedings of the 10th International Conference on Semantic Systems, SEMANTICS 2014, Leipzig, Germany, 4–5 September 2014, pp. 45–52 (2014). http://doi.acm.org/10.1145/2660517.2660518

22. Mendes, P.N., Jakob, M., García-Silva, A., Bizer, C.: DBpedia spotlight: shedding light on the web of documents. In: Proceedings of the 7th International Conference on Semantic Systems, I-Semantics 2011, pp. 1–8. ACM, New York (2011). http://doi.acm.org/10.1145/2063518.2063519

23. Mihalcea, R., Csomai, A.: Wikify!: linking documents to encyclopedic knowledge. In: Proceedings of the Sixteenth ACM Conference on Conference on Information and Knowledge Management, pp. 233–242 (2007)

24. Milne, D., Witten, I.H.: An effective, low-cost measure of semantic relatedness obtained from wikipedia links. In: Proceeding of AAAI Workshop on Wikipedia and Artificial Intelligence: An Evolving Synergy, pp. 25–30, July 2008

25. Milne, D., Witten, I.H.: Learning to link with wikipedia. In: Proceedings of the 17th ACM Conference on Information and Knowledge Management, pp. 509–518 (2008)

26. Moro, A., Raganato, A., Navigli, R.: Entity linking meets word sense disambiguation: a unified approach. Trans. Assoc. Comput. Linguist. **2**, 231–244 (2014)

27. Page, L., Brin, S., Motwani, R., Winograd, T.: The pagerank citation ranking: Bringing order to the web. Technical Report 1999-66, Stanford InfoLab, November 1999. http://ilpubs.stanford.edu:8090/422/

28. Plu, J., Rizzo, G., Troncy, R.: A hybrid approach for entity recognition and linking. In: ESWC 2015, 12th European Semantic Web Conference, Open Extraction Challenge, Portoroz, Slovenia, 31 May-4 June 2015 (2015). http://www.eurecom.fr/publication/4613

29. Ponzetto, S.P., Strube, M.: Knowledge derived from wikipedia for computing semantic relatedness. J. Artif. Intell. Res. (JAIR) **30**, 181–212 (2007)

30. Ratinov, L., Roth, D., Downey, D., Anderson, M.: Local and global algorithms for disambiguation to wikipedia. In: Proceedings of the 49th Annual Meeting of the Association for Computational Linguistics: Human Language Technologies, HLT 2011, vol. 1, pp. 1375–1384 (2011)

31. Shen, W., Wang, J., Han, J.: Entity linking with a knowledge base: issues, techniques, and solutions. IEEE Trans. Knowl. Data Eng. **27**(2), 443–460 (2015)

32. Singh, S., Subramanya, A., Pereira, F., McCallum, A.: Wikilinks: A large-scale cross-document coreference corpus labeled via links to Wikipedia. Technical report, UM-CS-2012-015 (2012)

33. Spitkovsky, V.I., Chang, A.X.: A cross-lingual dictionary for english wikipedia concepts. In: Chair, N.C.C., Choukri, K., Declerck, T., Doan, M.U., Maegaard, B., Mariani, J., Moreno, A., Odijk, J., Piperidis, S. (eds.) Proceedings of the Eight International Conference on Language Resources and Evaluation (LREC 2012), Istanbul, Turkey, May 2012

Evaluating Multiple Summaries Without Human Models: A First Experiment with a Trivergent Model

Luis Adrián Cabrera-Diego[1(✉)], Juan-Manuel Torres-Moreno[1,2],
and Barthélémy Durette[3]

[1] LIA, Université d'Avignon et des Pays de Vaucluse, Avignon, France
{luis-adrian.cabrera-diego,juan-manuel.torres}@univ-avignon.fr
[2] École Polytechnique de Montréal, Montréal, Canada
[3] Adoc Talent Management, Paris, France
durette@adoc-tm.com

Abstract. In this work, we extend the task of evaluating summaries without human models by using a trivergent model. In this model, three elements are compared simultaneously: a summary to evaluate, its source document and a set of other summaries from the same source. We present in this paper, a first pilot experiment using a French corpus from which we obtained promising results.

Keywords: Divergence of probability distributions · Trivergence of probability distributions · Text summarization evaluation

1 Introduction

Since the creation of the first Automatic Text Summarization (ATS) system, the evaluation of summaries has been a key but not less ambitious and controversial part of the research [6,14,17–19,21]. The evaluation of summaries consists in applying different techniques or measures to create a score that is correlated with the one a human evaluator would give [5].

According to [18], summaries can be evaluated intrinsically or extrinsically. In the first case, the condensed documents are compared against human models; both kinds of documents come from the same source document. In the second case, the summaries are evaluated with respect to their effect over other systems' performance, e.g. QA systems or search engines.

In the literature, we can find that most of the developed evaluation methodologies are intrinsic [6], for example, *Rouge* [11] and *Pyramids* [15]. However, the problem with these methods is that they have to face up to the definition of "ideal summary", which is hard to establish [19]. In other words, humans can generate summaries, either by extraction or abstraction, with different points of views. Thus, for frameworks like *Rouge* and *Pyramids* it is necessary to have several man-made summary models in order to represent correctly the perspective diversity.

© Springer International Publishing Switzerland 2016
E. Métais et al. (Eds.): NLDB 2016, LNCS 9612, pp. 91–101, 2016.
DOI: 10.1007/978-3-319-41754-7_8

To bypass the "ideal summary", some frameworks have been developed to compare the produced summaries against their source document. Among these, we can point out *SIMetrix*[1] [13] and *FRESA*[2] [17]. In both cases, the quality of summaries is determined by calculating the divergences of Kullback-Leibler [10] or Jensen-Shannon [12]. More specifically, the summary's words n-grams distribution is compared against the one of its source document.

We present, in this paper, a pilot study that uses a model where three elements are compared simultaneously: a summary to be evaluated, its source document and other summaries generated from the same source. Our goal is to evaluate summaries more accurately without the need of multiple man-made summary models.

This paper is divided in three sections. In Sect. 2 we explain the methodology used in this work. We introduce the results and their discussion in Sect. 3. The conclusions and future work are presented in Sect. 4.

2   Methodology

In this paper, we use the *Trivergence of probability distributions* [22]. It is a statistical measure that compares simultaneously three different probability distributions. This measure is based on the application of multiple divergences, like the one of Kullback-Leibler or Jensen-Shannon. For three probability distributions P, Q and R, the triple comparison can be done either by the multiplication or the composition of divergences.

In the multiplicative case, the trivergence τ_m is defined as in Eq. 1:

$$\tau_m(P||Q||R) = \delta(P||Q) \cdot \delta(P||R) \cdot \delta(Q||R) \tag{1}$$

Where three divergences $\delta(\bullet)$ are calculated separately. The resulting values of the divergences are then multiplied.

Regarding the composite trivergence τ_c, there are two divergences, one inside the other. In this case, the inner divergence is calculated first and then, its value, which it is a constant, is used for the outer divergence. The composite trivergence is defined as in Eq. 2:

$$\tau_c(P||Q||R) = \delta\left(P||\frac{\delta(Q||R)}{N}\right) \tag{2}$$

Where $\delta(\bullet)$ corresponds to the divergences and N to a normalization parameter. The normalization parameter N is used to make the inner divergence value closer to a probability figure. In other words, the inner divergence gives always a constant value which is frequently greater than 1. Thus, the division of this constant by a normalization factor allow us to make the value smaller and more similar to a probability figure. In our case, we defined N as the sum of the 3 distributions size ($|P| + |Q| + |R|$), however a different N could be set.

---

[1] http://homepages.inf.ed.ac.uk/alouis/.

[2] http://fresa.talne.eu.

Using as base the divergence of Kullback-Leibler, the inner divergence $\delta(Q||R)$ from Eq. 2 is defined as in Eq. 3:

$$KL(Q||R) = \sum_{\omega \in Q} q_\omega log \frac{q_w}{r_w} \tag{3}$$

Where ω are the words belonging to Q; q_ω and r_ω are the probabilities of ω to occur in Q and R respectively. The composite trivergence of Kullback-Leibler, thus, is defined as in Eq. 4:

$$\tau_c(P||Q||R) = \sum_{\sigma \in P} p_\sigma log \frac{p_\sigma}{\left(\frac{KL(Q||R)}{N}\right)} \tag{4}$$

Where σ are the words belonging to P; p_σ represents the probabilities of σ to occur in P; N is the normalization parameter. It should be said that we make use of $\left(\frac{KL(Q||R)}{N}\right)$ only if σ exists in Q, otherwise a smoothing value is used. We decided to do this, because a σ that does not exist in Q, should be considered as a unseen event.

As it was indicated previously, it is possible to have in some distributions unseen events, either in the multiplicative or composite trivergence. For example, not all the events in the distribution Q exist in distribution R. Therefore, we set a smoothing based on Eq. 5:

$$Smoothing = \frac{1}{|P| + |Q| + |R|} \tag{5}$$

Where $|P|$, $|Q|$ and $|R|$ corresponds to the respective distribution size of P, Q and R. Other smoothing methods can be used without having to change the trivergent model, e.g. Kat's back-off model [7] or Good-Turing estimation [4]. However, in order to use a more complex smoothing, it is necessary to take some considerations for the composite trivergence because its inner divergence is a constant value.

For our study, we considered the existence of a source document from where a set of summaries was generated. Thus, we defined Q as the source document, R as a summary to evaluate and P as all the summaries of the set excepting R. Every condensed document from the set can be evaluated with the trivergence. It is only necessary to interchange the summary of R by one from P.

In our experiments, two different divergences were used for the trivergent model: Kullback-Leibler (KL) and Jensen-Shannon (JS).

Despite [22] set a cardinality condition[3] for both trivergences, multiplicative and composite, in this experiment we decided to test all the possible model combinations[4]. The reason to test the different combinations was to verify whether

---

[3] The cardinality condition consists in computing the divergences from the smallest distribution to the largest one. For example, using as base Eqs. 1 and 2, the condition would be $|P| > |Q| > |R|$.

[4] We did not test the combinations given by the possible trivergence's associative property.

the performance of our methodology could change due to the divergences' commutative property. For example, the divergence of Kullback-Leibler is not symmetric, therefore $KL(P||Q) \neq KL(Q||P)$. This methodology is similar to the one used by [1], nonetheless, it was applied in an e-Recruitment context for ranking *résumés*.

Following the representation of documents used by other summarization evaluation frameworks [11,13,17], we decided to use 3 types of words n-grams: unigrams, bigrams and skip-bigrams (SU4). In consequence, we defined the score for each summary as the average trivergence according to the 3 types of words n-grams.

2.1 Corpus

For this pilot experiment, we created a small French corpus called *"Puces"*.[5] This corpus comprises one source document and a set of 170 extracts created by several humans and 10 ATS systems. The source document is composed of 30 sentences that belong to two topics: a new electronic microchips for computers and a fleas invasion in a military Swiss company. The topic mixture is due to the polysemy of the French word *puces*; one sense corresponds to microchips, while the other corresponds to fleas.

To produce the human summaries, each person had 15 min for reading the source document, then, they extracted the 8 sentences that represented better for them the subject. Regarding the ATS systems, they generated the summaries respecting the same size of 8 sentences. Among the ATS systems we used, we can name Essential Summarizer[6], Microsoft Word, Cortex [23], Artex [20] and Enertex [2]. All the documents of the corpus *Puces* were preprocessed before obtaining the n-grams. In first place we deleted the stop-words using the list available in *FRESA*. Then we stemmed the words with the French Snowball stemmer[7].

As the performance of the trivergence can change according to the data and the size of distribution P, we decided to test how this would affect our methodology. Thus, we set 12 different sizes, in terms of documents, for P (4, 8, 16, 32, 48, 64, 80, 96, 112, 128, 144, 160). For each size, we built 25 subcorpora by selecting randomly documents from *Puces*. All the 300 different subcorpora were used to determine R and P distributions. For example, when a subcorpus had 32 summaries, one of the extracts was used as R and the remaining 31 as P.

2.2 Evaluation

The application of the trivergence over each subcorpus generated a ranking, which was based, in turn, on the score obtained by each summary. Therefore, to

---

[5] The corpus can be downloaded from: http://dev.termwatch.es/~fresa/CORPUS/PUCES/.

[6] https://essential-mining.com.

[7] http://snowballstem.org.

evaluate our methodology, we calculated the correlation between the rankings given by *Rouge* and those obtained by the trivergent model. The evaluation with *Rouge* consisted in compare the summaries against 67 abstracts done by humans after reading the source document of *Puces*.

The correlation measures we used were: *Spearman's Rank-order Correlation Coefficient (Spearman's rho)* and *Kendall's tau-b Correlation Coefficient (Kendall's tau-b)*. These two correlation coefficients were chosen because they are specific tests to compare two ranks. Furthermore, they have been used in other researches with similar goals [5,13,17].

To be more specific, for each subcorpus, we calculated the correlation coefficients between the scores given by the trivergent model and *Rouge*. Then, considering the subcorpora size, we calculated the average correlation coefficients with their respective 95 % confidence interval bars.

We compared, as well, the trivergent model against two baselines under the same conditions. The baselines consisted in the average correlation between *Rouge* and the rankings given by *FRESA* and *SIMetrix*. From *FRESA*, we used its 3 measures: Kullback-Leibler, Symmetric Kullback-Leibler and Jensen-Shannon. In the case of *SIMetrix*, we tested its 4 measures based on divergences: Kullback-Leibler between source and summary, Kullback-Leibler between summary and source, Smoothed and Unsmoothed Jensen-Shannon. For SIMetrix, we needed to use its own English Porter stemmer as it was not possible to set a French one.

The statistical analysis was done with the programming language *R* [16] and its package *irr* [3].

3   Results and Discussion

In Fig. 1 we present the 3 trivergence combinations with the highest average *Kendall's tau-b*. We show in Fig. 1 as well, the measures coming from *FRESA* and *SIMetrix* with the best average *Kendall's tau-b* coefficients.

As it can be seen in Fig. 1, the correlation coefficients of the measures vary according to the subcorpora size. Small subcorpora give low average correlation coefficients and large confidence interval bars. While larger copora indicates more precisely the measures' performance, as the confidence interval bars become shorter. It can be seen, as well, that the result from the trivergence measures become more correlated to *Rouge* as we increase the copora size. This contrast with the outcomes coming from *FRESA* and *SIMetrix*, which are more stable no matter the corpus size.

Concerning the *Spearman's rho*, it can be seen in Fig. 2 that the trivergent model and baselines behave in the same way that for *Kendall's tau-b*. However, the correlations coefficients are higher. This difference should not be taken as a disagreement or discrepancy between the coefficients. In fact, the difference between these two coefficients is well known. Moreover, [8] states that the *Kendall's tau* will be about two-thirds of the value of *Spearman's rho* when the analyzed sample size is large.

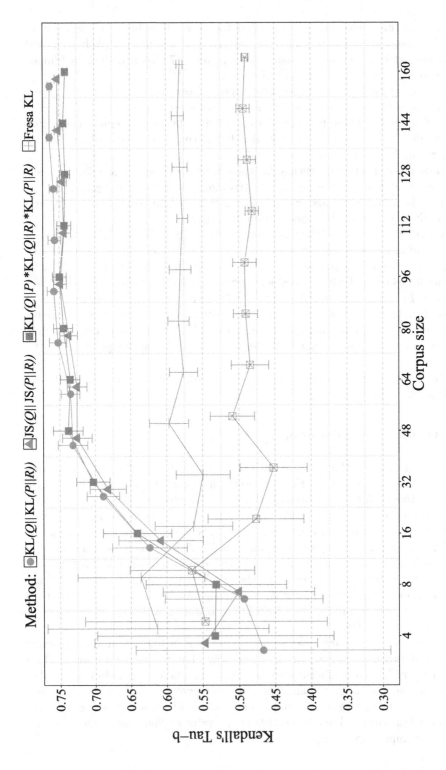

Fig. 1. Results according to the *Kendall's tau-b* correlation coefficient

Method: $\boxdot$ KL(Q‖KL(P‖R)) $\blacktriangle$ JS(Q‖JS(P‖R)) $\blacksquare$ KL(Q‖P)*KL(Q‖R)*KL(P‖R) $\boxplus$ Fresa KL

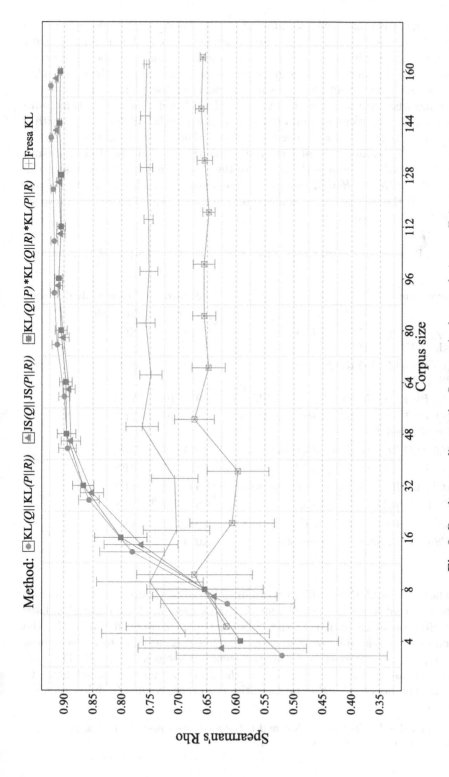

Fig. 2. Results according to the *Spearman's rho* correlation coefficient

In both correlation coefficients, for subcorpora greater than 16 documents, we observed that the trivergences measures become more correlated to *Rouge* with respect to the other frameworks. This means that the rankings of the trivergences become more similar to the ones given by *Rouge* as we increase the size of P.

To validate our results for the subcorpora of 32 documents, we decided to determine whether the *Kendall's tau-b* correlation coefficients of each measure were statistically different. To do this, we used a *Kruskal-Wallis test* [9] with a *Games-Howell post hoc test*.[8] For the *Kruskal-Wallis test*, we got a p-value = 3.88×10^{-14}, meaning that at least the correlation coefficients of one measure were statistically different with respect to the others. In Table 1, we present the results of the *Games-Howell post hoc test* by measure couples. It can be seen in Table 1 that the trivergent model measures are statistically different with respect to the baselines' measures.

Table 1. Results of the *Games-Howell post hoc test* for the subcorpora of 32 documents

Measure couples	Adjusted p-value	Statistically different ($\alpha = 0.05$)
KL($Q\|\|P$)· KL($Q\|\|R$)·KL($P\|\|R$) vs. KL($Q\|\|$KL($P\|\|R$))	0.915	NO
KL($Q\|\|P$)· KL($Q\|\|R$)·KL($P\|\|R$) vs. JS($Q\|\|$JS($P\|\|R$))	0.759	NO
KL($Q\|\|P$)· KL($Q\|\|R$)·KL($P\|\|R$) vs. *FRESA* KL	2.94×10^{-7}	YES
KL($Q\|\|P$)· KL($Q\|\|R$)·KL($P\|\|R$) vs. *SIMetrix* Unsmoothed JS	3.18×10^{-10}	YES
KL($Q\|\|$KL($P\|\|R$)) vs. JS($Q\|\|$JS($P\|\|R$))	0.995	NO
KL($Q\|\|$KL($P\|\|R$)) vs. *FRESA* KL	2.07×10^{-6}	YES
KL($Q\|\|$KL($P\|\|R$)) vs. *SIMetrix* Unsmoothed JS	1.36×10^{-9}	YES
JS($Q\|\|$JS($P\|\|R$)) vs. *FRESA* KL	8.34×10^{-6}	YES
JS($Q\|\|$JS($P\|\|R$)) vs. *SIMetrix* Unsmoothed JS	2.93×10^{-9}	YES
FRESA KL vs. *SIMetrix* Unsmoothed JS	0.020	YES

The good correlation coefficients obtained by our methodology may be related to the fact that as we increase the number of extracts in P, we make P to have a distribution more similar to the "ideal extract". Expressed differently, the sentences more relevant from the source document correspond to the sentences that appear more frequently in all the summaries belonging to P. This will affect, in consequence, the probability of the words n-grams; those belonging to relevant sentences have higher probabilities than those coming from less relevant sentences.

---

[8] We applied a *Kruskal-Wallis test* and a *Games-Howell post hoc test* as the results had heterogeneous variances.

If we make a comparison with *Rouge*, we generate the distribution of the "ideal summary by extraction" instead of building the distribution of the "ideal summary by abstraction". Moreover, we add in our methodology the comparison between the summary and its source document, as *Fresa* and *SIMetrix* do.

We found that for $KL(R||KL(Q||P))$, the inner divergence resulted in a negative constant. This made impossible to continue with the outer divergence as the logarithm is not defined for negative values. However, after a deep analysis, we found that the problem resides in the fact that the smoothing value was not small enough to make the divergence to be positive. The use of a smaller smoothing value or another smoothing method may solve the problem.

In addition, we observed that each sentence of the *Puces* source document appears at least in one summary, even when all the possible summaries are near of 6 millions due to the combinatorial $\binom{30}{8}$.

4   Conclusions and Future Work

In this paper we presented a first study about the trivergence of probabilities and its application in the evaluation of summaries without human models. The trivergence is a statistical measure where multiple divergences are used, either by multiplication or composition. The aim is to calculate the dissimilarity of 3 different distributions.

The tests were done over a small French corpus of summaries. To evaluate the trivergent model, we calculated the correlation of our rankings and 2 baselines with respect to *Rouge*. The results showed that the trivergence can have a better performance, in comparison to the baselines, when more than 16 summaries are analyzed. These outcomes are really favorable, especially for summarization challenges, where multiple artificial summaries must be evaluated. Moreover, our methodology could be a way to bypass the creation of abstracts by humans in order to evaluate ATS systems.

We consider feasible a reduction in the number of summaries that our evaluation methodology needs to have a good performance. A potential solution is to change the stemmer by a lemmatizer during the preprocessing step, as we use a highly inflected language. Another possible solution is to test other kinds of n-grams or to use TF-IDF to weight differently the n-grams.

As future work, we consider it fundamental to test our methodology in other languages and with other corpora like DUC/TAC, MultiLing and RPM2. As well, we want to continue testing and improving the trivergence of probabilities in other tasks and domains, for example, to rank *résumés* for the e-Recruitment.

References

1. Cabrera-Diego, L.A.: Automatic methods for assisted recruitment. Ph.D. thesis, Université d'Avignon et des Pays de Vaucluse, December 2015
2. Fernández, S., SanJuan, E., Torres-Moreno, J.M.: Textual energy of associative memories: performant applications of enertex algorithm in text summarization and topic segmentation. In: Gelbukh, A., Kuri Morales, A.F. (eds.) MICAI 2007. LNCS (LNAI), vol. 4827, pp. 861–871. Springer, Heidelberg (2007)
3. Gamer, M., Lemon, J., Singh, I.F.P.: irr: Various Coefficients of Interrater Reliability and Agreement (2012), rpackageversion0.84. https://CRAN.R-project.org/package=irr
4. Good, I.J.: The population frequencies of species and the estimation of population parameters. Biometrika 40(3–4), 237–264 (1953)
5. Hovy, E., Lin, C.Y., Zhou, L., Fukumoto, J.: Automated summarization evaluation with basic elements. In: Proceedings of the Fifth Conference on Language Resources and Evaluation (LREC 2006), pp. 604–611 (2006)
6. Jing, H., Barzilay, R., McKeown, K., Elhadad, M.: Summarization evaluation methods: experiments and analysis. In: AAAI Symposium on Intelligent Summarization, pp. 51–59 (1998)
7. Katz, S.M.: Estimation of probabilities from sparse data for the language model component of a speech recognizer. IEEE Trans. Acoustics Speech Signal Process. 35(3), 400–401 (1987)
8. Kendall, M.G., Stuart, A.: The Advanced Theory of Statistics, vol. 1, 2nd edn. Charles Griffin and Co., London (1948)
9. Kruskal, W.H., Wallis, W.A.: Use of ranks in one-criterion variance analysis. J. Am. Stat. Assoc. 47(260), 583–621 (1952)
10. Kullback, S., Leibler, R.A.: On information and sufficiency. Ann. Math. Stat. 22(1), 79–86 (1951)
11. Lin, C.Y.: Rouge: a package for automatic evaluation of summaries. In: Text Summarization Branches Out: Proceedings of the Association for Computational Linguistics 2004 Workshop, vol. 8 (2004)
12. Lin, J.: Divergence measures based on the Shannon entropy. IEEE Trans. Inf. Theory 37(1), 145–151 (1991)
13. Louis, A., Nenkova, A.: Automatically evaluating content selection in summarization without human models. In: Proceedings of the 2009 Conference on Empirical Methods in Natural Language Processing, vol. 1, pp. 306–314. Association for Computational Linguistics (2009)
14. Mani, I.: Summarization evaluation: an overview. In: Proceedings of the North American Chapter of the Association for Computational Linguistics (NAACL) Workshop on Automatic Summarization (2001)
15. Nenkova, A., Passonneau, R.: Evaluating content selection in summarization: the pyramid method. In: Susan Dumais, D.M., Roukos, S. (eds.) Proceedings of HLT-NAACL 2004, pp. 145–152. Association for Computational Linguistics, Boston (2004)
16. R Core Team: R: A Language and Environment for Statistical Computing. R Foundation for Statistical Computing, Vienna, Austria (2016). https://www.R-project.org/
17. Saggion, H., Torres-Moreno, J.M., da Cunha, I., SanJuan, E., Velázquez-Morales, P.: Multilingual summarization evaluation without human models. In: 23rd International Conference on Computational Linguistics (COLING 2010), pp. 1059–1067. Association for Computational Linguistics (2010)

18. Spärck-Jones, K., Galliers, J.R.: Evaluating Natural Language Processing Systems: An Analysis and Review. LNCS(LNAI), vol. 1083. Springer, New York (1996)
19. Steinberger, J., Ježek, K.: Evaluation measures for text summarization. Comput. Inform. **28**(2), 251–275 (2012)
20. Torres-Moreno, J.M.: Artex is another text summarizer (2012). arXiv:1210.3312
21. Torres-Moreno, J.M.: Automatic Text Summarization. Wiley, New York (2014)
22. Torres-Moreno, J.M.: Trivergence of Probability Distributions, at Glance. Computing Research Repository (CoRR) abs/1506.06205 (2015). http://arxiv.org/abs/1506.06205
23. Torres-Moreno, J.M., Velázquez-Morales, P., Meunier, J.G.: Condensés automatiques de textes. Lexicometrica. L'analyse de données textuelles: De l'enquête aux corpus littéraires, Special (2004)

PatEx: Pattern Oriented RDF
Graphs Exploration

Hanane Ouksili[1,2]($\boxtimes$), Zoubida Kedad[1], Stéphane Lopes[1],
and Sylvaine Nugier[2]

[1] DAVID Lab., Univ. Versailles St Quentin, Versailles, France
{hanane.ouksili,zoubida.kedad,stephane.lopes}@uvsq.fr
[2] Department STEP, EDF R&D, Chatou, France
sylvaine.nugier@edf.fr

Abstract. An increasing number of RDF datasets are available on the
Web. In order to query these datasets, users must have information about
their content as well as some knowledge of a query language such as
SPARQL. Our goal is to facilitate the exploration of these datasets. We
present in this paper PatEx, a system designed to explore RDF(S)/OWL
data. PatEx provides two exploration strategies: theme-based exploration and keyword search, and users can interactively switch between
the two. Moreover, the system allows the definition of patterns to formalize users' requirements during the exploration process. We also present
some evaluations performed on real datasets.

Keywords: RDF graph exploration · Theme discovery · Keyword
search

1 Introduction

A huge volume of RDF datasets is available on the Web, enabling the design
of novel intelligent applications. However, in order to use such datasets, the
user needs to understand them. The exploration of RDF datasets is deemed to
be a complex and burdensome task requiring a long time and some knowledge
about the data. Formal query languages are powerful solutions to express users
informational needs and to retrieve knowledge from an RDF dataset. However,
in order to write these structured queries, users must have some information
about the dataset schema. Besides, they must be familiar with a formal query
language that could be used to query this data, such as SPARQL.

To illustrate the problem and highlight our contributions, consider the
dataset provided by the AIFB[1] institute, which represents a research community
including persons, organizations, publications and their relationships. Assume
that the user is interested in the scientist named *"Rudi Studer"* and wants to
have all his cooperators in the domain of the Semantic Web only. The user has

---

This work is supported by EDF and french ANR project CAIR.

[1] http://www.aifb.kit.edu/web/Hauptseite/en.

© Springer International Publishing Switzerland 2016
E. Métais et al. (Eds.): NLDB 2016, LNCS 9612, pp. 102–114, 2016.
DOI: 10.1007/978-3-319-41754-7_9

first to submit several queries and manually browse the dataset in order to collect all the relevant properties, which will be used to formulate the queries that will provide the final answer. Our proposal to address this issue is to combine keyword search and theme browsing to guide the user during the identification of the relevant information. We argue that if the data is presented as a set of themes along with the description of their content, it is easier to target the relevant resources and properties by exploring the relevant themes only. The user can perform a theme-based exploration, then, she may select the desired theme, the field of Semantic Web in our example, and perform a keyword search with the query *"Studer Cooperator"*.

In this paper, we propose a system, called PatEx, which enables users to explore and to understand RDF datasets. Our approach does not require prior knowledge about the schema, or about a formal query language. Moreover, users can enrich the exploration process with some domain knowledge as well as their specific requirements. PatEx provides two exploration strategies: a theme browser and a keyword search engine, and the users can interactively switch between the two. To support the exploration process, we define the notion of pattern to formalize both users preferences and specific domain knowledge.

The paper is organized as follows. Section 2 presents an overview of the PatEx system. Section 3 defines the notion of pattern. In Sect. 4, we describe the way external knowledge is captured during the exploration. Sections 5 and 6 describe the approaches used to support the exploration strategies. Section 7 presents some evaluation results. In Sect. 8, we discuss related works and finally, we conclude the paper in Sect. 9.

2   The PatEx System: An Overview

Our system provides two complementary approaches to query RDF datasets: theme-based exploration and keyword search. Moreover, since two users can have different point of views, the system enables to specify some preferences which are formalized as patterns. An overview of the system's architecture is given in Fig. 1.

The user interacts with the system through the RDF GRAPH MANAGER. It enables to choose the RDF data to explore, to select the exploration strategy and to visualize the results. This module also enables an iterative exploration; for example, the user might select one or several identified themes from the initial RDF graph to refine the exploration. Moreover, it is possible to switch between the exploration modes, i.e. keyword search can be performed on a subgraph corresponding to one or several themes identified in the initial dataset.

The PATTERN MANAGER module captures user preferences and domain knowledge. It translates them into patterns evaluated on the selected RDF graph. The result is stored in the PATTERN STORE, then they are used to guide the exploration process. Theme-based exploration is performed by the THEME BROWSER. It receives the selected RDF graph from the RDF GRAPH MANAGER and uses the evaluated patterns from the PATTERN STORE. The result

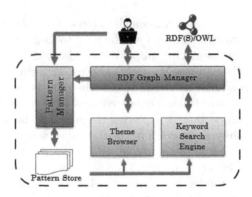

Fig. 1. Architecture of the PatEx System

of pattern evaluation on the RDF graph is mapped into graph transformations. Themes are then identified using a density based graph clustering algorithm executed on the enriched graph. Keyword queries are processed by the KEYWORD SEARCH ENGINE module which will look for elements that match the keywords, taking into account the result of patterns evaluation from the PATTERN STORE. The result is a set of subgraphs containing these elements, which are returned to the user through the RDF GRAPH MANAGER.

3   Pattern Definition

According to the knowledge available for a specific domain, or to the user's point of view, some equivalence relations can be defined between a property and a path in the dataset. In PatEx, we express these relations through patterns. The definition of patterns relies on the definition of *property expression* and *path expression* which are defined hereafter.

A *property expression* denoted *exp* is a 3-tuple (X, p, Y), where X and Y are either resources, literals or variables and p is a property. For example, $(X,$ *swrc:isAbout, Database*) is a *property expression* that represents triples having *Database* as object and *swrc:isAbout* as predicate.

A *path expression* describes a relation (possibly indirect) between resources. The relation is a path (a sequence of properties) between two resources. Note that a property expression is a special case of path expression where the path length is 1. More formally, A *path expression* denoted *exP* is defined as a 3-tuple (X, P, Y), where X and Y are either resources, literals or variables and P is a SPARQL 1.1 path expression[2]. For example, $(D.B., (owl{:}sameAs|\hat{}\,owl{:}sameAs\,)^{+}, Y)$ represents all the resources related to $D.B.$ by a sequence of *owl:sameAs* or its inverse.

A *pattern* represents the equivalence between two expressions, one property expression and one path expression. More formally, a *pattern* is a pair $[exp, exP]$ where *exp* is a property expression and *exP* is a path expression.

---

[2] http://www.w3.org/TR/sparql11-property-paths/.

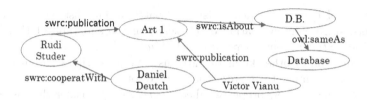

Fig. 2. RDF subgraph from AIFB data

As an example, consider the RDF graph of Fig. 2, the two resources *D.B.* and *Database* are related by the property *owl:sameAs* which indicates that they are equivalent. The value of the property *swrc:isAbout* for the resource *Art1* is *D.B.*, therefore, this property has also the value *Database*. To generalize this case, the following pattern is defined: $[(X, swrc:isAbout, Y),$ $(X, swrc:isAbout/(owl:sameAs|^owl:sameAs)^+, Y)]$. This pattern represents the fact that the property *swrc:isAbout* is equivalent to a path composed by a property *swrc:isAbout* followed by a sequence of *owl:sameAs*.

4   Expressing Knowledge Using Patterns

In order to capture the user intent and some domain knowledge during the exploration, we use the notion of pattern defined in Sect. 3.

In the AIFB dataset, the *swrc:cooperateWith* property between two researchers r_i and r_j states that they have already collaborated. But RDF graphs are often incomplete and this property is not always defined. If this property is missing for a pair of researchers but if we know that they wrote the same paper, we can infer that they have collaborated, even if the property *swrc:cooperateWith* is missing. The pattern manager module translates this knowledge into $Pattern_1$, shown in Table 1. Furthermore, it evaluates this pattern on the given RDF graph

Table 1. Examples of Patterns

Id	Pattern	Explanation of the *path expression*
1	$[(X, swrc:cooperateWith, Y), (X,$ $swrc:publication/^swrc:publication, Y)]$	The path is composed of the property *swrc:publication* followed by its inverse
2	$[(X, pat:related, Y), (X,$ $(owl:sameAs|^owl:sameAs)^+, Y)]$	The path is a sequence of *owl:sameAs* properties or its inverse
3	$[(X, pat:related, Y), (X, p, Y)]$	The path is composed of one property p
4	$[(X, pat:sameValue\_p, Y),(X, p/^p, Y)]$	The path is composed of the property p followed by its inverse

to find a set of triples L which satisfy the pattern. This list will be saved in the pattern store. The evaluation of $Pattern_1$ in the subgraph of Fig. 2 gives $L = \{< Rudi\ Studer,\ swrc:cooperateWith,\ Victor\ Vianu >\}$.

The evaluation of other kinds of patterns generates triples based on a new property. For example, to express that two resources are semantically close, we have introduced a new property called *pat:related*. Consider the case of two resources related by a sequence of *owl:sameAs* properties. We can state that they are semantically close as they represent in fact the same object. This is captured by the $pattern_2$ given in Table 1. The result of its evaluation is $L = \{< B.D.,$ *pat:related, Database* $>\}$.

5   Theme Identification

Theme identification aims at providing the users with a thematic view of an RDF graph taking into account their preferences. A set of themes is identified from an RDF graph, and for each theme, the labels that best reflect its semantic are extracted. Our approach relies on the idea that a theme corresponds to a highly connected area of the RDF graph.

Let G be the RDF graph and I a set of preferences representing the intent of the user. Theme identification is the task of identifying a set of k subgraphs $T = \{t_1, ..., t_k\}$ according to the user preferences I. Each subgraph $t_i = \{r_{i_1}, ..., r_{i_m}\}$ represents a theme and is composed of a set of resources.

We use a graph clustering algorithm which identifies highly connected areas in the graph in order to form clusters, each one corresponds to a theme. In our approach, we have used the MCODE graph clustering algorithm [1]. Our choice is based on three requirements for the chosen algorithm: (i) it has to exploit the graph density to form clusters, (ii) the number of clusters is not known and finally, (iii) it has to provide clusters which are not necessarily disjoint, as it is possible that one resource in our RDF graph belongs to two different themes. More details about the algorithm can be found in [14].

Users express the domain knowledge and their intent using an interactive interface. Then, this information is formalized by our system using patterns. In this paper, we will illustrate our approach using two specific kinds of preferences. The first one is **grouping two resources related by a property**. Some properties express a strong semantic link that should be used to assign resources to the relevant themes. For example, resources linked by the *owl:sameAs* property should obviously be assigned to the same theme, and this is always valid. Besides, a user may wish to give some domain properties more importance than others in a specific context, for example, in the AIFB dataset, if the user wants to have scientists which have already cooperated in the same theme, she will use the property *swrc:cooperateWith* in her preference. Preferences of this kind are translated into patterns by the pattern manager module. We define a new property in the pattern to express this information, namely the *pat:related* property. The structure of such pattern is illustrated by $Pattern_3$ in Table 1. This kind of preferences is taken into account by merging the two resources, forcing the clustering algorithm to assign them to the same theme.

The second type of preferences is **grouping resources according to values of a property**. In some cases, two resources are semantically close not because they are in the same highly connected area or related by a given property, but because they share the same value for a specific property. For example, if the user wants a theme to be related to a single country, she could state that resources having the same value for the *localisation* property should be grouped together. We introduce the property *pat:sameValue_p* in the pattern to express this information, where p is the considered property. The structure of such patterns is shown by $Pattern_4$ of Table 1. This preference is mapped on the graph by creating a complete graph with the resources resulting from the evaluation of the pattern.

Once the themes have been defined, the user is provided with a description of the content of each theme. In our approach, we provide these labels by extracting (i) the names of the classes having the highest number of instances, and (ii) the labels of the resources corresponding to nodes having the highest degree in the RDF graph.

6 Keyword Search

Another kind of exploration of an RDF graph is keyword search, where an answer to a keyword query is a set of connected subgraphs, each one containing the highest number of query keywords. Indeed, several elements (nodes or edges) may match a given keyword, therefore many subgraphs may satisfy the query. Our contribution is to use the patterns defined in Sect. 3 for the extraction of relevant elements for each keyword, even the ones which do not contain a query keywords but are relevant to the query.

Let G be an RDF graph and $Q = \{k_1, ..., k_n\}$ a keyword query. Keyword search on RDF graphs involves three key issues: (i) matching the keywords to the graph elements using an index and a mapping function; a set of lists $El = \{El_1, ..., El_n\}$ is returned, each list El_i contains graph elements el_{ij} which match the keyword k_i; we refer to these as *keyword elements*; (ii) merging keyword elements to form a subgraph, which provides a list of connected subgraphs, each one containing a single corresponding keyword element for a given keyword. Finally (iii) results have to be ranked according to their relevance.

Consider the RDF graph given in Fig. 2. Let *"Studer Cooperator"* be the keyword query issued to find the collaborators of the researcher named *Studer*. Using a mapping function and some standard transformation functions such as abbreviations and synonyms, two keyword elements will be identified: the node *Rudi Studer* and the property *swrc:cooperateWith*. Without using the patterns, the only answer to the query will therefore be the subgraph G_1 of Fig. 3 representing *Daniel Deutch*, a collaborator of *Rudi Studer*, both linked by the *swrc:cooperateWith* property. However, *Victor Vianu* is also a collaborator of *Rudi Studer* as they published a paper together (*Art1*) as is shown in Fig. 2. The subgraph G_2 of Fig. 3 will not be returned using the above process. In other words, the link *swrc:cooperateWith* between *Rudi Studer* and *Victor Vianu* can

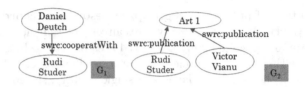

Fig. 3. Keyword search results

be replaced by the subgraph G_2. Our approach will return both results by using *Pattern*$_1$ of Table 1. We can see from this example that some relevant results could not be returned without taking into account the semantics provided by properties.

Note that if the keyword element is a property as in our example, the system will not extract all the occurrences of the property *swrc:cooperateWith*, it will consider instead that a property occurrence is relevant if either the object or the subject of the property is a keyword element. Otherwise, the result might include irrelevant answers. In our example, the system will return only cooperators of *Studer* and not cooperators of others researchers.

To construct results, we search for a connected subgraph which contains one fragment for each keyword k_i. We compute a Cartesian product between the different sets of relevant fragments to construct the combinations of relevant fragments. Each combination will be used to construct one possible result. Finally, we perform a bidirectional expansion search strategy [11] to join different relevant fragments and construct the result.

We have defined a ranking function to evaluate the relevance degree of each subgraph. Our approach combines two criteria: (i) *Compactness Relevance* and (ii) *Matching Relevance*. The former is computed with respect to the size of the subgraph and the defined patterns. Indeed, when using patterns, a long path can be replaced by one property which reduces its size. The latter is calculated using standard Information Retrieval metrics, such as TF-IDF in our case.

7   Experimental Evaluation

The PatEx system is implemented in Java with Jena API for the manipulation of RDF data and Lucene API for keyword search primitives. All the experiments were performed on a machine with Intel Core i5 2,40GHz CPU and 12GB RAM. In the following, we will discuss some of the performed evaluations for each exploration strategy.

7.1   Efficiency Evaluation

Theme Identification. We have used two subgraphs from DBpedia with different sizes and without literals, because they are not considered during the clustering step. We start the construction of each subgraph from a central resource

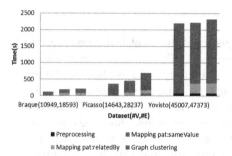

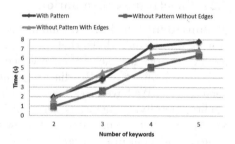

Fig. 4. Effect of the dataset size on the execution time (Color figure online)

Fig. 5. Effect of the query size on the execution time (Color figure online)

(eg. dbp:Picasso) and we recursively add all the adjacent resources. We have also used the *Yovisto*[3] dataset. We consider the execution time of each task: (1) mapping *pat:related* property, (2) mapping *pat:sameValue_p* property, (3) removing literals and finally (4) clustering. These times are first calculated without user preferences, then with one preference expressed as a *pat:related* property, and finally with the two kinds of preferences, the first expressed by *pat:related* property and the second expressed by *pat:sameValue_p* property. The results are shown in Fig. 4.

We observe that of all the tasks, clustering is the one which takes the longest execution time. The complexity of this part is polynomial and depends on the number of edges n, number of nodes m and the average nodes degree h: $complexity = O(nmh^3)$ [1]. Consequently, the execution time of the theme identification also depends on the same parameters. Besides, for the same dataset, the execution time changes with user preferences. This means that the execution time also depends on the number of resources involved in the user preference. When the user preference is mapped into a node merging transformation, the number of nodes and edges decreases which reduces the clustering time. However, when the user preference is mapped into the creation of complete graphs, the clustering time increases.

Keyword Search. We have used data from AIFB, which contains 8281 nodes and 29223 triples. We have computed execution times in three configurations: (i) the first one considers nodes content only, (ii) the second considers both the nodes and edges content, but without any condition during their extraction, and (iii) the third integrates patterns. We vary the query size from 2 to 5 keywords. For each size, the average execution time of 10 queries which include a wide variety of keywords for the three configurations are shown in Fig. 5. We observe that the execution time increases almost linearly with the number of keywords. The execution time of the configuration integrating patterns is higher than the others but remains in the same order of magnitude.

---

[3] https://datahub.io/fr/dataset/yovisto.

7.2   Effectiveness Evaluation

Theme Identification. To validate the quality of the identified themes, ground truth data is required. To this end, we have used DBpedia as it provides predefined categories for each resource which can be interpreted as themes. We have extracted a subset of DBpedia representing scientists, their research field, universities, etc. We have investigated the impact of user preferences on the results' quality. Identifying themes is performed for four scenarios. During the first one, we suppose that the user does not know the dataset and does not have any preferences. The goal in scenarios 2, 3 and 4 is the same and the themes represent information about scientists from the same domain. For this purpose, in scenario 2 we use the *dbo:field* property in the preference, which indicates the domain of scientists to group resources having the same value for this property by using $pattern_4$ of Table 1. In scenario 3, based on the assumption that a scientist and his student work in the same domain, we use the *dbo:doctoralAdvisor* property to express the preference, in order to group resources related by this property by using $pattern_3$ of Table 1. Finally, in scenario 4, we use the two previous preferences. The *Precision, Recall* and *F-score* achieved for these scenarios are shown in Fig. 6.

We can notice that user preferences have improved the precision but have reduced the recall. The recall decreases because more than one theme can correspond to one category. However, when we analyze the results, we can see that the themes represent the same category but each one focuses on a more specialized sub-category; for example, two of the provided themes correspond to mathematicians, but one describes statisticians and the other describes scientists in the specific field of mathematical logic. This is confirmed with the precision which is low for the first scenario but have increased significantly for the others. The discovered themes therefore focus on a specific area. Furthermore, if we compare the improvement of the precision in the different scenarios, we notice that it has the highest value in the second scenario. The precision is lower in the third scenario, this is because the hypothesis made about the definition of a theme in the third scenario is not necessarily true. For example, we can find in the dataset a mathematician who supervises another mathematician, but also a mathematician who supervises a statistician. We can also notice that user preferences have improved the F-score and therefore the overall quality of the identified themes. The only case in which the F-score did not improve with the preferences is the third scenario because, as explained above, the assumption of the user on the property *dbo:supervisor* is not always true.

Keyword Search. For these experiments, we have used 10 queries that are randomly constructed (with 2 to 5 keywords). We have compared the three configurations of the algorithm described in Sect. 7.1: (i) the first one considers nodes content only, (ii) the second considers both the nodes and edges content, but without any condition during their extraction, and (iii) the third integrates patterns. We have asked 4 users to assess the top-k results of each query using the three configurations on a 3-levels scale: 3 (perfectly relevant), 1 (relevant) and 0 (irrelevant). We have used NDCG@k (Normalized Discounted Cumulative Gain)

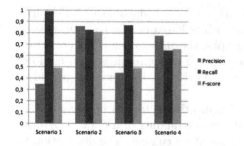

Fig. 6. Precision, Recall and F-score (Color figure online)

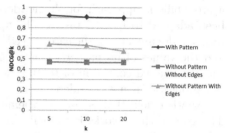

Fig. 7. Average NDCG@k: SWRC (Color figure online)

to evaluate the quality of the results. Figure 7 reports the average NDCG@k for the 10 considered queries with k=5, 10 and 20. We can see from this figure that the use of patterns in the algorithm allows to significantly outperform the two others configurations in terms of NDCG@k values at all levels. This means that, our approach finds more relevant results. Furthermore, since NDCG includes the ranking position of the results, the ranking algorithm of our approach is better because it includes patterns during the ranking step.

8   Related Works

In this section, we present an overview of previous works on RDF dataset exploration, theme identification and keyword search on RDF graphs.

Data graph exploration has attracted a great deal of interest. In [5,12], the authors present a survey of existing systems for data graphs exploration. However, the existing works are mainly interested in the visualization by using the structure (schema and the hierarchy of the data) [2] and statistics for data classification [18]. The data is then presented though a rich graphical interface component such as facets [9], histograms [18], sub-graphs [10], etc. The users browse the dataset manually using existing properties. We can notice that the semantic is not used in these works to guide the browsing process.

Some approaches have been proposed for theme identification in RDF graphs. In [4], an augmented graph is constructed by adding a new edge between resources which have a high similarity, calculated according to the common words in their labels. Then, a graph clustering algorithm to identify themes is presented. In [3], authors have proposed an approach based on the structural information of the graph. Density based graph clustering is applied on a summary graph where nodes correspond to classes and edges between classes are defined if their respective instances are connected in the RDF graph. Each cluster of classes is considered as a theme. Following the assumption that instances with strongly related types should be related similarly, instances are grouped according to the theme of their classes. This approach is based on the assumption that a schema of the dataset is provided, which is not often true. In [17], a probabilistic approach for theme identification in text documents (LDA) has been adapted to RDF datasets. The approach is based on the co-occurence of terms in documents. In the case of RDF graphs, the authors define each resource and the values of its properties as a document. The approach is then used on the constructed collection of textual documents. Hence, the semantic provided but properties is not considered.

Another field that is close to theme identification is topical profiling [7,13]. The goal of these works is to facilitate the understanding of a dataset by providing descriptive information (metadata) about its content. The description is given as labels that represent the existing topics in the dataset [7,13]. However, unlike our approach, these works return a set of topics that exist in the dataset, but they do not assign each node to one or more of these topics.

There are mainly two categories of approaches dealing with keyword search over RDF graphs. The first one uses information retrieval techniques. They define what a document is, for example, a triple [6] or a resource with its neighborhood [8]. Documents are indexed, then query keywords are mapped to the defined documents. Finally, a ranked list of documents is returned to the user. These works do not merge the relevant fragments corresponding to the query keywords.

In the second category, approaches use graphs to construct results [19,20]. To do so, they typically use a keyword-to-graph mapping function to identify a set of elements that contain the query keywords. Then, a connected "minimal" subgraph, which contains one identified element for each query keyword is returned as an answer. Since several elements can contain the same keyword, several results are returned to the user, and each approach defines a ranking function mainly based on the size of the subgraph, the matching relevance and the coverage. In these works, only the structure is considered during the construction of the result, unlike our approach, which integrates the semantics through patterns. In [15,16], the authors have predefined templates for the results based on the type of keyword elements. However, defining templates for the result might result in capturing some interpretations of the keywords only, which may not satisfy the user's needs.

9   Conclusion

In this paper, we have proposed an approach which combines thematic and keywords exploration in RDF data. To support the exploration process, we have defined the notion of pattern to formalize both users preferences and specific domain knowledge. We have implemented the two exploration strategies and proved that they reduce the user effort to explore, browse and understand an RDF(S)/OWL dataset without having any knowledge about the data. The experiments have shown that the result's quality is improved by including external knowledge formalized using patterns.

Future works include the definition of other types of patterns to express more than equivalence between property and path expressions. Besides, for the theme identification approach, we are currently working on the optimization of the clustering process in order to improve the efficiency of our approach. We will also address the problem of maintaining the themes over time as the dataset evolves.

References

1. Bader, G.D., Hogue, C.W.: An automated method for finding molecular complexes in large protein interaction networks. BMC Bioinform. **4**, 2 (2003)
2. Benedetti, F., Po, L., Bergamaschi, S.: A visual summary for linked open data sources. In Proceedings of the Posters & Demonstrations Track within ISWC, pp. 173–176 (2014)
3. Böhm, C., Kasneci, G., Naumann, F.: Latent topics in graph-structured data. In: CIKM, pp. 2663–2666 (2012)
4. Castano, S., Ferrara, A., Montanelli, S.: Thematic clustering and exploration of linked data. In: Ceri, S., Brambilla, M. (eds.) Search Computing III. LNCS, vol. 7538, pp. 157–175. Springer, Heidelberg (2012)
5. Dadzie, A.-S., Rowe, M.: Approaches to visualising linked data: a survey. Semant. Web **2**, 89–124 (2011)
6. Elbassuoni, S., Blanco, R.: Keyword search over RDF graphs. In: CIKM, pp. 237–242 (2011)
7. Fetahu, B., Dietze, S., Pereira Nunes, B., Antonio Casanova, M., Taibi, D., Nejdl, W.: A scalable approach for efficiently generating structured dataset topic profiles. In: Presutti, V., d'Amato, C., Gandon, F., d'Aquin, M., Staab, S., Tordai, A. (eds.) ESWC 2014. LNCS, vol. 8465, pp. 519–534. Springer, Heidelberg (2014)
8. Guha, R., McCool, R., Miller, E.: Semantic search. In: WWW, pp. 700–709 (2003)
9. Heim, P., Ertl, T., Ziegler, J.: Facet graphs: complex semantic querying made easy. In: Aroyo, L., Antoniou, G., Hyvönen, E., ten Teije, A., Stuckenschmidt, H., Cabral, L., Tudorache, T. (eds.) ESWC 2010. LNCS, vol. 6088, pp. 288–302. Springer, Heidelberg (2010)
10. Heim, P., Lohmann, S., Stegemann, T.: Interactive relationship discovery via the semantic web. In: Aroyo, L., Antoniou, G., Hyvönen, E., ten Teije, A., Stuckenschmidt, H., Cabral, L., Tudorache, T. (eds.) ESWC 2010, Part I. LNCS, vol. 6088, pp. 303–317. Springer, Heidelberg (2010)
11. Kacholia, V., Pandit, S., Chakrabarti, S., Sudarshan, S., Desai, R., Karambelkar, H.: Bidirectional expansion for keyword search on graph databases. In: VLDB, pp. 505–516 (2005)
12. Marie, N., Gandon, F.L.: Survey of linked data based exploration systems. In: IESD (2014)
13. Meusel, R., Spahiu, B., Bizer, C., Paulheim, H.: Towards automatic topical classification of LOD datasets. In: LDOW co-located with WWW, vol. 1409 (2015)
14. Ouksili, H., Kedad, Z., Lopes, S.: Theme identification in RDF graphs. In: Ait Ameur, Y., Bellatreche, L., Papadopoulos, G.A. (eds.) MEDI 2014. LNCS, vol. 8748, pp. 321–329. Springer, Heidelberg (2014)
15. Pradel, C., Haemmerlé, O., Hernandez, N.: Swip: a natural language to SPARQL interface implemented with SPARQL. In: Hernandez, N., Jäschke, R., Croitoru, M. (eds.) ICCS 2014. LNCS, vol. 8577, pp. 260–274. Springer, Heidelberg (2014)
16. Rahoman, M., Ichise, R.: Automatic inclusion of semantics over keyword-based linked data retrieval. IEICE Trans. **97–D**, 2852–2862 (2014)

17. Sleeman, J., Finin, T., Joshi, A.: Topic modeling for RDF graphs. In: LD4IE co-located with ISWC, pp. 48–62 (2015)
18. Voigt, M., Tietz, V., Piccolotto, N., Meibner, K.: Attract me! How could end-users identify interesting resources? In: WIMS, pp. 1–12 (2013)
19. Wang, H., Zhang, K., Liu, Q., Tran, T., Yu, Y.: Q2Semantic: a lightweight key-word interface to semantic search. In: Bechhofer, S., Hauswirth, M., Hoffmann, J., Koubarakis, M. (eds.) ESWC 2008. LNCS, vol. 5021, pp. 584–598. Springer, Heidelberg (2008)
20. Yang, S., Wu, Y., Sun, H., Yan, X.: Schemaless and structureless graph querying. VLDB Endow. 7, 565–576 (2014)

Multilingual Protest Event Data Collection with GATE

Vera Danilova[1,2,3(✉)], Svetlana Popova[3,4], and Mikhail Alexandrov[1,2]

[1] Russian Presidential Academy (RANEPA), Moscow, Russia
vera.danilova@e-campus.uab.cat, malexandrov@mail.ru
[2] Autonomous University of Barcelona, Barcelona, Spain
[3] ITMO University, St. Petersburg, Russia
spbu.svp@gmail.com
[4] Saint Petersburg State University, St. Petersburg, Russia

Abstract. Protest event databases are key sources that sociologists need to study the collective action dynamics and properties. This paper describes a finite-state approach to protest event features collection from short texts (news lead sentences) in several European languages (Bulgarian, French, Polish, Russian, Spanish, Swedish) using the General Architecture for Text Engineering (GATE). The results of the annotation performance evaluation are presented.

Keywords: Protest event data · Event features · Information extraction · GATE

1 Introduction

Protest activity reflects peoples satisfaction, confidence and belief in leader. It has been a hot issue in the recent years. *"Presidents, prime ministers and assorted rulers, consider that you have been warned: A massive protest can start at any time, seemingly over any issue, and can grow to a size and intensity no one expected. Your countrys image, your own prestige, could risk unraveling as you face the wrath of the people"* - wrote Frida Ghitis, special to CNN, in June 2013 referring to the dramatic events that took place in Turkey and Brazil. November of the same year marked the beginning of the civil unrest in Ukraine, which lead to the coup detat and grew into an international conflict and a civil war.

Research related to the protest phenomenon, contentious collective action more broadly, is of prime interest to social scientists and governmental workers. For decades sociologists from the institutions, such as Berkman Center for Internet and Society, University of Illinois at Urbana-Champaign Cline Center for Democracy and others have been studying regularities in contentious collective action and accumulating statistics for protest prediction and the analysis of its origins, dynamics and aftermath from protest event data (single events, small event sets and, since recently, big data).

Under partial support of the Government of the Russian Federation Grant 074-U01.

E. Métais et al. (Eds.): NLDB 2016, LNCS 9612, pp. 115–126, 2016.
DOI: 10.1007/978-3-319-41754-7_10

Earlier protest studies have been based on the manual analysis of newspapers data. Since 90's, automatic approaches have been applied to the protest database population and coding that mostly apply natural language processing techniques.

News media remains the most used source for protest event data collection. Its advantages are accessibility and good temporal coverage, while the biased view of events is known to be its main pitfall. Since recently, social media data is being used to study the connections between real and virtual protest activity, between protest-related news reports often controlled by the government and social media discussions.

Event selection and event coding are two main subtasks within the computational approach to protest database population, where, giving a rough definition, *event selection* is the automatic collection of relevant articles describing events in a given domain or domains, and *event coding* is the procedure of automatic code assignation to certain pieces of information (event type, actor, location, etc.).

This article provides additional details on our system for multilingual protest event data collection described in [3] and continues the evaluation of protest event features' (Event_Type, Event_Reason, Event_Location and Event_Weight) annotation performance.

2   Related Work

The pioneer systems for automatic socio-political event data collection, such as KEDS (Kansas Event Data System) [13] and its adapted versions, used keywords to speed up the search of relevant texts in digital archives and extracted events by matching the entries of ontology-based lexicons.

A description of the active projects in socio-political event data collection, such as El:DIABLO[1], W-ICEWS[2], SPEED (Social, Political, and Economic Event Database [7]), GDELT (Global Database of Events, Language and Tone [15]), is given in [3]. Event extraction in these systems also relies on the use ontology-based lexicons. SPEED employs its own ontology of destabilizing events, and the rest of the projects use CAMEO (Conflict and Mediation Event Observations [14]), which covers only 4 types of protest events. Event annotation and coding in the above systems is based on the application of relation extraction techniques (ICEWS) or the simple identification of event trigger and surrounding entities (other projects). Geocoding in these systems is performed using either special external tools or by quering an ontology followed by the application of heuristics. The characteristics of the above systems are shortly described in Table 1.

A description of the prototypes for protest event data collection from English-language news that experiment with machine learning techniques is presented in [3]. An overview of the general approaches to event extraction is given in [8].

---

[1] Event/Location: Dataset In A Box, Linux-Option: http://openeventdata.github.io/eldiablo/.

[2] World-Wide Integrated Crisis Early Warning System: http://www.lockheedmartin.com/us/products/W-ICEWS/iData.html.

Table 1. The main characteristics of machine coding-based systems. *Supervision* denotes the control of the output by an expert (experts). *Customizable* indicates whether the end user can modify the program in accordance with his needs.

Project	EL:DIABLO	W-ICEWS (2001-present)	SPEED (1945-2005)	GDELT 2.0 (1979-present)
Supervision	Unsupervised	Unsupervised	Supervised	Unsupervised
Customizable	+	-	-	-
Language Coverage	English	English, Spanish, Portuguese	English	Google Translate
Focus	Interstate conflict mediation	Interstate conflict mediation	Civil unrest	Interstate conflict mediation
Sources	around 160 websites	over 6000 news feeds	over 800 news feeds	thousands of news feeds
Ontology	CAMEO	CAMEO	SSP	CAMEO
Dictionaries	customizable	own, entities & generic agents	-	WordNet[a] & NER-enhanced
Data acquisition	own web scraper	Factiva[b], OSC[c]	Heritrix[d], BIN[e]	Google Translate
Geocoder	Penn State GeoVista project coder[f], UT/Dallas coder[g]	Country-Info.txt[h]	GeoNames[i]	CountryInfo.txt
Actor & Event Coding	CoreNLP[j], PETRARCH[k]	BBN Serif, JabariNLP[l]	Apache open NLP[m], EAT[n]	CoreNLP, PETRARCH

[a] WordNet thesaurus: http://wordnet.princeton.edu
[b] Commercial Factiva Search Engine: http://new.dowjones.com/products/factiva/
[b] http://en.wikipedia.org/wiki/Open_Source_Center
[d] The Internet Archive's open-source crawling project: https://webarchive.jira.com/wiki/display/Heritrix/Heritrix
[e] Automatic Document Categorization for Highly Nuanced Topics in Massive-Scale Document Collections [9]
[f] Geovisualization and Spatial Analysis of Cancer Data: http://www.geovista.psu.edu/grants/nci-esda/software.html
[g] Geospatial Information Sciences (GIS) project by the University of Texas at Dallas: https://github.com/mriiiron/utd-geocoding-locator
[h] CountryInfo.txt includes about 32 000 entries on 240 countries and administrative units: country names, synonyms, major city and region names, national leaders: https://github.com/openeventdata/CountryInfo
[i] Free Geographical Database: http://www.geonames.org/
[j] Java-based NLP tools including conditional random fields-based NER: https://github.com/stanfordnlp/CoreNLP
[k] Open Event Data Alliance Software: https://openeventdata.github.io/
[l] See [1]
[m] A maximum entropy and perceptron-based machine learning NLP toolkit: http://opennlp.apache.org/
[n] See [7]

A short survey of the approaches to multilingual event extraction from text has been done in [2]. A detailed survey and discussion of all of the above approaches and systems has been presented in [16].

A recent project that considers future protest detection in news and social media in several languages (Spanish, Portuguese and English) called EMBERS is presented in [12]. It uses a commercial platform[3] for multilingual text pre-processing, TIMEN [11] for date normalization and probabilistic soft logic for location assignment (in case of news and blogs). The key algorithm for multilingual planned protest detection applies extraction patterns obtained in a

---

[3] Rosette Linguistics Platform: http://www.basistech.com/text-analytics/rosette/.

semi-automatic way. The algorithm of pattern generalization from a seed set relies on syntactic parsing. The resulting patterns include two or more word lemmas for an event type term (e.g., *demonstration, protest, strike, march*, etc.) and an indicator of a planned action (future-tense of the verbs like *to organize, to plan, to announce, to prepare*, etc.), language specification and separation threshold. EMBERS is geared towards the identification of planned events and is reported to be quite accurate in this task. Authors consider the portion of events that has not been captured during the experiments as referring to spontaneous events. The data obtained by this system (the fact of a planned protest, location and date) is insufficient to populate protest event databases.

As compared to the previously mentioned systems, our prototype has the following advantages:

- detects events at the sentence level and enriches them with protest-specific features;
- determines the event weight (combines size, duration, violence involvement, regular character, intensity of a protest);
- uses a protest event ontology;
- uses simple generic multilingual patterns (no need of large labeled datasets for learning);
- employs few external tools;
- does not use deep linguistic processing.

We opted for the GATE framework[4], because it is a well-known free-source architecture with a large community of contributors that provides resources for multilayer text annotation in multiple languages (incl. semantic annotation). Also, it performs fast processing of simple cascaded grammars on big corpora and allows easy incorporation of external tools.

3   Data

A multilingual corpus that includes news reports in Bulgarian, French, Polish, Russian, Spanish and Swedish has been collected using Scrapy[5]. These languages have been selected, because the authors have a good command of them. English has been currently excluded, because the instruments for this language are being developed by other groups of researchers. In the nearest future we will add English for performance comparison purposes. The feasibility of the use of news reports lead sentences (titles and subtitles) as the unit of analysis has been proved by a number of studies, [10,17] to name a few.

The corpus has been crawled from the following number of news-oriented sites per language: *12* (Bulgarian), *10* (French), *12* (Polish) *35* + Livejournal (Russian), *16* (Spanish), *14* (Swedish).

The crawled data (URL, title, subtitle, metadata, text body, time) is saved in a JSON file (Fig. 1). Next, duplicates, near-duplicates, as well as messages

---

[4] General Architecture for Text Engineering: http://gate.ac.uk.
[5] Python-based Scrapy crawling framework: http://scrapy.org.

containing stopwords in the title are filtered out. Stopwords list has been constructed by an expert on the basis of the corpus analysis. The resulting corpus includes *13710* messages.

URL	Title	Subtitle	Metadata	Text_body	Time
http://www.reuters.com/article/2014/08/22/us-yemen-protests-idUSKBN0GM12C20140822	Tens of thousands of Yemeni Houthis protest against govt in capital		Saudi Arabia, Yemen, Abd-Rabbu Mansour Hadi, Ali Abdullah Saleh, Mohammed Al-Sayaghi	Tens of thousands of Yemenis massed in the capital Sanaa on Friday in a protest called by the Shiite Houthi movement, which wants the government to reverse a decision on cutting fuel subsidies and resign...	Fri Aug 22, 2014 4:39pm

Fig. 1. An entry of the obtained corpus (output.json).

A detailed description of data collection and filtering is given in [3,16].

4   Knowledge Resources

The following features are selected to be extracted in the current version:

- **Event_Type** (the *what* of an event: rally, march, boycott, strike, picketing, etc.),
- **Event_Reason** (the *position* of a protesting group towards an *issue*: *for* a cause (expressions of support incl. conmemorations and demands) or *against* a cause),
- **Event_Location** (the *where* of an event: names of countries, cities and physical settings),
- **Event_Weight** (an attribute that defines the *importance* of an event and takes into account the values of the following slots: *Event_Duration, Event_Intensity, Event_Iteration, Event_Size, Violence_Use*).

These features have been selected, because they most frequently define the given event (protest action) in the studied examples. Event_Weight allows to select events featured by press as long-term, intensive, repeating, large-scale and violent.

The following knowledge resources have been developed for the extraction of the above features:

(i) *Ontology* that includes a classification of protest forms, subevents (events connected by temporal and/or causal relations with the main event) and event properties;
(ii) Multilingual *lexicons* that correspond to ontology classes;
(iii) JAPE[6] *patterns* for event features annotation that take into account lexical, syntactic, semantic and discourse properties.

---

[6] Java Annotation Patterns Language:
http://en.wikipedia.org/wiki/JAPE_(linguistics).

A quantitative description of lexicons and grammar patterns that correspond to each of the features is given in Table 2.

Table 2. Quantitative characteristics of the knowledge resources (gazetteers and patterns) corresponding to the event features.

Feature name	Gazetteer lists	Gazetteer entries	Patterns
Event_Type	20	252	1
Event_Reason	10	225	7
Event_Location	6	8841	4
Event_Weight	15	364	14

Ontology. The ontology has been organized manually on the basis of the analysis of a number of resources, specifically:

- previous domain-specific ontologies (CAMEO, SSP)
- DBpedia[7] and WordNet ontologies
- WordReference, Oxford dictionary, Wiktionary, Merriam Webster dictionaries
- Wikipedia free encyclopedia
- terminology-related discussions in social scientists' blogs (Jay Ulfelder, Bad Hessian, Anthony Boyle, etc.)
- linguistic studies [5]
- 5000 multilingual documents from our corpus

The resulting structure has been formalized using Protégé[8]. Figure 2 gives a general idea of the ontology contents (depth 2).

The central event (protest) contains its own hierarchy. It is defined by a set of properties (features), such as geotemporal attributes, cause, significance (size, duration, violence involvement, regular character, intensity), status (ongoing, past, planned, N/A), actors (participants). To the left of the main event there are preceding subevents, such as *appeal to protest, change of time and/or place, threat of protesting* or other, and to the right - subevents that can follow the main event, such as *property damage, response of authorities or other people groups, violent acts* or other events. The ontology is described in detail in [16].

Gazetteers. Gazetteer lists have been constructed on the basis of the corpus data and search engine output (Google Search) analysis and correspond to the ontology classes. Each list contains one element (one-word or multiword) per line. `MajorType`, `MinorType` and language features are separated by a colon as follows: `huelga:Protest_Action:Strike:es`.

---

7 DBpedia: http://dbpedia.org.
8 Protégé ontology editor: http://protege.stanford.edu.

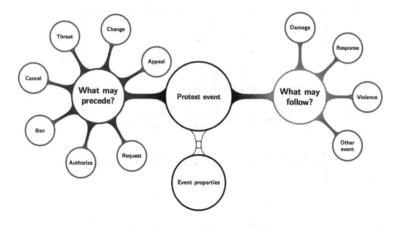

Fig. 2. A schematic representation of the semantic resource.

In order to improve the annotation of locations, we populate gazetteer lists with the entries from DBpedia (612845 elements) via Large Knowledge Base gazetteer (LKB) by Ontotext[9] using SPARQL queries. The problem with the direct text annotation with LKB gazetteer PR is that it is not flexible (preprocessed (e.g., lemmatized) text cannot be taken as input). Therefore, it is not able to perform high-quality annotation of texts in inflected languages. Ontotext has developed another gazetteer[10] that has this functionality, but it is not available for download.

Patterns. Patterns are two-level: morphosyntactic and informational. A schematic representation of a pattern (type I) for the annotation of event reason is presented in Fig. 3.

Here, in the phrase *"Demonstration in solidarity to the New Greek Government"* we identify Event_Type (*"demonstration"*) and Event_Reason that includes the *Position* of the protesting group (here, *Position = Support*) towards an *Issue* (here, *Issue = "new_greek_government"*) on the basis of the pattern. Here, the event type can be represented by a noun phrase or a verb, the position - by a simple or complex preposition, and the issue - by a (nested) noun phrase.

The system uses cascaded finite-state transducers that are connected to the main GATE pipeline using Jape Plus Extended processing resource (PR). The sequence of grammars is indicated in the `main.jape` file. Each grammar includes one or more rule/action pairs, where the left hand side describes a pattern or a sequence of patterns and the right hand side manipulates the annotations.

---

[9] Gazetteer LKB: https://confluence.ontotext.com/display/KimDocs37EN/Large+Knowledge+Base+(LKB)+gazetteer.

[10] Linked Data Gazetteer PR: https://confluence.ontotext.com/display/SWS/Linked+Data+Gazetteer+PR.

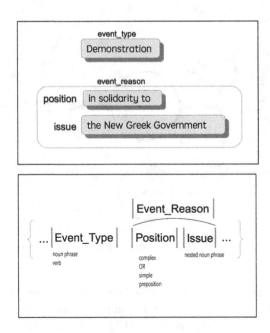

Fig. 3. A schematic representation of the Event_Reason pattern type I.

5   GATE Processing

Gate pipeline is a sequence of PRs that perform multilayer annotation of unstructured text and optionally write the output to a file of a specified type (in our case, a CSV). We also use ad hoc PRs for gazetteer/ontology population. The processing stages are outlined in Table 3.

At the pre-processing stage, the following PRs get activated: Document Reset (reset of old annotations prior to a new iteration), Document Normalizer (filtering of certain punctuation marks), GATE Unicode Tokenizer (simple tokenization), modified ANNIE Sentence Splitter (a modified version that does not take into account split annotations inside headlines).

At the next stage, part of speech tagging is performed via Generic Tagger PR connected to the TreeTagger[11], which covers all of the considered languages but Swedish. Swedish text is labeled separately with The Stockholm Tagger[12]. A GATE plugin for Swedish called SVENSK is described in the literature [6], however, it is not available for download.

The ontology lookup is performed using a complex of PRs. LKB creates annotations on the basis of a SPARQL query to DBpedia. OntoGazetteer is connected

---

[11] TreeTagger: http://www.ims.uni-stuttgart.de/projekte/corplex/TreeTagger/.
[12] The Stockholm Tagger:
http://www.ling.su.se/english/nlp/tools/stagger/stagger-the-stockholm-tagger-1.
98986.

to the in-house ontology and it is run with several auxiliary gazetteers: its default Hash Gazetteer, BWPgazetteer[13] (uses the Levenshtein distance calculation) and Flexible Gazetteer (token, lemma or other annotation can be used as input).

Grammar-based annotation of informational blocks is done via Jape Plus Extended PR. As compared to the standard Jape Transducer PR, Jape Plus Extended allows the use of additional constraints[14] without including them into the right hand side.

Gazetteers are populated using Gazetteer List Collector PR. When checked by the expert, the updated versions are used in the following iterations. Currently, we do not directly populate the in-house ontology, however, this functionality is available via a special grammar.

The results can be exported via Configurable Exporter PR to any format (in our case, CSV), specified in the corresponding configuration file.

Table 3. GATE pipeline stages with the corresponding processing resources.

Stage	Processing resource
Pre-processing	Document Reset
	Document Normalizer
	GATE Unicode Tokenizer
	modified ANNIE sentence splitter
Part-of-speech tagging	Generic Tagger [**TreeTagger**], Swedish Stagger
Ontology-based gazetteer lookup	Large KB Gazetteer [DBpedia]
	Ontogazetteer [in-house ontology]
	Hash Gazetteer
	BWPgazetteer
	Flexible Gazetteer
Grammar processing	Jape Plus Extended
Gazetteer population	Gazetteer List Collector
Export to CSV	Configurable Exporter

6   Evaluation

The evaluation of protest-related documents' selection performance is described in [3]. The annotation quality has been evaluated as follows.

Datasets. For the purposes of annotation performance evaluation, 2 datasets have been formed on the basis of the crawled headlines. The development set

---

[13] BWP Gazetteer: https://sourceforge.net/projects/bwp-gazetteer/.
[14] https://code.google.com/archive/p/gateplugin-japeutils/wikis/Constraints.wiki.

contains 575 information-rich multilingual headlines (90-100 per language) that are used both for runtime settings tuning, patterns enhancement and performance evaluation. The test set includes 3000 multilingual headlines (500 per language) randomly selected from the main dataset.

Metrics. Protest features annotation performance has been evaluated on both the development and test sets using the standard Precision and Recall metrics. Precision is the fraction of retrieved documents that are relevant, and Recall is the fraction of relevant documents that are retrieved[15].

$$Precision = \frac{|G \bigcap C|}{|G|}, Recall = \frac{|G \bigcap C|}{|C|}, \tag{1}$$

where G is the total amount of sequences retrieved from the reports' lead sentences for a given scenario slot, and C is the amount of documents that contain text spans, which are approved by the expert as relevant representatives of the same slot.

Comparison to the Gold Standard. The annotation quality has been evaluated as follows. The Gold Standard annotations have been created in GATE by a multilingual expert. The development and test sets have been divided into language-specific subsets. The number of false positive, false negative and true positive annotations has been counted automatically (GATE's Annotation Diff function). An annotation is considered a true positive if it covers and does not lie beyond the natural language representation of a semantic component (for Event_Reason Issue feature and Event_Location) and is associated with a correct class/instance in the concept hierarchy (Event_Type, Event_Location, Event_Weight, Event_Reason Position feature). An annotation is considered a false positive if it does not cover or lies beyond the natural language representation of a semantic component. In case an extra function word is spanned by a given Event_Type or Event_Reason annotation, it will nonetheless be considered as a true positive (partially correct). A false negative is a span of text corresponding to the natural language representation of a target semantic component that has not been covered by the target annotation.

Results. The evaluation results are shown in Tables 4 and 5. Here, ET stands for Event_Type, ER - Event_Reason, EL - Event_Location, EW - Event_Weight. The lowest Precision and Recall values are observed for the Event_Weight feature annotation in the test set, because of the insufficiency of the corresponding gazetteer data.

---

[15] http://nlp.stanford.edu/IR-book/html/htmledition/evaluation-of-unranked-retrieval-sets-1.html.

Table 4. Annotation quality evaluation on the development set

Language	No. of Docs	No. of Annots	Precision				Recall			
			ET	ER	EL	EW	ET	ER	EL	EW
Bulgarian	93	244	1	0.97	0.93	1	1	0.97	0.82	1
French	85	218	1	0.94	0.92	0.98	1	0.98	0.87	0.88
Polish	93	231	1	0.96	0.96	0.98	1	0.72	0.90	0.90
Russian	107	293	1	0.98	1	0.92	1	0.98	0.89	0.80
Spanish	89	218	1	1	0.83	0.98	1	0.94	0.78	1
Swedish	108	195	1	0.98	1	1	1	0.95	0.65	0.88

Table 5. Annotation quality evaluation on the test set (500 documents per language)

Language	No. of Annots	Precision				Recall			
		ET	ER	EL	EW	ET	ER	EL	EW
Bulgarian	1053	0.99	0.98	0.95	0.87	1	0.97	0.87	0.86
French	1093	0.92	0.92	0.91	0.66	0.99	0.90	0.91	0.62
Polish	893	0.91	0.96	0.90	0.90	1	0.84	0.84	0.67
Russian	1031	0.93	0.95	0.97	0.67	0.99	0.90	0.89	0.75
Spanish	1023	0.96	0.93	0.90	0.90	0.99	0.88	0.93	0.83
Swedish	906	0.99	0.97	0.95	0.80	1	0.95	0.91	0.54

7   Conclusions

This paper presents a strategy of the protest features' annotation in the unstructured text using GATE for multilingual protest event data collection and shows the annotation quality evaluation results. The system first crawls and filters the data (title, subtitle, time, text body, metadata, URL) on the documents related to protests and processes the lead sentences with GATE. The output is a CSV with event features that correspond to the input documents.

Information extraction from lead sentences obviously does not allow to obtain complete descriptions of events of interest and find the relationships with other events, however, obtaining maximum useful data from titles and subtitles can speed up the algorithm of protest event extraction in general. Also, in [4], we have proved that the use of the obtained event features is beneficial in that it enhances short-text (headlines) clustering accuracy.

This work continues as follows: (i) improvement of protest-related documents selection (experiments with introducing additional constraints into queries, clustering similar events, etc.), (ii) selection of resources (PoS taggers, morphological analyzers) to enhance multilingual text pre-processing, e.g., to be able to annotate noun plurality, animacy, etc., for the whole set of languages, as well as to reduce the error rate (TreeTagger yields satisfactory results within the current implementation, however, there is a number of significant drawbacks [16]),

(iii) multilingual data annotation (for feature classifier training, evaluation and comparison of results).

References

1. Boschee, E., Weischedel, R., Zamanian, A.: Automatic information extraction. In: Proceedings of the International Conference on Intelligence Analysis (2005)
2. Danilova, V., Alexandrov, M., Blanco, X.: A survey of multilingual event extraction from text. In: Métais, E., Roche, M., Teisseire, M. (eds.) NLDB 2014. LNCS, vol. 8455, pp. 85–88. Springer, Heidelberg (2014)
3. Danilova, V.: A pipeline for multilingual protest event selection and annotation. In: Proceedings of TIR15 Workshop (Text-based Information Retrieval), pp. 309–314. IEEE (2015)
4. Danilova, V., Popova, S.: Socio-political event extraction using a rule-based approach. In: Meersman, R., et al. (eds.) OTM 2014. LNCS, vol. 8842, pp. 537–546. Springer, Heidelberg (2014)
5. Dobrovolskiy, D., Pöppel, L.: Lexical synonymy within the semantic field POWER. In: Chahine, I.K. (ed.) Current Studies in Slavic Linguistics, pp. 281–295 (2013)
6. Eriksson, M., Gambäck, B.: SVENSK: a toolbox of Swedish language processing resources. In: Proceedings of the 2nd International Conference on Recent Advances in Natural Language Processing (RANLP), Tzigov Chark, Bulgaria, pp. 336–341 (1997)
7. Hayes, M., Nardulli, P.F.: SPEEDs Societal Stability Protocol and the Study of Civil Unrest: an Overview and Comparison with Other Event Data Projects, October 2011. Cline Center for Democracy University of Illinois at Urbana-Champaign (2011)
8. Hogenboom, F., Frasincar, F., Kaymak, U., de Jong, F.: An overview of event extraction from text. In: Workshop on Detection, Representation, and Exploitation of Events in the Semantic Web (DeRiVE 2011), vol. 779, pp. 48–57. CEUR Workshop Proceedings (2011)
9. Leetaru, K.: Automatic Document Categorization for Highly Nuanced Topics in Massive-Scale Document Collections: The SPEED BIN Program. Cline Center for Democracy University of Illinois at Urbana-Champaign, March 2011
10. Lejeune, G.: Structure patterns in information extraction: a multilingual solution? In: Advances in Method of Information and Communication Technology AMICT 2009, Petrozavodsk, Russia, vol. 11, pp. 105–111 (2009)
11. Llorens, H., Derczynski, L., Gaizauskas, R.J., Saquete, E.: TIMEN: an open temporal expression normalisation resource. In: LREC, vol. 34, pp. 3044–3051 (2012)
12. Muthiah, S., Huang, B., Arredondo, J., Mares, D., Getoor, L., Katz, G., Ramakrishnan, N.: Planned protest modeling in news and social media. In: AAAI 2005, pp. 3920–3927 (2015)
13. Schrodt, P.A.: KEDS: Kansas Event Data System, version 1.0 (1998)
14. Schrodt, P.A.: CAMEO: Conflict and mediation event observations. event and actor codebook. Department of Political Science, Pennsylvania State University, Version: 1.1b3, March 2012
15. Schrodt, P.A.: GDELT: global data on events, location and tone. In: Workshop at the Conflict Research Society, Essex University, 17 September 2013
16. Danilova, V.: Linguistic support for protest event data collection. Ph.D. Thesis, Autonomous University of Barcelona, 27th November 2015. http://www.tdx.cat/handle/10803/374232
17. Wunderwald, M.: Event Extraction from News Articles (Diploma Thesis), Dresden University of Technology, Department of Computer Science (2011)

A Multi-lingual Approach to Improve Passage Retrieval for Automatic Question Answering

Nouha Othman[1](✉) and Rim Faiz[2]

[1] Institut Supérieur de Gestion de Tunis, LARODEC,
Université de Tunis, Tunis, Tunisia
othmannouha@gmail.com
[2] IHEC Carthage, LARODEC, Université de Carthage, Tunis, Tunisia
rim.faiz@ihec.rnu.tn

Abstract. Retrieving topically relevant passages over a huge document collection is deemed to be of central importance to many information retrieval tasks, particularly to Question Answering (QA). Indeed, Passage Retrieval (PR) is a longstanding problem in QA, that has been widely studied over the last decades and still requires further efforts in order to enable a user to have a better chance to find a relevant answer to his human natural language question. This paper describes a successful attempt to improve PR and ranking for open domain QA by finding out the most relevant passage to a given question. It uses a support vector machine (SVM) model that incorporates a set of different powerful text similarity measures constituting our features. These latter include our new proposed n-gram based metric relying on the dependency degree of n-gram words of the question in the passage, as well as other lexical and semantic features which have already been proven successful in a recent Semantic Textual Similarity task (STS). We implemented a system named PRSYS to validate our approach in different languages. Our experimental evaluations have shown a comparable performance with other similar systems endowing with strong performance.

Keywords: Question answering · Passage retrieval · Support vector machine · N-gram · Text similarity measures

1 Introduction

Question answering (QA) is a complex form of information retrieval (IR) characterised by information needs expressed by human natural language questions. Unlike classical IR, where full documents are considered relevant to the information request, in QA, the user is rather interested in a precise and concise information which should be directly generated as a response to a his natural language question. Today, the explosive rise in the amount of data and information has made the need to direct and precise information more than ever

© Springer International Publishing Switzerland 2016
E. Métais et al. (Eds.): NLDB 2016, LNCS 9612, pp. 127–139, 2016.
DOI: 10.1007/978-3-319-41754-7_11

necessary, hence why QA has attracted increasing interest in recent years [18], thanks to the QA tracks in the yearly TREC[1] and CLEF[2] campaigns.

We emphasize that the domain of a QA system (QAS) can be either closed dealing with questions under a specific domain such as in [1], or open dealing with general questions in various domains without any limitation such as in [8,15,18]. In this work, we focus on the second category as the techniques employed are not tailored toward a specific domain.

Fundamentally, a typical QA system (QAS) can be viewed as a pipeline composed of four main modules [20] namely, question analysis, document retrieval, passage retrieval and answer extraction each of which has to deal with specific issues and challenges. One significant module that serves as the building block of efforts is Passage Retrieval (PR) module, which allows reducing the search space from a massive collection of documents to a fixed number of passages. Undoubtedly, a correct answer to a posted question can be found only when it already exists in one of the retrieved passages. Additionally, it has been proved that the performance of the PR module significantly affects that of the whole system [20].

In fact, the issue of retrieving a relevant passage to a given natural language question over a sizable document collection has been a lively topic of research in recent years, such as in [5,11]. However, most existing QASs are still unable to provide correct answer to any given question from a large repository even though it contains the correct answers. Therefore, we assume that if we improve the PR engine of QASs, we could enhance their performance and increase the number of correct passages retrieved containing the relevant answers to the given question.

In the present paper, we propose a multi-lingual approach to improve PR and ranking tasks for an open domain QAS. Our approach is based on a Ranking SVM model which combines different textual similarity measures. These latter involves our new proposed n-gram similarity measures, relying on the dependency degree of n-gram words of the question in the passage, as well as other powerful lexical and semantic features which have already shown promise in the Semantic Textual Similarity task (STS) at *SEM 2013 [4]. Our approach attempts to automatically deliver the top one relevant passage in return for the user's natural language question.

The remainder of this paper is organized as follows. In Sect. 2, we provide an overview of major related work on PR. Our approach is then introduced in Sect. 3 where we detail its different steps. Section 4 is advocated to describing our experimental study carried out on the CLEF dataset to validate our proposed approach and comparing our results with those of similar solutions performing the same task. Finally, we draw our conclusion and our main future work directions in Sect. 5.

---

[1] http://trec.nist.gov/.

[2] http://www.clef-initiative.eu/.

2   Related Work

There is an abundance of work on PR which is usually considered as the kernel of most QASs and it attempts to reduce the search space for finding out an answer. Indeed, PR was initially explored to overcome the shortcomings of document retrieval and then integrated into QA, basically because a passage unit is more appropriate than a document to rank for QA [2]. Through an early quantitative evaluation performed by [20] of the existing PR algorithms for QA based on lexical matching between the question and the passage, it was deduced that most proposed PR algorithms based on lexical matching process each question term as an independent symbol and take into account neither the order of words nor their dependency relations. In order to address this drawback, there have been several attempts to consider term dependencies. There are further numerous works that include syntactic matching such as [7] drawing syntactic relations dependencies instead of keywords to match passages with the given query. Nonetheless, the major limitation of syntactic matching is the need of a syntactic parser which requires adaptation and its performance significantly affects that of the entire QAS. Other works relied on semantics such as [19] in order to answer specific question types. Although semantic matching based approaches allow to capture relevant answers, they are difficult to adapt to multilingual tasks as they require semantic resources such as FrameNet which is currently available only in English and do not cover neither all domains nor all terms. Also, it is claimed that such approaches are often more time consuming than bag-of-words ones. It is important to note that there have been further subsequent works combining both semantic and syntactic approaches in the context of PR such as [17] in order to take advantage of both approaches. Otherwise, another powerful method was proposed to go beyond the above simple lexical matching, relying on n-gram structures which refer to contiguous term sequences extracted from a given text. N-grams is a statistical model that has been widely employed in natural language processing, since it take into account the simple dependency between terms rather than independent symbols. Within this context, [14] conducted a probabilistic method for web-based QA, where the web documents are shortened into passages to be ranked using an n-gram score based on tf-idf. In addition, [6] proposed a PR engine for QA based on an n-gram similarity measure where the passages containing more question n-grams and longer ones are favored when comparing the extracted candidate passages using a keyword technique. Besides, the same previous approach was followed by [5] but the only difference is about the n-gram model applied which considers the passage n-grams occurring in the question as well as their proximity. By and large, our proposed n-gram measure is different from the cited n-gram based ones insofar as we put more focus on common n-grams between the passage and the question and we proceed differently to determine the n-grams.

We emphasize that passage ranking is viewed as a crucial subtask to improve the PR performance and ensure the relevance of the retrieved passages. Some proposed methods were based on knowledge such as [2,3], where knowledge about question and answers is represented in the form of rules and inferences are derived from a knowledge base. Nevertheless, the major disadvantage of such methods is the need of a huge number of inferences and manual design of features. Other works like [16] resort to pattern-matching approaches where patterns are pre-defined context evaluation rules designed and extracted from the query and the passages and they are often related to a specific domain and depend on the question type. Such approaches work well only if it is possible to reach high answer redundancy such as in the web. They also usually are extremely complex and need a large amount of training data. Furthermore, the words context was used as a simple intuition for passage ranking, such as [21] who introduced a machine learning-based QA framework that integrates context-ranking model which re-rank the passages retrieved from the initial retrievers to find out the relevant answers. This model utilizes a Support Vector Machine (SVM) to incorporate contextual information of proper names, semantic features and syntactic patterns and the position of each term in the context window to predict whether a given candidate passage is relevant to the question type. Nonetheless, the performance of this context model highly dependents on the question classification. Inspired by the success of SVM application in the latter work, we resorted to SVM to combine different lexical and semantic text features with our proposed n-gram measure to boost the PR engine and further ensure the relevance of the retrieved passages.

3   Proposed Approach Overview

The general architecture of our approach is illustrated in Fig. 1. Its basic principle is to extract from a given document collection a set of candidate passages that are most likely to fit the user's query, filter and rank them using a Ranking SVM model that incorporates powerful similarity measures in order to return the top ranking passage. It is noteworthy that we can not only rely on our n-gram measure to guarantee high relevance, since it is just based on simple dependencies between terms, thus, we attempt to improve the performance of our approach by adding other effective lexical and semantic features.

A crucial step that should precede retrieving passages is question processing which attempts to generate a formal query by preprocessing the question entered by the user and extracting the useful terms by applying text cleaning, question words elimination, tokenization, stop words removal and stemming. The formal query can be defined as: $Q = \{t_1, t_2, ..., t_q\}$ where t represents a separate term of the query Q and q denotes the number of query terms.

3.1   Candidate Passage Extraction:

At this stage, all passages containing the question terms should be extracted as candidate passages. In advance, the document collection should be segmented into paragraphs called passages, which must preferably be neither too short nor too long as, short excerpts may not include the answer while long ones may contain additional information that can skew the response or require more

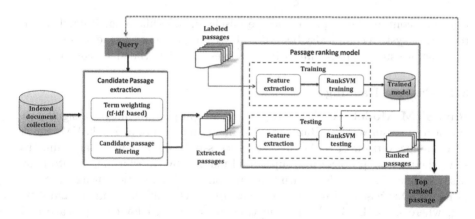

Fig. 1. Architecture of our proposed passage retrieval approach

robust extraction module. To index our passages, for each one, we save its id, number, text and its document name. Then, we extract their terms, apply the stemming and remove all the stopwords. A passage P can be formally defined as: $P = \{t_1, t_2, ..., t_p\}$ where t denotes a separate passage term and p is the number of passage terms. Subsequently, we assign a frequency to each passage term in order to calculate the maximum frequency as well as the term weights. Once the passages are indexed, their terms will be stored in an inverted index.

Query and Passage Term Weighting: To calculate the weight of the query terms, we are based on the formula used in [6]: $w(t_i, q) = 1 - \frac{log(n(t_i))}{1 + log N}$ where n is the number of passages containing the term t and N is the total number of passages. This formula does not take into account the frequency of words but it only considers their discriminating power between passages. Therefore, a term occurred in a single passage will have a maximum weight as the ratio value in the formula is low. Candidate passages that contain at least one of the query terms can be determined by looking for the query terms in the inverted index, where for each term the list of related passages is recorded and taking the intersection of these passages. The candidate passages are defined as follows: $Pc = \{P_1, P_2, ..., P_n\}$ where P_i represents a candidate passage and n their number. Notice that the weight of the candidate passages terms is calculated in the same manner as that of the query terms (using formula [6]).

Candidate Passage Filtering: In order to filter the candidate passages, we need to calculate the textual similarity between each passage and the question by means of a metric that only considers common words between the query and the passage, defined as follows: $s(p, q) = \frac{\sum_{t_i \in P \cap Q} w(t_i, q)}{\sum_{t_i \in Q} w(t_i, q)}$. The candidate passages are then ranked according to their similarity scores. Their number (n) will be reduced to (nb) which should be a happy medium between a big and a small number, because a big nb does not meet our main goal of reducing the search space mentioned above, but it has the advantage of not excluding passages that may contain the response and are misclassified, while a small number can reduce the system complexity but with a big chance to ignore some passages. For example, we suggest to set nb to 100 passages to satisfy the given constraints.

3.2 Passage Ranking

RankSVM Model: At this second stage, we propose to integrate significant similarity measures into a Ranking SVM model referred to as RankSVM, which is a ranking version of SVM that was successfully applied in IR and aims to solve the ranking problem in a supervised manner by transforming it into pairwise classification and then learn a prediction ranking function using the SVM principle. Our passage ranking model consists of two phases: training and testing, where in both, the different metrics are computed for each passage and entered into the RankSVM classifier to be ranked given their scores. During the former phase, a set of annotated passages entered in the passage ranking model where each one is labeled either Right (R) or Wrong (W) while in the testing phase, the passages are unlabeled since they are the output of the first stage. It is worth noting that most incorporated features have already shown promise in the Semantic Textual Similarity task (STS) at *SEM 2013 track [4] which requires participants to determine the similarity degree between pairs of text sentences. Among the proposed features, we will apply WordNet-based Conceptual Similarity, Named Entity Overlap, Edit distance, Syntactic Dependencies and instead of the N-gram based similarity applied in this work and was not working very well, we resort to that introduced by ourselves. The given features were adapted to the context of QA, that is to say, the sentence pairs become pairs of passage-question. Note that we have selected lexical and semantic features in order to further ensure answer relevance since, at this stage, features based on term frequencies are not sufficient for retrieving relevant passages.

Feature Extraction: In what follows, we start by introducing our proposed n-gram passage similarity measure, then we briefly remind the other employed features fully detailed in [4].

NGpsim: our n-gram based passage similarity measure: Our basic intuition is to just focus on the common n-grams between the question and the passage. Thus, we begin by identifying the common terms between the question and a passage to derive their corresponding n-grams. We build the vector $\overrightarrow{CTqp}$ of common terms between the question and the passage, by browsing through the question terms and check for each of them if it is also a passage term to add it in the vector. This vector is defined as follows:

$$\overrightarrow{CTqp} \begin{pmatrix} t_1 \; p_1Q \; [p_{11},..,p_{1m}] \\ t_2 \; p_2Q \; [p_{21},..,p_{2m}] \\ .. \quad .. \\ t_n \; p_nQ \; [p_{n1},..,p_{nm}] \end{pmatrix}$$

where t_i is the ith term in common between the question and the passage with $i=\{1..n\}$, n is the number of question terms, p_iQ is the position of ith term in the question, p_{ij} is the jth position of the ith term in the passage and $j=\{1..m\}$. m is the passage terms number. Thereafter, we construct the n-gram vectors of the question $\overrightarrow{NGQ}$ and the passage $\overrightarrow{NGP}$ by browsing the vector $\overrightarrow{CTqp}$ and grouping the terms having successive positions in the question and the passage as follows: $\overrightarrow{NGQ} = \{ngQ_1, ngQ_2, ..., ngQ_q\}$ and $\overrightarrow{NGP} = \{ngP_1, ngP_2, ..., ngP_p\}$ where q is the number of query n-grams and p it that of the passage n-grams. The weight of each question n-gram is calculated on the basis of its length and the sum of its term weights using formula 1:

$$w(ngQ) = l \times \sum_{t_i \in terms(ngQ)} w(t_i, q) \tag{1}$$

where l denotes the number of terms included in the question n-gram (ngQ). Indeed, the weights sum when multiplied by the n-gram length can favor adjacent words over the independent ones in the similarity calculation. We suppose that grouped terms are more significant and less ambiguous than separate ones. Hence, a term that belongs to an n-gram should have a greater weight than an independent one while for the passage n-grams, they are weighted regarding their similarity degree with those of the question. We give a cumulative weight to the passage by browsing through the question n-grams and at each one either its full weight or a lower one is added to the passage weight. More concretely, if a question n-gram fully occurs in the passage, its whole weight will be added to the total weight, while if it is divided into smaller n-grams named subn-grams, a lower weight will be added to the cumulative weight which should be fixed according to the number of subn-grams in the passage. Note that there are three possible cases:

- Case 1: The n-gram of the query is one of the passage n-grams:
$ngQ_i = ngP_j, ngP_j \in NGP$
- Case 2: It is made by combining a number n of the passage n-grams:
$ngQ_i = \cup ngP, ngP \in NGP$
- Case 3: It is included in one of the passage n-grams:
$ngQ_i \in ngP_j, ngP_j \in NGP$

Let w be the weight to add to the passage when we browse through the question n-grams ngQ. In the cases 1 and 3, ngQ exists in the passage, so the additional weight w is set to: $w(ngP) = w(ngQ) = l \times \sum_{t_i \in terms(ngQ)} w(t_i, q)$ where l denotes the length of the n-gram ngQ and $w(t_i, q)$ is the weight of its term t_i. In the case 2, ngQ is divided into subn-grams in the passage, let sng be the number of these subn-grams. In this case, the additional weight w is computed as follows: $w(ngP) = \frac{w(ngQ)}{sng} = \frac{l}{sng} \times \sum_{t_i \in terms(ngQ)} w(t_i, q)$. Our n-gram based passage similarity measure named NGpsim between a passage and a question is equal to the ratio between the weight of the passage and that of the question previously computed. Recall that the weight of a passage is a sum of partial weights calculated at each step to be added to the total passage weight, while the query weight is calculated on the basis of its length and the sum of the weight of its terms. We can define NGpsim as follows :

$$NGpsim(p,q) = \frac{w(P)}{w(Q)} = \frac{\sum_{i=1}^{q} \frac{l_i}{sng_i} \times \sum_{t \in terms(ngQ_i)} w(t,q)}{l(Q) \times \sum_{t_i \in (Q)} w(t_i, q)} \qquad (2)$$

where q is the number of the question n-grams, $l(Q)$ is the number of the question terms and $w(t_i, q)$ is the weight of a question term. For instance, let us consider the following terms of a question Q and those of a passage P:

$Q(terms)=$ **Commission, Member, States**, *draw, guidelines*, **Television, Frontiers, Directive**.

$P(terms)=$ **Commission**, *encourage, audiovisual, benefits, media, service, providers*, **Member, States**, *ensure, implementation, new*, **Television, Frontiers, Directive**, *once, adopted, Directive*

The corresponding vector $\overrightarrow{CTqp}$ between the given Q and P is set to:

$\overrightarrow{CTqp}(Q,P)=$ *Commission, Member, States, Television, Frontiers, Directive*. From which we can generate:

$\overrightarrow{NGQ(Q)}$ =[Commission, Member, States][Television, Frontiers, Directive] and $\overrightarrow{NGP(P)}$ =[Commission][Member, States][Television, Frontiers, Directive].

In this example, $\overrightarrow{NGQ}$ is composed of two n-grams so, we have two partial passage weights to calculate. The first question n-gram is divided into two subn-grams in the passage so, *sng* equals 2 while the second one exactly equals a passage n-gram so, *sng* equals 1.

In a nutshell, $NGpsim$ is different from the existing n-gram based measures insofar as instead of extracting all n-grams for all n possible values of the query and the passage, as in [5,6], or all the n-grams of size n, as in [14], we just extract common n-grams between the question and the passages with different lengths. So, no additional step is needed to select common n-grams from all the extracted n-grams. Also, for the weight of n-grams, we consider both the sum of the terms weight and their lengths like in [14] while [5,6] consider only the sum of the terms weight.

WordNet-based Conceptual Similarity: First, both the question q and a passage p are analysed to extract all the corresponding WordNet synsets. For each synset, we keep only noun synsets and group them respectively into the set of synsets of the question named C_q and that of the passage named C_p. If the synsets belong to another POS categories such as verb, adjective, pronoun, we seek their derivationally related forms to obtain a related noun synset and we put it in the corresponding set of synsets. Given C_p and C_q the sets of concepts contained in a passage p and the question q, with $|C_p| \geq |C_q|$, WordNet-based Conceptual Similarity between p and q is set to:

$$ss(p,q) = \frac{\sum_{c_1 \in C_p} \max_{c_2 \in C_q} s(c_1, c_2,)}{|C_p|} \qquad (3)$$

where $s(c_1, c_2)$ is a conceptual similarity measure calculated using a variation of the Wu-Palmer formula, called ProxiGenea defined by formula 4 as follows:

$$pg_3(c_1, c_2) = \frac{1}{1 + d(c_1) + d(c_2) - 2d(c_0)} \tag{4}$$

where c_0 denotes the most specific concept that is present both in the synset path of c_1 and c_2 and d denotes the function returning the depth of a given concept which constitutes the number of nodes between a concept and the root.

Named Entity Overlap: A Named Entity Recognizer is used such as that implemented by [9], which is composed by 7 classes trained for MUC: Organization, Person, Money, Time, Location, Percent, Date. Thereafter, a per-class overlap measure is computed considering the class of each named entity. For instance, "Manchester" as a Location does not correspond to "Manchester" as an Organization. This similarity measure is calculated as:

$$O_{NER}(p, q) = \frac{2 * |N_p \cap N_q|}{|N_p| + |N_q|} \tag{5}$$

where N_p and N_q are the sets of NEs detected in p and q.

Edit distance: Edit distance is a similarity measure that quantifies how dissimilar two text fragments are by calculating the minimum number of operations needed to transform one string into the other. Numerous variants of edit distance using different operations exist. One of the most common one, that we have applied, is called Levenshtein distance. The edit distance is defined as follows:

$$sim_{ED}(p, q) = 1 - \frac{Lev(p, q)}{max(|p|, |q|)} \tag{6}$$

where $Lev(p, q)$ is the Levenshtein distance between the passage and the query.

4   Evaluation

4.1   Datasets, Tools and Metrics:

In order to evaluate the proposed approach, we implemented a system named PRSYS using the open source system JIRS[3] (JAVA Information Retrieval System) described in [10] for indexing and search and SVM^{light}[4] for passage ranking, adapting them to our requirements. For our experiments, we selected the dataset proposed in the ResPubliQA2010 exercise [12] of CLEF claimed to be the most suitable ones to validate our approach as it is based on passages. The aim of this exercise is to output either paragraphs or exact answers to a pool of 200 complex questions from two test collections. The document collection

---

[3] http://sourceforge.net/projects/jirs/.
[4] http://svmlight.joachims.org/.

includes a subset of JRC-Acquis[5] as well as a small portion of the Europarl[6] collection made up of approximately 50 parallel documents. Note that we used all the questions and the full english collection because we utilized the english versions of major tools deemed to achieve higher performance like the english version of WordNet Lexical Database, but we can certainly validate our approach in other languages by integrating multilingual tools. But beforehand, we should evaluate the performance to our NGpsim measure. To this end, we used the dataset provided in the ResPubliQA 2009 exercise [13] of CLEF where the aim is to retrieve passages from the corpus to answer a question picked out from a set of 500 different questions falling into five types: factual, definition, reason, purpose and procedure. The corpus includes only JRC-Acquis, where for each language, roughly 10700 documents are used containing over 1 million passages. In our experiments, we employ that of English, French and Spanish and all the question pool.

To evaluate our *NGpsim* measure, we are mainly based on the following measures which were widely used in QA:

- The accuracy: is the percentage of correctly answered questions.
- The Mean Reciprocal Rank (MRR): is the multiplicative inverse of the rank position of the first correct answer.
- The number of questions having correct passages ranked first.

To evaluate the performance of our PR approach, we used the following metrics proposed by CLEF:

- The c@1 measure: is the main CLEF evaluation metric for passage and answer selection tasks, defined as: $c@1 = \frac{1}{n}(n_R + n_U \frac{n_R}{n})$ where n_R denotes the number of correctly answered questions, n_U is the number of questions unanswered and n is the total number of questions.
- The number of: unanswered questions (#NoA), questions correctly answered (#R), questions wrongly answered (#W), unanswered questions where a right candidate answer is discarded (#NoA R) and questions unanswered with wrong candidate answer (#NoA W).

Note that we have set a threshold value for the final score to be 0.15 as it is the most common used value for ranking. Thus, we answer the question only if the highest score value exceeds 0.15, otherwise, no answer is outputted to the user, as we think that returning no response is better that delivering a wrong one.

4.2 Experimental Results

We report the results of our PR engine when using only NGpsim feature in order to evaluate its performance. Obviously, in this first evaluation we have not need RankSVM yet as we test just one similarity measure.

Table 1 shows that NGpsim measure has yielded significant results for the three languages. It is important to note that we have compared our system

---

[5] https://ec.europa.eu/jrc/en/language-technologies/jrc-acquis.
[6] http://www.europarl.europa.eu/.

Table 1. Experimental PR results using only NGpsim

Language	English		Spanish		French	
System	PRSYS	NLEL	PRSYS	NLEL	PRSYS	NLEL
MRR	0.453	0.377	0.413	0.373	0.411	0.367
Accuracy	0.851	0.782	0.811	0.762	0.805	0.742
Number of questions having correct passages in the first 10 positions	298	274	284	267	282	265
Number of questions whose correct passage is in first position	186	161	174	154	171	153

results to those obtained by NLEL System [6] which is based on an n-gram PR model and it was ranked first in the CLEF 2009 QA track for French and Spanish and second for English language. PRSYS when just based on NGpsim measure succeeded in outperforming NLEL for the 3 languages on all criteria. A significant number of questions was answered correctly with a good difference equals 24 for English and 17 questions more than NLEL for both Spanish and French languages. We also got more answers in the first position with a difference between 18 and 25 questions. Moreover, we got a greater accuracy and MRR values for all languages as the number of correct answers is higher in the first positions and lower in the last ones. Thus, we can so admit that our propose NGpsim measure is more efficient than that of NLEL.

We move on to evaluate our PR approach combining different features other than NGpsim, we compared the results obtained by our system run with those reported by the best participant systems in the CLEF paragraph selection task described in [12]. From Table 2, we see that PRSYS gives better results than all other ones performing the same task in terms of both accuracy and $c@1$. Besides, the fact that our $c@1$ value is greater than the accuracy score proves that the use of our NoA criterion is justified and has enabled to get higher $c@1$. We have

Table 2. Comparison between PRSYS and similar systems

System	c@1	Accuracy	#R	#W	#NoA	#NoA R	#NoA W
PRSYS	**0.85**	**0.77**	**154**	**23**	**23**	**0**	**0**
uiir101PSenen	0.73	0.72	143	54	3	0	3
bpac102PSenen	0.68	0.68	136	64	0	0	0
dict102PSenen	0.68	0.67	117	52	31	17	14
elix101PSenen	0.65	0.65	130	70	0	0	0
nlel101PSenen	0.65	0.64	128	68	4	2	2
uned102PSenen	0.65	0.65	129	71	0	0	0

also observed that most unanswered and incorrectly answered questions where opinion ones which are very complex. Furthermore, we can reason out that the combination of different lexical and semantic features has allowed to generate relevant passages with high accuracy and c@1 values.

5 Conclusion

Getting directly a precise answer to a human natural question constitutes a big challenge mainly in the database and information systems field due to the ever increasing amount of data and information. In this paper, we have proposed a multi-lingual approach for retrieving passages for open domain QA using a RankSVM classifier to combine different state-of-the-art features other than our proposed n-gram measure. Our experiments and analysis have shown promise and demonstrated that our approach is competitive and our new similarity measure is efficient outperforming the system ranked first in the PR task of CLEF. Interesting perspectives emerge then to further strengthen our proposed solution. We mainly look forward to extending our experiments on larger datasets to decide on the threshold value for the ranking final score and better refining the ranking model by incorporating other features such as a syntactic one but without increasing the program complexity. We also look forward to making our system find a precise answer rather than a full passage.

References

1. Abacha, A.B., Zweigenbaum, P.: MEANS: a medical question-answering system combining NLP techniques and semantic web technologies. IPM **51**(5), 570–594 (2015)
2. Araki, J., Callan, J.: An annotation similarity model in passage ranking for historical fact validation. In: Proceedings of the 37th International ACM SIGIR Conference on Research and Development in IR, pp. 1111–1114. ACM (2014)
3. Bilotti, M.W., Elsas, J., Carbonell, J., Nyberg, E.: Rank learning for factoid question answering with linguistic and semantic constraints. In: Proceedings of the 19th ACM International Conference on Information and KM, pp. 459–468. ACM (2010)
4. Buscaldi, D., Le Roux, J., Flores, J.J.G., Popescu, A.: Lipn-core: semantic text similarity using n-grams, wordnet, syntactic analysis, esa and information retrieval based features. In: Proceedings of the 2nd Joint Conference on LCS, p. 63 (2013)
5. Buscaldi, D., Rosso, P., Gómez-Soriano, J.M., Sanchis, E.: Answering questions with an n-gram based passage retrieval engine. JIIS **34**(2), 113–134 (2010)
6. Correa, S., Buscaldi, D., Rosso, P.: NLEL-MAAT at ResPubliQA. In: Peters, C., Di Nunzio, G.M., Kurimo, M., Mandl, T., Mostefa, D., Peñas, A., Roda, G. (eds.) CLEF 2009. LNCS, vol. 6241, pp. 223–228. Springer, Heidelberg (2010)
7. Cui, H., Sun, R., Li, K., Kan, M.Y., Chua, T.S.: Question answering passage retrieval using dependency relations. In: Proceedings of the 28th Annual International ACM SIGIR Conference, pp. 400–407. ACM (2005)
8. Fader, A., Zettlemoyer, L., Etzioni, O.: Open question answering over curated and extracted knowledge bases. In: Proceedings of the 20th ACM SIGKDD International Conference on Knowledge Discovery and Data Mining, pp. 1156–1165. ACM (2014)

9. Finkel, J.R., Grenager, T., Manning, C.: Incorporating non-local information into information extraction systems by gibbs sampling. In: Proceedings of the 43rd Annual Meeting on Association for Computational Linguistics, pp. 363–370. ACL (2005)

10. Gómez, J.M., Buscaldi, D., Rosso, P., Sanchis, E.: JIRS language-independent passage retrieval system: a comparative study. In: Proceedings of the 5th International Conference on NLP (ICON-2007), pp. 4–6 (2007)

11. Keikha, M., Park, J.H., Croft, W.B., Sanderson, M.: Retrieving passages and finding answers. In: Proceedings of the 2014 Australasian Document Computing Symposium, p. 81. ACM (2014)

12. Peñas, A., Forner, P., Rodrigo, Á., Sutcliffe, R.F.E., Forascu, C., Mota, C.: Overview of respubliqa 2010: question answering evaluation over european legislation. In: CLEF 2010 LABs and Workshops, Notebook Papers (2010)

13. Peñas, A., Forner, P., Sutcliffe, R., Rodrigo, Á., Forăscu, C., Alegria, I., Giampiccolo, D., Moreau, N., Osenova, P.: Overview of ResPubliQA 2009: question answering evaluation over European legislation. In: Peters, C., Di Nunzio, G.M., Kurimo, M., Mandl, T., Mostefa, D., Peñas, A., Roda, G. (eds.) CLEF 2009. LNCS, vol. 6241, pp. 174–196. Springer, Heidelberg (2010)

14. Radev, D., Fan, W., Qi, H., Wu, H., Grewal, A.: Probabilistic question answering on the web. JASIST 56(6), 571–583 (2005)

15. Ryu, P.M., Jang, M.G., Kim, H.K.: Open domain question answering using wikipedia-based knowledge model. IPM 50(5), 683–692 (2014)

16. Severyn, A., Nicosia, M., Moschitti, A.: Building structures from classifiers for passage reranking. In: Proceedings of the 22nd ACM International Conference on CIKM, pp. 969–978. ACM (2013)

17. Shen, D., Lapata, M.: Using semantic roles to improve question answering. In: Proceedings of EMNLP/CoNLL, pp. 12–21 (2007)

18. Sun, H., Ma, H., Yih, W.t., Tsai, C.T., Liu, J., Chang, M.W.: Open domainquestion answering via semantic enrichment. In: Proceedings of the 24th International Conference on WWW, pp. 1045–1055 (2015)

19. Tari, L., Tu, P.H., Lumpkin, B., Leaman, R., Gonzalez, G., Baral, C.: Passage relevancy through semantic relatedness. In: TREC (2007)

20. Tellex, S., Katz, B., Lin, J., Fernandes, A., Marton, G.: Quantitative evaluation of passage retrieval algorithms for question answering. In: Proceedings of the 26th Annual International ACM SIGIR Conference, pp. 41–47. ACM (2003)

21. Yen, S.J., Wu, Y.C., Yang, J.C., Lee, Y.S., Lee, C.J., Liu, J.J.: A support vector machine-based context-ranking model for question answering. JIS 224, 77–87 (2013)

Information Extraction on Weather Forecasts with Semantic Technologies

Angel L. Garrido[1]([✉]), María G. Buey[1], Gema Muñoz[1],
and José-Luis Casado-Rubio[2]

[1] IIS Department, University of Zaragoza, Zaragoza, Spain
{algarrido,mgbuey,gmunoz}@unizar.es
[2] Spanish Meteorological Service (AEMET), Madrid, Spain
jcasador@aemet.es

Abstract. In this paper, we describe a natural language application which extracts information from worded weather forecasts with the aim of quantifying the accuracy of weather forecasts. Our system obtains the desired information from the weather predictions taking advantage of the structure and language conventions with the help of a specific ontology. This automatic system is used in verification tasks, it increases productivity and avoids the typical human errors and probable biases in what people may incur when performing this task manually. The proposed implementation uses a framework that allows to address different types of forecasts and meteorological variables with minimal effort. Experimental results with real data are very good, and more important, it is viable to being used in a real environment.

Keywords: Weather forecast · Information extraction · Ontologies

1 Introduction

Nowadays, weather forecasts rely on mathematical models of the atmosphere and oceans in order to predict the weather based on current weather conditions. Several global and regional forecast models are used in different countries worldwide, which make use of different weather observations as inputs. Powerful supercomputers are used to work with vast datasets and they perform necessary complex calculations to deliver weather forecasts. These predictions are numerical, and are hardly interpretable by those who are not experts in the field. In order to convert these numerical results into information understandable by everyone, forecasters make their own interpretation of the mathematical models and create graphics, maps, and texts in natural language to explain the weather conditions of the atmosphere which may occur in the next few hours or days.

However, these interpretations are prone to errors that can be produced by both mathematical models and humans (or in some cases by software). Therefore, comparing weather forecasts with the data coming from actual observations

This research work has been supported by the CICYT project TIN2013-46238-C4-4-R, and DGA-FSE. Our gratitude to Dr. Eduardo Mena and AEMET.

© Springer International Publishing Switzerland 2016
E. Métais et al. (Eds.): NLDB 2016, LNCS 9612, pp. 140–151, 2016.
DOI: 10.1007/978-3-319-41754-7_12

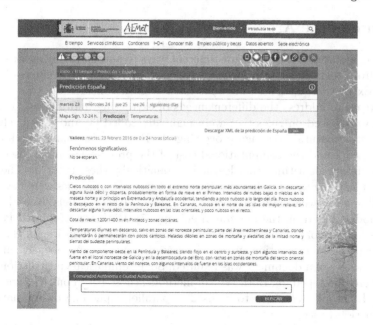

Fig. 1. A sample of a meteorological service's web page offering worded forecasts.

is a very interesting task, which can provide useful information to meteorological services. The problem is that many forecasts are published using natural language (a sample is shown in Fig. 1), and thus the verification is not trivial.

Therefore, meteorological services must convert by hand these worded weather forecasts into verifiable data, and then compare these data with actual observations of the different meteorological stations in the forecast area. Since we are talking about a closed domain environment, with great influence of specific knowledge to interpret texts, machine learning based approaches may not be the ideal to get good results [1]. On the other hand, the simple use of regular expressions can also be insufficient because the predictions often use ambiguous terms. Hence, we have investigated to improve information extraction on weather forecasts expressed in natural language, by using an ontology which guides the extraction process using a new methodology, based mainly on having different extraction methods stored in the ontology, and applying the most appropriate depending on the case. We have also developed an application that deals with real data in collaboration with the Spanish Meteorological Service (AEMET). Despite the fact that experimental dataset is written in Spanish, our approximation is generic enough to be applied to forecasts in other languages.

The remainder of the paper is organized as follows: Sect. 2 studies the state of the art related to information extraction based on ontologies close to this meteorological context. The information extraction process with a complete example is explained in Sect. 3. Section 4 interprets the results of our first experiments with real data. Finally, Sect. 5 summarizes the key points of this work, provides our conclusions, and explores future work.

2   State of the Art

Regarding methodologies, there are different ways to automatically extract information from texts in natural language, all framed within the context of Information Extraction [2]. These methods have evolved considerably over the last twenty years to address the different needs of extracting information. The first systems were based on rules that were manually coded [3] and which relied on an exhaustive natural language processing [4]. However, as manual coding was a tedious work and the computational cost of the process was high, researchers began to design algorithms that learned automatically these rules [5]. Then, the statistical learning was developed, where two types of techniques were developed in parallel: generative models based on hidden Markov models [6] and conditional models based on the maximum entropy [7]. Both were replaced later by the global conditional models, known as Conditional Random Fields [8]. Then, as the scope of the extraction systems increased, a more comprehensive analysis of the structure of a document was required. So, grammatical construction techniques were developed [9]. Both type of methods, those based on rules and those statistical methods, continue to be used in parallel depending on the nature of the extraction tasks.

Recently, the influence of ontologies has increased, and they are widely used as resources to allow exchange knowledge between humans and computers. An ontology [10] is a formal and explicit specification of a shared conceptualization that can be used to model human knowledge and to implement intelligent systems. Ontologies are sometimes used to model data repositories [11,12], or they can be used to classify elements [13–15]. And in other cases, they are used to guide extraction data processes from heterogeneous sources [16]. When the used method is based on the use of ontologies in an information extraction system, it belongs to the OBIE (Ontology Based Information Extraction) system group [17]. The characteristics of these systems are: (1) they process semi-structured texts in natural language, (2) perform the information extraction process guided by one or more ontologies, and (3) the output is formally represented according to the information of the ontologies used in the system. In this context, there are different approaches. Some oriented to automatic labeling of content: those which process the documents looking for instances of a given ontology [18], or those that obtain structured information from semi-structured resources [19]. There are also other existing works that aim to build an ontology from the processed information [20]. Most of these approaches try to provide methods and extraction techniques for general purposes or for certain domains.

However, our work focuses on a particular domain (weather forecasts) and it highly depends on the context, so the application of these methods turns complicated or limited in many ways. Moreover, our work is clearly different from them because it uses embedded information extraction procedures within the ontology, i.e. the ontology itself knows how to extract the information. The system exploits an ontological model defined for the domain to detect important events in the extraction stage, and then, the model is used to evaluate different treatment options of information that may exist.

Regarding weather issues, predictions are the result of numerical and statistical methods which try to anticipate the weather that is going to affect an area. There are many approaches to make these predictions understandable by everybody, i.e. to translate them into natural language format [21], but we have not found any approach focused on describing the semantics and the linguistic information about different atmospheric variables in order to extract them from texts, or the formal structure of predictions. Moreover, there are many approaches which focus on determining and verifying the actual effectiveness of the weather predictions. However, to the best of our knowledge, there are not works specially dedicated to the treatment of worded weather forecasts that also perform an evaluation of the accuracy of the interpretations of predictions made by people or automatic systems.

3   Information Extraction Process

Next, we describe our proposed system, called AEMIX, which identifies and extracts information contained in weather forecasts expressed in natural language format. AEMIX aims at verifying that forecasts match the real observation data in a specific date. Some screenshots of the software and a figure explaining the complete process can be found in the webpage of our research group[1].

On the one hand AEMIX receives a set of weather forecasts, and in the other hand, a spreadsheet with the corresponding actual data from all the observation stations. The first step consists of a preprocess, which cleans the texts and stores all the textual information into a database. Then, AEMIX fragments the forecast into several paragraphs according to the atmospheric variable described. The three most relevant variables to verify are: the temperature, the precipitation, and the wind. Each of these sets of sentences is analyzed with the aid of an ontology in order to facilitate the extraction process. The words are converted into numerical information, and finally they are also stored in the database. With this database, AEMIX will be able to execute verification tasks, but this issue is out of the scope of this work.

For each meteorological variable we need: (1) a set of attributes, and (2) information about the data required to be found. For instance, if we refer to precipitation, the attributes are the typology, the quantification, and the temporal evolution. Respectively, the information required could be {*drizzle, rain, snow, hail*}, {*weak, moderate, heavy, very heavy, torrential*}, and {*persistent, frequent, intermittent*}. Certain attributes of atmospheric variables can not be verified, because observations of them are sparse or non-existent (snow).

As we mentioned before, the extraction process is guided by a proprietary ontology designed by us containing the knowledge about different meteorological variables, and how to identify and extract them in the text of a forecast. A fragment of the ontology can be appreciated in Fig. 2. This sample shows an excerpt of the ontology which focuses on extracting information about precipitation, one of the variables studied. The ontology includes references to the extraction

---

[1] http://sid.cps.unizar.es/SEMANTICWEB/GENIE/Genie-Projects.html.

methods used during the process. One advantage of this architecture is that new atmospheric variables can be added dynamically in a quick and easy way.

The stages of the process are explained in detail with examples in the following subsections.

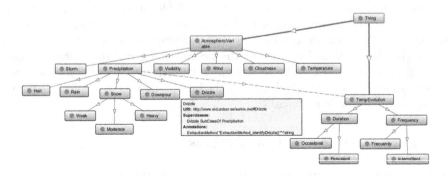

Fig. 2. A partial sample of the AEMIX ontology model. It shows the information about the atmospheric variable "Precipitation", and details about the extraction of the information about "Drizzle"

3.1 Stage 1: Adjusting the Input Texts

In this first stage, we perform a number of different tasks which clean weather forecasts before the data extraction. The four key elements in this stage are: (1) date and time, (2) time period when the weather forecast is valid, (3) geographical scope of validity of the prediction, and (4) the textual content of the forecast. Following, we show an example of a Spanish weather forecast[2], downloaded from the Spanish Meteorological Service (AEMET) website[3]:

"ZCZC FPSP85 LECR 260600 SPANISH METEOROLOGICAL SERVICE WEATHER FORECAST FOR GALICIA REGION DAY JANUARY 26th 2015 AT 10:00 OFFICIAL TIME FORECAST VALID FROM 00 UNTIL 24 HO FRIDAY 28. CLOUD OR OVERCAST WITH WEAK SHOWERS, AND POSSIBILITY OF LOCAL MODERATE STORMS, LESS LIKELY IN THE SOUTHWEST. SNOW LEVEL AROUND 600–800 M. MINIMUM TEMPERA-TURE DECREASING SLIGHTLY, AND MAXIMUM UNCHANGED. WEAK FROST INLAND. STRONG NORTHEAST WINDS IN THE NORTH COAST OF FISTERRA AND WEAK IN THE REST. NNNN"

Once AEMIX has completed this stage, the output result is:

– *Date and time:* 01/26/2015 - 10:00 AM.
– *Range or Type of weather forecast:* FP85[4].

----

[2] For clarity's sake, we show the examples translated to English.
[3] http://www.aemet.es/.
[4] FP85 is the tag used by AEMET to indicate that this is a two-day weather forecast.

- *Geographical information:* Galicia[5].
- *Weather forecast text:* *"CLOUD OR OVERCAST WITH WEAK SHOWERS, AND POSSIBILITY OF LOCAL MODERATE STORMS, LESS LIKELY IN THE SOUTHWEST. SNOW LEVEL AROUND 600–800 M. MINIMUM TEMPERATURE DECREASING SLIGHTLY, AND MAXIMUM UNCHANGED. WEAK FROST INLAND. STRONG NORTHEAST WINDS IN THE NORTH COAST OF FISTERRA AND WEAK IN THE REST."*

This information is stored in the system database together with observation data from a standardized spreadsheet for the same date, also retrieved from the corporate website of the meteorological service.

3.2   Stage 2: Text Analyzer

At this stage, AEMIX queries the ontology to obtain the expected text structure, according to the type of weather forecast. With this information:

1. AEMIX studies the weather forecast and it finds the different possible meteorological variables by using pattern matching.
2. According to the information provided by the ontology, the system uses the most adequate method for fragmenting the texts into groups of sentences, taking into account the heterogeneity of this kind of texts. For example, the information about a given variable could be located in two (or more) different sentences. Or the opposite, the same sentence may contain information about more than one variable.
3. AEMIX returns a set of tuples (`<V,S>`) for each variable. V stands for the type of an atmospheric variable, namely: temperature, precipitation, storms, visibility, cloudiness or wind. S represents a list of sentences referred to the same variable in the forecast. In those cases, where the same sentence appears linked to two (or more) variables, the next stage will be responsible for filtering the relevant information for each one.

Following the previous example, at this stage we obtain the next set of tuples:

- *V1=* cloudiness, *S1=* *"CLOUD OR OVERCAST"*.
- *V2=* precipitation, *S2=* *"WEAK SHOWERS, AND POSSIBILITY OF LOCAL MODERATE STORMS. SNOW LEVEL AROUND 600–800 M"*.
- *V3=* temperature, *S3=* *"MINIMUM TEMPERATURE DECREASING SLIGHTLY, AND MAXIMUM UNCHANGED. WEAK FROST INLAND"*.
- *V4=* wind, *S4=* *"STRONG NORTHEAST WINDS IN THE NORTH COAST OF FISTERRA AND WEAK IN THE REST."*.

We have designed different methods, based on patterns rules and machine learning techniques, in order to analyze the text and identify meteorological variables, and then to cut off the text and provide the tuples. The advantage of our modular development is we can interchange these methods easily, only modifying the ontology.

---

[5] Galicia is a Spanish region.

3.3   Stage 3: Data Extractor

At the third stage is where the information extraction is properly located. There-fore, it aims at converting the tuples (`<V,S>`), obtained in the previous stage from the worded weather forecast, to a numerical format. For doing this, the system queries the ontology to obtain the most appropriate methods to extract the desired data from the sentences. The ontology has been previously populated with different methods which can be applied on the input text according to the related variable. These methods, mainly based on pattern rules, are readily exchangeable for testing through the ontology.

In summary, through the knowledge of the atmospheric variable features depicted on the ontology, the system identifies the sentence format, and conse-quently, it is able to use regular expressions to recover the accurate data. At the end of this stage, AEMIX returns a list of tuples of elements (`<Attribute, Value>`). The *attributes* represent each of the features of an atmospheric vari-able. For instance, regarding temperature, the features could be "minimum", "maximum", or "frosts". The *values* are is the quantity which establishes the determination of a variable. As an example, in the case of wind, the possible values of the attribute *"direction"* are {"N"}, {"NE"}, {"SE"}, {"S"}, {"SW"}, {"W"} or {"NW"}.

Following the running example, at this stage we obtain the next set of tuples with extracted data referring, for instance, to temperature (V3):

1. *Attribute:* minimum —— *Value:* "DECREASING SLIGHTLY".
2. *Attribute:* maximum —— *Value:* "UNCHANGED".
3. *Attribute:* frosts —— *Value:* N/A[6].

The system uses lexical patterns and morpho-syntactic analyzers to locate these attributes. As soon as AEMIX has extracted the textual information, it converts them into numerical information using the rules provided by the ontol-ogy. If we go ahead with the aforementioned example, rules can be similar to these:

– *DECREASING SLIGHTLY* = {-5, -2}
– *UNCHANGED* = {-2, +2}[7]

These rules have been previously established and stated into the ontology, inferred through a writing style guide of weather forecasts. An example of this style guide can be downloaded from our research group website[8]. After consulting the ontology, AEMIX obtains for each prediction a set of atmospheric variables identified in the forecast, and for each variable a set of tuples (`<Attribute, Value>`). Therefore, the system achieves exactly:

---

[6] The system returns a N/A (Not Applicable) value when there is no possibility of performing the verification process. For example, frosts: there are no observational data related to frost.

[7] This means that if it is said in the weather prediction temperature values remain unchanged, verification values will be actually valid between 2 degrees up or down.

[8] http://sid.cps.unizar.es/SEMANTICWEB/GENIE/Genie-Projects.html.

1. *Atmospheric variable:* cloudiness.
   - *Attribute:* adjetive —— *Value:N/A.*
2. *Atmospheric variable:* precipitation.
   - *Attribute:* type —— *Value:*{0, 2}.
   - *Attribute:* storms —— *Value:N/A.*
   - *Attribute:* snow-level —— *Value:N/A.*
3. *Atmospheric variable:* temperature.
   - *Attribute:* minimum —— *Value:*{-5, -2}.
   - *Attribute:* maximum —— *Value:*{-2, 2}.
   - *Attribute:* frosts —— *Value:N/A.*
4. *Atmospheric variable:* wind.
   - *Attribute:* direction —— *Value:*{"NE"}.
   - *Attribute:* speed —— *Value:*{41, 70}.
   - *Attribute:* location —— *Value:*{(42.9, -9.35), (42.9, -9.05), (43.9, -7.7), (43.6, -7.5)}.
5. *Atmospheric variable:* wind.
   - *Attribute:* direction —— *Value:*{"NE"}.
   - *Attribute:* speed —— *Value:*{6, 20}.
   - *Attribute:* location —— *Value:Rest.*

As we can see, if the attributes are geographical ("location"), the values are a set of geographical points defining the area of interest, or a specific word, like "Rest", which is translated as *the set of the stations in the region included in the forecast, and not mentioned yet.* When no location is indicated, it implies that the forecast is referred to the region as a whole. We can also appreciate that there are two items for the wind, because there are two different forecasts according to the location. The first one for the north coast of Fisterra, and the second one for the rest of the Galicia region. Besides, there are several attributes with the value "N/A" (not applicable). In these cases, since observational data are not available, AEMIX does not make the effort of extracting them because there is no way of verifying (and we have to remember that this is the final aim of the project). It is important to point out that numerical data units are not specified in the extracted data, but they are defined within the ontology itself, related to attributes that require them by their nature.

Finally, all these data are stored in an appropriate format in the database, ready for the next stage: the verification against the observation data stored in the first stage.

3.4   Stage 4: Verification

The aim of this stage is to verify the accuracy of the forecast, compared with observation data. Verification is a wide and complicated field in meteorology. The techniques to apply depend on the meteorological variable to study (continuous variables, such as temperature, and discrete ones, such as the presence of thunderstorms, are treated differently), its statistical properties (temperature and precipitation behave in a very different way), and the type of forecast (deterministic or probabilistic). Besides, factors such as the representativity and density of observations in a region must be taken into account.

Hence many different scores are used to summarize the quality of forecasts. Selecting one or the other depends on the specific problem we are dealing with. For example, for the temperature in our worded forecasts the bias and the root mean square error will be the most useful scores, because they can give us more insight into the mistakes made by human forecasters. Anyway, we are not going into more detail because this task is out of the scope of this paper.

4   Experiments

Tests on the system have been conducted on actual data provided by the afore-mentioned Spanish Meteorological Agency (AEMET). We used a sample of 2,828 worded weather forecasts corresponding to one year of predictions (2011) over Galicia region, and the corresponding observation data from 58 observation stations. We worked with temperature and precipitation variables during the same year, accounting a total of 77,339 observation registers. Previously, we identified the forecaster's linguistic uses to depict the principal zones of the Galicia region, and we related them with the corresponding geographical areas though the graphical interface of AEMIX. We assessed the expected maximum and minimum temperature on the forecasts, and for precipitation we monitored the amount of rain collected throughout the day also regarding the prevision. We focus on data from a month (February 2011). Regarding technical environment, we have used Java and Firebird, Freeling[9] as the NLP tool, and SVM-light[10] for machine learning tasks. The ontology have been implemented with OWL.

We tested the performance of the system using well-known measures in the field of the Information Extraction (*precision, recall,* and *F-measure*). In order to calculate them, we took into account for each forecast: the number of attributes to be extracted, the number of extracted attributes, and the number of attributes that had been retrieved properly. The baseline was the application of a set of extraction rules based on regular expressions without using our ontology-based extraction approach. Though the percentage of hits are quite high (near 77 %), AEMET needs at least 90–95 % of accuracy, so the set of rules is not good enough. The errors are mainly due to the incorrect use of the symbolic pattern rules when applying them on certain sentences widely separated from the standard use of language in weather forecasts.

We can see the results of the extraction process with the baseline method-ology (Experiment 1) in Fig. 3. We can appreciate the influence of the use of the ontologies to guide the extraction, leading to a small but essential improve-ment of the process (Experiment 2). The results clearly show that while the temperature is the easiest meteorological variable to extract, the precipitation and geographical areas are a bit more complex to deal with. This is due to the

---

[9] http://nlp.lsi.upc.edu/freeling/.
[10] http://svmlight.joachims.org/.

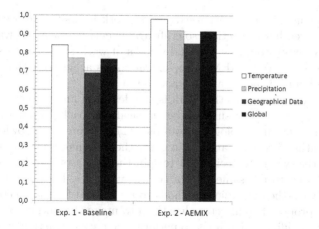

Fig. 3. F-measure results from the experiments 1 (Baseline) and 2 (AEMIX). Both experiments are based on data from AEMET forecasts and observation for February 2011.

greater semantic ambiguity of the expressions used to describe both precipitation aspects and geographical areas.

With our new enhancements, AEMIX achieved better results (an F-measure above 90 %, and hence valid). The rules to extract the data from the forecasts are the same in experiments 1 and 2, but in the second experiment the decision to apply one or another rule is led by the ontology, and its use avoids most errors. With these experiments we have proved that if we have a semantic aid to guide the extraction we can get an important improvement on the task of extracting data from the weather forecasts.

5   Conclusion and Future Work

In this work, we have dealt with information extraction tasks applied to a particular type of texts: weather forecasts expressed in natural language. The lack of homogeneity of text structure, the number of atmospheric variables, the ambiguous area descriptions, and the presence of elements with similar adjectives, make difficult the application of known techniques. Besides, we needed a high degree of precision in order to correctly carry out a verification process.

We have presented AEMIX, a natural language application which extracts information from worded weather forecasts with the aim of verifying them against actual observation data collected from meteorological stations. The usefulness of this work is to avoid that this work is done by hand by experts, which is expensive, time-consuming, and subjected to many human errors. The authors have performed manual experiments, and rates of up to 10 % of human errors were found. Additionally, we realized that there is often a kind of internal psychological bias that makes the human reviewers give for good weather predictions, which actually only approach to observations. An automatic process, however, is rigorous and is not subject to moods, fatigue, humor, etc.

Usually doing this information extraction work by hand can assume about 5 min for each weather prediction, so crafting the entire process shown in experiments (2,828 texts) would be almost 18 weeks of work from one person, which is highly unfeasible. With AEMIX the extraction is achieved without effort: in less than one hour all the data is extracted and ready.

The main contribution of this work is the design of an ontology-driven approach which improves the results of the classical approach. In our solution, the motivation of using an ontology is two-fold: (1) to represent knowledge about aspects of weather forecasts, and (2) to guide the automatic information extraction, helping the system to decide how to split the forecasts in order to group the sentences referred to the same meteorological variables. Besides, this ontology is populated with the extraction methods to be applied for each meteorological variable. The proposed architecture has the advantage of allowing to incorporate several and very different methods to perform the data extraction with minimal effort, and it avoids errors due to ambiguous language. The first tests seem to indicate that using semantics tools to guide the extraction process improves the results obtained by other approaches, and therefore the application can be used in a real environment with severe restrictions in terms of effectiveness.

With regard to the information extraction through semantic techniques, we want to advance in the following working lines: (1) to check the system operation by testing more extensively in new geographical locations, and by analyzing other atmospheric variables, (2) to explore tools that automate the construction of the ontology; its development has been completed manually in this work with the help of experts, but it would be interesting to do it automatically, given a forecast style guide, and (3) to check rigorously the generality of the method with other languages than Spanish. Moreover, and this is another reason of having used ontologies, we want to explore their capacity to store rules and axioms to improve the extraction mechanism.

References

1. Sarawagi, S.: Information extraction. Found. Trends Databases **1**(3), 261–377 (2008)
2. Russell, S., Norvig, P.: Artificial Intelligence: A Modern Approach. Artificial Intelligence. Prentice-Hall, Upper Saddle River (1995)
3. Appelt, D.E., Hobbs, J.R., Israel, D., Tyson, M.: Fastus: a finite-state processor for information extraction from real-world text. In: 13th International Joint Conferences on Artificial Intelligence (IJCAI 1993), vol. 93, pp. 1172–1178 (1993)
4. Grishman, R.: Information extraction: techniques and challenges. In: Information Extraction a Multidisciplinary Approach to an Emerging Information Technology, pp. 10–27 (1997)
5. Soderland, S.: Learning information extraction rules for semi-structured and free text. Mach. Learn. **34**(1–3), 233–272 (1999)
6. Seymore, K., McCallum, A., Rosenfeld, R.: Learning hidden Markov model structure for information extraction. In: AAAI 1999 Workshop on Machine Learning for Information Extraction, pp. 37–42 (1999)

7. McCallum, A., Freitag, D., Pereira, F.C.: Maximum entropy Markovmodels for information extraction and segmentation. In: 27th International Conference on Machine Learning (ICML 2000), vol. 17, pp. 591–598 (2000)
8. Lafferty, J., McCallum, A., Pereira, F.C.: Conditional random fields: probabilistic models for segmenting and labeling sequence data (2001)
9. Viola, P., Narasimhan, M.: Learning to extract information from semi-structured text using a discriminative context free grammar. In: 28th International ACM SIGIR Conference on Research and Development in Information Retrieval, pp. 330–337 (2005)
10. Gruber, T.R.: A translation approach to portable ontology specifications. Knowl. Acquisition 5(2), 199–220 (1993)
11. Mika, P.: Ontologies are us: a unified model of social networks and semantics. In: Gil, Y., Motta, E., Benjamins, V.R., Musen, M.A. (eds.) ISWC 2005. LNCS, vol. 3729, pp. 522–536. Springer, Heidelberg (2005)
12. Barbau, R., Krima, S., Rachuri, S., Narayanan, A., Fiorentini, X., Foufou, S., Sriram, R.D.: Ontostep: enriching product model data using ontologies. Comput. Aided Des. 44(6), 575–590 (2012)
13. Vogrinčič, S., Bosnić, Z.: Ontology-based multi-label classification of economic articles. Comput. Sci. Inf. Syst. 8, 101–119 (2011)
14. Garrido, A.L., Gómez, O., Ilarri, S., Mena, E.: An experience developing a semantic annotation system in a media group. In: Bouma, G., Ittoo, A., Métais, E., Wortmann, H. (eds.) NLDB 2012. LNCS, vol. 7337, pp. 333–338. Springer, Heidelberg (2012)
15. Garrido, A.L., Buey, M.G., Ilarri, S., Mena, E.: GEO-NASS: a semantic tagging experience from geographical data on the media. In: Catania, B., Guerrini, G., Pokorný, J. (eds.) ADBIS 2013. LNCS, vol. 8133, pp. 56–69. Springer, Heidelberg (2013)
16. Kara, S., Alan, Ö., Sabuncu, O., Akpınar, S., Cicekli, N.K., Alpaslan, F.N.: An ontology-based retrieval system using semantic indexing. Inf. Syst. 37(4), 294–305 (2012)
17. Wimalasuriya, D.C., Dou, D.: Ontology-based information extraction: an introduction and a survey of current approaches. J. Inf. Sci. 36(3), 306–323 (2010)
18. Cimiano, P., Handschuh, S., Staab, S.: Towards the self-annotating web. In: 13th International Conference on World Wide Web, pp. 462–471 (2004)
19. Buitelaar, P., Cimiano, P., Frank, A., Hartung, M., Racioppa, S.: Ontology-based information extraction and integration from heterogeneous data sources. Int. J. Hum. Comput. Stud. 66(11), 759–788 (2008)
20. Getman, A.P., Karasiuk, V.V.: A crowdsourcing approach to building a legal ontology from text. Artif. Intell. Law 22(3), 313–335 (2014)
21. Goldberg, E., Driedger, N., Kittredge, R.: Using natural-language processing to produce weather forecasts. IEEE Expert 9(2), 45–53 (1994)

Keyword Identification Using
Text Graphlet Patterns

Ahmed Ragab Nabhan[1,2(✉)] and Khaled Shaalan[3,4]

[1] Faculty of Computers and Information, Fayoum University, Faiyum, Egypt
ahmed.nabhan@gmail.com
[2] Member Technology, Sears Holdings, Hoffman Estates, USA
[3] The British University in Dubai, Dubai, United Arab Emirates
[4] School of Informatics, University of Edinburgh, Edinburgh, UK
khaled.shaalan@buid.ac.ae

Abstract. Keyword identification is an important task that provides useful information for NLP applications including: document retrieval, clustering, and categorization, among others. State-of-the-art methods rely on local features of words (e.g. lexical, syntactic, and presentation features) to assess their candidacy as keywords. In this paper, we propose a novel keyword identification method that relies on representation of text abstracts as word graphs. The significance of the proposed method stems from a flexible data representation that expands the context of words to span multiple sentences and thus can enable capturing of important non-local graph topological features. Specifically, graphlets (small subgraph patterns) were efficiently extracted and scored to reflect the statistical dependency between these graphlet patterns and words labeled as keywords. Experimental results demonstrate the capability of the graphlet patterns in a keyword identification task when applied to MEDLINE, a standard research abstract dataset.

Keywords: Word graphs · Pattern analysis · Graph features · Machine learning · MEDLINE

1 Introduction

Keywords are important meta data that is useful for many document processing tasks including indexing and retrieval, clustering, and classification. Keywords can act as relevancy indicators about documents that are retrieved given a search query. This meta data can be provided by authors, or by manual or automatic methods to provide better search and access experience when using scientific databases. Lists of keywords, as document fields, are of high quality considering they are provided manually either by authors or assigned by librarians. Although these lists are typically limited in size (three to five items), there is an opportunity to utilize them through supervised learning methods that aim to identify statistical dependencies between text features and word labels (e.g. keyword vs. non-keyword).

ⓒ Springer International Publishing Switzerland 2016
E. Métais et al. (Eds.): NLDB 2016, LNCS 9612, pp. 152–161, 2016.
DOI: 10.1007/978-3-319-41754-7_13

The problem of automatic extraction of keywords within text has been addressed through both NLP and graph-based methods. Several resources including manually-generated keywords, lexical, and syntactic annotations have been used to identify keywords within texts [1–5]. Machine learning techniques have been developed to extract useful features for the task of keywords identification. Witten et al. (1999) proposed a Naive Bayes method for learning features relevant to keyword extraction, see also [6]. Li et al. developed a semi-supervised approach to learning an indicator by assigning a value to each important phrase in the document title and to propagate that value to other phrases throughout the document [7]. Tomokiyo (2003) proposed an approach to key phrase extraction based on language models (cf. [8]).

Automatic methods for keyword identification utilize lexical, syntactic, semantic, and presentation features of texts. These features can be used to learn rules for identifying keywords in testing data. Counting-based methods (e.g. TF-IDF [9]) are built on lexical features (e.g. stemmed, lemmatized, or surface forms of the words). Morpho-Syntactic features (e.g. part-of-speech tagging information, base-phrase chunks) can be used as a filter to exclude words (e.g. adverbs and prepositions) or phrases (e.g. verb phrases) unlikely to be keywords.

Graph-based document representation can enable powerful methods such as random walks and frequent subgraph mining to capture relevant features for many NLP tasks. Ranking-based graph methods have been proposed to quantify the importance of a word in a document relative to neighboring words. TextRank [10], LexRank [11], and NeRank [12] are examples of methods that employed graph properties such as vertex centrality and ranking for keyword extraction and text summarization. Mining of frequent subgraphs and subtrees were demonstrated to be useful for document categorization [13]. Graph-based term weighting methods were proposed for information retrieval applications [14].

In this paper, a graph pattern mining method is proposed to identify keywords in word graphs constructed from text abstracts. Word graphs provide a flexible representation that enables exploration of complex, non-local features that can span multiple sentences. The method combined lexical features with graph substructures (e.g. graphlets) to explore contexts of keywords and identify siginificant patterns. Then, these graph-based features were used by a machine learning based classifier for testing data.

2   Research Methods

The problem being addressed in this study can be defined as follows. Let a text document be mapped to a word graph where vertices represent words, and edges indicate word co-occurrences in sentences. That is, if the words w_i and w_j appear next to each other in a sentence, an edge is created to link the two vertices representing w_i and w_j. Given a set of word graphs $D = \{G_1, G_2, ..., G_n\}$ where each graph G_i represents a text document, the problem is to structurally analyze elements in D to extract significant graph substructures (e.g. graphlets, paths) in the neighborhood of each word vertex. Then, a binary classifier can be used to classify word vertices in a test corpus as keywords or non-keywords based on their graph substructure features.

Vertex labels in word graphs can represent content at some linguistic level. For instance, at the very basic level, a vertex can correspond to a primitive lexical unit (e.g. surface, lemmatized or stemmed word form), syntactic unit (e.g. a phrase) or a sentence (in which case a graph can represent a whole corpus). Edges in these graphs can represent a lexical relationship (e.g. neighborhood, lexical similarity) or syntactic relationships (e.g. subject-verb-object relationships). In this study, text abstracts were mapped to graphs where vertices represented words and edges between a pair of vertices were created if the pair of vertices appeared next to each other in a sentence. The edges did not have weights and the vertices had two features: a lexical label and a part-of-speech tag.

2.1   Notations

A graph is a structure $G = (V, E)$, where $V = \{v_1, v_2, ..., v_n\}$ is a set of vertices and $E \subseteq V \times V$ is the set of edges. A graph is said to be undirected iff $\forall e = (v_i, v_j) \in E, \exists e' = (v_j, v_i) \in E$. A graph $G(V, E)$ is isomorphic to graph $G'(V', E')$ if there exists a bijective function $f: V \rightarrow V'$ such that $e = (v_i, v_j) \in E$ iff $e'(f(v_i), f(v_j)) \in E'$. Graph automorphism is a symmetry property of a graph so that its vertex set can be mapped into itself while preserving edge-vertex connectivity. A subgraph $H(V', E')$ of G is defined such that $V' \subseteq V$ and $E' \subseteq E$. The size of a graph or a subgraph is defined as the number of items in the vertex set.

Graphlets are small-sized nonisomorphic subgraphs (typically $2 \leq |V| \leq 5$) within a given larger graph [15]. A graph can be characterized by its collection of graphlets [16]. Graphlet automorphism allows for modeling the relationship of a graphlet and its component vertices. *Automorphism orbits* are defined by distances of a vertex of interest (pivot) and the rest of vertices in a graphlet. Each vertex reachable from a pivot vertex p in a number of edges d are said to be in $d - orbit$.

2.2   The Graph-Based Method

In this study, we present a graph-based method for identification of keywords in research abstracts. Text preprocessing operations (e.g. removing punctuation marks and stop words, lower-casing) were applied. Text were processed to extract bigram patterns and these patterns were used to represent edges with end points representing vertices. The NLTK toolkit [17] part-of-speech tagger provided syntactic information that were used during classification to apply the graphlet methods only to vertices with nouns and adjectives part of speech tags. These helped in increasing the true negative (non-keywords) score.

The method relied on graphlets that were extracted from a word graph representation of the abstracts. For each word vertex, a graph search algorithm was designed to extract 3-graphlet and 4-graphlet subgraphs such that this word was a pivot vertex. The relationships between the pivot vertex (a candidate keyword)

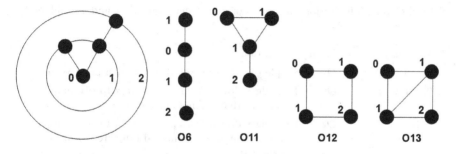

Fig. 1. An example 4-graphlet with four possible orbit forms of the configuration $< 0, 1, 1, 2 >$.

and its neighbors are defined based on orbit positions and inter-edges between these non-pivot vertices.

Graphlets can be enumerated systematically and thus are efficient to extract. *Automorphism orbits* allow for efficient comparison and counting of graphlets and thus they can be efficiently explored and scored. A string representation of graphlet determines (1) the layout of neighboring vertices in each orbit, and (2) an indicator of the configuration (e.g. how other non-pivot vertices are linked). Illustration of orbit configurations are given in Figs. 1 and 2. For instance, a 4-graphlet orbit 0 1 2 2 (Fig. 2) represents a pivot vertex at the origin (0) and the next neighbor is at distance 1, and then two vertices are connected to that neighbor at a two-edges distance from the pivot. In this case, two graphlet forms can be possible: (1) a form where the neighbors-of-a-neighbor can be linked by one edge, or (2) another form where there are no edges between those neighbors-of-a-neighbor.

In this paper, we followed the methodology proposed by Vacic et al. [16] for enumerating and counting graphlets using the configuration-form scheme. Table 1 shows the graphlet orbit configuration and form schemes used in this study. During feature extraction, graphlet patterns were discovered systematically by applying each scheme in Table 1, starting with a pivot vertex (the word for which the features were extracted). Two trivial graphlets (namely: 1-graphlet and 2-graphlets) were not used in our study.

Figure 1 shows four varieties of graphlets of the same configuration, only differing by the way non-pivot vertices are linked. The feature extraction algorithm stores and retrieves graphlets using the configuration string (e.g. $< 0, 1, 2, 2 >$), a named form (e.g. O7), together with labels (words surface forms) of the vertices ordered by the configuration string. Hash function values were generated using these compact string representations of graphlets and were used to read from and write to the feature table where features frequency were stored.

In this study, 3-graphlet and 4-graphlet subgraphs were used as features for classification of words as keywords vs. non-keywords. A minimum support value of 5 was used to reduce feature space size. Graphlets of size 5 were not used due to their somewhat large number of orbit configurations that would increase

Table 1. Orbit configuration and forms of 2-graphlet, 3-graphlet and 4-graphlet patterns

Orbit configuration	Form	Description
$< 0, 1 >$	O_1	Simple edge from pivot vertex to non-pivot vertex
$< 0, 1, 1 >$	O_3	Two edges drawn from pivot vertex to two non-pivot vertices that are not connected
$< 0, 1, 1 >$	O_4	Two edges drawn from pivot vertex to two non-pivot vertices that are connected (all three forms a triangle)
$< 0, 1, 2 >$	O_2	A simple path connecting pivot vertex to two other non-pivot vertices
$< 0, 1, 1, 1 >$	O_8	A star-like shape where pivot vertex is in the middle and no edges between non-pivot vertices
$< 0, 1, 1, 1 >$	O_{10}	A pivot vertex is in the middle and there exists one edge between a pair of the non-pivot vertices
$< 0, 1, 1, 1 >$	O_{14}	A pivot vertex is in the middle and there exists two edges between two pair of the non-pivot vertices
$< 0, 1, 1, 1 >$	O_{15}	A pivot vertex is in the middle and all non-pivot vertices connected to each other
$< 0, 1, 1, 2 >$	O_6	A simple path connecting four vertices and the pivot is any non-terminal vertex in that path (Fig. 1)
$< 0, 1, 1, 2 >$	O_{11}	A pivot vertex is connected to two non-pivot and a third vertex is connected to either of the non-pivot vertices (Fig. 2)
$< 0, 1, 1, 2 >$	O_{12}	A pivot vertex is connected to two non-pivot and a third vertex is connected to both of the non-pivot vertices (Fig. 1)
$< 0, 1, 1, 2 >$	O_{13}	A pivot vertex is connected to two non-pivot and a third vertex is connected to both of the non-pivot vertices (Fig. 1)
$< 0, 1, 2, 2 >$	O_7	A pivot vertex is connected to one non-pivot which is connected to two non-pivot vertices that are not connected (Fig. 2)
$< 0, 1, 2, 2 >$	O_9	A pivot vertex is connected to one non-pivot which is connected to two non-pivot vertices that are connected via an edge (Fig. 2)
$< 0, 1, 2, 3 >$	O_5	A simple path connecting the four vertices where the pivot is at either ends

the running time of the algorithm. A conditional probability model was used to quantify the correlation between graphlet features and class labels (keywords vs. non-keywords). Let $g_1, g_2, ..., g_n$ be a set of graphlets containing a given word v_i (as a pivot vertex) and that word has a class label c_k. The problem of assigning

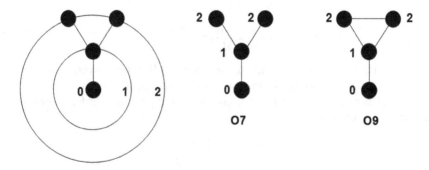

Fig. 2. An example 4-graphlet with two possible orbit forms of the configuration $< 0, 1, 2, 2 >$. There are two cases with the outer vertices on orbit 2: first case is where no edges between these vertices and second case where there is an edge linking them.

a class label c_k to a word v_i given a set of graphlets with v at the origin can be stated as estimating the probability value $P(c_k|v_i, g_1, g_2, ..., g_n)$, as a function of graphlet features. Using Bayes' rule,

$$
\begin{aligned}
P(c_k|v_i, g_1, g_2, ..., g_n) &\propto P(c_k) \times P(v_i, g_1, g_2, ..., g_n|c_k) \\
&= P(c_k) \times P(v_i|c_k) \times P(v_i, g_1, g_2, ..., g_n|c_k) \\
&= P(c_k) \times P(v_i|c_k) \times \prod_{i=1}^{n} P(g_i|c_k),
\end{aligned}
\tag{1}
$$

assuming graphlet features are conditionally independent on class label c_k. The probability value $P(c_k)$ represents prior information about class distribution in the data set. The probability model $P(v_i|c_k)$ can be viewed as a unigram probability of word vertex v_i. During initial phases of the study, there was no observed performance improvements of using the unigram probability $P(v_i|c_k)$ and thus this probability value was not used in the final model. Finally, a label was assigned to a word such that it maximized the probability value that is a function of graphlet features:

$$
\hat{y} = \underset{k \in \{1,2\}}{\mathrm{argmax}} P(c_k) \times \prod_{i=1}^{n} P(g_i|c_k)
\tag{2}
$$

Using the above model, a Naive Bayes classifier was used to decide on whether or not a test word vertex represented a keyword, given graphlets that were explored for that test word vertex.

2.3 TF-IDF Method

In this study, TF-IDF scores were used in the baseline system. TF-IDF is a frequency-based method that takes into account word as well as document frequencies [18]. For a given word i in a document j, the w_{ij} score was computed as follows.

$$w_{ij} = \frac{n_{ij}}{d_j} log_2 \frac{N}{n_i}, \tag{3}$$

where n_{ij} is the frequency of word i in document j, $|dj|$ is total words in document j, N is the total number of documents in the corpus, and n_i is the size of set of document containing word i. After computing w_{ij} for all words, a normalized score was computed as

$$W_{ij} = \frac{w_{ij}}{\sqrt{\sum_i wij}} \tag{4}$$

The TF-IDF score was computed for words after applying the syntactic filter where words other than nouns and adjectives were excluded. Words were ranked according to the score and top five high scored words were labeled as keywords.

3   Experiments and Results

3.1   Data Description

The dataset of the present study was drawn from MEDLINE research abstract database. MEDLINE is a high quality medical publication database supported by the National Library of Medicine (NLM) in the United States. One of the fields of an abstract record is the Other Term (OT) field that indicates author-supplied list of keywords. Not all MEDLINE abstracts have that field and the data was collected such that only abstracts with non-empty OT fields were included in the test dataset. A corpus of around 10,000 abstracts was collected. Text pre-processing steps were applied. For each abstract, a word graph was constructed and 3-and 4-graphlets were extracted for every word that pass the syntactic filter (only nouns and adjectives were considered). Other word types (e.g. adverbs, conjunctions) were included in the graphs to serve as hubs to connect other words, but no specific graphlets are generated for these category of words.

3.2   Performance Evaluation

A 5-fold cross validation was conducted for two systems. A system based on TF-IDF method was implemented as a baseline. The evaluation of both experimental and baseline systems were based on top five scoring candidate keywords. For the experimental model, for each document, the five top scoring keywords in the positive class (keyword) were considered if ($\hat{y}$) probability score is higher than that value of the negative class (non-keyword) according to Eq. 2. When, for a test document, the top five model scores given positive class labels were lower than scores corresponding to negative class labels, no keyword was generated.

A frequent pattern mining system was implemented to extract 3- and 4-graphlets based on a minimum support value of five (a graphlet has to appear in five documents to be selected as a feature). A probability model of graphlets conditionally independent on word class label (keyword vs. non-keyword) was build

Table 2. Keyword classification accuracy

System	Recall	Precision
TFIDF	0.55	0.73
graph-based	0.6	0.79

Table 3. A sample of graphlet patterns

Orbit Configuration	Form	Vertex Labels
$< 0, 1, 2 >$	O2	vascular,endothelial,growth
$< 0, 1, 2, 2 >$	O7	control,and,cervical,prevention
$< 0, 1, 2, 2 >$	O7	cervical,cancer,early,screening
$< 0, 1, 2, 2 >$	O7	overall,survival,(pfs),(os)
$< 0, 1, 1, 2 >$	O6	cancer,early,screening,detection

by counting pattern occurrences and normalization. A uniform prior probability (0.5) was used for $P(c_k)$. A small probability value $\epsilon = 1 \times 10^{-5}$ was used for unseen graphlet parameters. Recall and precision scores were computed based on the five top scoring candidate keywords in each test document. Performance results for the two systems are shown in Table 2.

As a by-product of the feature extraction process, a number of significant patterns were discovered. Here, we present some of the graphlet patterns that achieved high scores. A number of these patterns is shown in Table 3.

4   Discussion and Conclusions

This study addressed the problem of keyword identification in scientific abstracts using a novel graph-based method. An important advantage of the proposed method is the use of efficient enumeration of significant graphlet patterns in word graphs. A Bayes' rule-based model was applied to measure the statistical dependency between significant graphlet patterns and word class labels (e.g. keywords vs. non-keywords labels). The pivot vertices in the graphlets feature set represent candidate keywords. The proposed method was compared to standard, frequency-based method (TF-IDF) and experimental results showed competitive performance in terms of Recall and Precision.

This study advances the state-of-the-art of keyword identification methods by introducing an efficient method for exploring significant patterns in word graphs. Graph-based representation of text has an important advantage over traditional methods of text analysis. These graphs allow access to a wider context of the word vertex and hence enables capturing of non-local dependencies that might not be straightforward to obtain through syntactic and lexical analysis.

A major contribution of this study is an efficient algorithm for mining significant patterns in word graphs. While graph-based algorithms are known to be computationally expensive, the methods presented in this study enabled calculation of pattern frequency without resorting to testing of graph isomorphism. This was possible through usage of an orbit-based scheme representation of graphlets. This graphlet-based method is different from sequence-based methods (e.g. n-gram models) in two aspects. First, one graphlet pattern may contain words from multiple sentences in a given utterance (e.g. a text abstract), whereas sequence-based methods only contain words that co-occur within sentences. Second, sequence-based methods cannot capture some subtle topological patterns of a candidate keyword (e.g. as illustrated by the patterns shown in Fig. 1).

The presented method utilized a simple text structure, that is of words co-occurrences, to identify vertices and edges of text graphs. This simplicity of graph representation ignores the syntactic structure of sentences. One way to capture syntactic relationships is to identify graph vertices at a multi-word or phrase level (e.g. noun phrases). However, this might affect graph topology in a way that can prevent capturing of some significant, micro-level (e.g. word level) graph substructures. Another limitation of the presented method is lack of edge information (i.e. edges were unlabeled, unweighted). One potential future enhancement of the work would include, in addition to adjacency, some information such as bigram frequency and syntactic dependency.

While the method was designed to solve the problem of identification of word vertices that are keywords, as a by-product, a set of significant graphlet patterns are produced. These patterns can be useful in other tasks such as exploration, text categorization, and document clustering. The proposed probability model allows for quantification of pattern importance and thus these patterns are already measured as significant. Future work includes designing a text categorization method that rely on text graphlets.

References

1. Andrade, M.A., Valencia, A.: Automatic extraction of keywords from scientific text: application to the knowledge domain of protein families. Bioinformatics **14**(7), 600–607 (1998)
2. Matsuo, Y., Ishizuka, M.: Keyword extraction from a single document using word co-occurrence statistical information. Int. J. Artif. Intell. Tools **13**(1), 157–169 (2004)
3. Hammouda, K.M., Matute, D.N., Kamel, M.S.: CorePhrase: keyphrase extraction for document clustering. In: Perner, P., Imiya, A. (eds.) MLDM 2005. LNCS (LNAI), vol. 3587, pp. 265–274. Springer, Heidelberg (2005)
4. Hasan, K.S., Ng, V.: Automatic keyphrase extraction: a survey of the state of the art. In: ACL, pp. 1262–1273 (2005)
5. Daiber, J., Jakob, M., Hokamp, C., Mendes, P.N.: Improving efficiency and accuracy in multilingual entity extraction. In: Proceedings of the 9th International Conference on Semantic Systems, pp. 121–124. ACM (2013)
6. Witten, I.H., Paynter, G.W., Frank, E., Gutwin, C., Nevill-Manning, C.G.: KEA: practical automatic keyphrase extraction. In: Proceedings of the Fourth ACM Conference on Digital Libraries, pp. 254–255. ACM (1999)

7. Liu, F., Pennell, D., Liu, F., Liu, Y.: Unsupervised approaches for automatic keyword extraction using meeting transcripts. In: Proceedings of Human Language Technologies, pp. 620–628. Association for Computational Linguistics (2009)

8. Tomokiyo, T., Hurst, M.: A language model approach to keyphrase extraction. In: Proceedings of the ACL 2003 Workshop on Multiword Expressions: Analysis, Acquisition and Treatment, vol. 18, pp. 33–40. Association for Computational Linguistics (2003)

9. Zhang, Y., Zincir-Heywood, N., Milios, E.: Narrative text classification for automatic key phrase extraction in web document corpora. In: Proceedings of the 7th Annual ACM International Workshop on Web Information and Data Management, pp. 51–58. ACM (2005)

10. Mihalcea, R., Tarau, P.: TextRank: Bringing Order into Texts. Association for Computational Linguistics, Stroudsburg (2004)

11. Erkan, G., Radev, D.R.: LexRank: graph-based lexical centrality as salience in text summarization. J. Artif. Intell. Res. **22**, 457–479 (2004)

12. Bellaachia, A., Al-Dhelaan, M.: Ne-rank: a novel graph-based keyphrase extraction in twitter. In: Proceedings of the 2012 IEEE/WIC/ACM International Joint Conferences on Web Intelligence and Intelligent Agent Technology, vol. 01, pp. 372–379. IEEE Computer Society (2012)

13. Jiang, C., Coenen, F., Sanderson, R., Zito, M.: Text classification using graph mining-based feature extraction. Knowl. Based Syst. **23**(4), 302–308 (2010)

14. Blanco, R., Lioma, C.: Graph-based term weighting for information retrieval. Inf. Retrieval **15**(1), 54–92 (2012)

15. Pržulj, N.: Biological network comparison using graphlet degree distribution. Bioinformatics **23**(2), e177–e183 (2007)

16. Vacic, V., Iakoucheva, L.M., Lonardi, S., Radivojac, P.: Graphlet kernels for prediction of functional residues in protein structures. J. Comput. Biol. **17**(1), 55–72 (2010)

17. Bird, S.: NLTK: the natural language toolkit. In: Proceedings of the COLING/ACL on Interactive Presentation Sessions, pp. 69–72. Association for Computational Linguistics (2006)

18. Aizawa, A.: An information-theoretic perspective of TF-IDF measures. Inf. Process. Manage. **39**(1), 45–65 (2003)

Disentangling the Structure of Tables in Scientific Literature

Nikola Milosevic[1]([✉]), Cassie Gregson[2], Robert Hernandez[2],
and Goran Nenadic[1,3]

[1] School of Computer Science, University of Manchester,
Kilburn Building, Oxford Road, M13 9WJ Manchester, UK
{nikola.milosevic,g.nenadic}@manchester.ac.uk
[2] AstraZeneca Ltd, Riverside, Granta Park, Cambridge CB21 6GH, UK
[3] Health EResearch Centre, Manchester, UK

Abstract. Within the scientific literature, tables are commonly used to present factual and statistical information in a compact way, which is easy to digest by readers. The ability to "understand" the structure of tables is key for information extraction in many domains. However, the complexity and variety of presentation layouts and value formats makes it difficult to automatically extract roles and relationships of table cells. In this paper, we present a model that structures tables in a machine readable way and a methodology to automatically disentangle and transform tables into the modelled data structure. The method was tested in the domain of clinical trials: it achieved an F-score of 94.26 % for cell function identification and 94.84 % for identification of inter-cell relationships.

Keywords: Table mining · Text mining · Data management · Data modelling · Natural language processing

1   Introduction

Tables are used in a variety of printed and electronic documents for presenting large amounts of factual and/or statistical data in a structured way [1,21,25]. They are a frequent option in written language for presenting large, multi-dimensional information. For example, in experimental sciences, tables are usually used to present settings and results of experiments, as well as supporting information about previous experiments, background or definitions of terms. Tables are, in particular, widely used in the biomedical domain. However, while there have been numerous attempts to automatically extract information from the main body of literature [11,20,27], there have been relatively few attempts to extract information from tables.

One of the main challenges in table mining is that the existing models used to represent tables in the literature are focused on visualisation, rather than on content representation and mining. For example, tables in PubMedCentral (PMC)[1] are presented in XML with tags describing rows, cells, header and body

---

[1] http://www.ncbi.nlm.nih.gov/pmc/.

© Springer International Publishing Switzerland 2016
E. Métais et al. (Eds.): NLDB 2016, LNCS 9612, pp. 162–174, 2016.
DOI: 10.1007/978-3-319-41754-7_14

of the table. However, these tags are used only for formatting and there is no guarantee that cells labelled as headers are also semantically headers of the table. Therefore, the table layout structure and relationships between cells make preprocessing and decomposition necessary before machine understanding tasks can be performed.

Hurst [10] introduced five components of table processing: *graphical* (a basic graphical representation of the table, e.g. bitmap, rendered table on screen or paper), *physical* (a description of the table in terms of physical relationships between its basic elements when rendered on a page), *functional* (the purpose of areas of the table with respect to the use of the table by the reader), *structural* (the organisation of cells as an indication of the relationships between them), and *semantic* (the meaning of the text in the cell). Following Hurst, we differentiate five steps of table processing:

1. **Table detection** locates the table in document.
2. **Functional analysis** detects and marks functional areas of tables such as navigational (e.g. headers, stubs, super-row) and data cells.
3. **Structural analysis** determines the relationships between the cells. For each cell in the table, it finds related header(s), stub(s) and super-row cells.
4. **Syntactic analysis** looks at the value of the cells at the syntactic level, for example, by identifying whether the value is a numeric expression.
5. **Semantic analysis** determines the meaning of data and attempts to extract and represent specific information from tables.

In this paper, we focus on functional and structural analysis of tables in clinical literature available openly in PMC. The aim is to facilitate further syntactic and semantic processing of tables by providing a model to capture necessary information about cells' functions, relationships and content.

2 Background

There are three main areas of table processing in literature: (1) table detection, (2) functional analysis and (3) table mining applications, which include information retrieval, information extraction, question answering and knowledge discovery.

Table detection is a hard problem for graphical document formats because it may involve visual processing of the document in order to find visual structures recognisable as tables [2]. Other formats can be also challenging. For example, table tags exist in HTML, but they are often used for formatting web page layout. Previous work focused on detecting tables from PDF, HTML and ASCII documents using Optical Character Recognition [13], machine learning algorithms such as C4.5 decision trees [17] or SVM [19,22], and heuristics [26].

Functional analysis examines the purpose of areas of the table. The aim here is to differentiate between cells containing data and cells containing navigational information, such as headers, stubs or super-rows (see Fig. 1). To solve this problem, several machine learning methods like C4.5 decision trees [5,12],

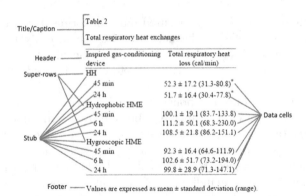

Fig. 1. Table components (PMC29053): Cell – the basic grouping within a table; Caption – a textual description of the table's content; Footer – additionally describes data or symbols used in the table; Column – a set of vertically aligned cells; Row – a set of horizontally aligned cells. Header – top-most row (or set of several top-most rows) that defines data categories of data columns; Stub or column header – typically the left-most column of the table that categorizes the data in the rows. Super-row – groups column headers and data by some concept.

conditional random fields [23] and hierarchical clustering [9] were used on web and ASCII text tables in general domain. Hurst also developed a system that is able to perform text mining analysis according to his model [10]. His system was composed of several components performing smaller tasks such as detection, functional analysis and structural analysis of tables by using rule based and/or machine learning approaches.

There are several *applications* that use tables. For example, the BioText Search engine [6,8] performs information retrieval from text, abstracts, figures and tables in biomedical documents. Wei et al. [23] created a question-answering system that looked for answers in tables, using CRF and information retrieval techniques. Few attempts were made to extract information using linked data and databases [16,24] with machine learning methods trained on a standardized set of tables [21]. There have been several approaches to semantically annotate columns using external resources, such as search engines [18] or linked data resources [14]. *WebTables* used a hypothesis that tables in web documents could be viewed as relational database tables [4]. They created a huge corpus database that consists of 154 million distinct relational tables from the world wide web that can be used for schema auto-complete, attribute synonym finding or joint graph traversal in database applications. Most of these approaches restricted themselves to simple tables because they lacked functional or structural processing steps.

Doush and Pontelli [7] created an spreadsheet *ontology* that included a model of table for screen reader's purposes. Their model considered tables that can be found in spreadsheets (usually simple matrix tables). They differentiate between data and header cells, but they do not include other possible cell roles such as stub and super-row cells.

3   Table Model

Since current table models focus mainly on visualization for human readers, we here propose a new model for computational processing which is comprised of two components:

- Table types: common table structural types that determine the way of reading the table;
- Data model: models the table structure and data in a way that data can be automatically processed by the machine (including visualisation).

3.1   Table Types

We define three main structural table types with several sub-types based on the table's dimensionality:

- **One-dimensional (list) tables** are described by a single label. The label is usually placed in the header (see Fig. 2 for an example). One-dimensional tables may have multiple columns, representing the same category, where multi-column structure is used for space saving purposes.
- **Two-dimensional** or *matrix tables* have data arranged into a matrix categorised usually by two labels: a column header and row header (stub). In our model, these tables may have multiple layers of column or row headers (see Fig. 4 for an example).
- **Multi-dimensional tables** contain more than two dimensions. We identify two types of multi-dimensional tables:
  - **Super-row tables** contain super-rows that group row headers below them (see example in Fig. 1). A super-row table can have multiple layers of super-rows, forming a tree-like structure. This structure is typically visually presented with an appropriate number of white spaces in front of each stub's label.
  - **Multi-tables** are tables composed of multiple, usually similar tables, merged into one table. In some cases, headers of concatenated tables inherit some categorisation from the header of the first table.

3.2   Data Model

The proposed data model captures necessary information for semantic understanding of tables to facilitate further processing and knowledge gathering. We have extended the spreadsheet ontology for tables [7] by adding entities that are not specific for navigation in screen readers, so it contains information that can aid text mining from tables and visualization.

The model has article, table and cell layers (see Fig. 3), which are arranged in a tree-like instantiation, with the *Article* node as the top element, containing the article information (i.e. title, reference, authors, text) and a list of tables. The article layer also stores where tables are mentioned within the document.

Table 2

Examples of the objectives of the cards

- Describe a situation before interpretation
- Devise the interpretation of a situation as a hypothesis
- Search for different interpretations of the same situation
- Identify the cognitive and behavioural consequences of the different hypotheses
- Search for a link between the interpretation given for a situation and a personal real-life experience
- Put the hypotheses in hierarchical order in terms of their probability
- Search for arguments for or against a hypothesis
- Conceive a way of testing a given hypothesis in reality

Fig. 2. Example of a list table (PMC161814)

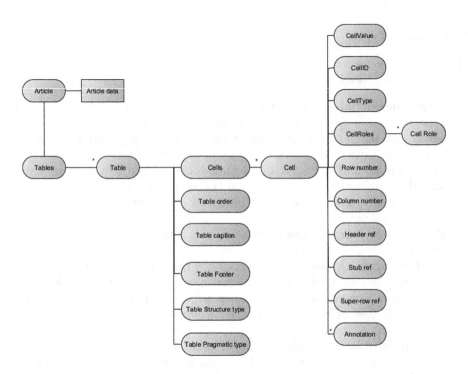

Fig. 3. Hierarchy of the proposed table model

Caption, footer, order of the table in the document and its structural type (dimensionality of the table, as defined above) are stored in the *table layer*. The table node also contains a list of table's cells. Finally, in the *cell layer*, the model stores the information about each cell including its value, function and position in the table. In the cell layer, the model also stores information regarding structural references to the navigational cells (headers, stubs and super-rows). The references to navigational cells are set by the ID of the closest cell in the navigational area. If navigational cells contain multiple layers, we apply a cascading style of cell referencing, where lower layers (closer to the data cell) reference the higher order layer (see Fig. 4). In this layer, the model further captures any

Table 1

Reported and audited outcome data

Trial phase Audited	Pre-intervention		Post-intervention	
	Intervention	Control	Intervention	Control
Patients seen	307	209	418	237
Referrals	56	39	80	63
Number of practices	27	25	27	25

Fig. 4. Example of cascading referencing of the header relationships (PMC270060). The cell with the value 56 is linked to the header "Intervention", which is linked to the super-header "Pre-intervention".

possible annotations of the cell content, which might be added during the table processing. For example, annotations may be syntactic, giving information about the type of value inside the cell, or semantic, mapping to a knowledge source (e.g. ontology or thesauri such as UMLS [3]). For each annotation, we record the span positions of annotated parts in content, concept names, ids in the lexicon or ontologies with which the text was annotated, name of the annotation knowledge source, its version, and environment description.

We note that spanning cells in our model are split and the content of a cell is copied to all cells that were created in the splitting process. Column, row numbers and cell ids are assigned after the splitting of spanning cells.

4 Methodology for Automatic Table Disentangling

We propose a methodology that automatically performs the functional and structural analysis of tables in PMC documents. The method uses a set of heuristic rules to disentangle tables and transform them into the previously described table model.

4.1 Identification of Functional Areas

The aim of functional analysis is to identify functional areas (headers, stubs, super-rows, data cells) within the table.

Header Identification. In most PMC documents, headers are marked using a *thead* XML tag. If *thead* tag exists, we assume that it is correctly labelled. For tables that do not have a header annotated, we examine syntactic similarity of the cells over the column. This is performed by using a window that takes five cells from the column and checks whether the content has the same syntactic type (i.e. string, single numeric value (e.g. 13) numeric expression (e.g. 5 ± 2), or empty). If all cells are of the same syntactic type, we assume that the table does not have a header. However, if the cells are syntactically different, for example, the first 2 cells are strings while the rest are numeric, we move the window down until it reaches the position where all cells in it have the same syntactic type.

The cells above the window are marked as header cells. The window size of five cells is chosen based on experimental experience. We have encountered tables that had up to four rows of headers, so the window size needs to be large enough to capture syntactic type differences. The algorithm then marks as header only rows with all cells marked as headers.

Another heuristic for determining header rows is to check whether some of the first row cells spans over several columns. If they do, we assume that the header contains the next rows, until we reach first one with no spanning cells.

Headers in multi-tables are usually placed between horizontal lines. Only the first header is usually marked with *thead* tags. If multiple cells between the lines have content, these cells are marked as header cells. However, if only one cell has content, these cells are classified as super-rows.

Stub Identification. Stubs or column headers are usually cells in the left most column. However, if cells in the left-most columns are row-spanning, the stub contains the next columns, until the first column with no spanning cells is identified.

Super-row Identification. Super-rows are rows that group and categorise stub labels. They can have multiple layers. In order to recognise super-rows, our method uses the following heuristics:

- A super-row can be presented as cells that span over the whole row. If these cells have regular content, they are labelled as super-rows.
- A tables may have multiple layers of super-rows. Authors usually present sub-group of relationships with leading blank spaces (indentation) at the beginning of the grouped elements. The number of blank spaces determines the layer of categorisation (i.e. the first layer has usually one blank space, the second has two, etc.). In other words, indentation level visually structures the super-row and stub layers. The row with a label that has less blank spaces than the labels in a stub below, is categorising them, and is therefore considered their super-row. Since there can be multiple levels of super-rows, we used a stack data structure in order to save the associated super-rows of the currently processed cell.
- In PMC documents, it is usual for spanning cells to be presented as a column with multiple cells where only one cell has content (usually the leading one). Rows with only one cell with content are labelled as super-rows.

4.2   Identification of Inter-cell Relationships

Once the functional areas are detected, we further attempt to identify relationships between cells. Using the detected functional areas, the method classifies tables into one of the four structural classes (one-dimensional, matrix, super-row, multi-table). This classification is based on a set of rules about the functional areas of the table. For example, if the table contains multiple headers, it is classified as multi-table. If it contains super-rows it is a super-row table. If table has only one dimension, it is a list table. Otherwise, it is matrix table.

Depending on the class, the method decides which relationships to search for. For example, data cells in one-dimensional tables can contain only headers, in matrix tables they contain relationships with stubs and headers, while in super-row tables they contain an additional relationships with the super-rows, which may be cascading with multiple layers. Data cells are related to header cells above, stub cells on the left and super-rows above. Navigational cells are related to the higher layers of navigational cells as defined in the cascading referencing model.

Detected functions and relationships of cells can be stored in an XML file and mySQL database according to our model.

5   Results and Evaluation

5.1   Data Set

We collected a data set by filtering PMC data for clinical trial publications. We mapped MEDLINE citations, with clinical trial publication type ("Clinical Trial", "Clinical Trial, Phase I", "Clinical Trial, Phase II", "Clinical Trial, Phase III" and "Clinical Trial, Phase IV"), to the PMC and extracted full text articles. The data set contains 2,517 documents in XML format, out of which 568 (22.6 %) had no tables in XML (usually containing only reference to a scanned image). The data set contains a total of 4,141 tables, with 80 cells per table on average.

Clinical trial publications are rich in tables, containing, on average, 2.4 tables per article. Biomedical literature, as a whole, contains on average 1.6 tables per article (30 % less than clinical literature).

5.2   Table Disentangling Performance

The system was able to process 3,573 tables from the data set (86 %). Table 1 presents the numbers of tables identified as belonging to different types. It is interesting that matrix tables make over 55 % of tables, along with super-row tables (over 42 %), while list and multi-table are quite rare (around 2 %).

We performed the evaluation of the functional and structural analyses on a random subset of 30 articles containing 101 tables. The evaluation sample contains tables from each table type and has been evaluated manually. The detailed information about the evaluation data set and the performance on structural table type recognition is given in Table 1.

The results for the functional and structural analyses are presented in Tables 2 and 3. Associations to the right roles and navigational relationships (headers, stubs, super-rows) were considered true positives (TP). Association to the non-existing roles or relationships were considered false positives (FP), while missing association were considered false negatives (FN). For the functional analysis, the method archived an F-score of 94.26 %, with the lowest performance on the identification of super-row areas for the multi-tables. Our results are comparable and better than previously reported. For example, Hurst [10] combined

Table 1. Overview of the dataset and accuracy of the recognition of structural table types

	Overall	List tables	Matrix tables	Super-row tables	Multi-table
Number of tables	3573	27 (0.76 %)	1974 (55.24 %)	1517 (42.46 %)	55 (1.54 %)
Number of evaluated	101	6	51	27	17
Accuracy of structural type recognition	92.07 %	100 %	96 %	96.3 %	70.6 %

Table 2. Evaluation of functional table analysis

	TP	FP	FN	Precission	Recall	F-Score
Cell role – header	**1041**	**35**	**260**	**96.70 %**	**80.00 %**	**87.60 %**
List	1	0	0	100.00 %	100.00 %	100.00 %
Matrix	469	9	0	98.10 %	100.00 %	99.00 %
Super-row	275	18	20	93.85 %	93.22 %	93.53 %
Multi-table	296	8	240	97.36 %	55.22 %	70.47 %
Cell role – stub	**1250**	**87**	**22**	**93.49 %**	**98.27 %**	**95.82 %**
List	0	0	7	N/A	N/A	N/A
Matrix	407	1	3	99.75 %	99.26 %	99.51 %
Super-row	488	17	4	96.63 %	99.10 %	97.89 %
Multi-table	355	69	8	83.72 %	97.79 %	90.22 %
Cell role – super-row	**414**	**102**	**66**	**80.23 %**	**86.25 %**	**83.13 %**
List	12	7	0	63.15 %	100.00 %	77.42 %
Super-row	359	26	27	93.24 %	93.00 %	93.12 %
Multi-table	43	63	37	40.57 %	53.75 %	46.24 %
Cell role – data	**3709**	**167**	**41**	**95.69 %**	**98.91 %**	**97.27 %**
List	31	7	6	81.57 %	83.78 %	82.66 %
Matrix	1438	1	12	99.93 %	99.17 %	99.55 %
Super-row	1517	11	21	99.28 %	98.63 %	98.95 %
Multi-table	723	148	2	83.00 %	99.72 %	90.60 %
Overall	**6414**	**391**	**389**	**94.25 %**	**94.28 %**	**94.26 %**

Naive Bayes, heuristic rules and pattern based classification archiving F-score of around 92 % for functional analysis. Similarly, Tengli et al. [21] reported F-score of 91.4 % for the table extraction task in which they recognised labels and navigational cells, while Wei et al. [23] reported an F-measure of 90 % for detecting headers using CRF. Cafarella et al. [4] detected navigational cells with precision

Table 3. Evaluation of structural table analysis

	TP	FP	FN	Precission	Recall	F-Score
References – header	**5402**	**768**	**47**	**87.55 %**	**99.13 %**	**92.98 %**
List	7	0	0	100.00 %	100.00 %	100.00 %
Matrix	2076	15	3	99.30 %	99.85 %	99.60 %
Super-row	2501	61	6	97.61 %	98.63 %	98.95 %
Multi-table	818	692	38	54.17 %	95.56 %	69.15 %
References – stub	**4982**	**147**	**0**	**97.10 %**	**100.00 %**	**98.55 %**
Matrix	1788	14	0	99.22 %	100.00 %	99.61 %
Super-row	2057	95	0	95.58 %	100.00 %	97.74 %
Multi-table	1137	38	0	96.70 %	100.00 %	98.35 %
References – Super-row	**1663**	**78**	**269**	**95.52 %**	**86.07 %**	**90.55 %**
List	29	0	6	100.00 %	82.85 %	90.62 %
Super-row	1456	66	215	95.66 %	87.13 %	91.12 %
Multi-table	178	12	42	93.68 %	80.91 %	86.82 %
Overall	**12047**	**993**	**316**	**92.38 %**	**97.44 %**	**94.84 %**

and recall not exceeding 89 % and Jung et al. [12] reported 82.1 % accuracy in extracting table headers.

For the task of structural analysis, the system achieved an F-score of 94.84 %. For comparison, Hurst's system performed with 81.21 % recall and 85.14 % precision. It is also important to note that input data in Hurst's system were perfectly formatted, while the PMC data is often not. To the best of our knowledge, there is no other system that attempted to performing the combined task of functional and structural table analysis.

During the error analysis, we identified misleading mark-up and complex tables unique to a specific paper in the evaluation set as the main reasons for errors. In PMC documents, XML mark-up features such as spanning cells, head tags, and breaking lines are often misused to make tables look visually appealing. Although we have applied some heuristics that can overcome some of the issues, some of the misleading XML labelling remains challenging. Furthermore, there are tables that are not only complex in structure, but their structure is unique to specific paper, and thus difficult to generalise. Our method made significant number of errors on multi-tables, since it is challenging to determine whether a row is a new header or just an emphasized row or super-row just by analysing XML structure. Errors in wrongly recognizing headers or super-rows cause high amount of false links in structural analysis, since relationships in the subsequent rows will be wrongly annotated. However, multi-tables are relatively rare, so this did not heavily affect the overall results.

6   Conclusion

In this paper we have presented a model to computationally represent tables found in scientific literature. We also presented a domain-independent methodology to disentangle tables and add annotations about functional areas and relationships between table cells. The evaluation has shown that the table structure can be identified with high F-scores (above 94 %) which is encouraging. Even though we performed evaluation on the PMC clinical trial documents, the proposed approach can be extended to HTML or any other XML-like format. The Implementation of the method is available at https://github.com/nikolamilosevic86/TableAnnotator.

Although the results for these steps are encouraging, there are still a number of challenges in table mining, mainly in the semantic analysis of the cell content and the methods to query and retrieve table data.

The proposed model can serve as a basis to support applications in information retrieval, information extraction and question answering. We have already performed several information extraction experiments [15] and in the future we are planning to develop a general methodology for information extraction from tables in biomedical literature that uses the presented approach as its basis. Our methodology can be also used as a basis for semantic analysis and querying of tables. In addition, the model can aid systems in accessibility domain. For example, screen readers for visually impaired people could enable easy navigation through tables by providing information about cell's relationships and functions.

Acknowledgments. This research is funded by a doctoral funding grant from the Engineering and Physical Sciences Research Council (EPSRC) and AstraZeneca Ltd.

References

1. Alley, M.: The Craft of Scientific Writing. Springer Science & Business Media, New York (1996)
2. Attwood, T.K., Kell, D.B., McDermott, P., Marsh, J., Pettifer, S., Thorne, D.: Utopia documents: linking scholarly literature with research data. Bioinformatics **26**(18), i568–i574 (2010)
3. Bodenreider, O.: The unified medical language system (UMLS): integrating biomedical terminology. Nucleic Acids Res. **32**(Suppl 1), D267–D270 (2004)
4. Cafarella, M.J., Halevy, A., Wang, D.Z., Wu, E., Zhang, Y.: Webtables: exploring the power of tables on the web. Proc. VLDB Endowment **1**(1), 538–549 (2008)
5. Chavan, M.M., Shirgave, S.: A methodology for extracting head contents from meaningful tables in web pages. In: 2011 International Conference on Communication Systems and Network Technologies (CSNT), pp. 272–277. IEEE (2011)
6. Divoli, A., Wooldridge, M.A., Hearst, M.A.: Full text and figure display improves bioscience literature search. PloS One **5**(4), e9619 (2010)
7. Doush, I.A., Pontelli, E.: Non-visual navigation of spreadsheets. Univ. Access Inf. Soc. **12**(2), 143–159 (2013)

8. Hearst, M.A., Divoli, A., Guturu, H., Ksikes, A., Nakov, P., Wooldridge, M.A., Ye, J.: Biotext search engine: beyond abstract search. Bioinformatics **23**(16), 2196–2197 (2007)
9. Hu, J., Kashi, R., Lopresti, D., Wilfong, G.: A system for understanding and reformulating tables. In: Proceedings of the Fourth IAPR International Workshop on Document Analysis Systems, pp. 361–372 (2000)
10. Hurst, M.F.: The interpretation of tables in texts. Ph.D. Thesis, University of Edinburgh (2000)
11. Jensen, L.J., Saric, J., Bork, P.: Literature mining for the biologist: from information retrieval to biological discovery. Nat. Rev. Genet. **7**(2), 119–129 (2006)
12. Jung, S.W., Kwon, H.C.: A scalable hybrid approach for extracting head components from web tables. IEEE Trans. Knowl. Data Eng. **18**(2), 174–187 (2006)
13. Kieninger, T., Dengel, A.R.: The T-Recs table recognition and analysis system. In: Lee, S.-W., Nakano, Y. (eds.) DAS 1998. LNCS, vol. 1655, pp. 255–270. Springer, Heidelberg (1999)
14. Limaye, G., Sarawagi, S., Chakrabarti, S.: Annotating and searching web tables using entities, types and relationships. Proc. VLDB Endowment **3**(1–2), 1338–1347 (2010)
15. Milosevic, N., Gregson, C., Hernandez, R., Nenadic, G.: Extracting patient data from tables in clinical literature: Case study on extraction of BMI, weight and number of patients. In: Proceedings of the 9th International Joint Conference on Biomedical Engineering Systems and Technologies (BIOSTEC 2016), vol. 5, pp. 223–228 (2016)
16. Mulwad, V., Finin, T., Syed, Z., Joshi, A.: Using linked data to interpret tables. In: Proceedings of the First International Conference on Consuming Linked Data, vol. 665, pp. 109–120. CEUR-WS.org (2010)
17. Ng, H.T., Lim, C.Y., Koo, J.L.T.: Learning to recognize tables in free text. In: Proceedings of the 37th Annual Meeting of the Association for Computational Linguistics on Computational Linguistics, pp. 443–450. ACL (1999)
18. Quercini, G., Reynaud, C.: Entity discovery and annotation in tables. In: Proceedings of the 16th International Conference on Extending Database Technology, pp. 693–704. ACM (2013)
19. Son, J.W., Lee, J.A., Park, S.B., Song, H.J., Lee, S.J., Park, S.Y.: Discriminating meaningful web tables from decorative tables using a composite kernel. In: 2008 IEEE/WIC/ACM International Conference on Web Intelligence and Intelligent Agent Technology, WI-IAT 2008, vol. 1, pp. 368–371. IEEE (2008)
20. Spasić, I., Livsey, J., Keane, J.A., Nenadić, G.: Text mining of cancer-related information: review of current status and future directions. Int. J. Med. Inf. **83**(9), 605–623 (2014)
21. Tengli, A., Yang, Y., Ma, N.L.: Learning table extraction from examples. In: Proceedings of the 20th International Conference on Computational Linguistics, pp. 987–994. ACL (2004)
22. Wang, Y., Hu, J.: A machine learning based approach for table detection on the web. In: Proceedings of the 11th International Conference on World Wide Web, pp. 242–250. ACM (2002)
23. Wei, X., Croft, B., McCallum, A.: Table extraction for answer retrieval. Inf. Retrieval **9**(5), 589–611 (2006)
24. Wong, W., Martinez, D., Cavedon, L.: Extraction of named entities from tables in gene mutation literature. In: Proceedings of the Workshop on Current Trends in Biomedical Natural Language Processing, pp. 46–54. ACL (2009)

25. Yesilada, Y., Stevens, R., Goble, C., Hussein, S.: Rendering tables in audio: the interaction of structure and reading styles. In: ACM SIGACCESS Accessibility and Computing, pp. 16–23. No. 77–78. ACM (2004)
26. Yildiz, B., Kaiser, K., Miksch, S.: pdf2table: a method to extract table information from pdf files. In: IICAI, pp. 1773–1785 (2005)
27. Zhu, F., Patumcharoenpol, P., Zhang, C., Yang, Y., Chan, J., Meechai, A., Vongsangnak, W., Shen, B.: Biomedical text mining and its applications in cancer research. J. Biomed. Inf. **46**(2), 200–211 (2013)

Extracting and Representing Higher Order Predicate Relations between Concepts

Sanjay Chatterji$^{(\boxtimes)}$, Nitish Varshney, Parnab Kumar Chanda, Vibhor Mittal, and Bhavi Bhagwan Jagwani

Samsung R&D Institute India, Bangalore, India
{sanjay.chatt,nitish.var,p.chanda,vibhor.m21,b.jagwani}@samsung.com

Abstract. In a text, two concepts can hold either direct or higher order relationship where function of some concepts is considered as another concept. Essentially, we require a mechanism to capture complex associations between concepts. Keeping this in view, we propose a knowledge representation scheme which is flexible enough to capture any order of associations between concepts in factual as well as non-factual sentences. We utilize a five-tuple representation scheme to capture associations between concepts and based on our evaluation strategy we found that by this we are able to represent 90.7 % of the concept associations correctly. This is superior to existing pattern based methods. A use case in the domain of content retrieval has also been evaluated which has shown to retrieve more accurate content using our knowledge representation scheme thereby proving the effectiveness of our approach.

1 Introduction

Knowledge representation is a well researched topic. Popular methods of knowledge representations include predicate logic (propositional or zero-order, first-order, higher order), semantic network and frame. Predicate logic is very popular when we wish to capture the expressiveness of a natural language text ranging from less expressive (propositional) to more expressive (higher order) logic. Predicate logic is useful when we wish to capture concepts in terms of functions and arguments. In higher order predicate logic, the arguments can themselves be functions(predicates) thereby allowing us to capture more complex expressions. Semantic network is a graphical way of representing concepts and the relationship between concepts. This is a very intuitive way of knowledge representation and is almost similar to how humans tend to store, retrieve and manipulate knowledge. In semantic network, the associations between concepts are arbitrary: without contextual information. It is hard to interpret without a context. Consider the text: *The ideal voltmeter is a potentiometer.* From here, semantic network captures the relation as (potentiometer is-A voltmeter). Similarly, from other texts it may find that (Micro-ammeter is-A voltmeter) and (Electrostatic-voltmeter is-A voltmeter). Using these relations we cannot infer what is an "ideal" voltmeter unless the context *ideal* is taken into consideration while capturing the relations.

© Springer International Publishing Switzerland 2016
E. Métais et al. (Eds.): NLDB 2016, LNCS 9612, pp. 175–186, 2016.
DOI: 10.1007/978-3-319-41754-7_15

To this end, we propose a knowledge representation scheme that utilizes the advantage of both higher order predicate logic and semantic network. With this we can represent arbitrary relation between concepts using complex terms. A *Complex term* is defined a unit of knowledge which encapsulates one or more concept terms and their direct or hierarchical relations or functions. For example, the complex term "The ideal voltmeter is a potentiometer" holds a binary relation between a concept term and a function of another concept term, which can be represented as *is-A(potentiometer, ideal(voltmeter))*. To capture these complex relationships we have introduced a five-tuple knowledge representation scheme that gives us enough flexibility to capture from the simplest relations to complex higher order relations or functions between concept terms. Using these higher order functions, we are able to derive more relations by inferencing. Again, these representations allow us to capture the nonfactual or conditional relations that exist in a sentence. This can be used in answering the nonfactual questions.

Our work is presented as follows. In Sect. 2, we discuss some related work. In Sect. 3, we present the system architecture and describe the methodology which takes natural language text as input, generates tuples and stores in a relational database. Additionally, we also use clustering to normalize the extracted relations that help us in inferring relations under similar context. Further, we also present a mechanism for visualizing the relationship between concepts following a hierarchical graphical structure. In Sect. 4, we describe the evaluation strategy for our representation scheme. We present the accuracy of our tuple representation and also evaluate a use case in the domain of content retrieval for education text materials utilizing our proposed knowledge representation scheme.

2   Related Work

Semantic roles in different parts of a sentence are annotated by the Semantic Role Labeling (SRL) tools. The LTH-SRL of [1] is a SRL tool that gives predicate/argument structure of input sentences based on NomBank [2] and PropBank [3] semantic role structures. Given an input sentence this tool returns the POS and lemma of each word, existing dependency relations between words, verbal predicates with their arguments (as annotated in PropBank) and nominal predicates with their arguments (as annotated in NomBank).

There are four general schemes for representing knowledge, namely, logic, semantic network, production rules and frames. There are a few comparative studies on different representation schemes [4,5]. Each of these schemes has specific usage in some particular application. Some uses of different representation schemes are discussed by Clark [6]. Theoretically, these four knowledge representation schemes are equivalent. Anything which can be represented by one scheme can also be represented by another scheme.

Logical representation schemes can be arranged in terms of their expressiveness. Propositional logic is the most restrictive logical representation scheme [7] which is able to represent sentences which can either be true or false.

This scheme uses propositions, connectives, brackets and True/False symbols and the syntactic rules of the language. First order predicate logic is more expressive as along with symbols and rules of propositional logic it also allows constants, variables, predicates, functions, quantifiers and semantic rules of the language [8]. Ali and Khan [5] have given first order predicate logic representations of different syntactic relations of linguistic phrases. Their algorithm splits the sentences into phrases and based on the patterns of the phrase categories and some other factors they have identified 36 different types of representations. They are able to capture 80 % of sentences in these representation types.

Logical relations between concepts can best be represented graphically using Ontology. Uschold and King [9] and Gomez and Perez [10] gave a skeleton methodology for building ontology. These techniques are then extended and refined in multiple ways [11–13]. The Suggested Upper Merged Ontology [14] is the largest formal ontology in existence today. Though, it began as just an upper level ontology encoded in first order logic but subsequently its logic has been expanded to also include some higher-order elements of the logical relations.

3   Overview of the Work with System Architecture

In simple natural text, Ali and Khan [5] have manually identified different patterns of grammatical categories of words and phrases. In complex natural text, each argument of a relation may have a hierarchy of such patterns. It is very difficult, if not impossible, to manually create all such hierarchy of patterns in complex natural text. In this paper, we are trying to extract all levels of syntactic and semantic relations semi-automatically and represent them using a common higher order predicate logic knowledge representation framework. We use the term *relation* for indicating both function and relation to make it consistent. Such relations can be unary function of a concept term, binary relation between two concept terms or hierarchy of functions and relations of concept terms of any depth.

There are several steps in this task. The first step is to collect documents of a domain and run Semantic and Syntactic Role Labeling (SSRL) tools. The concept terms are identified using the index of the books of that domain. Then, we extract different levels of relations between the concept terms and filter out the relations where no concept term is present. The filtered relations are called complex terms.

We design a unique representation scheme for representing such hierarchy of relations between concept terms in a five-tuple format and store them in a database. The relations in the database are clustered into groups such that similar types of relations are in same group. We store the cluster information for each relation in the database. We have also represented these higher order predicate logic relationships using our higher order graphical network. The system architecture is shown in Fig. 1. Each step of the system is described in details in the following subsections.

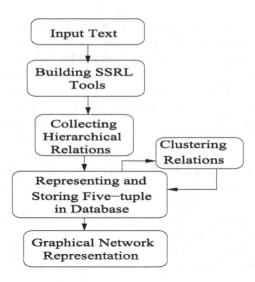

Fig. 1. System architecture for extracting and representing higher order predicate logic.

3.1 Building Syntactic and Semantic Role Labeling (SSRL) Tools

The LTH SRL parser [1] identifies the predicate relations and their arguments. But, each predicate relation is returned as an independent relation. Similarly, the syntactic parser also identifies the relations between the words. Olney et al. [15] have used the LTH SRL and syntactic parser outputs for identifying the relations between the concepts and stored them in ConceptMap. They have considered all the predicate relations given by LTH SRL parser and extracted is-a, type-of and location relations from syntactic parser. They have also represented these relations independently, as the ConceptNet does not have the power to represent hierarchical relations. We have used this technique to extract the same set of predicate relations and their arguments.

Hearst [16] has written some grammatical rules for identifying the hyponym relations between the phrases. They have identified six lexico-syntactic patterns that capture all possible hyponymy relations in simple English phrases. We have extracted these relations and combined them with the is-a relation. We have also combined the prepositions and other words (upto a certain distance) following the relation as suggested by [17].

Most of these extractors extract relations from factual sentences. There is scope in identifying more rules for extracting relations especially from non-factual sentences. By observing these gaps, we have implemented a few rules for extracting If-Then relation, Mathematical relation and Of-Nmod relation. The If-Then relation is extracted from the compound sentences with the following patterns. Here, X and Y should be complete phrases which are related by the If-Then relation.

Patterns of If-Then relation:

- If X Then Y
- If X, Y
- When X, Y
- Y when X

The Mathematical relations include proportional, directly proportional, inversely proportional, reverse, ratio, etc. Such relations are extracted from simple sentences with the following patterns. Here, X and Y are related by the corresponding Mathematical relation.

Patterns of Mathematical relation:

- X is proportional/directly proportional/inversely proportional to Y
- X is reverse of Y
- ratio of X and Y
- X divided by Y

We have also created rules for extracting OF and Nmod (Of-Nmod) relations. The corresponding patterns are given below. The Nmod relation is based on the occurrence of the Nominal MODifier (NMOD) dependency relation between the concepts. We have shown the dependency relation in angle bracket. In the following patterns, X, Y and Z are nouns or noun phrases. The phrase without concept term is the relation of the phrases containing the concept term.

Patterns of Of-Nmod relation:

- X of Y
- X of Y and Z
- X <NMOD> Y

3.2   Collecting Hierarchical Relations from SSRL Tools

Now, to get the hierarchy of relations, we have considered the starting and ending points of the arguments of each relation. If a relation (R1) and all its arguments are contained in an argument of another relation (R2) then we consider R1 is inside R2. We combine each such pair or relations, iteratively. One example sentence containing three predicate relations is shown below. First the relations 'call' and 'IN' are combined and then the combined relation is combined with 'word' and 'derive_from'.

Example 1:

- Input Sentence: The word magnet is derived from an island in Greece called Magnesia.
- Predicate and Arguments given by LTH-SRL: derive_from(word magnet, island in Greece called Magnesia), call(island in Greece, Magnesia)
- Predicate and Arguments extracted using [15]: IN(island, Greece)
- Predicate and Arguments extracted using our NMOD rule: word(magnet)

- Hierarchical relation combining derive_from and call: derive_from(word magnet, call(island in Greece, Magnesia))
- Hierarchical relation combining derive_from, call and word: derive_from (word(magnet), call(island in Greece, Magnesia))
- Hierarchical relation combining derive_from, call, word and IN: derive_from (word(magnet), call(IN(island, Greece), Magnesia))

3.3   Representing Hierarchical Relations Using Five-Tuple and Storing in Database

In complex natural text, a complex term is related to another complex term by a relation. We represent such knowledge in the database using five-tuple (C1, T1, C2, T2, R), where C1 and C2 are complex terms of type T1 and T2, respectively. R is the relation between the complex terms. The type of the complex term can be a concept term ($T = 0$) or a unary (T=1) or binary (T=2) relation of complex terms. We use underscore (_) where the corresponding field is not applicable. Each five-tuple itself is a complex term.

In the five tuple representation, the innermost complex term is represented first. We attach a global unique identifier to this complex term as a reference. This unique identifier is reused across the five-tuples where the particular complex term is used. The complex term "Modulus of elasticity is called Young's Modulus" is represented by the following five-tuples.

4. (elasticity, 0, _, _, Modulus)
5. (4, 1, Young's Modulus, 0, call)

The fourth five-tuple means 'Modulus' is the unary relation of the concept term 'elasticity'. The fifth five-tuple means 'call' is the binary relation of 4th complex term and the concept term 'Young's Modulus'. Similarly, the complex sentence of Example 1 is represented by the following five-tuples.

6. (magnet, 0, _, _, word)
7. (island, 0, Greece, 0, IN)
8. (7, 2, Magnesia, 0, call)
9. (6, 1, 8, 2, derive_from)

The sixth and seventh five-tuples mean that the 'word' is the unary relation of the concept term 'magnet' and the 'IN' is the binary relation of the concept terms 'island' and 'Greece', respectively. The eighth five-tuple means 'call' is the binary relation of 7th complex term and the concept term 'Magnesia'. The ninth five-tuple means 'derive_from' is the binary relation of 6th and 8th complex terms.

3.4   Relation Clustering

Consider a factual statement "Ribosome is responsible for protein synthesis in cells". With the use of the SSRL tool, we are able to capture a relation represented as *responsible(Ribosome, IN(protein synthesis, cells))*. Using this representation, we should be easily able to reason with queries which ask for the role, function or purpose of Ribosomes. For this purpose, we should interpret responsible, role, function and purpose under same type. We can achieve this by clustering the contextually similar relations stored in the database.

Olney et al. [15] have manually clustered the extracted relations from Biology book and manually mapped the clusters with a fixed set of relation names. They did not consider goal driven relationships and logical patterns as these are not appropriate for the biology domain. But, it is not possible to manually judge the type of logical relations. We have discussed below the steps of a semi-automatic process we have followed for identifying the types of logical relations stored in the database.

We have first collected the pre-trained vectors trained on a part of Google News dataset. The model contains 300-dimensional vectors for 3 million words and phrases. With the help of this model, we have calculated the vector for each relation word using word2vec tool[1]. Then, the cosine similarities of the vectors for each pair of logical relation words are used to measure the similarity for clustering the logical relations. We have used the WEKA tool [18] for this clustering implementation.

We further expanded each cluster using Roget's thesaurus[2] to cover words that may not be seen in the corpus vocabulary. For an input relation word, which may not still exist in the relation clusters we use the following heuristic to find the semantically closest cluster. We compute the average vector for each cluster and the vector of the unseen relation word. We assign the unseen relation word to the cluster whose average vector has the highest cosine similarity with the vector of the unseen relation word.

3.5 Graphical Network Representation

We have also designed a scheme for representing the higher order predicate relationships graphically. There are two levels of representation. At the first level, similar to the ConceptNet representation scheme [19], we have considered concept terms as nodes. When there is a logical relation between two terms then we give an edge between the corresponding concepts. The first level representation of the sentence in Example 1 is shown in Fig. 2.

There can be multiple types of relations between concept terms. We are not able to show the exact relation at the first level graph. The exact logical relation between the concepts is represented at the second level. The second level graphical representation of the sentence in Example 1 is shown Fig. 3. Here, White nodes indicate concept terms, Gray nodes indicate unary relation and Black nodes indicate binary relation.

---

[1] http://deeplearning4j.org/word2vec.html.
[2] http://www.thesaurus.com/.

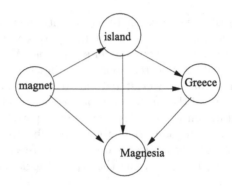

Fig. 2. First level graphical representation of the sentence in Example 1.

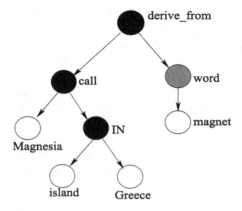

Fig. 3. Second level graphical representation of the sentence in Example 1.

4   Evaluation

In this section we will first discuss the strategy we follow to evaluate the effectiveness of our work. It will also help in understanding the real need of such representation scheme. We have discussed this strategy using three real example statements, three questions and their five-tuple representations. Finally, the actual results are displayed.

4.1   Evaluation Strategy

The above stated knowledge representation mechanism can best be used in suggesting content for a question posed by a student. Our evaluation strategy is based on this usecase. Consider the following three statements.

- Statement 1: Glass is made up of silica.
- Statement 2: When glass is heated it melts.
- Statement 3: When glass falls it breaks.

Now, consider the following questions.

- Question 1: What is the temperature at which glass will melt?
- Question 2: What happens when you heat glass?
- Question 3: What will happen if glass is warmed up at high pressure?

If we represent the above three statements by the simple two-tuple representations (glass, silica), (glass, melt) and (glass, break) then given any question on 'glass' the content retrieval engine based on this knowledge base will return all three statements as output. So, it is clear that we have to take relation names into consideration, for reasoning.

Now, suppose we represent them using triplets make(glass, silica), heat(glass, melt) and fall(glass, break) as is usually done in ConceptNet and Ontology. Then for each of the first two questions which are represented as the triplets ?(glass,melt) and heat(glass, ?), respectively, the second statement will be given as a relevant content. For the third question which is represented as warm_up(glass, ?) it will not be able to suggest any content. This is because, *warm_up* and *heat* did not match.

Consider, [S 10 - S 16] proposed five-tuple representations for the above three statements. These numbers are preceded by S to denote statement. The above three questions are also processed via the same pipeline as that of the text and similar five-tuple representations were generated which are shown using [Q 17 - Q 22]. These numbers are preceded by Q to denote question. Here, the question mark (?) denotes an unknown concept which needs to be inferred following our hierarchy of relations present in our knowledge base.

S 10. (glass, 0, silica, 0, make)
S 11. (glass, 0, _, _, heat)
S 12. (glass, 0, _, _, melt)
S 13. (11, 1, 12, 1, if-then)
    {heat(glass) ==> melt(glass)}
S 14. (glass, 0, _, _, fall)
S 15. (glass, 0, _, _, break)
S 16. (14, 1, 15, 1, if-then)

Q 17. (glass, 0, ?, _, temperature)
Q 18. (17, 2, 12, 1, if-then)
Q 19. (11, 1, ?, _, if-then)
Q 20. (pressure, 0, _, _, high)
Q 21. (glass, 0, 20, 1, warm)
Q 22. (21, 2, ?, _, if-then)

Using these five-tuple representations we are able to return some statement for a question. For example, the question five-tuple $Q19$ has a direct match with the statement five-tuple $S13$. Therefore, Statement 2 is returned as the relevant content for the Question 2.

4.2   Results and Discussions

We generate our knowledge base from K-12 physics book of NCERT[3], India, using our above stated methodology. There are 20,000 sentences in our dataset. We have also downloaded and cleaned the index of that book to be used as concept list. We, then, choose a set of 100 questions whose information need can be answered from the dataset. The questions are also processed via the same pipeline and the similar five-tuple representations were generated for them. We extracted the top 1000 relations from our corpus. We generated around 90 clusters using our clustering mechanism.

We are able to represent 90.7 % of the relations that exist between the concepts in natural sentences. This result shows that it is superior to the existing first order predicate logic based representation scheme of [5] which is able to represent 80 % of the sentences.

We compare our content search system with standard TF-IDF based Baseline System which returns ordered results based on keyword match and TF-IDF ranking scheme. We have considered the top three results for comparison. Then, we have incorporated relation clusters in the TF-IDF based search engine to prepare the Modified System. In this Modified System recall is high as the system is able to search the relevant contents where the exact relation name is not present (similar relation name is present). For example, we have put *heat* and *warm* under one cluster. Hence, our representation scheme returns the second statement as a relevant content for only the 2nd and 3rd questions in Sect. 4.1. Here, we do not care the "high pressure" phrase in Question 3.

Then, we have used the five-tuple representation to build our Final System. We have used these five-tuples for reasoning. In this Final System, precision @ K (K=3) is increased as the query is more restrictive. Therefore, the result contains lesser irrelevant contents. For example, for the third question in Sect. 4.1 our final system does not suggest any content. The results of the TF-IDF based Baseline system, modified system and final advanced system for content search task are shown using Precision, Recall and F-measure values in Table 1.

Table 1. Evaluation result of the baseline TF-IDF based and our advanced content retrieval engine.

System	Precision	Recall	F-measure
Baseline system with TF-IDF based ranking	0.72	0.67	0.69
Modified system after adding relation clustering	0.68	0.79	0.73
Our final system after reasoning through five-tuples	0.86	0.74	0.79

---

[3] ncert.nic.in/ncerts/textbook/textbook.htm.

5   Conclusion

In this paper, we have presented a knowledge representation scheme for capturing higher order relations between concepts. We have developed an approach where we use syntactic and semantic analysis to extract such higher order relations. For visual interpretation of the complex associations between concepts, we have presented a hierarchical graphical structure. Finally, we have demonstrated the effectiveness of our representation scheme by evaluating a use case in the domain of content retrieval for education materials.

The future research work is focused on the following fields: (1) Increasing the set of rules to extract more relations (2) Exploring the possibility to extend the use case from trivial content retrieval to more focused question-answering system.

Acknowledgments. This research was supported by Samsung R & D Institute India - Bangalore. We thank our colleagues Mr. Tripun Goel, Mr. Krishnamraju Murali Venkata Mutyala, Mr. Chandragouda Patil, Mr. Ramachandran Narasimhamurthy, Mr. Srinidhi Nirgunda Seshadri, Dr. Shankar M. Venkatesan for providing insight and expertise that greatly assisted the research. We thank them for comments to improve the manuscript.

References

1. Johansson, R., Nugues, P.M.: Dependency-based syntactic-semantic analysis with PropBank and NomBank. In: Proceedings of CoNNL, pp. 183–187 (2008)
2. Meyers, A., Reeves, R., Macleod, C., Szekely, R., Zielinska, V., Young, B., Grishman, R.: The nombank project: an interim report. In: Meyers, A. (ed.) HLT-NAACL 2004 Workshop: Frontiers in Corpus Annotation, pp. 24–31. Association for Computational Linguistics, Boston, Massachusetts (2004)
3. Palmer, M., Gildea, D., Kingsbury, P.: The proposition bank: an annotated corpus of semantic roles. Comput. Linguist. **31**, 71–106 (2005)
4. Colton, S.: Knowledge representation (2007). http://tinyurl.com/gozhw5t. Accessed 21 Feb 2016
5. Ali, A., Khan, M.A.: Selecting predicate logic for knowledge representation by comparative study of knowledge representation schemes. In: International Conference on Emerging Technologies, ICET 2009, pp. 23–28 (2009)
6. Clark, P.: Knowledge representation in machine learning. Mach. Hum. Learn. 35–49 (1989)
7. Brachman, R.J., Levesque, H.J., Reiter, R.: Knowledge Representation. MIT Press, Cambridge (1992)
8. Mylopoulos, J.: An overview of knowledge representation. SIGART Bull. 5–12 (1980)
9. Uschold, M., King, M.: Towards a methodology for building ontologies. Citeseer (1995)
10. Gmez-Prez, A., Fernndez-Lpez, M., de Vincente, A.: Towards a method to conceptualize domain ontologies. In: Proceedings of ECAI-1996 Workshop on Ontological Engineering (1996)

11. Uschold, M.: Converting an informal ontology into ontolingua: some experiences. Technical report, University of Edinburgh Artificial Intelligence Applications Institute AIAI TR (1996)
12. Uschold, M.: Building ontologies: towards a unified methodology. Technical report, University of Edinburgh Artificial Intelligence Applications Institute AIAI TR (1996)
13. Fernndez-Lpez, M., Gmez-Prez, A., Juristo, N.: Methontology: from ontological art towards ontological engineering. In: Proceedings of Symposium on Ontological Engineering of AAAI (1997)
14. Pease, A., Niles, I., Li, J.: The suggested upper merged ontology: a large ontology for the semantic web and its applications. In: Working Notes of the AAAI-2002 Workshop on Ontologies and the Semantic Web, vol. 28 (2002)
15. Olney, A.M., Cade, W.L., Williams, C.: Generating concept map exercises from textbooks. In: Proceedings of the 6th Workshop on Innovative Use of NLP for Building Educational Applications, IUNLPBEA 2011, pp. 111–119. Association for Computational Linguistics, Stroudsburg (2011)
16. Hearst, M.A.: Automatic acquisition of hyponyms from large text corpora. In: Proceedings of the 14th Conference on Computational Linguistics, COLING 1992, vol. 2, pp. 539–545. Association for Computational Linguistics, Stroudsburg (1992)
17. Fader, A., Soderland, S., Etzioni, O.: Identifying relations for open information extraction. In: Proceedings of the Conference on Empirical Methods in Natural Language Processing, EMNLP 2011, pp. 1535–1545. Association for Computational Linguistics, Stroudsburg (2011)
18. Hall, M., Frank, E., Holmes, G., Pfahringer, B., Reutemann, P., Witten, I.H.: The weka data mining software: an update. ACM SIGKDD Explor. Newsl. 11, 10–18 (2009)
19. Speer, R., Havasi, C.: Representing general relational knowledge in conceptnet 5. In: Chair, N.C.C., Choukri, K., Declerck, T., Doan, M.U., Maegaard, B., Mariani, J., Moreno, A., Odijk, J., Piperidis, S. (eds.) Proceedings of the Eight International Conference on Language Resources and Evaluation (LREC 2012). European Language Resources Association (ELRA), Istanbul (2012)

Towards Evaluating the Impact of Anaphora Resolution on Text Summarisation from a Human Perspective

Mostafa Bayomi[1(✉)], Killian Levacher[1], M. Rami Ghorab[3],
Peter Lavin[1], Alexander O'Connor[2], and Séamus Lawless[1]

[1] ADAPT Centre, Knowledge and Data Engineering Group, School of Computer
Science and Statistics, Trinity College Dublin, Dublin, Ireland
{bayomim,killian.levacher,peter.lavin,
seamus.lawless}@scss.tcd.ie
[2] ADAPT Centre, School of Computing, Dublin City University, Dublin, Ireland
alexander.oconnor@dcu.ie
[3] IBM Analytics, IBM Technology Campus, Dublin, Ireland
rami.ghorab@ie.ibm.com

Abstract. Automatic Text Summarisation (TS) is the process of abstracting key content from information sources. Previous research attempted to combine diverse NLP techniques to improve the quality of the produced summaries. The study reported in this paper seeks to establish whether Anaphora Resolution (AR) can improve the quality of generated summaries, and to assess whether AR has the same impact on text from different subject domains. Summarisation evaluation is critical to the development of automatic summarisation systems. Previous studies have evaluated their summaries using automatic techniques. However, automatic techniques lack the ability to evaluate certain factors which are better quantified by human beings. In this paper the summaries are evaluated via human judgment, where the following factors are taken into consideration: informativeness, readability and understandability, conciseness, and the overall quality of the summary. Overall, the results of this study depict a pattern of slight but not significant increases in the quality of summaries produced using AR. At a subject domain level, however, the results demonstrate that the contribution of AR towards TS is domain dependent and for some domains it has a statistically significant impact on TS.

Keywords: Text summarisation · Anaphora resolution · TextRank

1 Introduction

Natural Language Processing (NLP) has different tasks [1, 2]. One of these tasks is Automatic Text Summarisation (TS) that has been the subject of a lot of interest in the NLP community in recent years [2]. The goal of automatic summarisation is to process

M.R. Ghorab—Postdoctoral Researcher at Trinity College Dublin at the time of conducting this research.

E. Métais et al. (Eds.): NLDB 2016, LNCS 9612, pp. 187–199, 2016.
DOI: 10.1007/978-3-319-41754-7_16

the source text to produce a shorter version of the information contained in it then present this version in a way that suits the needs of a particular user or application. Text summaries attempt to provide concise overviews of content, and can allow a reader to make a quick and informed decision regarding whether a document contains the information they seek, and thus whether it would be worth the time and effort required to read the entire document. The rapid growth of the Web has resulted in a massive increase in the amount of information available online, which in turn has increased the importance of text summarisation.

Various techniques have been proposed in the literature for the automatic summarisation of text, some of which are supervised, while others are unsupervised. Supervised techniques involve the need for an existing dataset of example summaries [3]. In contrast, unsupervised techniques do not rely upon any external knowledge sources, models or on linguistic processing and interpretation to summarise text [4].

There are two primary approaches to automatic summarisation. Extractive methods work by selecting a subset of existing words, phrases, or sentences from the original text to form the summary. In contrast, abstractive methods build an internal semantic representation and then use natural language generation techniques to create a summary that is closer to what a human might generate. Such a summary may contain words that are not included in the original text. These approaches are outlined in more detail in Sect. 2, below.

Throughout the evolution of automatic text summarisation, attempts have continuously been made to improve the quality of the summaries produced. As a result, diverse NLP techniques have been combined together in an attempt to deliver advancements in the process, with mixed results. While some combinations of techniques demonstrated positive impact [5], some did not [6].

One approach that has been shown to deliver a positive impact on summarisation is Anaphora Resolution (AR). A number of studies have combined AR with the summarisation algorithm, and their results have shown a positive impact on summaries [6, 7].

Various systems were developed to carry out automatic evaluation, of the produced summaries, taking into consideration some measures and aspects in the evaluation process, such as comparing the summary to other (ideal) summaries created by humans [9]. An example of these measurements involves counting the number of overlapping units such as n-gram, word sequences, and word-pairs between the computer-generated summaries and the ideal summaries created by humans. The problems with these evaluations are: (1) the necessity for existing human summaries and the fact that (2) automatic systems lack evaluation measurements, such as conciseness and informativeness. Metrics such as these are often missing from evaluation systems, as they are difficult to quantify automatically and need to be assessed by human beings. Our research attempts to take these factors into consideration as part of the evaluation.

The contribution of this paper involves investigating the impact of AR on text summarisation, both in general, and with respect to different domains, by using human evaluation to judge the generated summaries. The evaluation includes factors that cannot be quantified by automatic systems, and therefore require a human to judge. These factors are: informativeness, readability & understandability, conciseness and the overall quality of the summary.

The summarisation system proposed in this paper builds upon TextRank [4] for the generation of summaries. In the evaluation carried out for this study, a comparison is conducted between two summarisation approaches. In the first approach, summarisation is performed directly on the original text. In the second approach, however, the anaphora of the original text is resolved before summarisation is performed.

2 Related Work

Many advances have been achieved in the field of text summarisation, which generate summaries of a variety of documents in different repositories, such as scientific repositories and meeting recordings [10]. The two main approaches to text summarisation consist of abstraction [11] and extraction [12] methods. Abstraction involves "understanding" the meaning of a text in order to build an internal semantic representation of it and subsequently use natural language generation techniques to create a summary that is close to what a human would generate. The challenge faced by abstraction techniques is that they rely upon the requirements for semantics and training data to automatically interpret and "understand" the meaning of the content in order to summarise the original text. As a result of this, summarisation techniques based upon abstraction have so far met with limited success [13]. On the other hand, extraction approaches require the identification of the most important sentences (or passages), in the text in order to extract the summary. Thus, the extraction method is deemed more feasible [4] and is selected as the approach to be used in this paper.

Various methods have been discussed in the literature to improve extractive summarisation systems. Early systems depended upon simple approaches, such as: *Sentence Location*, where the importance of the sentence is determined by: (1) its location in the text (being at the beginning or end of the text); (2) the emphasis of the text (e.g. being a heading); and (3) the formatting of the text (e.g. a sentence highlighted in bold would be considered more important than other sentences) [14]. The *Cue Phrase* approach focuses on words that are used to determine the appropriate sentences for the summary such as "this paper", "propose", and "concluding" [15]. The *most frequent words* approach extracts a summary by determining the most important concepts by exploring the term distribution of a document [16]. Another approach that extracts summaries by measuring the length of the sentences is the *Sentence Length* approach, which automatically ignores sentences that have a length that falls under a predefined threshold [15].

Robust graph-based methods for NLP tasks have also been gaining a lot of interest. Arguably, these methods can be singled out as key elements of the paradigm-shift triggered in the field of NLP. Graph-based ranking algorithms are essentially a way of deciding the importance of a vertex within a graph, based on global information recursively drawn from the entire graph. The basic idea implemented by a graph-based ranking model is that of "voting" or "recommendation". When one vertex links to another, it is basically casting a vote for that vertex. The higher the number of votes for a vertex, the higher the importance of that vertex. Moreover, the importance of the vote is determined by the importance of the vertex casting the vote. One of the most significant algorithms in this spectrum is the TextRank algorithm [4], which is based on the PageRank algorithm [17].

TextRank is an unsupervised graph-based ranking model that summarises text by extracting the most important sentences of the text after ranking them according to the number of overlapping words between them. TextRank can be applied to tasks ranging from automated extraction of keyphrases, to extractive summarisation and word-sense disambiguation, and it works as follows:

Formally, let $G = (V, E)$ be a directed graph with the set of vertices V and set of edges E, where E is a subset of $V \times V$. In this model, the graphs are built from natural language texts where V represents a sentence from the text, and E is the edge that represents the connection between two sentences. Each edge acts as a weight that represents the similarity between two sentences. For a given vertex V_i, let $In\ (V_i)$ be the set of vertices that point to it (predecessors), and let $Out\ (V_i)$. be the set of vertices that vertex V_i points to (successors). The score of a vertex V_i is defined as follows:

$$S(V_i) = (1 - d) + d * \sum\nolimits_{j \in In(V_i)} \frac{1}{Out(V_i)} S(V_i) \tag{1}$$

where d is a damping factor that can be set between 0 and 1, which has the role of integrating into the model the probability of jumping from a given vertex to another random vertex in the graph. The factor d is usually set to 0.85.

TextRank has been demonstrated to perform well as an algorithm for extracting summaries [4], however, combining it with other NLP approaches may enhance the quality of these summaries. These additional approaches can be employed in the text pre-processing step, or furthermore in the ranking process. In this research we combine TextRank with other approaches such as Sentence Location and Anaphora Resolution. Furthermore, we combine different approaches in the text pre-processing step such as stopword removal and stemming.

Information about anaphoric relations can be beneficial for text summarisation. Anaphora is a reference which points back to some previous item, with the 'pointing back' word or phrase called an anaphor, and the entity to which it refers, or for which it stands, its antecedent [6]. For example: "*The boy is clever, he is the first in the class*". Here, "*he*" is the reference (anaphor) that points back to its antecedent "*The boy*".

Anaphoric information is used to enrich the latent semantic representation of a document, from which a summary is then extracted. Different algorithms for text summarisation are based on the similarity between sentences. AR can be used to identify which discourse entities are repeatedly mentioned, especially when different mentioning forms are used. Previous work involving the inclusion of AR in summarisation, reports some increase in performance. Vodolazova et al. [5] achieved an improvement of around 1.5 % over their summarisation system based on the lexical Latent Semantic Analysis (LSA) by incorporating anaphoric information into it. The performance was tested on the DUC 2002 data using the ROUGE evaluation toolkit [9]. The authors also mentioned two strategies for including anaphoric relations: (1) *addition*, when anaphoric chains are treated as another category of term for the input matrix construction; (2) *substitution*, when each representative of an anaphoric chain in the text is substituted by the chain's first representative. The evaluation results show that the substitution approach performs significantly worse than the addition approach and in some tests even worse than the same system without including AR [5]. Although the addition strategy is better, in some cases the output summary becomes incoherent. This happens when a pronoun appears in the summary and its antecedent does not. As a result, for coherency, in this research the addition strategy is used, and after that the anaphora chain is tracked. If a pronoun appears in the summary without its antecedent, the pronoun is substituted by its antecedent.

Various researchers have investigated how the quality of automatically generated summaries can be evaluated. Vodolazova et al. [12] attempted to analyse the impact of shallow linguistic properties of the original text on the quality of the generated summaries. Nenkova et al. [18] focused on how sentence structure can help to predict the linguistic quality of generated summaries.

Summarisation evaluation methods can be broadly classified into two categories [19]: *extrinsic* evaluation, where the summary quality is judged on the basis of how helpful summaries are for a given task, and *intrinsic* evaluation which is based on an analyses of the summary, and involves a comparison between the original text (the source document) and the generated summary to measure how many ideas of the source document are covered by the summary. Intrinsic evaluations can then be divided into: *content evaluation* and *text quality evaluation*. Content evaluation measures the ability to identify the key topics of the original document, while text quality evaluation judges the readability, grammar and coherence of automatic summaries [8].

Text Summarisation evaluation can be classified into two approaches: Automatic evaluation and Human evaluation. Many evaluation studies [4, 5, 8] used automatic evaluation systems such as ROUGE [9]. ROUGE is a tool which includes several automatic evaluation methods that measure the similarity between summaries. Other studies have asked human judges to evaluate the quality of generated summaries generated using their approach [20].

The problems with the automatic evaluation are that (1) they first of all necessitate the existence of human summaries for comparison with the generated summaries; additionally (2) automatic systems lack evaluation measurements such as conciseness and informativeness because these criteria are difficult to quantify automatically and need to be assessed by humans. We take these factors into consideration as part of the evaluation in this research where we adopted the intrinsic method and used human evaluation to judge the quality of summaries generated by two different approaches.

3 Design

In order to evaluate the impact of Anaphora Resolution from a human perspective a system has been developed. The system consists of the following modules:

Summarisation:

- *Summarisation Algorithm*: The task of summarisation is an unsupervised extractive task, so TextRank is used as the summarisation module.
- *Stemming*: TextRank measures the similarity between sentences by calculating the number of overlapping words between them. Therefore, to ensure that words that

share the same root (i.e. ones which are deemed very close to each other in meaning), words are stemmed before comparing them to each other. The Lancaster stemmer was selected to be used in this research. This follows on the successful approach of Augat and Ladlow [21], where they compared the performance of three stemmers: WordNet stemmer, Lancaster stemmer, and Porter stemmer; of the three, the Lancaster stemmer performed best.

– *Stopword Removal*: As stopwords are generally assumed to be of less, or no, informational value, the system performs stopword removal on the original text before stemming and summarisation.

Anaphora Resolution. The AR system used is the Stanford Deterministic Coreference Resolution System [22] which implements a multi-pass sieve coreference resolution (or anaphora resolution) system. The system relates pronouns to their nominal antecedents.

Addition Approach. As it has been shown by Vodolazova et al. [5], that the substitution approach performs significantly worse than the addition approach, the addition approach is used in this system.

Sentence Location. The sentence location approach is applied to extract the first sentence from the original document to be used as the first sentence in the summary, even if TextRank did not extract it. The selected dataset for this research is Wikipedia, and since the first sentence in any Wikipedia article states the definition of the article, the sentence location can provide valuable information to the summarisation process.

Chain Tracking. The Anaphora resolution system produces an anaphoric chain. This chain consists of an antecedent (first representative, a name for example) and the anaphor that refers to that antecedent (and may be more than one anaphor). This chain is used to check the coherence of the summary produced by the system, by checking that the anaphoric expressions contained in the summary still have the same interpretation that they had in the original text, and if it does not have the same interpretation, the anaphor is substituted by its antecedent.

The system implemented for this research carries out the text processing in the following order:

1. Anaphora in the original text is resolved.
2. Stopwords are removed.
3. Stemming is applied.
4. TextRank is executed on the new text.
5. Sentence ranks are produced.
6. The highest ranked sentences are selected and extracted from the original text.
7. The first sentence is selected to be used in the summary (if it was not already selected by TextRank).
8. If a pronoun is found in the summary without its antecedent, it is replaced by the antecedent.

4   Evaluation

4.1   Dataset

In order to investigate the impact of AR on summaries generated from different subject domains, 70 Wikipedia abstracts were selected from various subject domains. The abstract is the first section from the Wikipedia article. The abstracts have different length (from around 180 words to more than 560 words.) Wikipedia was chosen as an open source of content in multiple domains. The domains were selected at random and the abstracts were randomly chosen from within the following domains: Accidents, Natural Disasters, Politics, Famous People, Sports, and Animals. Two summaries were generated for each abstract, one summary was generated without applying Anaphora Resolution and the second summary was generated by applying Anaphora Resolution before summarisation[1].

When we tried to generate the two summaries for each article, we noticed in 45 articles that the two summaries were identical, which reflected the possibility that AR may have had no impact on the produced summaries.

We also noticed that in producing the two summaries, the identical summaries were often produced from the shorter articles. This means that, the length of text in the article proportional to the impact of AR on the produced summaries (i.e. the shorter the article, the less the impact of AR on the summary, and vice versa).

Hence, 45 abstracts were removed and the final selected set of abstracts, for the experiment, are 25 articles of the original document-set where each domain has 5 articles.

4.2   Experimental Setup

To conduct the comparative evaluation between the two summaries, a web application was built. The abstracts were divided into groups; each group had five abstracts that came randomly from different domains. When the first user logged in, they were randomly assigned a group. The next user was then randomly assigned a group from the remaining unassigned groups. This continued until all the groups were assigned to users. The process was then repeated for the next set of users who logged in to the system. This ensured an even spread of assessment. Each article in the group that is presented to the user was followed by the two generated summaries. The users were not aware of which summary was produced using which technique. Each participant in the experiment was asked to evaluate each summary according to the following characteristics:

1. **Readability and Understandability**: Whether the grammar and the spelling of the summary are correct and appropriate.
2. **Informativeness**: How much information from the source text is preserved in the summary.

---

[1] In our experiment, the first summary is marked as TR and the second summary is marked as TR + AR.

3. **Conciseness**: As a summary presents a short text, conciseness means to assess if this summary contains any unnecessary or redundant information.
4. **Overall**: The overall quality of the summary.

An open call for participation in the experiment was made on several mailing lists and through social media. The selected participants' ages were ranging from 24 to 50 years old and they were from different backgrounds.

The users were asked to evaluate each characteristic on a six-point Likert scale (ranging from one to six, where one is the lowest quality and six is the highest). For the sake of data completeness, each participant was asked to fill in answers for all the evaluation characteristics. A text box was provided in case the user had any comments regarding the quality of the summaries or regarding the difference between them. The users were allowed to leave this box empty in case they did not have any comments. After each summary was evaluated individually, the user was asked to indicate which summary they preferred. In most cases this characteristic was automatically set by the system based upon the user's individual evaluation of the two summaries, while still allowing for manual adjustment if the user so wished. However if the user evaluated the two summaries equally, then neither summary was preselected, and they had to manually make a selection.

5   Results and Discussion

Thirty-eight users participated in the experiment. Each user evaluated at least four abstracts in different domains and also added comments about the two summaries. The final results were analysed regarding:

(a)   The general impact of Anaphora Resolution on Text Summarisation.
(b)   The domain-specific impact of Anaphora Resolution on Text Summarisation.

5.1   General AR Impact

Table 1 reports the mean scores of the user evaluations, the variance, and p-value for each characteristic. The results show that the difference between the two approaches (TR and TR + AR) is not statistically significant, which coincides with results obtained by Mitkov et al. [6]. However, from the results obtained in our research, we can notice a pattern. Although the difference is not significant, the differences are constantly positive, which would suggest the combined approach does have a positive influence overall. Nevertheless, this pattern will need to be confirmed with future experiments and more detailed analyses.

As part of the analysis, the users' comments were also examined. In general, the comments show that the majority of the users preferred the second summary (TR + AR). The users' comments also showed that AR has no significant impact on the quality of the produced summaries. This is demonstrated by the following sample comment: *"I noticed that usually the difference between the two summaries is just one or two sentences. This is why it is hard to judge which one is better than the other. They are very close to each other with just a small difference."*

Table 1. General Anaphora resolution impact

Criteria / Approach		TR	TR + AR
Readability& Understand-ability	Mean	4.8625	4.8688
	Variance	0.00625	
	P-Value	0.9291 (> 0.05)	
Informativeness	Mean	4.3875	4.4563
	Variance	0.0688	
	P-Value	0.4671 (> 0.05)	
Conciseness	Mean	4.2375	4.375
	Variance	0.1375	
	P-Value	0.1276 (> 0.05)	
Overall	Mean	4.35625	4.4
	Variance	0.04375	
	P-Value	0.6195 (> 0.05)	
Preferable	Mean	0.8375	1.09375
	Variance	0.2563	
	P-Value	0.1596 (> 0.05)	

5.2 Domain Specific AR Impact

Table 2 reports the mean scores of the user evaluations, the variance, and p-value for each characteristic in each of the different domains tested. In this experiment, abstracts were divided by domains to investigate the intuition that language characteristics differ by subject domain, and thus to assess if the impact of AR on summarisation also differed.

The impact is shown to be statistically significant in the *"Politics"* domain for Conciseness (p-value = **0.02**) and Readability (p-value = **0.05**). Also in the *"Animals"* domain, the difference between the two summarisation approaches is statistically significant with respect to two characteristics: Informativeness (p-value = **0.001**) and User Preference (p-value = **0.02**). In the *"Politics"* and *"Animals"* domains, AR has a positive impact on the generated summaries in all characteristics. In support of this argument, user comments confirmed this finding. The following is a sample of user testimonies: *"First summary did not summarise first part of the topic as efficient as the second summary."* Where the first summary here refers to the TR summary and the second summary is the TR + AR summary.

In contrast, in the *"Accidents"* domain, the summaries generated with TR only are more preferable and are evaluated more positively with respect to all characteristics than summaries with AR.

Regarding the *"Famous People"* domain, although it was expected that AR would have a positive impact on its articles as they have more pronouns than others, the results actually show a negative impact of AR in all characteristics (variance ranges from −0.17 to −0.72). User comments on these domains show that the summary that was generated with TR only is more preferable than TR + AR. One of these comments is: *"The first summary* [the TR summary] *included an extra piece of information about*

Table 2. Domain specific Anaphora resolution impact

Domain / Criteria		C1[a]	C2[b]	C3[c]	C4[d]	C5[e]
Accidents	TR	5.21	4.50	4.71	4.50	0.93
	TR+AR	5.00	4.36	4.29	4.14	0.79
	Varience	-0.21	-0.14	-0.43	-0.36	-0.14
	P-Value	0.27	0.61	0.16	0.29	0.81
Natural Disasters	TR	5.00	4.60	4.33	4.56	0.82
	TR+AR	4.96	4.61	4.51	4.60	1.22
	Varience	-0.04	-0.09	0.18	0.04	0.40
	P-Value	0.75	0.64	0.27	0.79	0.28
Politics	TR	4.61	4.33	3.56	4.11	0.72
	TR+AR	5.00	4.33	4.44	4.28	1.11
	Varience	0.39	0.00	0.89	0.17	0.39
	P-Value	**0.05**	0.99	**0.02**	0.62	0.44
Famous People	TR	4.79	4.21	4.34	4.31	1.28
	TR+AR	4.62	4.17	4.21	4.21	0.55
	Varience	-0.17	-0.03	-0.14	-0.10	-0.72
	P-Value	0.38	0.89	0.53	0.63	0.07
Sports	TR	4.63	4.43	4.27	4.30	0.73
	TR+AR	4.80	4.37	4.30	4.37	1.10
	Varience	0.17	-0.07	0.03	0.07	0.37
	P-Value	0.28	0.71	0.84	0.71	0.38
Animals	TR	4.96	4.13	4.13	4.21	0.50
	TR+AR	4.92	4.96	4.42	4.54	1.67
	Varience	-0.04	0.83	0.29	0.33	1.17
	P-Value	0.80	**0.001**	0.18	0.12	**0.02**

[a]Readability & understand-ability, [b]Informativeness, [c]Conciseness, [d]Overall, [e]Preferable

travels which was omitted in the second [the TR + AR summary]." Which means that the TR summary is more informative than the TR + AR summary.

It was noted from the results in the *"Natural Disasters"* and *"Sports"* domains that the impact of AR varies from one characteristic to another. For example, the second summary (AR + TR) in both domains is more preferable than the first (positive variance, 0.40 and 0.37 respectively), however, the first summary is slightly more readable than the second one. A user comment shows that: *"These two are tight. The first is more readable."*

6 Conclusion and Future Work

This paper presented an alternative approach to evaluate the impact of Anaphora Resolution on Text Summarisation. Various researchers have attempted to evaluate this impact automatically, however, automatic evaluation always lack evaluation characteristics that are difficult to effectively measure automatically and need to be evaluated manually. In this research we measured this impact by asking humans to evaluate two summaries, one of which was generated without Anaphora Resolution (AR), while the second was generated with Anaphora Resolution. The results showed that, in general, AR has a slight but not significant impact on the quality of the summaries produced. Further experimentation on the domain-specific impact of AR showed that AR is domain dependent and its impact varies from one domain to another. The results showed that AR has a statistically significant impact on TS for some criteria.

In the experimental setup stage, it was noticed that the length of the article and the density of the anaphoric references have an influence upon the summaries produced.

Future work will therefore be carried out in order to expand upon these findings. It will investigate the relation between text characteristics, such as the length, the style, the domain, and the number of anaphors in the original text, and their impact upon the generated summaries.

Acknowledgements. This research is supported by Science Foundation Ireland through the CNGL Programme (Grant 12/CE/I2267) in the ADAPT Centre (www.adaptcentre.ie) at Trinity College Dublin.

References

1. Bayomi, M., Levacher, K., Ghorab, M.R., Lawless, S.: OntoSeg: a novel approach to text segmentation using ontological similarity. In: Proceedings of 5th ICDM Workshop on Sentiment Elicitation from Natural Text for Information Retrieval and Extraction, ICDM SENTIRE. Held in Conjunction with the IEEE International Conference on Data Mining, ICDM 2015, Atlantic City, NJ, USA, 14 November 2015
2. Lawless, S., Lavin, P., Bayomi, M., Cabral, J.P., Ghorab, M.: Text summarization and speech synthesis for the automated generation of personalized audio presentations. In: Biemann, C., Handschuh, S., Freitas, A., Meziane, F., Métais, E. (eds.) NLDB 2015. LNCS, vol. 9103, pp. 307–320. Springer, Heidelberg (2015)

3. Cruz, F., Troyano, J.A., Enríquez, F.: Supervised TextRank. In: Salakoski, T., Ginter, F., Pyysalo, S., Pahikkala, T. (eds.) FinTAL 2006. LNCS (LNAI), vol. 4139, pp. 632–639. Springer, Heidelberg (2006)

4. Mihalcea, R., Tarau, P.: TextRank: bringing order into texts. In: Proceedings of EMNLP 2004, pp. 404–411. Association for Computational Linguistics, Barcelona, Spain (2004)

5. Vodolazova, T., Lloret, E., Muñoz, R., Palomar, M.: A comparative study of the impact of statistical and semantic features in the framework of extractive text summarization. In: Sojka, P., Horák, A., Kopeček, I., Pala, K. (eds.) TSD 2012. LNCS, vol. 7499, pp. 306–313. Springer, Heidelberg (2012)

6. Mitkov, R., Evans, R., Orăsan, C., Dornescu, I., Rios, M.: Coreference resolution: to what extent does it help NLP applications? In: Sojka, P., Horák, A., Kopeček, I., Pala, K. (eds.) TSD 2012. LNCS, vol. 7499, pp. 16–27. Springer, Heidelberg (2012)

7. Ježek, K., Poesio, M., Kabadjov, M.A., Steinberger, J.: Two uses of anaphora resolution in summarization. Inf. Process. Manag. **43**(6), 1663–1680 (2007)

8. Steinberger, J., Ježek, K.: Evaluation measures for text summarization. Comput. Inform. **28**(2), 251–275 (2012)

9. Lin, C., Rey, M.: ROUGE : a package for automatic evaluation of summaries. In: Text Summarization Branches Out: Proceedings of ACL-2004 Workshop, vol. 8 (2004)

10. Murray, G., Renals, S., Carletta, J.: Extractive summarization of meeting recordings. In: Proceedings of Interspeech 2005 - Eurospeech, 9th European Conference on Speech Communication and Technology, Lisbon, Portugal, 4–8 September 2005

11. Fiszman, M., Rindflesch, T.C.: Abstraction Summarization for Managing the Biomedical Research Literature (2003)

12. Vodolazova, T., Lloret, E., Muñoz, R., Palomar, M.: Extractive text summarization: can we use the same techniques for any text? In: Métais, E., Meziane, F., Saraee, M., Sugumaran, V., Vadera, S. (eds.) NLDB 2013. LNCS, vol. 7934, pp. 164–175. Springer, Heidelberg (2013)

13. Nenkova, A., Mckeown, K.R.: Automatic summarization. In: Proceedings of 49th Annual Meeting of the Association for Computational Linguistics: Tutorial Abstracts of ACL 2011. Association for Computational Linguistics (2011)

14. Edmundson, H.P.: New methods in automatic extracting. J. ACM (JACM) **16**(2), 264–285 (1969)

15. Teufel, S., Moens, M.: Sentence extraction as a classification task. In: Proceedings of ACL, vol. 97 (1997)

16. Luhn, H.P.: The automatic creation of literature abstracts. IBM J. Res. Dev. **2**(2), 159–165 (1958)

17. Page, L., Brin, S., Motwani, R., Winograd, T.: The PageRank citation ranking: bringing order to the web. Stanford InfoLab (1999)

18. Nenkova, A., Chae, J., Louis, A., Pitler, E.: Structural features for predicting the linguistic quality of text. In: Krahmer, E., Theune, M. (eds.) Empirical Methods. LNCS, vol. 5790, pp. 222–241. Springer, Heidelberg (2010)

19. Sparck Jones, K., Galliers, J.R., Walter, S.M.: Evaluating Natural Language Processing Systems: An Analysis and Review. LNCS, vol. 1083. Springer, Heidelberg (1996)

20. Saggion, H., Lapalme, G.: Concept identification and presentation in the context of technical text summarization. In: Proceedings of 2000 NAACL-ANLP Workshop on Automatic Summarization, pp. 1–10. Association for Computational Linguistics, Stroudsburg, PA, USA (2000)

21. Augat, M., Ladlow, M.: An NLTK package for lexical-chain based word sense disambiguation (2009)
22. Lee, H., Peirsman, Y., Chang, A., Chambers, N., Surdeanu, M., Jurafsky, D.: Stanford's multi-pass sieve coreference resolution system at the CoNLL-2011 shared task. In: Proceedings of 15th Conference on Computational Natural Language Learning: Shared Task, pp. 28–34. Association for Computational Linguistics, Stroudsburg, PA, USA (2011)

An LDA-Based Approach to Scientific Paper Recommendation

Maha Amami[1,2]([✉]), Gabriella Pasi[1], Fabio Stella[1], and Rim Faiz[3]

[1] Department of Informatics, Systems and Communication,
University of Milano Bicocca, Milan, Italy
{amami,pasi,stella}@disco.unimib.it
[2] LARODEC, ISG, University of Tunis, Tunis, Tunisia
[3] LARODEC, IHEC, University of Carthage, Tunis, Tunisia
rim.faiz@ihec.rnu.tn

Abstract. Recommendation of scientific papers is a task aimed to support researchers in accessing relevant articles from a large pool of unseen articles. When writing a paper, a researcher focuses on the topics related to her/his scientific domain, by using a technical language.

The core idea of this paper is to exploit the topics related to the researchers scientific production (authored articles) to formally define her/his profile; in particular we propose to employ topic modeling to formally represent the user profile, and language modeling to formally represent each unseen paper. The recommendation technique we propose relies on the assessment of the closeness of the language used in the researchers papers and the one employed in the unseen papers. The proposed approach exploits a reliable knowledge source for building the user profile, and it alleviates the cold-start problem, typical of collaborative filtering techniques. We also present a preliminary evaluation of our approach on the DBLP.

Keywords: Content-based recommendation · Scientific papers recommendation · Researcher profile · Topic modeling · Language modeling

1 Introduction

In the last years a big deal of research has addressed the issue of scientific papers recommendation. This problem has become more and more compelling due to the information overload phenomenon suffered by several categories of users, including the scientific community. Indeed, the increasing number of scientific papers published every day implies that a researcher spends a lot of time to find publications relevant to her/his research interests. In particular, recommender systems serve in this context the purpose of providing the researchers with a direct recommendation of contents that are likely to fit their needs.

Most approaches in the literature have addressed this problem by means of collaborative filtering (CF) techniques, which evaluate items (in this case papers) based on the behavior of other users (researchers), by exploiting the

© Springer International Publishing Switzerland 2016
E. Métais et al. (Eds.): NLDB 2016, LNCS 9612, pp. 200–210, 2016.
DOI: 10.1007/978-3-319-41754-7_17

rates assigned by other users to the considered items. However, generally, CF approaches assume that the number of users is much larger than the number of items [9].

This is verified in applications like movie recommendations, where there are usually few items and several users. For instance, the MovieLens 1M[1] dataset contains 1,000,209 ratings from 6,040 users and 3,706 movies [8]. Moreover the users are clients who are very likely to interact with the system several times, often consuming similar items; therefore ratings are quite easy to obtain. Hence, CF recommendation models can make accurate recommendations for most users in e-commerce domains.

On the contrary, in the domain of scientific papers recommendation, there are usually less users than papers, which results in the data sparsity problem. In fact usually there are few users who select the same papers, and thus finding similar users only based on explicit ratings of papers is a difficult task.

A second general issue that affects CF is the cold start problem, which occurs when a new item to be recommended has not been rated by any user. Furthermore, when a new user is introduced, the system is not able to provide recommendations. A possible strategy is to force the user to rate a minimum number of items before starting to use the system.

Content-based approaches recommend items based on both the items content and a profile that formally represents the user interests [12]. Content-based approaches exploit the items metadata and content to provide recommendations based on users preferences represented in the user profile. However, content-based recommendation approaches need reliable sources of the user interests. More precisely they rely on a user model (the profile) that specifies the user topical preferences; these preferences must be captured either by means of an explicit user involvement or by means of the analysis of various kinds of interactions of the user with the system (implicit feedback). In this way the cold start problem could be shifted from items to users [6,9].

To improve the recommendations some systems apply hybrid approaches [4], which combine content-based and collaborative filtering techniques.

By means of hybrid approaches [11,17,19] the cold start problem for both new users and items can be alleviated. For instance, in [17] the authors suggest to combine ratings and content analysis into a uniform model based on probabilistic topic modeling.

However, in [1] the authors show that producing accurate recommendations depends on the choice of the recommendation algorithm, and mainly it depends on the quality of the information about users. The noise injected in a user profile affects the accuracy of the produced recommendations.

In this paper we make the assumption that the users (researcher) scientific corpus (publications co-authored by the researcher) is a reliable source of data and information. In fact, in the task of writing articles, a researcher focuses on a set of topics related to her/his scientific investigations, and s/he uses a technical language related to those topics. These core topics play an important role in the

---

[1] www.grouplens.org/.

selection of unseen papers. The rationale behind the approach we propose in this paper is to make use of the researchers scientific corpus to formally define her/his profile: in this way the user model will exploit the core concepts contained in the articles authored by the researcher.

Formally, the approach we propose relies on topic modeling: the profile of a researcher is a topic model obtained by applying LDA to the abstract of a sample of the articles written by the researcher. Topic models provide the identification of core topics from a provided text collection; each topic is formally represented by means of a probability distribution of n-grams, i.e. sequences of n words. We propose then to formally represent each unseen article (to be recommended) by means of a language model. Then the topics generated by topic modeling from the authors collection and the language models of the unseen papers are compared to assess their similarity, which is employed by the recommendation mechanism.

The outline of this paper is the following. In Sect. 2 we review the related works based on LDA-recommendation models with a brief discussion of their limitations. The proposed content-based recommendation model is presented in Sect. 3. The evaluation of our recommendation model is presented in Sect. 4, with a description of the employed dataset and a discussion of the obtained results. In Sect. 5 we draw some conclusions and outline our future work.

2   Related Work

Several content-based recommender systems formally represent the user profile as a bag of words, represented by a vector. For example in [13] both researchers and unseen papers are represented as vectors in an n-dimensional space of terms, and the cosine similarity measure is applied to determine the relevance of a paper to a user profile.

In [5] both user profiles and unseen papers are represented as trees of concepts from the ACM's Computing Classification System (CCS); the recommender system matches the concepts in the user profile to each concept in the paper representation by means of a tree matching algorithm. Based on this technique the unseen papers in a scientific library are recommended to a user (researcher). A limitation of this approach is that the considered concepts are limited and too general to be able to well distinguish different topics.

Latent Dirichlet allocation (LDA) [3] has been employed as a technique to identify and annotate large text corpora with concepts, to track changes in topics over time, and to assess the similarity between documents. The purpose of this algorithm is the analysis of texts in natural language in order to discover the topics they contain and to represent them by means of probability distributions over words. For real world tasks, LDA has been successfully applied to address several tasks (e.g., analysis of scientific trends [2], information retrieval [18], and scholarly publication search engines[2]).

---

[2] Rexa.info.

In [10] the authors have proposed an LDA-based method for recommending problem-related papers or solution-related papers to researchers, in order to satisfy user-specific reading purposes. Here the LDA algorithm was used with a fixed number of topics to generate the document-topic distributions from the entire corpus. In this paper any comparative evaluation with an appropriate baseline is provided.

In [17] the authors have proposed an extension of LDA for recommending scientific articles called collaborative topic regression (CTR). This hybrid approach combines collaborative filtering based on latent factor models and content analysis based on topic models. Here matrix factorization and LDA are merged into a single generative process, where item latent factors are obtained by adding an offset latent variable to the document-topic distribution. Like CF approaches, this method is able to predict articles already rated by similar users. However, it performs differently on unseen papers, and it can make predictions to the ones that have similar content to other articles that a user likes.

In [19] the authors have proposed a hybrid recommender which is a latent factor model called CAT. It incorporates content and descriptive attributes of items (e.g., author, venue, publication year) to improve the recommendation accuracy.

However, both the CTR and the CAT systems suffer two limitations. First, they have problems when asked to make accurate predictions to researchers who have only few ratings. Second, they may not effectively support tasks that are specific to a certain research field such as recommending citations. For instance, a biological scientist who is interested in data mining applications to biological science might desire the recommender systems to support tasks such as recommending unseen papers on data mining techniques. While both CAT and CTR are using a rich source of metadata to generate recommendations, they are subject to certain limitations that affect their effectiveness in modeling citation patterns. The citation context, defined as a sequence of words that appear around a particular citation [14] is not highlighted in the learned models.

3 The Proposed Approach

In this section we introduce our LDA-based approach to scientific paper recommendation to address the issues pointed out in Sect. 2.

The rationale behind our approach is that the generation of the researcher profile should rely on the content generated by the researcher her/himself, as it exposes the topics of interests of the researcher, as well as the technical language s/he uses to generate her/his articles. The researcher profile is then conceived as a mixture of topics extracted by the LDA algorithm from the researcher past publications. To estimate if an unseen article could be of interests to the researcher, a formal representation of the unseen article by a language model is provided, which is then compared with each topic characterizing the researcher profile (we remind that a topic is formally represented as a probability distribution over the considered vocabulary, as also a language model is).

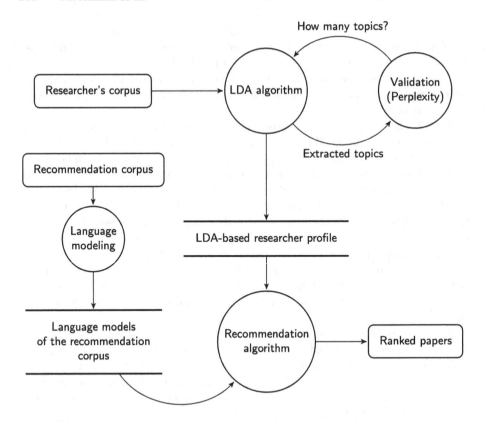

Fig. 1. Process flow of our recommendation model

3.1   Notation and Approach

We denote by Q_i the set of articles (co-)authored by a researcher A_i, and by $D = \{d_1, d_2, \ldots, d_M\}$ the set (consisting of M papers) that contains the articles unseen by researcher A_i and that could be potentially interesting to her/him. We refer to Q_i as to the researcher's corpus and to D as to the recommendation corpus. Furthermore, let W_i be the vocabulary employed in the extended set of articles, i.e., the set including both the researcher's corpus and the recommendation corpus ($Q_i \cup D$). W_i contains then the set of words w occurring in the corpus Q_i of researcher A_i and/or in the recommendation corpus D. The researcher profile is formally represented as a topic model obtained by applying the LDA algorithm to the texts constituted by the abstracts of the considered articles authored by the considered researcher (as it will be better explained in Sect. 3.2). We remind that formally each topic is represented by a probability distribution over the considered vocabulary $p_i(w|k)$ where w ranges over the vocabulary W_i and k is a topic among the K_i topics. In particular, we denote by $p_i(w|1 : K_i)$ the topic model associated with researcher A_i.

The objective of the content-based filtering algorithm is to recommend to researcher A_i the top m papers from the recommendation corpus D. We address this task by means of the following sub-tasks (see Fig. 1); *(i)* to define the topic model representing the researcher's interests: as previously outlined this is done by applying the LDA algorithm to the researcher's corpus Q_i, *(ii)* to validate the topics extracted by the LDA algorithm form the researcher's corpus, *(iii)* to evaluate if a given article d_j from corpus D has to be recommended to researcher A_i; to this purpose the distance between the validated topics and the language model of the article to be recommended is computed, and *(iv)* to rank papers belonging to D in descending order of similarity and recommend, to researcher A_i, the first m papers in the provided ranking. In Sect. 3.2 the above sub-tasks will be detailed.

3.2   Generation of the Researcher Profile and Topic Validation

As explained in Sect. 3.1, sub-tasks *(i)* and *(ii)* are performed by applying LDA to the texts extracted from the researcher's corpus Q_i for each researcher A_i. In particular, to select in a distinct way the optimal number of topics K_i for each researcher A_i the researcher's corpus Q_i is cross validated. More specifically, the optimal number of topics K_i for a given researcher A_i is selected by optimizing the cross validated perplexity, where perplexity [16] measures the uncertainty in predicting the occurrence of a word when using a given model. In topic modeling, the perplexity measures how well the topics extracted by LDA using the training papers (in-sample papers, i.e., a portion of Q_i), allows to predict the occurrence of words belonging to validation papers (out-of-sample papers, i.e., the papers belonging to Q_i which are not used by LDA to extract topics). Perplexity is defined as follows [7]:

$$Perplexity(Q_i^{out-of-sample}) = exp\{-\log \frac{p(Q_i^{out-of-sample}|p_i(w|1:K_i))}{|Q_i^{out-of-sample}|}\} \quad (1)$$

where $Q_i^{out-of-sample}$ is the set of out-of-sample papers belonging to Q_i.

3.3   The Recommendation Algorithm

Step (iii) of the proposed procedure consists of computing, for each researcher, the similarity between her/his K_i validated topics and the language model computed for each article in the recommendation corpus D. Formally, we propose to define the similarity between the profile of the researcher A_i and the article $d_j \in D$ as the maximum value among the K_i similarity values between the language model of article d_j and the K_i topics (probability distributions over words, as illustrated in Sect. 3.2) associated with the profile of author A_i. As each topic is represented as a probability distribution over words (as produced by the LDA algorithm), the similarity between a topic in the LDA-based researcher profile and the language model representing the article to be recommended is defined by exploiting the *symmetrized Kullback Leibler divergence* between the above probability distributions. The language model associated with the unseen article $d_j \in D$ is computed as follows:

$$p(w|d_j) = \frac{nocc(w, d_j) + \frac{\mu nocc(w,Q)}{\sum_w nocc(w,Q)}}{\sum_w nocc(w, d_j) + \mu} \tag{2}$$

where $nocc(w, Q_i)$ is the number of occurrences of word w in Q_i, and μ is the hyperparameter of the Dirichlet distribution. Indeed, Formula (2), which is known as *Bayesian smoothing using Dirichlet priors*, does not incur the black swan paradox; i.e. a word w not occurring in the researcher's corpus Q_i is assigned a null probability value.

Given the topic distribution $p_i(w|k)$, i.e., the probability distribution over words w associated with topic k, extracted by LDA using the corpus of papers (co-)authored by researcher A_i, and the language model $p(w|d_j)$ associated with paper $d_j \in D$, we compute the symmetrized Kullback Leibler divergence between the topic k and the paper d_j as follows:

$$SKL(k, j) = \frac{1}{2} \sum_{w \in W_i} p(w|d_j) \log \frac{p(w|d_j)}{p(w|k)} + \frac{1}{2} \sum_{w \in W_i} p(w|k) \log \frac{p(w|k)}{p(w|d_j)} \tag{3}$$

Then, for each researcher A_i and paper $d_j \in D$, we find the topic k^* which minimizes the symmetrized Kullback Leibler divergence (3) across all the K_i validated topics associated with researcher A_i. Then, the similarity between researcher A_i and paper d_j is defined as follows:

$$Similarity(i, j) = \frac{1}{SKL(k^*, j)} \tag{4}$$

where we assume $SKL(k^*, j) \neq 0$ for each paper $d_j \in D$. Formula (4) corresponds to an optimistic computation of the similarity between researcher A_i and paper $d_j \in D$. Indeed, we are assuming that each researcher is summarized by a single topic k^*, i.e. the topic which is the most similar to the language model associated with the considered paper $d_j \in D$. Once we have computed for each paper $d_j \in D$ the similarity (4), we rank papers of the recommendation corpus D in descending order of similarity and use the ranking to recommend papers to researcher A_i. We then apply the same procedure to all researchers to implement step *(iv)* of the proposed procedure.

4   Evaluation of the Effectiveness of the Proposed Recommendation Approach

In this section we describe the experimental evaluations that we have conducted to verify the effectiveness of our approach. We first present the dataset and pre-processing step followed by details of the experimental procedure.

4.1 Dataset

We used the dataset of ArnetMiner[3], which contains 1.5 million papers from DBLP and 700 thousand authors. We have preprocessed this dataset to select only papers with complete titles and abstracts [15]; we denote this reduced set by $\mathcal{L}$ with $|\mathcal{L}| = 236,012$. To the purpose of our evaluations we have randomly selected 1,600 authors. We denote the set of the considered authors as $U = \{A_1, \ldots, A_{1,600}\}$ with $|U| = 1,600$ and we denote by $Q_A = \{q_1, \ldots, q_{N_A}\}$ the set of papers written by author A_i with $N_A \geq 10$.

We build the profile for each author based on her/his scientific production, namely the papers s/he (co-)authored by using the MALLET topic model API[4]. We have applied the following pre-processing steps to the titles and abstracts of the author's scientific production. First, we eliminated any words occurring in a standard stop list. Then, we converted the abstracts to a sequence of unigrams. To the purpose of defining a feasible test set, we have assumed that citations in papers written by a researcher A_i represent her/his preferences. We denote the test set by C, and its cardinality is defined as:

$$|C| = \sum_{A \in U} |C_A| = 24,539$$

where $C = \{d_j \in \mathcal{L} | \exists q_i \in Q, (q_i \to d_j) \text{ and } d_j \notin Q\}$ where $q_i \to d_j$ means q_i cites d_j.

4.2 Metrics

Two possible metrics to quantitatively assess the effectiveness are precision and recall. However, as the unrated papers (false positives) in the test set are unlabeled. It is not possible to establish if they are known by the user. This makes it difficult to accurately compute precision.

Hence, the measure we have used to assess the effectiveness of the proposed algorithm is recall. In particular, we have performed a comparative study to evaluate our approach with respect to the CAT and CTR algorithms. The recall quantifies the fraction of rated papers that are in the top-m of the ranking list sorted by their estimated ratings from among all rated papers in the test set. For each researcher A_i:

$$Recall@m = \frac{|N(m; A_i)|}{|N(A_i)|} \tag{5}$$

where $|.|$ denotes the cardinality of a set, $N(A_i)$ is the set of items rated by A_i in the test set and $N(m; A_i)$ is the subset of $N(A_i)$ contained in the top-m list of all papers sorted by their estimated relevance to the user model.

The recall for the entire system can be summarized using the average recall from all researchers.

---

[3] aminer.org/citation.

[4] mallet.cs.umass.edu/index.php.

4.3   Parameters

To the purpose of our experiments, we have selected the optimal number of topics K_i for each researcher by optimizing the cross validated perplexity as described in Sect. 3.2 with 5-cross validations. We used the left-to-right method defined in [16] to compute the perplexity.

The value of μ in the language model presented in Sect. 3.3 is a value determined empirically, and it is set to $\mu = 0.000001$.

4.4   Results

We compare the average Recall@m results for 1,600 researchers produced by our approach to the results produced by the state-of-the-art LDA-based recommender systems CTR and CAT. We report the averaged results of 5 repeated experiments to measure the performance of the different methods. Figure 2 shows the comparison of the average Recall@m values for 1,600 researchers with $m \leq 100$. Our approach achieves better Recall@m values than CTR and CAT systems. For $m = 40$, our approach performs better with a 60.4 % improvement over CTR and CAT.

As explained in Sect. 2, CTR and CAT systems are not able to make accurate recommendations to researchers who use few metadata (rates) to create the user profile.

The advantage of our approach does not only consist in being as accurate as or better than other hybrid approaches, but in employing a researcher profile that is only based on past publications and therefore using few metadata (only content item) and alleviate the cold start problem for new items. Furthermore,

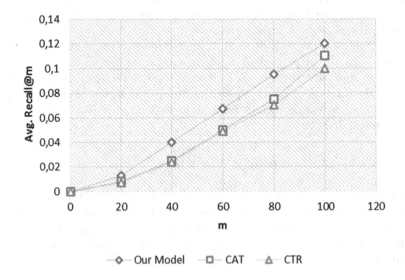

Fig. 2. Comparison of average Recall@m results for 1600 researchers

our approach offers ways to better explain researchers why a specific paper is recommended.

5 Conclusion and Future Work

In this paper we have proposed a fully content-based approach to the recommendation of scientific papers based on the researchers corpus. The researcher profile is built upon the topics generated by LDA algorithm on the researchers publications corpus. The profile built by this technique is easily interpretable, and it can explain the recommendation results. Our preliminary experiments show that our approach is performing well compared to the LDA-state-of-the art models, which make use of several metadata.

As a future work we aim to extend our work to include various attributes from the citation graph such as unseen article's recency, and the author's impact factor to improve the recommendation results.

References

1. Bellogín, A., Said, A., de Vries, A.P.: The magic barrier of recommender systems – no magic, just ratings. In: Dimitrova, V., Kuflik, T., Chin, D., Ricci, F., Dolog, P., Houben, G.-J. (eds.) UMAP 2014. LNCS, vol. 8538, pp. 25–36. Springer, Heidelberg (2014)
2. Blei, D., Lafferty., J.: Dynamic topic models. In: Proceedings of the 23rd International Conference on Machine Learning, pp. 113–120 (2006)
3. Blei, D.M.: Probabilistic topic models. Commun. ACM **55**, 77–84 (2012)
4. Burke, R.: Hybrid recommender systems: survey and experiments. User Model. User-Adap. Inter. **12**(4), 331–370 (2002)
5. Chandrasekaran, K., Gauch, S., Lakkaraju, P., Luong, H.P.: Concept-based document recommendations for CiteSeer authors. In: Nejdl, W., Kay, J., Pu, P., Herder, E. (eds.) AH 2008. LNCS, vol. 5149, pp. 83–92. Springer, Heidelberg (2008)
6. De Nart, D., Tasso, C.: A personalized concept-driven recommender system for scientific libraries. In: The 10th Italian Research Conference on Digital Libraries, pp. 84–91 (2014)
7. Griffiths, T.L., Steyvers, M.: Finding scientific topics. Proc. Natl. Acad. Sci. **101**, 5228–5235 (2004)
8. Herlocker, J.L., Konstan, J.A., Terveen, L.G., Riedl, J.T.: Evaluating collaborative filtering recommender systems. ACM Trans. Inf. Syst. (TOIS) **22**(1), 5–53 (2004)
9. Jannach, D., Zanker, M., Felfernig, A., Friedrich, G.: Recommender Systems: An Introduction. Cambridge University Press, New York (2010)
10. Jiang, Y., Jia, A., Feng, Y., Zhao, D.: Recommending academic papers via users' reading purposes. In Proceedings of the 6th ACM International Conference on Recommender Systems, Dublin, Ireland, pp. 241–244 (2012)
11. Liu, Q., Chen, E., Xiong, H., Ding, C.H., Chen, J.: Enhancing collaborative filtering by user interest expansion via personalized ranking. IEEE Trans. Syst. **42**(1), 218–233 (2012)
12. Lops, P., De Gemmis, M., Semeraro, G.: Content-based recommender systems: state of the art and trends. In: Ricci, F., Rokach, L., Shapira, B., Kantor, P.B. (eds.) Recommender Systems Handbook, pp. 77–105. Springer, Heidelberg (2010)

13. Philip, S., John, A.O.: Application of content-based approach in research paper recommendation system for a digital library. Int. J. Adv. Comput. Sci. Appl. **5**(10), 37–40 (2014)
14. Sugiyama, K., Kan, M.A.: Exploiting potential citation papers in scholarly paper recommendation. In: Proceedings of the 13th ACM/IEEE-CS Joint Conference on Digital libraries, pp. 153–162 (2013)
15. Tang, J., Zhang, J., Yao, L., Li, J., Zhang, L., Su, Z.: ArnetMiner: extraction and mining of academic social networks. In: Proceedings of the Fourteenth ACM SIGKDD International Conference on Knowledge Discovery and Data Mining, pp. 990–998 (2008)
16. Wallach, H.M., Murray, I., Salakhutdinov, R., Mimno, D.: Evaluation methods for topic models. In Proceedings of the 26th Annual International Conference on Machine Learning, pp. 1105–1112 (2009)
17. Wang, C., Blei, D.M.: Collaborative topic modeling for recommending scientific articles. In: Proceedings of the 17th ACM SIGKDD International Conference on Knowledge Discovery and Data Mining, pp. 448–456 (2011)
18. Wei, X., Croft, B.C.: LDA-based document models for Ad-hoc retrieval. In: Proceedings of the 29th Annual International ACM SIGIR Conference on Research and Development in Information Retrieval, pp. 178–185 (2006)
19. Zhang, C., Zhao, X., Wang, K., Sun, J.: Content + attributes: a latent factor model for recommending scientific papers in heterogeneous academic networks. In: de Rijke, M., Kenter, T., de Vries, A.P., Zhai, C.X., de Jong, F., Radinsky, K., Hofmann, K. (eds.) ECIR 2014. LNCS, vol. 8416, pp. 39–50. Springer, Heidelberg (2014)

Short Papers

Topic Shifts in StackOverflow:
Ask it Like Socrates

Toni Gruetze[(✉)], Ralf Krestel, and Felix Naumann

Hasso Plattner Institute, Potsdam, Germany
{toni.gruetze,ralf.krestel,felix.naumann}@hpi.de

Abstract. Community based question-and-answer (Q&A) sites rely on well-posed and appropriately tagged questions. However, most platforms have only limited capabilities to support their users in finding the right tags. In this paper, we propose a temporal recommendation model to support users in tagging new questions and thus improve their acceptance in the community. To underline the necessity of temporal awareness of such a model, we first investigate the changes in tag usage and show different types of collective attention in StackOverflow, a community-driven Q&A website for computer programming topics. Furthermore, we examine the changes over time in the correlation between question terms and topics. Our results show that temporal awareness is indeed important for recommending tags in Q&A communities.

1 Tags in Q&A Communities

During the last two decades, various popular question-and-answer (Q&A) platforms, such as Ask.com, Experts-Exchange, Quora, and Yahoo! Answers have emerged. These platforms provide their users with the opportunity to ask questions and challenge other users of the community to share their knowledge based on these questions. Since 2008 StackOverflow (SO) is the de facto standard question-and-answer website for topics in computer science and programming. In SO, users earn reputation through active participation in the community, e.g., by posing well-received questions or providing helpful answers. After achieving a high reputation level, a user is allowed to up-vote or comment on questions and answers or even edit posts of other users.

Vasilescu et al. showed that experienced programmers (i.e., active GitHub committers) ask fewer questions and give more answers [9]. Anderson et al. investigated the correlation between delay of and reputation for a given answer [1]. Wang et al. show that the majority of questioners in SO asks only one question during their membership [10]. Only little research has been done on how to support these "newbies" on posing well-received questions. A qualitative study discusses efficient ways on how to give code examples for well-received questions [7]. Bazelli et al. establish a connection between extroversion and openness used in the questions and high reputation points of users [2].

Another means of improving a question is by categorizing the post appropriately and help a skilled responder to find it. To this end SO supports tags to

© Springer International Publishing Switzerland 2016
E. Métais et al. (Eds.): NLDB 2016, LNCS 9612, pp. 213–221, 2016.
DOI: 10.1007/978-3-319-41754-7_18

categorize questions. Questioners are forced to choose at least one computer science topic for their new post so that other users, who subscribed topics of their interest and profession, have the chance to find the post. Hence, an appropriately tagged question has a higher probability of getting useful answers.

Depending on the social platform, tags are used with different temporal characteristics. In platforms like Twitter many topics, such as sport events or disasters, are discussed in a very narrow time span (e.g., the Superbowl final: #SuperBowl or the Paris attacks of November 2015: #PrayForParis). Lehmann et al. study different classes of collective attention on Twitter [5] and Gruetze et al. evaluate the influence of such temporal changes for the tag recommendation task [4]. Based on the example of Quora, Maity et al. show that the topics discussed in general-purpose Q&A platforms underly strong temporal dynamics (e.g., political topics) [6]. The topics posted on SO (i.e., programming languages, databases, or operating systems) seem to be more static and lead to the assumption that they slowly evolve over time.

In contrast to this first intuition, we show that tag usage in SO indeed underlies strong temporal effects. In the following, we discuss four different types of temporal topic popularity patterns in SO. We further show how the likelihood for question terms can strongly change over time for SO topics. Finally, we show that recommender systems that incorporate these temporal changes outperform their static counterparts and thus improve the support for SO users in asking well-received questions to receive appropriate answers and thus earning the Socratic badge, which is awarded for asking a well-received questions.

2   Tags in StackOverflow

As the basis for the following analysis, we use the Stack Exchange dataset containing approximately 10 million questions posted between August 1st 2008 and July 31st 2015 on StackOverflow. The questions were posted by over 4.5 million users and categorized using over 40 thousand unique tags. The dataset is licensed under Attribution-ShareAlike 3.0 and available online.[1]

Figure 1b shows the number of questions and answers provided by the SO community. The steady growth of posts emphasizes the increase of popularity and spread of this Q&A platform. We presume that the large 2014 drop in the number of answers and questions originates from the Mighty Mjölnir community update that allows experienced users to close duplicated questions.

The temporal dynamics of discussed topics (i.e., applied tags) is shown in Fig. 1c. While the number of distinct topics per month increases nearly linearly over time, the number of new topics, i.e., tags that were not used before, does not. During the first two years, the number of new topics is relatively high but strongly decreasing. This shows that the platform is gaining coverage of the relevant topics for developers. As of mid 2010 the number of new topics is relatively

---

[1] https://archive.org/details/stackexchange.

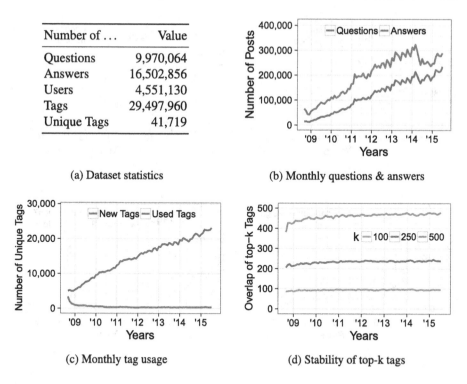

Number of ...	Value
Questions	9,970,064
Answers	16,502,856
Users	4,551,130
Tags	29,497,960
Unique Tags	41,719

(a) Dataset statistics

(b) Monthly questions & answers

(c) Monthly tag usage

(d) Stability of top-k tags

Fig. 1. SO dataset statistics including monthly usage charts (Color figure online)

stable between 200 and 400 per month. We believe that these topics are real new topics, i.e., topics that emerge at this point in time, such as a new programming languages (e.g. Dart (2011)) or new frameworks (e.g. Apache Spark (2013)). The continuing increase of distinct tags per month originates from a gain in topic coverage and is asymptotic to the total number of topics in SO.

Figure 1d shows the stability of the most questioned topics, i.e., the top 100, 250, and 500 tags. The top-100 topics are very stable such that, starting from 2010, these topics overlap with the top-100 of the previous month by approximately 95 %. The number of overlapping top 250 and 500 topics is large too. Beginning with 2011, both values show a relatively constant overlap of about 92 %.

All three statistics show that the community, as well as the coverage of relevant topics for programmers, is constantly increasing. Because the community is growing faster than the newly emerging topics, we expect the number of distinct topics discussed per month to converge in the near future. However, we show that the topics discussed in SO — while making the impression to be well covered by a significant number of questions and answers — still underly significant temporal effects, such that the consideration of up-to-the-minute statistics significantly increase the understanding of the topics in SO.

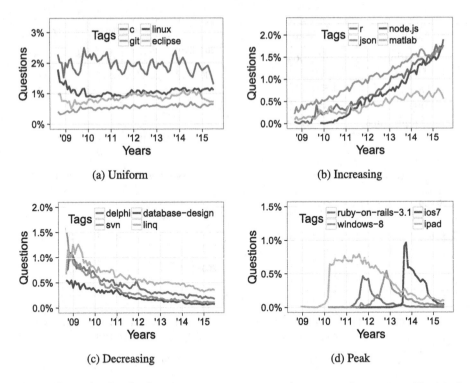

(a) Uniform

(b) Increasing

(c) Decreasing

(d) Peak

Fig. 2. Examples for the four most common temporal tag attention pattern. Depicted in each figure, the percentage of questions that were assigned to a specific topic. (Color figure online)

3   Temporal Tagging Behavior in StackOverflow

In comparison to other social media platforms, such as Twitter and Facebook, SO users exhibit a differing temporal tagging behavior.

Lehmann et al. show different classes of collective attention for tags in Twitter that span a period of one day to one week [5]. Similarly, we are able to classify most **types of attention** in SO into four classes, namely uniform, increasing, decreasing, and peak. Examples for all four classes are shown in Fig. 2 and are next discussed in more detail:

Uniform: Figure 2a shows a uniform usage pattern of the topics with a relatively stable expected monthly usage value. Topics with the uniform usage ratio pattern are growing as fast as the platform (Fig. 1b). While showing a stronger variance of monthly percentages, the topic C still provides a constant expected value of around 2 %. The other topics stay relatively constant over the seven year time span with around 1 % (Linux and Eclipse) or 0.5 % (Git).

Increasing: Figure 2b depicts topics with an increasing attention on the SO platform. The two programming languages R and MATLAB gain in popularity.

Both languages, which have a strong focus on mathematics and statistics, are commonly used in the area of data science and data analytics. Until mid 2009 the number of MATLAB questions was higher than R related posts. Afterwards, the number of R questions follows a stronger growth. Notably, the number of questions concerning R is higher than the ones concerning C at the end of the data period.

Decreasing: Figure 2c depicts the decrease in spreading of the topics (halved popularity). For instance, the number of questions about the proprietary and expensively licensed programming framework Delphi decreases. Simultaneous, the number of questions regarding open source alternatives, such as .NET and Java, increases. The decreasing spread of question regarding SVN shows that more and more software development projects switch to other version control solutions (e.g., Git).

Peak: Figure 2d shows the spreading of discussion topics following a peak pattern, which behave like a mixture of increasing and decreasing pattern happening in a short time frame. The peak attention pattern is a typical scheme for topics covering particular versions of frameworks, platforms, etc., as shown for iOS7. However, the pattern can also appear for non-version topics such as iPad, which was hyped on SO in spring 2010, the time of the official release of the device. While being a major concern of developers also shortly after the release, the community lost interest, such that, 2 years and 9 month later the relative frequency of questions with this topic halved.

Topic Shifts in SO

Besides the changes of topic popularity, rather the community also produces **topic shifts**, i.e., temporal changes of the vocabulary used for questions associated with a specific topic.

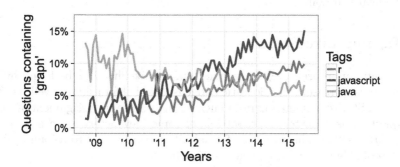

Fig. 3. Topic shift example for tags on questions containing the term 'graph' (Color figure online)

For instance, Fig. 3 contrasts three different topics in the context of posts containing the term 'graph' in the question text. Due to the ambiguity of the term, different meanings might be covered, e.g., a chart or infographic, a structure connecting a set of objects by links (graph-theory), or a plot of a mathematical function. These meanings are expressed by the assignment of different tags. For example, the percentage of tag Java is decreasing over time, due to the fading importance of graph theoretic questions. In contrast, the number of questions about R, which is usually used for plotting of mathematical functions and calculating statistics, increase. The questions assigned to Javascript typically cover visualization of graph data.

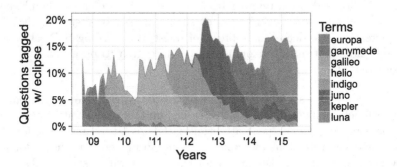

Fig. 4. Topic shift example for keywords on questions tagged with eclipse (Color figure online)

Figure 4 presents the frequency of selected terms in questions tagged with Eclipse. Specifically, the terms represent the version names of the IDE during the examined period. Evidently, the number of questions follow a temporal pattern that is strongly correlated with the yearly release cycle of the IDE. Similar effects can be observed for other topics, such as Android with version names from Cupcake to Lollipop.

4   Tagging Like Socrates

By assigning appropriate tags to a question, the likelihood to be found by the community is increased. The Socratic badge is awarded for asking a well-received question in SO. To model a tag recommender, we phrase the assignment of topics to a question as a posterior probability of topic parameters given a question: $P(\theta_{topic}|q)$. As shown in [4], such a formalization can be used to recommend the k-most probable topics for a given question by:

$$rec_k(q) = \arg\max_{\{t_1..t_k\} \subset T} \sum_{i=1}^{k} P(\theta_{t_i}) \cdot P(q|\theta_{t_i})$$

where T refers to the set of known topics. As shown in Fig. 2, the collective attention for topics in SO underlies strong temporal changes. In our experiments, we model the prior probability $P(\theta_{t_i})$ as frequency of posts with this topic from the past, including all recent posts. The likelihood $P(q|\theta_{t_i})$ is estimated based on nouns found in the question texts for a given topic (e.g., such as the version names for Eclipse questions shown in Fig. 4). Note, we could have used more word classes or even all words, however, tests showed a comparable recommendation quality for models based on nouns as for models with all textual information while minimizing the vocabulary size. To further facilitate the probability estimation, we make the assumption of independence between the probabilities of terms (as the Naïve Bayes model). This plain model greatly facilitates the update of its parameters and thus enables the efficient incorporation of current topic changes, such that the model parameters can be updated for every newly posed question. In contrast, more complex methods, such as dynamic topic models, are based on more expensive parameter inference methods. In particular, they require a temporal partitioning of the data (i.e., the parameter estimation can be executed only for time slices, e.g. yearly) [3].

Note, this recommender definition is rather simple and does not include features like tag co-occurrence, code snippet information, or user preferences. As shown in previous research, such features yield a significant improvement of the recommendation quality [8,11]. However, due to the focus on the temporal aspects, we did not use this knowledge for the experiments and do not discuss these features here and rather focus on the improvements due to temporal patterns.

To measure the influence of topic shifts in SO for tag recommendations, we evaluated four different configurations of the recommender: the first model follows a static learn once strategy, which is based on the topic knowledge derived from all posts created in the year 2010. The second model is updated annually, this yearly model recommends based on the posts of the previous calendar year (e.g., recommendations in 2011 are based on the posts of 2010). Third, the live model is continuously updated based on the latest SO posts. This model is based on all posts that were created in the one year period before a question q was submitted. Finally, we define the simple baseline that is only based on the prior probability $P(\theta_{t_i})$ and thus always recommends the k most commonly used tags for all posts independent from the question text ($\forall t_i, t_j \in T : P(q|\theta_{t_i}) = P(q|\theta_{t_j})$). To compare the performance of all four models, we computed top-k recommendations for all questions Q from the years 2011–2015. Based on these recommendations, we measured recall in comparison to the actual used tags assigned to the question.

$$recall@k = \frac{1}{|Q|} \sum_{q \in Q} \frac{|rec_k(q) \cap act_{Tags}(q)|}{|act_{Tags}(q)|}$$

Results

Figure 5a shows the recall as the function of k for all four approaches. As expected, the baseline model performs worst. This is due to the ignorance of question contents. For $k = 5$ the static model was already able to recommend 39 % of the actually used tags correctly. The yearly model improved this performance

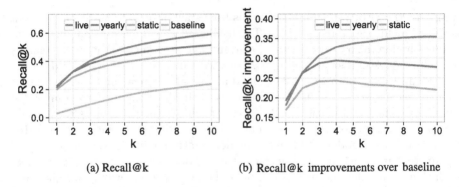

(a) Recall@k   (b) Recall@k improvements over baseline

Fig. 5. Recall@k statistics for different recommendation models (Color figure online)

by 5 percent points, whereas the live model doubled these improvements and achieved a recall of 49 %. This is a respectable performance given that the static baseline is already able to correctly identify 2 out of 5 tags given the set possible set of over $40k$ tags. The significant increase of ten percentage points is based on the additional temporal information and shows the potential even for a seemingly static corpus such as SO. Considering the top-10 recommendations, the live model is able to improve on recall by 10 percentage points (up to 59 %), whereas the gap to the static (and yearly) model is increased by 6 (respectively 5) percentage points.

To investigate the adaptation of the models with respect to topic shifts, Fig. 5b shows the recall improvement of the content based models (static, yearly, and live) in comparison to the baseline. The improvements are significant for all competitors, however, only the live model is able to constantly increase the improvements for increasing k values. This underlines the superior adaptation to topic shifts due to the fast update of the model parameters.

5   Conclusion

We studied the temporal topic changes in the Q&A platform StackOverflow. We examined four different types of popularity patterns of topics over time to show that indeed there are topic shifts. We showed that the likelihood for question terms can strongly change over time for SO topics. Finally, we showed that the tag recommender that are able to incorporate these temporal changes are able to significantly outperform their static counterparts.

In future work we will survey further Q&A platforms and compare the identified temporal effects. We will investigate whether other features, such as user preferences, code snippets, named entities, or tag co-occurrences, are changing over time and can help to improve tag recommendation for new questions. Another interesting application for our findings would be trend prediction in SO and thus for the computer programming field.

Acknowledgments. This research was funded by the German Research Society, DFG grant no. FOR 1306. We thank the StackOverflow community for sharing the valuable knowledge targeted in this work. Finally, we wish to acknowledge the anonymous reviewers for their detailed and helpful comments to the manuscript.

References

1. Anderson, A., Huttenlocher, D., Kleinberg, J., Leskovec, J.: Discovering value from community activity on focused question answering sites: a case study of Stack Overflow. In: Proceedings of the ACM SIGKDD Conference on Knowledge Discovery and Data Mining, pp. 850–858 (2012)
2. Bazelli, B., Hindle, A., Stroulia, E.: On the personality traits of StackOverflow users. In: Proceedings of the International Conference on Software Maintenance (ICSM), pp. 460–463 (2013)
3. Blei, D.M., Lafferty, J.D.: Dynamic topic models. In: Proceedings of the International Conference on Machine Learning (ICML), pp. 113–120 (2006)
4. Gruetze, T., Yao, G., Krestel, R.: Learning temporal tagging behaviour. In: Proceedings of the International Conference on World Wide Web (WWW Companion), pp. 1333–1338 (2015)
5. Lehmann, J., Gonçalves, B., Ramasco, J.J., Cattuto, C.: Dynamical classes of collective attention in Twitter. In: Proceedings of the International Conference on World Wide Web (WWW), pp. 251–260 (2012)
6. Maity, S., Sahni, J.S.S., Mukherjee, A.: Analysis and prediction of question topic popularity in community Q&A sites: a case study of Quora. In: Proceedings of the International AAAI Conference on Web and Social Media (ICWSM) (2015)
7. Nasehi, S.M., Sillito, J., Maurer, F., Burns, C.: What makes a good code example? A study of programming Q&A in StackOverflow. In: Proceedings of the International Conference on Software Maintenance (ICSM), pp. 25–34 (2012)
8. Stanley, C., Byrne, M.D.: Predicting tags for StackOverflow posts. In: Proceedings of the IEEE International Conference on (ICCM), pp. 414–419 (2013)
9. Vasilescu, B., Filkov, V., Serebrenik, A.: StackOverflow and GitHub: associations between software development and crowdsourced knowledge. In: Proceedings of the International Conference on Social Computing (SocialCom), pp. 188–195 (2013)
10. Wang, S., Lo, D., Jiang, L.: An empirical study on developer interactions in Stack-Overflow. In: Proceedings of the Annual ACM Symposium on Applied Computing (SAC), pp. 1019–1024 (2013)
11. Wang, S., Lo, D., Vasilescu, B., Serebrenik, A.: EnTagRec: an enhanced tag recommendation system for software information sites. In: Proceedings of the International Conference on Software Maintenance and Evolution (ICSME), pp. 291–300 (2014)

Semantic-Syntactic Word Valence Vectors for Building a Taxonomy

Oleksandr Marchenko[✉]

Faculty of Cybernetics, Taras Shevchenko National University of Kyiv,
Glushkova ave. 4D, Kyiv 03680, Ukraine
rozenkrans@yandex.ua

Abstract. The paper describes the use of semantic-syntactic word valence vectors as context vectors in the Formal Concept Analysis for building high-quality taxonomies. Research and experiments have confirmed a significant improvement in the quality of constructed taxonomies when in tensor models the number of dimensions is increased while generating semantic-syntactic word valence vectors. The increased arity of the tensor model gives a more accurate description of the multidimensional semantic and syntactic relations and allows allocating more commutative semantic-syntactic properties of the words that are used in the Formal Concept Analysis.

Keywords: Ontologies · Taxonomies · Formal concept analysis · Non-negative tensor factorization

1 Introduction

Taking into account that taxonomy is a fundamental hierarchical basis for designing ontological knowledge bases, the development of methods for automated taxonomy building is a relevant and extremely popular trend in computational linguistics and artificial intelligence in general.

Formally, the task is to build a hierarchical graph from the input set of Nouns N using the relation of *hyponymy/hyperonymy (subclass/superclass)*.

Two basic classes of methods are developed for automated taxonomy building via text corpora processing, namely the methods based on clustering words with semantic proximity measures and the methods of the set theory for word-concepts ordering. Both classes of methods work with the vector space model that presents words or terms as corresponding feature vectors obtained when processing and analyzing a certain text corpus.

The first class of the methods is characterized by semantic proximity usage to determine the distance between vectors of words to figure out how similar they are semantically and whether they can be placed into one cluster. To this end, the cosine measure between vectors of words can be used. The methods of this class are divided, in their turn, into agglomerative (bottom-up clustering)

© Springer International Publishing Switzerland 2016
E. Métais et al. (Eds.): NLDB 2016, LNCS 9612, pp. 222–229, 2016.
DOI: 10.1007/978-3-319-41754-7_19

and divisive methods (top-down clustering). The best representatives of these approaches are described in [1–3].

The other group, the set theory methods, performs taxonomy building by establishing a partial order on the set of word-concepts using the relation of inclusion among sets of features. One of the best representatives of this trend is the Formal Concept Analysis (FCA) [4]. The Formal Concept Analysis approach allows generating taxonomies of higher quality than hierarchies built by agglomerative/divisive clustering algorithms [5].

We propose using semantic-syntactic word valence vectors as context vectors in the Formal Concept Analysis for building high-quality taxonomies.

The remainder of the paper is structured as follows. Section 2 contains an overview of the Formal Concept Analysis. Section 3 presents the description of a method for generating semantic-syntactic word valence vectors. Section 4 describes experiments of the use of semantic-syntactic word valence vectors as context vectors in the Formal Concept Analysis for building high-quality taxonomies, and demonstrates constructed taxonomies' quality estimates calculated in experiments. Section 5 is the conclusion of the paper.

2   The Formal Concept Analysis

The Formal Concept Analysis works with the data structure known as a context table $\mathcal{K}$. The context table contains linguistic context data about the word usage in a certain subject area, along with a certain basic set of lexemes in the text corpora processed and analyzed with the purpose to create the table $\mathcal{K}$. To cite an example, it can be a matrix containing data on the compatibility of an input set of Nouns N with a basic set of qualitative Adjectives. In general terms, one can say that this method works with a table $Objects \times Attributes$. In [5] the context $\mathcal{K}$ of the FCA method is a table that stores information about the compatibility of Noun terms with a certain set of Verbs describing $Predicate - Object$ relation. Noun-terms are presented in the context $\mathcal{K}$ as vectors where non-zero elements indicate connectivity of a Noun to the corresponding Verbs. The values are usually the frequencies of the joint appearance of $Predicate - Verb \times Object - Noun$ pairs in the text corpus. There are also options for using binary matrices with the values 1 and 0 only.

It is worthwhile noting that in order to create the context $\mathcal{K}$ only the words with stable connections are selected, i.e. only Verbs with stable connections to at least one of the Nouns from N are selected to make a set V, i.e. a Verb can get into V if at least one frequent pair $(Verb, Noun)$ with $Predicate \times Object$ relation exists in a text corpus: $\forall v \in V \, \exists n \in N : \vartheta(v, n) \geq Th$, where V is the set of Verbs, N is the set of Noun-terms, $\vartheta(v, n)$ is the frequency of co-occurrences of (v, n) in the text corpus, Th is a threshold.

The FCA uses the partial order theory to analyze relationships between objects G and their features M. The FCA singles out a feature set B from the formal context $\mathcal{K}$, where $B \subseteq M$, and a bijective relation exists between B and a set of objects $A \subseteq G$. Such a correlated pair is called a *formal concept*

(A, B). Formal concepts are partially ordered $(A1, B1) \leq (A2, B2) \Leftrightarrow (A1 \subseteq A2)$ and $(B2 \subseteq B1)$. Items $o1, o2 \in G$ are conceptually clusterized if $\{o1, o2\} \subseteq A$, where (A, B) is a formal concept in $\mathcal{K}$.

The FCA builds a lattice of formal concepts from the binary representation of objects. The following two rules are applied to interpret the lattice as taxonomy:

1. One concept of taxonomy c_B labeled with B is introduced for each formal concept (A, B) if $|A| \geq 2$. The concepts are arranged according to the order in the lattice.

2. For each object $o \in G$ one taxonomy concept c_o labeled with o is introduced. The concepts are arranged in such a way that $c_o \leq c_B$, where (A, B) is a formal concept and $o \in A$, and there is no formal concept (A', B') such that $(A', B') \leq (A, B)$ and $o \in A'$.

Among the advantages of the FCA method, one should note the quality of the taxonomies built. They are much better than the taxonomies constructed by the clustering methods [5]. The taxonomies obtained with the FCA method are transparent, i.e. they are easy to interpret. The main drawback of the FCA method is the exponential time complexity $O(2^n)$, where n is the number of Noun-terms that have to be organized in a taxonomy. In practice, however, the FCA demonstrates a near-linear time due to the excessive sparseness of the Noun-term vector representation.

3   Using Semantic-Syntactic Word Valence Vectors in the FCA Method

We propose to employ semantic-syntactic word valence vectors as term-context vectors in the Formal Concept Analysis.

Semantic-syntactic word valence vectors are generated with non-negative tensor factorization (NTF) of words compatibility tensors. Figure 1 shows such three-dimensional tensor for storing the frequencies of word combinations for the *Subjects* × *Predicates* × *Objects* structures derived from the analysis of a large text corpus.

A large and sparse tensor is formed as a result of syntactic-frequency analysis of text corpus. The non-negative factorization method [6,7] is applied to the tensor to obtain a more concise and convenient presentation.

The basic idea of the method is to minimize the sum of squared differences between the original tensor and its factorized model. For the three-dimensional tensor $T \in R^{D1 \times D2 \times D3}$ it is described by the equation:

$$\min_{x_i \in R^{D1}, y_i \in R^{D2}, z_i \in R^{D3}} \left\| T - \sum_{i=1}^{k} x_i \circ y_i \circ z_i \right\|_F^2,$$

where k is the rank of factorization and $\circ$ denotes the outer product of vectors. The Frobenius norm is used.

In the NTF method, a restriction of being non-negative is added, thereby turning the model into

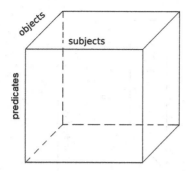

Fig. 1. A three-dimensional tensor for storing the frequencies of word combinations for *Subjects* × *Predicates* × *Objects* structures.

$$\min_{x_i \in R^{D1}_{\geq 0}, y_i \in R^{D2}_{\geq 0}, z_i \in R^{D3}_{\geq 0}} ||T - \sum_{i=1}^{k} x_i \circ y_i \circ z_i||^2_F.$$

The model can be represented graphically as the following illustration Fig. 2. This decomposition consists of the sum of the k outer products of vectors triples (according to the number of tensor dimensions).

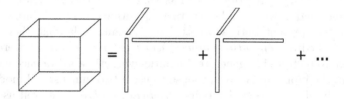

Fig. 2. A graphical representation of NTF as the sum of the k outer products.

Figure 3 shows the NTF decomposition of frequency estimates of word compatibilities for *Subject* × *Predicate* × *Object* triples into three matrices.

Generalized models of selectional preference and some elements of frame model semantics (verb subcategorization frame) are extracted from text corpora using the NTF model for analysis of three-dimensional (S, V, O) *Subject* × *Predicate* × *Object* relations. As a result, each *Subject* − *Noun*, *Predicate* − *Verb* and *Object* − *Noun* gets its own vector (the size of k) from one of the matrices. The original value of tensor T for triplet (S, V, O) T_{svo} can be restored from the factorized model via calculation of the sum:

$$T_{svo} = \sum_{i=1}^{k} s_i * v_i * o_i$$

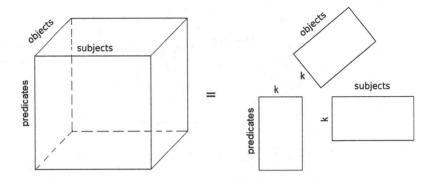

Fig. 3. Graphical representation of the NTF for a linguistic tensor.

Sentences are transformed into triples, e.g. *the hunter shot a wolf* →
(*hunter, shot, wolf*). To calculate the frequency for the triple (*hunter, shot, wolf*) one needs to find the vector-subject for the word *hunter*, and then the vector-predicate for the word *shot*, and finally, the vector-object for the word *wolf*. Then the frequency for the combination of these words is calculated according to the formula given above. If the score is higher than the threshold, one can infer that such a triple is possible in the language.

Vectors from the matrices of the factorized tensor describe commutative properties of words. They prescribe the links that a given word creates, i.e. with which lexemes and in what kind of syntactic positions these links are generated. These vectors represent both syntactic and semantic relations; therefore, the vectors are called as *semantic-syntactic word valence vectors*. Several tensor models have to be tested to generate semantic-syntactic word valence vectors for Nouns from the input set N. We propose to use a two-dimensional model based on the matrix of frequencies for pairs *Predicate* × *Object*, as well as a three-dimensional model to represent triples *Subject* × *Predicate* × *DirectObject*, and a four-dimensional model to represent both *Subject* × *Predicate* × *DirectObject* and *Subject* × *Predicate* × *DirectObject* × *IndirectObject* in text corpora.

4   Experiments

Method. As the basic algorithm for the Formal Concept Analysis, the method described in [8] was chosen. As context vectors in the FCA method, the semantic-syntactic word valence vectors for Nouns are chosen in the syntactic positions of the Object. To this end, the following tensors are collected and factorized:

1. a two-dimensional model is the matrix M: *Predicate − DirectObject*;
2. a three-dimensional model is the tensor T: *Subject − Predicate − DirectObject*;
3. a four-dimensional model is the tensor P: *Subject − Predicate − DirectObject − IndirectObject*.

To build the tensor, text articles from the English Wikipedia have been selected. For each word from the input noun set N, all the articles of the English Wikipedia containing this word are included into the training set $TSet$.

Also, we should include the articles from the English Wikipedia that are connected with the articles from the training set $TSet$ via the *category* link or some other kinds of direct links.

The training set $TSet$ is filtered. The texts of $TSet$ are clusterized using the Latent Semantic Analysis [9]. As the final training set, the largest cluster of texts is selected. It allows minimizing the influence of a polysemy on the model. The largest cluster contains texts of the correct topic domain because the maximal topical intersection of all texts from the $TSet$ should cover all the texts on the corresponding topic (the topic of future taxonomy made of set N). So words of N inside any document of the largest cluster are mainly used with proper meanings and with appropriate contexts.

This is followed by the stage of the assembling matrix M and tensors T and P.

The texts from the training set are parsed with the Stanford parser [10] and the values of tensor elements are set. For example, if a syntactic structure of a sentence under analysis shows that the word *hunter* is a *Subject*, the word *hunted* is a *Predicate*, and the word *boar* is a *Direct Object*, then the value $T[i, j, k]$ is increased by 1. It is assumed that i is the index of the word *hunter*, j is the index of the word *hunted*, k is the index of the word *boar*, and $T[i, j, k] = T[i, j, k] + 1$. At the end of the processing, the element $T[i, j, k]$ will contain the frequency of the triple (*hunter hunted boar*) for the phrases *the hunter hunted a boar* in the training set.

If a particular item is absent in the triplet or quadruplet, then the missing element is filled with the empty word $\oslash$. For example, the four-dimensional model *Subject* × *Predicate* × *DirectObject* × *IndirectObject* for the sentence *the hunter hunted a boar* has no Indirect Object and it is replaced by a sentence *the hunter hunted a boar* $\oslash$. Where more than one element is missing in the sentence structure, no data are recorded in tensor. All elements that do not exceed a certain threshold are zeroed in order to keep only stable values and to get rid of the noise. After collecting data, the matrix and tensor factorizations are performed. For the two-dimensional matrix model, the method of Lee-Seung [11] is employed. For the NTF of three-dimensional and four-dimensional tensors, the Parafac method is used [12]. For all the nouns from the set N the vectors from the Object matrices are collected. These semantic-syntactic Noun valence vectors are used by the FCA method [8] as context vectors for constructing taxonomies of the Noun set N.

Typically, the quality of taxonomy building algorithm is assured by the methods of comparison of generated taxonomies. Usually, the comparison is made with taxonomies that are constructed manually.

To ensure the availability of hand-crafted taxonomy the input noun set N was formed as shown below. As a reference taxonomy, a subnet from Word-Net [13] is taken starting from a certain node N_0 as a root together with all its descendants up to the distance of 4 (sometimes 5) using relation $is - a$.

The first words of the synsets are taken from all nodes of the subnet in order to form an input set N. To calculate estimates of the structural proximity among the hand-crafted taxonomies and the derived taxonomies, the method described in [5] is applied.

Results. There are three series of experiments: with the set N corresponding to $N_0 = transport$; with the set N corresponding to $N_0 = food$; with the set N corresponding to $N_0 = profession$. The FCA method uses the following vectors as the context:

1. as a baseline, the vectors from the incidence matrix of the words in pairs $Verb - Noun$ (for this case taxonomies were built and evaluated as described in [5] with the optimal value of the threshold t=0.005);
2. the semantic-syntactic word valence vectors for $Noun - Objects$ from the two-dimensional factorized matrix model;
3. the semantic-syntactic word valence vectors for $Noun - Objects$ from the three-dimensional factorized tensor model;
4. the semantic-syntactic word valence vectors for $Noun - Objects$ from the four-dimensional factorized tensor model.

Each time four taxonomies T_1, T_2, T_3, T_4 are built by FCA algorithm [8] as a result. Generated taxonomies have been examined for structural similarity to the reference WordNet taxonomies. The estimates are presented in Table 1.

Table 1. Estimates of structural similarity of the built taxonomies to standard taxonomies

N	T_1	T_2	T_3	T_4
$N_0 = transport$	67.39 %	70.21 %	76.47 %	78.73 %
$N_0 = food$	65.98 %	68.29 %	73.41 %	75.12 %
$N_0 = profession$	68.01 %	70.77 %	76.81 %	79.37 %

These estimates give the strong evidence of the sustainable advantage of semantic-syntactic word valence vectors over simple context vectors from the incidence matrix of words when the valence vectors are used in the Formal Concept Analysis method. There is a notable accuracy increase for the taxonomies built on semantic-syntactic word valence vectors when the number of tensor dimensions increases. It takes place because the multidimensional tensor model can capture subtle sophisticated connections between predicates ($Verbs$) and their arguments ($Nouns$) that are unavailable for simple two-dimensional models. Even the non-negative factorization of the two-dimensional matrix gives a notable improvement compared to the case when simple context vectors are used by the FCA algorithm because the Non-negative Matrix Factorization reliably extracts clear and stable relations from the incidence matrix. The further increase of the dimension number in the tensor model allows accumulating a larger amount of complex multidimensional semantic and syntactic data and allocating more semantic-syntactic relations that are used as *attributes* in the FCA method thereby improving the quality of the generated taxonomies.

5 Conclusion

The paper describes the use of semantic-syntactic word valence vectors as context vectors in the Formal Concept Analysis method for automatic construction of high-quality taxonomies. The research and experiments have confirmed a significant improvement in the quality of the constructed taxonomies with increasing number of tensor model dimensions while generating semantic-syntactic word valence vectors. The tensor model arity increase gives a more accurate description of the multidimensional semantic and syntactic relations and allows identifying more commutative semantic-syntactic properties of words used in the Formal Concept Analysis for building taxonomies of a higher quality.

References

1. Caraballo, S.A.: Automatic construction of a hypernym-labeled noun hierarchy from text. In: Proceedings of the 37th Annual Meeting of the ACL-99, pp. 120–126 (1999)
2. Hindle, D.: Noun classification from predicate-argument structures. Proceedings of the Annual Meeting of the ACL-90, pp. 268–275 (1990)
3. Pereira, F., Tishby, N., Lee, L.: Distributional clustering of english words. In: Proceedings of the 31st Annual Meeting of the ACL-1993, pp. 183–190 (1993)
4. Ganter, B., Wille, R.: Formal Concept Analysis: Mathematical Foundations. Springer, New York (1999)
5. Cimiano, P., Hotho, A., Staab, S.: Comparing conceptual, divisive and agglomerative clustering for learning taxonomies from text. In: Proceedings of the European Conference on Artificial Intelligence-2004, pp. 435–439 (2004)
6. Van de Cruys, T.: A non-negative tensor factorization model for selectional preference induction. J. Natural Lang. Eng. **16**(4), 417–437 (2010)
7. Van de Cruys, T., Rimell, L., Poibeau, T., Korhonen, A.: Multi-way tensor factorization for unsupervised lexical acquisition. In: Proceedings of COLING-2012, pp. 2703–2720 (2012)
8. Vychodil, V.: A new algorithm for computing formal concepts. In: Proceedings of the 19th European Meeting on Cybernetics and Systems Research, pp. 15–21 (2008)
9. Deerwester, S., Dumais, S.T., Furnas, G.W., Landauer, T.K., Harshman, R.: Indexing by latent semantic analysis. J. Am. Soc. Inform. Sci. **41**(6), 391–407 (1990)
10. Stanford Parser version 3.6.0 (2015). http://nlp.stanford.edu/software/lex-parser. shtml#Download
11. Lee, D.D., Seung, H.S.: Algorithms for Non-Negative Matrix Factorization. NIPS (2000). http://hebb.mit.edu/people/seung/papers/nmfconverge.pdf
12. Cichocki, A., Zdunek, R., Phan, A.-H., Amari, S.-I.: Nonnegative Matrix and Tensor Factorizations: Applications to Exploratory Multi-way Data Analysis and Blind Source Separation. Wiley, Chichester (2009)
13. Miller, G., Beckwith, R., Fellbaum, C., Gross, D., Miller, K.: Introduction to WordNet: An On-line Lexical Database. http://wordnetcode.princeton.edu/5papers.pdf

Generating Tailored Classification Schemas for German Patents

Oliver Pimas[1]([✉]), Stefan Klampfl[1], Thomas Kohl[2], Roman Kern[1], and Mark Kröll[1]

[1] Know-Center GmbH Graz, Graz, Austria
{opimas,sklampfl,rkern,mkroell}@know-center.at
[2] SKF GmbH Schweinfurt, Schweinfurt, Germany
thomas.kohl@skf.com

Abstract. Patents and patent applications are important parts of a company's intellectual property. Thus, companies put a lot of effort in designing and maintaining an internal structure for organizing their own patent portfolios, but also in keeping track of competitor's patent portfolios. Yet, official classification schemas offered by patent offices (i) are often too coarse and (ii) are not mappable, for instance, to a company's functions, applications, or divisions. In this work, we present a first step towards generating tailored classification. To automate the generation process, we apply key term extraction and topic modelling algorithms to 2.131 publications of German patent applications. To infer categories, we apply topic modelling to the patent collection. We evaluate the mapping of the topics found via the Latent Dirichlet Allocation method to the classes present in the patent collection as assigned by the domain expert.

Keywords: Patent classification · German patents · keyword extraction · LDA

1 Introduction

Since patent applications and patents reflect a company's know-how and value, it is important (i) to keep track of published patent applications of others, and (ii) to use it as a source for technology intelligence. Official classification such as the International Patent Classification (IPC)[1] offered by patent offices are not applicable to functions, applications, or divisions of a company.

In this work, we present a first step towards automatically generating patent schemas tailored to a company's needs. We approach this unsupervised learning scenario by applying keyword extraction as well as clustering techniques to 2131 publications of German patent applications. First, we evaluate three keyword extraction algorithms to identify representative candidates for customized categories. Based on a domain expert's feedback, we adjust the keyword extraction,

---

[1] http://www.wipo.int/classifications/ipc/en/.

© Springer International Publishing Switzerland 2016
E. Métais et al. (Eds.): NLDB 2016, LNCS 9612, pp. 230–238, 2016.
DOI: 10.1007/978-3-319-41754-7_20

for instance, to account for characteristics of the German language. To evaluate these adjustments, we compare rank positions of the extracted key terms and the ones provided by the domain expert. Second, we use Latent Dirichlet Allocation (LDA) to identify topics within the patents. Ideally, the resulting topics reflect a company's customized patent classes. We compare the resulting topics with the domain expert's class assignment. A visual inspection shows that LDA is capable of identifying some of the expert's classes but does not yield an optimal coverage.

2 Key Term Extraction

Observing a domain expert classifying patents shows heavy reliance on specific trigger words, which we call *weak signals*. We thus compare and evaluate three extraction methods, integrating the domain expert's feedback.

Our dataset is a collection of 2131 publications of German patent applications, registered between 2003-01-01 and 2012-12-31. We only process the patent description and claim and do not exploit any information specific to this data set. As a first step towards a tailored classification schema we aimed at extracting weak signals in the form of key terms, which would then serve as candidates for custom categories. We generated such key terms by employing and comparing three different approaches: (i) *Tf-idf*, (ii) *TextSentenceRank*, and (iii) *GriesDP*. Each of these methods yields a comparable score that allows a ranking of the resulting key terms. In all approaches we restricted these terms to the stems of common and proper nouns. For display purposes we associate the stems with the most frequently occurring word form.

Tf-Idf (term frequency-inverse document frequency) is a common weighting schema used in information retrieval [9]. It reflects the importance of a word by normalizing the occurrences of a term in a document by the number of occurrences in the corpus. Several variants of the tf-idf measure exist; here we assign to each term t and document d a weight

$$w(t, d) = f_{t,d} \cdot \log \left(1 + \beta \cdot \frac{N}{1 + n_t} \right) \tag{1}$$

where $f_{t,d}$ is the raw frequency of term t in document d, N is the total number of documents, and n_t is the number of documents, the term t occurs in. β is a configurable trade-off parameter that weights the influence of the inverse document frequency against the term frequency; here we choose $\beta = 1$.

To arrive at a global ranking of terms t, we have to aggregate these scores in (1) over documents d. Simply choosing the maximum weight for each term over documents would often yield top-ranking terms that are very specific and occur only in single documents. Instead, we empirically chose the 95th percentile, $TF(t) = P_{95} \{w(t, d)\}_d$, where $\{\cdot\}_d$ denotes the distribution over documents. In this way, we skip the highest 5 % of weights for each term, and thus arrive at key terms that are specific for a broader set of documents.

The **TextSentenceRank** algorithm [10] is a graph based ranking method to natural language text, returning a list of relevant key terms and/or key sentences ordered by descending scores. It builds a graph where vertices correspond to sentences or tokens and connects them with edges weighted according to their similarity (textual similarity between sentences or textual distance between words). Furthermore, each sentence vertex is connected to the vertices corresponding to the tokens it contains.

The score $s(v_i)$ of a vertex v_i is given by

$$s(v_i) = (1 - d) + d \sum_{j \in I(v_i)} \frac{w_{ij}}{\sum_{v_k \in O(v_j)} w_{jk}} s(v_j), \tag{2}$$

where d is a parameter accounting for latent transitions between non-adjacent vertices (set to 0.85 as in [2,7]), w_{ij} is the edge weight between v_i and v_j, and $I(v)$ and $O(v)$ are the predecessors and successors of v, respectively. The scores can be obtained algorithmically by an iterative procedure, or alternatively by solving an eigenvalue problem on the weighted adjacency matrix. The TextSentenceRank algorithm is an extension to the original TextRank algorithm [7], which could either compute key terms or key sentences. For the application at hand, we concatenate all patents into a single document and extract the key terms together with their scores as our ranked list of key terms.

GriesDP. The third method is the normalized dispersion measure proposed by Gries in [3], which we call *GriesDP*:

$$DP(t) = \frac{1}{1 - \frac{1}{n}} \sum_{i=1}^{n} \frac{|s_i - v_{t,i}|}{2}. \tag{3}$$

It compares the relative sizes of n disjoint corpus segments s_i (here we choose the document boundaries) with the actual relative frequencies $v_{t,i}$ of a given term t inside these segments. The values of the normalized measure $DP(t)$ lie between 0 and 1. Small values indicate that t is distributed across the segments as one would expect given the corpus sizes, whereas large values indicate that t occurs much more in certain subparts of the corpus than in others. Terms with small dispersion values can also be interpreted as corpus-specific stop words. Here, we chose the terms with the largest dispersion values as our list of key terms.

Comparison of Extraction Algorithms. Table 1 compares the top 10 key terms generated by each of the three extraction algorithms, Tf-idf, TextSentenceRank, and GriesDP. The ranking is determined by the scores assigned by the respective algorithm. Across algorithms these scores vary in terms of their value range. As expected, all of the algorithms yield mostly technical terms specific to the given application domain. Interestingly, however, the actual ranked lists of key terms appear to be quite different for different extraction methods. Perhaps the most apparent difference is that TextSentenceRank yields much more general terms, such as "Richtung" (direction) or "Beschreibung" (description), compared to Tf-idf and GriesDP, which tend to extract words describing

Table 1. Top 20 key terms generated by each of the three extraction algorithms ranked by the assigned score. Score values are rounded to two or three decimal places, respectively.

Tf-idf		TextSentenceRank		GriesDP	
Radialkanal	223.33	radial	0.189	Axialsteg	0.994
Druckmittelübertragungselement	184.01	Außenring	0.162	Lagerscheibe	0.993
Wälzlageraußenring	171.43	Wälzlager	0.148	Klauenring	0.993
Dichtsteg	160.42	Wälzkörper	0.137	Zwischenplatte	0.990
Ringkanalöffnung	157.27	Innenring	0.113	Funktionsring	0.989
Funktionsring	146.26	Käfigring	0.108	Rohrabschnitt	0.988
Kreisringscheibe	146.26	Beschreibung	0.102	Befestigungsschenkel	0.987
Rollenkörper	124.25	Richtung	0.100	Radialborden	0.986
Montageschiene	111.66	Ausführungsform	0.099	Rampe	0.986
Distanzstück	110.09	Lagerring	0.098	Kreisringscheibe	0.984

specific mechanical parts. GriesDP and Tf-idf share a number of terms in their top 20 list, although at quite different positions.

2.1 Incorporation of Expert Feedback into Extraction Metrics

To evaluate the extracted key terms as suggestions, a domain expert was asked to select around 1000 relevant terms. We found that among the top 1000 terms obtained by each of the algorithms, Tf-idf yielded 52, TextSentenceRank matched only 1, and GriesDP extracted 7 of the terms that were marked as relevant by the expert. This distinguishes Tf-idf as the most promising algorithm for the application at hand.

In the following we aim at modifying this measure as to incorporate insights obtained from the expert feedback:

1. Terms selected by the expert are on average longer than the average length across all extracted key terms from the patents. Figure 1 shows that there is a considerable difference in the distributions of lengths, both in terms of the number of characters and the number of syllables.
2. Terms selected by the expert typically consist of compound words. Since in German compounds are agglutinated into single words, this is also related to the first insight.
3. Terms selected by the expert tend to occur more in the beginning of the document. Figure 2 compares the distributions of relative document positions of relevant and non-relevant phrases. It can be seen that relevant phrases occur more often in the first quarter and less often in the second quarter of the documents.

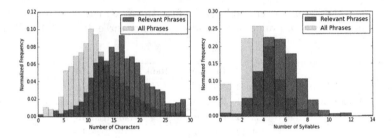

Fig. 1. Comparison of key phrase lengths in terms of characters and syllables between all extracted phrases (pink) and phrases marked relevant by the expert. (Color figure online)

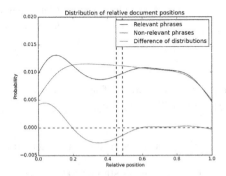

Fig. 2. Comparison of the distributions of relative document positions of relevent phrases (blue) and non-relevant phrases (green). For display purposes, the distributions were smoothed by kernel density estimation. The difference of the distributions is shown in red. The blue and green dashed lines denote the average relative document positions of relevant (0.449) and non-relevant phrases (0.486), respectively. (Color figure online)

We incorporated these insights into an *modified Tf-idf* metric, which combines the original Tf-idf metric, $TF(t)$ (1), with additional scores in a weighted manner. More precisely,

$$TF_{mod}(t) = w_{orig} \cdot TF(t) + w_{comp} \cdot S_{comp}(t) + w_{relComp} \cdot S_{relComp}(t) + w_{loc} \cdot S_{loc}(t), \tag{4}$$

where S_{comp}, $S_{relComp}$, and S_{loc} are scores capturing the influence of the term's components, the related compounds, and the relative location in the document, respectively. The scores are weigthed by $w_{loc} = 0.7$ and $w_{orig} = w_{comp} = w_{relComp} = 0.1$, which were heuristically chosen. Both the scores S_{comp} and $S_{relComp}$ take into account the composition of terms into a larger compound terms, which is a particular property of the German language. The component score S_{comp} of a term t computes the average Tf-idf score of its components $t_1, \ldots, t_n$,

$$S_{comp}(t) = \frac{1}{n} \sum_{i=1}^{n} TF(t_i). \tag{5}$$

For example, the component score of "Kreisringscheibe" (circular disk) would be computed from the Tf-idf scores of "Kreis" (circle), "Ring" (ring), and "Scheibe" (disk). The related compound score $S_{relComp}$ of a term t is the average score of all compounds that share at least one component with t,

$$S_{relComp}(t) = \frac{1}{n} \sum_{i=1}^{n} \frac{1}{m_i} \sum_{j=1}^{m_i} TF(c_j), \tag{6}$$

where $c_1, \ldots, c_{m_i}$ are compound words that consist of the i-th component of t. For "Kreisringscheibe" all compounds consisting of the components "Kreis", "Ring", and "Scheibe" would be considered, e.g., "Lagerscheibe" (bearing disk) or "Klauenring" (guide ring).

The relative location based score S_{loc} captures the characteristics of the distribution of relevant phrases over the relative position in the patent document (Fig. 2). It collects all occurrences of a term t in the dataset and computes their positions within the document, normalized to a value between 0 and 1. The contribution of each position to the overall score is then given by the value of the difference D between the probability density functions of the relevant and non-relevant phrases (i.e., the red curve in Fig. 2):

$$S_{loc}(t) = \sum_{d} \sum_{i} D(l_{t,d}^{i}), \tag{7}$$

where $l_{t,d}^{i}$ is the relative location of the i-th occurrence of term t in document d. Terms that occur more often in the first quarter of the document are boosted, while terms that occur more often in the second quarter of the document are penalized. Evenly distributed terms receive a neutral score, since $\int_{l} D(l)\,dl = 0$.

Table 2. Performance of the Tf-idf measures and modification scores evaluated on the position of relevant phrases among the top 1000 extracted terms. "Average rank" is the mean ranking position of relevant phrases; "Weighted average rank" computes the average by additionally weighting each term with its respective score value.

Measure	Average rank	Weighted average rank
TF (1)	543.81	332.82
S_{comp} (5)	604.77	481.67
$S_{relComp}$ (6)	451.30	293.78
S_{loc} (7)	497.18	284.06
TF_{mod} (4)	434.27	302.42

To evaluate the Tf-idf modifications we measured the rank positions of the relevant terms provided by the expert feedback. These values are shown in Table 2, which compares both the average rank $\frac{1}{n} \sum_{i} r_i$ of n positions $r_1, \ldots, r_n$, and the average rank weighted by the respective scores $s_1, \ldots, s_n$, $\frac{\sum_{i} s_i r_i}{\sum_{i} s_i}$, among the top 1000 key terms extracted by the algorithm in question. The weighted average rank also accounts for differences in the assigned score values, even if the ranking positions are similar. It can be seen that the modified Tf-idf algorithm, TF_{mod} (4), provides a considerable improvement over the standard Tf-idf method, TF (1), in terms of both performance measures. On average, relevant phrases are ranked more than 100 positions higher. Table 2 also evaluates the individual scores S_{comp}, $S_{relComp}$, and S_{loc}. It indicates that the scores based on the relative location of the phrases within a document, S_{loc}, and the related compounds, $S_{relComp}$, provide the largest potential for improvement. On the other hand, the component score S_{comp} alone would not be beneficial for the ranking of relevant terms. It is worth noting here, that the average rank of the modified Tf-idf measure, which combines all four other variables, is better than those of all of the other measures alone.

3   Category Inference

In order to automate the classification process we try to model the latent topics present in a patent collection. Companies often organise patents according to their applications or function, or to fit a company's divisions or areas. We expect such categories to share common concepts or ideas.

3.1   Latent Dirichlet Allocation (LDA) on Patent Collections

Following this hypothesis, we apply LDA [1], an unsupervised learning algorithm, that models topics as a certain probability distribution over a limited vocabulary. We filter the vocabulary by removing stop words and only considering noun phrases when computing LDA. As described in Sect. 2.1, a domain expert selected around 1000 relevant terms from our suggestions. This expert also assigned 63 classes to those relevant terms.

To evaluate how well the topics map this target classes, we set $k = 63$ and look at the words with the highest probabilities $P(w|\theta)$ for a certain topic θ. If a topic θ most likely produces relevant terms from a single class c, θ corresponds to c.

Figure 3 shows four latent topics uncovered by LDA, each represented by the 25 terms it most likely produces. Terms that were annotated as relevant by the expert are highlighted in different colors based on the class assigned by the expert. Thus, extracted topics that contain relevant terms highlighted in a single color can be mapped to the respective class. For example, topics 13 and 14 can be mapped to the classes *gearing* and *clutches*, correspondingly. On the other hand, different highlighting colors within an extracted topic means that it produced relevant terms from different classes with a high probability, therefore we cannot map it to one of the classes provided by the expert.

Topic 8:	Topic 9:	Topic 13:	Topic 14:
lagerring	elektromotor	welle	reibungskupplung
arbeitskammer	hebel	maschinen	betätigungselement
druckmittel	stößel	schaltmuffe	anpressplatte
rotor	druckstift	verzahnung	hebelfederzungen
gehäuse	antriebsstrang	getriebes	anpresskraft
druckmittelleitung	schaltgetriebe	modulationseinheit	trägerzungen
arbeitskammern	anker	gangrad	doppelkupplung
wälzkörperreihe	hebelende	planetenträger	spanndrahtring
außenumfang	druckstifts	getriebe	druckring
druckspeicher	aktor	schallemission	einrückkraft
lageranordnung	ritzel	schaltgabel	betätigung
stator	druckstifte	hauptantrieb	zustand
verriegelungskulisse	fahrzeug	maschine	gegenplatte
lagerrings	elektromagneten	getriebeeingangswelle	richtung
innenumfang	kardanwelle	ansicht	hebelfeder
schaltventil	ausführungsform	übersetzung	kupplungsdeckel
spät	abtriebsorgan	dämpfungselement	wesentlichen
anschlagstellung	drehmoment	lagerung	form
pendelrollenlager	getriebeeingangswelle	antriebsstranges	betätigungselementes
verriegelungsstift	abtriebswelle	verbindung	kupplungsscheibe
wälzkörperreihen	elektromotors	schlittens	phase
brennkraftmaschine	ausgestaltung	schlitten	auflagedrahtring
rotors	fahrzeugs	schiene	kraftfahrzeugmotors
richtung	verbrennungsmotor	getriebeelement	steifigkeit
druckmittelleitungen	magnetgehäuse	gehäuse	schließen

Fig. 3. Selection of 4 latent topics uncovered by LDA.

4 Related Work

Patents contain a valuable body of knowledge useful to many communities such as industry, business or research. Not only do they document technological advancements of our society, ideas expressed in patents can also reveal academic as well as business trends (cf. [4]). To gain access to this knowledge, techniques and methods from many areas including natural language processing, information retrieval, and machine learning are being applied to patents (cf. [11]). The broad spectrum of tasks when it comes to patents include Patent Retrieval, Patent Classification, Patent Visualization, Patent Valuation, and Cross-Language Mining (cf. [12]).

Key terms play an important role in the classification of patents as well. Larkey (cf. [6]), for instance, presents a patent classification system, which employs representative terms and phrases that are weighted by frequency and structural information. Kim et al. (cf. [5]) focused on structural elements of a patent as well; specifically they used so-called semantic elements such as claims, purposes, or effects. Pesenhofer et al. (cf. [8]) extracted key terms from respective Wikipedia pages, instead of the patents themselves.

5 Conclusion and Future Work

This work is inspired by a domain expert's observation during manually assigning patents to company related classes. We deploy three different key term extraction algorithms to identify relevant class descriptors, improving results by taking German language characteristics into account. To infer a classification schema, we apply topic modelling, which provides us with a flat structure of categories. Results of this work inform the development of a patent research application for SKF[2], a leading global technology provider that aims at facilitating and thus improving in-house retrieval processes.

---

[2] http://www.skf.com/group/index.html.

Acknowledgements. The authors thank Stefan Falk for helping with the experiments. The Know-Center is funded within the Austrian COMET Program under the auspices of the Austrian Ministry of Transport, Innovation and Technology, the Austrian Ministry of Economics and Labour and by the State of Styria. COMET is managed by the Austrian Research Promotion Agency FFG.

References

1. Blei, D., Ng, A., Jordan, M.: Latent dirichlet allocation. J. Mach. Learn. **3**, 993–1022 (2003)
2. Brin, S., Page, L.: Reprint of: the anatomy of a large-scale hypertextual web search engine. Comput. Netw. **56**(18), 3825–3833 (2012)
3. Gries, S.T.: Dispersions and adjusted frequencies in corpora. Int. J. Corpus Linguist. **13**(4), 403–437 (2008)
4. Jung, S.: Importance of using patent information. Most intermediate training course on practical intellectual property issues in business (2003)
5. Kim, J.H., Choi, K.S.: Patent document categorization based on semantic structural information. Inform. Process. Manage. **43**(5), 1200–1215 (2007)
6. Larkey, L.: Some issues in the automatic classification of us patents. Massachusetts Univ Amherst Department of Computer Science (1997)
7. Mihalcea, R., Tarau, P.: Textrank: bringing order into texts. In: Conference on Empirical Methods in Natural Language Processing. Barcelona, Spain (2004)
8. Pesenhofer, A., Edler, S., Berger, H., Dittenbach, M.: Towards a patent taxonomy integration and interaction framework. In: Proceedings of the Workshop on Patent Information Retrieval (2008)
9. Salton, G., McGill, M.J.: Introduction to Modern Information Retrieval (1986)
10. Seifert, C., Ulbrich, E., Kern, R., Granitzer, M.: Text representation for efficient document annotation. J. Univers. Comput. Sci. **19**(3), 383–405 (2013)
11. Tseng, Y.H., Lin, C.J., Lin, Y.I.: Text mining techniques for patent analysis. Inform. Process. Manage. **43**(5), 1216–1247 (2007)
12. Zhang, L., Li, L., Li, T.: Patent mining: a survey. SIGKDD Expl. Newslett. **16**(2), 1–19 (2015)

A Platform Based ANLP Tools
for the Construction of an Arabic
Historical Dictionary

Faten Khalfallah[✉], Handi Msadak, Chafik Aloulou,
and Lamia Hadrich Belguith

Faculty of Economic Sciences and Management, MIRACL, Sfax, Tunisia
{Faten.khalfallah105,Handi.msadak,Chafik.aloulou,
lamia.belguith}@gmail.com

Abstract. In this paper, we provide for the linguists a method to facilitate the creation of a standard Arabic historical dictionary in order to save the lost period and to be up to date with other languages. In this method, we propose a platform of Automatic Natural Language Processing (ANLP) tools which permits the automatic indexing and research from a corpus of Arabic texts. The indexation is applied after some pretreatments: segmentation, normalization, and filtering, morphological analysis. The prototype that we've developed for the generation of standard Arabic historical dictionary permits to extract contexts from the entered corpus and to assign meaning from the user. The evaluation of our system shows that the results are reliable.

Keywords: Historical dictionary · Platform · ANLP · Arabic · Corpus · Indexation · Morphological analyzer · Stemming

1 Introduction

The need to a better comprehension of the actual language, its culture recognition in its period and a concentration to the linguistics toward the fast evolution of the language nowadays oblige linguists to develop huge references in the different languages supporting the big volume of vocabularies and presenting its evolution by the time, it's the historical dictionary. Fortunately, there are many linguists who recognized this problem since a long time and tried to collect the history of their languages by developing historical dictionaries such the French historical dictionary and the English one. But, unfortunately, because of its complexity and semi algorithmic nature of its morphology employing numerous rules and constraints on inflexion, derivation and cliticization, the Arabic language is not yet stored although its importance and richness. That's why, we will propose in this paper a solution that can help linguists to save the time already passed and build their own database presenting the Arabic vocabularies from its birth and describing its evolution historically and geographically.

E. Métais et al. (Eds.): NLDB 2016, LNCS 9612, pp. 239–248, 2016.
DOI: 10.1007/978-3-319-41754-7_21

2 State of the Art

A dictionary is a reference database containing a set of words of a language or a domain of activity, generally represented under the alphabetic order. In general, a dictionary indicates the root, the definition, the spelling, the sense and the syntax of the entry [1, 2]. The general structure of a dictionary can have the form of: {Key = Description}, where the keys are generally words from the language and the descriptions are sets of words representing the definitions, the explanations or the correspondences synonym, antonym, translation, etymology) [1, 3]. There are two main types of dictionaries: the classical and the electronic, their content is necessarily similar but the main difference between them is not only in the presented information but also in its use, the content display, and the research capacity.

In fact, Arabic dictionaries started from an early time, exactly in the year 718 and was developed from a version to another till the present. The problem is that some entries which exist in a dictionary and don't in another one because of the evolution of the period. Moreover, it is not practical for the linguists to use the separated dictionaries in order to look for the evolution of a word sense from different dictionaries. So the majority of the old dictionaries have been neglected and some entries have been disappeared from a dictionary to another.

Historical dictionaries are available for many languages (French, English, Swedish, German,…), but unfortunately, there isn't an Arabic one until now although the different initiatives in different regions in the world which didn't exceed the collect phase although our awareness that the fact of having a historical dictionary will be beneficial for the most of researchers in different fields (linguistics, history, etymology …). That's why, as researchers, we found that it's necessary to think about an Arabic historical dictionary which will collect all the entries from the first appearance until the present.

The idea of historical dictionaries has started in the 19th century just after the appearance of the historical analysis method. Multiple international projects were launched in different countries aiming to create a historical dictionary. The main objective of the idea was to describe the evolution of word in language. We can take as an example the Arabic word قطار *kitar* which had 3 main senses from the year 39 BD till nowadays. First, it meant "المطر the rain" till the year 283, then "مجموعة من الإبل a set of camels" till 1307 to become "قطار train" until now.

In the following part, we will enumerate some of them.

2.1 DHLF: Historical Dictionary of the French Language

It was an etymological and historical dictionary written by Marianne Tomi, Tristan Hordé, Chantal Tanet and Alain Rey and edited by LE ROBERT dictionaries [4]. Its last version appeared in 2012 containing more than 50000 entry and up to 40000 paper which offer a complete synthesis about the evolution of the French words. This dictionary is characterized by its simplicity and clearness, it is based on the dictionaries build by the linguists and the lexicologists in the XVII, XVIII and XIX centuries. This dictionary didn't use any Automatic Natural Language Processing (ANLP) process. It permits just to have access to its contents in sequentially. After the access to this dictionary, we noticed that it presents for each entry its morphology, the appearance date, the origin, the sense and its evolution across the time.

2.2   DWB: German Historical Dictionary

The DWB has been the most important German dictionary since the XVI century. 32 volumes were published between 1838 and 1971. It contains the history of each word by some citations. The use of such dictionary can recognize the ancient use and the modern use of each entry [4]. After the study of this dictionary, we noticed that it doesn't include automatic language processing for its creation, but it gives access to other dictionaries in its results. It displays for each entry the different senses, the different papers containing the entry in addition to its date and links to search in other dictionaries.

2.3   OED (English Historical Dictionary)

Oxford English Dictionary is a reference dictionary for the English Language which is published by the Oxford University. This project was launched in 1857 by the poet Richard Chenevix Trench [6] who has shown the different problems of the English dictionaries and announced his motivation to begin to work on a historical dictionary in a conference. The first edition of this dictionary was published in 1928 containing about 414825 paper. This dictionary was able to be online since 1992 but there is no ANLP [4, 5]. As it is described in the different previous works, the creation of a historical dictionary requires a long time to classify the right information in its right place. That's why, as computer scientists, we decided to develop a Platform using the Automatic Natural Language Processing tools to guide the linguists to the creation of a historical dictionary for the Arabic language to save the lost time and be able to update the dictionary easily.

3   Proposed Method

Our project consists on developing a platform of ANLP tools in order to help linguists create a historical dictionary of the Arabic language. Thus, we've been inspired from the different outputs of the historical dictionaries already done and we've noticed that such a platform requires fundamental steps as follows: The corpus collection, the Natural Language Processing and the creation of the index which will facilitate the access to the corpus for the research. The Fig. 1 describes the proposed steps.

The Arabic language needs appropriate processing and analysis to be fully exploited. Indexation and research are efficient as long as Arabic content has been automatically processed through complex stages: mining, segmentation, normalization, morphosyntactical analysis, stemming, tagging, etc. By combining power and accuracy, this analysis enables highly structured indexing and high performance in research. The following figure resumes the proposed system tasks (Fig. 2).

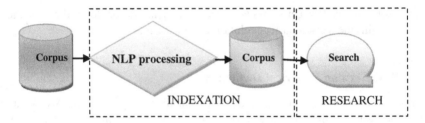

Fig. 1. General schema of the proposed platform

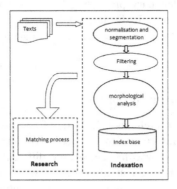

Fig. 2. System tasks

3.1   Indexation

The indexation will facilitate the access to a document content. In this section, we will present our proposed method detailing the different pretreatments.

(A) *Corpus creation*

Our corpus is collected from the LDC and it is a set of heterogeneous texts from different domains (literature, poetry, prose, dictionaries, etc.). It contains 30 texts from different periods from 700 to 1300, for every period, we reserved 5 texts. These texts will be treated in order to be indexed in units, that's why a segmentation task is necessary.

(B) *Segmentation*

The segmentation [13] is a process used for the morphological treatment of the language allowing to divide a text to a set of words called tokens. It's necessary for our system in order to identify the fields (author, title, date, place, …) and store the records. Since our target is to segment an Arabic text in words, we will just explain the Arabic words limits rules. The approach consists on the analysis of the punctuation marks, the special characters, and spaces and a list of particles as the coordination and conjunctions which are considered as segmentation criteria and act as Arabic words separators. Our method permits to browse the text character by character and eliminate them.

Moreover, this method permits to eliminate all the numbers and the non-Arabic characters. Unfortunately, the outputs of this step are not clean because there are a lot of words which don't bring any interest to the indexation process, that's why, it's necessary to filter them.

(C) *Filtering*

After segmenting the texts, it is important to clean the texts by deleting all the stop words. This step is a delicate task because Arabic is an inflected language and highly differentiable. And since, the stop words are non significant words and very commune in multiple texts, it's not necessary to index them or to look for them in the research phase. The stop words (i.e. بسبب, عليها, بعض, ف, ثم, etc.) are generally linguistic tools used to link or coordinate the sentences. And in order to eliminate these stop words, we used a data base in a txt file format which contains a list of non significant words. The filtered texts now contains multiple words having different forms but the same root, that's why, it is really recommended to provide a morphological analysis to study the words bare and unify them by the same descriptive.

(D) *Morphological analysis*

The morphological analysis consists on determining for each word, the list of possible morphosyntactic characteristics (category, gender, number, time, etc.) and their canonical forms, their roots, etc. For that, a stemming phase [13] is useful before starting the indexing process since it represents one of the most important treatments for the Arabic language, in the concept of information research which aims to find inflected forms of a word from its form represented in the document by the application of affixes truncation [4, 6, 12]. We have to remember that the form of Arabic words can have four affixes categories: the enclitic, prefixes, suffixes and proclitic. Besides, an Arabic word can have a more complicated if all the affixes are linked to its standard form. Therefore, we've identified a solution to this problem by the affixes removal approach which is called the light stemming [7, 11] allowing a remarkable way to produce an excellent information retrieval and allows a word to truncate at both extremities. In fact, various versions of Light Stemmer were developed for Arabic and have followed the same steps. The idea was to delete, first of all, the character «و» ("and") for the light2, light3, light8 and light10 [8, 9] if the rest of the word is composed of 3 characters or more. Although the deletion of the character «و» is important, it triggers a problem because it exists a lot of commune Arabic words starting by this character, thus, the length criteria here is stricter than the definite articles. Second, remove all the definite articles in the case where it stills two or more characters. Finally, access the list of suffixes once in the order from right to left while removing all that at the end of the word if there are two or more characters. As it is described previously, the light stemmers don't delete any string which will be considered as Arabic prefixes.

(E) *Index generation*

In the Indexation process, the analyzed words will be stored in a data structure called inversed index which is already stored in a file system or in memory as a set of index files. This index is a form of content representation which represents a relation which links each document to the set of keywords or descriptors. In addition, it permits to facilitate the access to the documents content [15]. In fact, we've represented our index in a table which contains the term frequencies (TF) of a given word in each document that contains the term and the texts which have appeared. Thus, we can determine for a text the list of words that can contain, and for a word the list of texts in which it appears from an IndexReader, using the term of interest.

3.2   Research

An information research system has an objective to output the most pertinent documents comparing with the request. This process consists on corresponding the list of generated index by an indexation system with the user request.

The research phase can be resumed by the schema presented in Fig. 3.

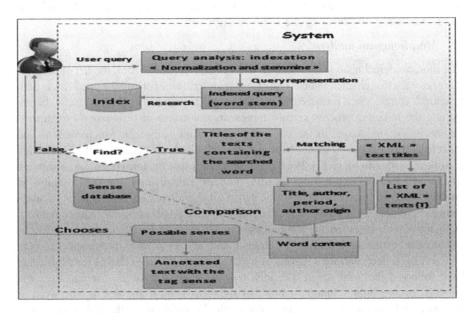

Fig. 3. Research system architecture

The research information can't be accomplished only by the user intervention. To be able to do a research, the user makes a query as a single word in everyday language. The system analyzes the content of the request (single word) and converts it into part of the indexing language after standardized and lemmatized.

Then, it compares the term of the query with the inversed index: if the result of the comparison returns true, the system will return the text titles containing the query word. Then, the system compares his titles with the titles of the "xml" files in the corpus and returns the texts with their titles, authors, authors origins and period and it extracts the contexts of this word from each text. For the extraction of these contexts, we've proposed a method which returns the left and the right context of a word. In fact, this method targets, first, to place the text content in a table (each word in a cell). Second, it permits to browse the table displaying every time the wished number of apparition left and right of the searched word.

4 Implementation and Evaluation

The aim of our application is to develop an ANLP tools platform enabling to build an Arabic historical dictionary from a corpus. The objective of this dictionary is to collect the vocabulary of the standard Arabic language and to describe its evolution (temporal and geographical aspects) which gives a reflection of civilizations and represents the witness of the mentalities evolution.

This historical dictionary will include a set of simple entries which will be analyzed separately. As we described in our proposed method, we need a segmentation tool to subdivide texts into separate simple words, filtering tool to eliminate the stop words from the corpus, a morphological analyzer including a stemmer, an indexation tool to match between the set of texts and different useful information and a research tool allowing to search in the indexed set of texts. Apache Lucene is a suitable technology for nearly any application that requires full-text search, especially cross-platform. We've used this tool for the needed tasks since it is a high-performance, full-featured text search engine library written entirely in Java.

In fact, Lucene stores the input data in a data structure called inverted index, which is already stored in the files system or in the memory as a set of index files. It allows users to execute some quick researches by keywords and to find documents responding a given query. Before adding the textual data to the index, it is treated by an Arabic analyzer which permits the conversion of textual data in fundamental units of research called terms. During the analysis, the text is set to various treatments: words extraction, removal of the stop words and the punctuation signs, words stemming. For our case, we've used the Arabic analyzer. The text analysis is done before the indexation and the query analysis. It converts the textual data in tokens, and these tokens are added to lucene index as terms. After finishing the indexation phase, we move to the research phase where we examine the index to find the words then obtain the documents containing these words.

Our corpus of texts contains 30 texts from different periods starting from 700 to 1300, for every period, we reserved 5 texts. The size of our corpus is 155 MO counting 2 millions words. The index is presented in a table in which it's displayed for every word its stem, root, possible vowellation, Part Of Speech, number of apparition of each word in a text and the title of the corresponding text. "The first version of our platform is composed of three main zones, the first is called The current text النصّ الجاري" in which the text will be uploaded to display, the second is called "Research operation عمليّـة البحث" which represents the research possibilities (with criteria, by entering a word to search, etc.) and the last zone which contains four parts "update texts تحديث النصوص", "update senses تحديث معاني الكلمات", "words senses معاني الكلمات" and "search results نتيجة البحث".

Our platform permits to add a new word from a new text with its relative information (sense, period and location) in the sense database, it permits also to add the tag "sense" in the xml texts next to the corresponding word and according to its context. The user can intervene to give a sense to the word, so that the sense data base can be updated according to the treated texts. The platform permits to access to all the different positions of the word in the corpus. So we can display the evolution of the use of the words.

To evaluate our work, we've chosen 100 words for testing and we've calculated the recall, the precision and the F-measure. The table below summarizes some of the result.

The Table 1 resumes the evaluation results of the research phase, the average of the F-measure obtained for the 100 words is equal to 90 %. The advantage of our system is in the fact of the indexation, because this step is done only one time for the entered texts of the corpus and will just take time when there is a corpus update. The response time in the research phase depends on the size of the text and of course on the computer characteristics. The following table shows some statistics after a research done on the word "Allah الله".

The Table 2 proves that the response time depends on the size of the text. The bigger is the text size, the longer is the response time.

Table 1. System evaluation according to the pertinence

Words	Doc	Returned	Not returned	Precision	Recall	F-measure
كتب kataba	Pertinent	25	2	92.60 %	100 %	96.16 %
	Non Pertinent	0	3			
ضاحكا dhahikan	Pertinent	6	1	85.71 %	100 %	92.30 %
	Non Pertinent	0	23			
قدمه kadamuhu	Pertinent	27	2	93.10 %	96.45 %	94.75 %
	Non Pertinent	1	0			
وجد Wajada	Pertinent	30	0	100 %	100 %	100 %

Table 2. Response time variation

Word	Texts	Text size (KO)	Words number in the text	Number of apparition in the text	Response time
الـلّـه	الشافية في علم التصريف	97	12 428	23	0 s
	عيون الحكمة	125	13 392	9	0 s
	تحفة القادم	270	30 623	151	1 s
	عيون الانباع	353	36 245	228	1 s
	كتاب الموازنة	581	64 892	203	1 s
	معجم الشعراء	792	90 617	763	4 s
	مفردات ألفاظ القرآن	853	86 092	1 130	5 s
	الجامع لأحكام القرآن	1 251	142 053	2 412	11 s

90 % of F-measure is an acceptable rate for our first system trial but we have to search for the different causes that didn't let the system provide 100 %. When we focused on the causes, we found that the main problem leading to this result is the scripts and the rich and complex morphology of the Arabic language which lack short vowels and other diacritics that distinguish words and grammatical functions. These linguistic complexities result in significant recognition difficulties. There are different flexion forms of the entry and various vowelling style of a simple root which increases retrieval effectiveness over the use of words or stems.

5 Conclusion

Our system aims to generate automatically an Arabic historical dictionary from a corpus through ANLP tools. Various tasks are done to the entered corpus for the indexation and the research: segmentation, filtering, morphological Analysis. Our system still need semantic resources to assign the sense to the entries according to their period. As future work, we will try to ameliorate the performance of our system using semantic analyzers, and grammatical rules to extract the suitable contexts. We are going to link our system to some electronic dictionaries from different periods to display for each entry its details from different sources spanning different periods. Besides, we propose to include the semantic analysis to recognize automatically for each word its signification without the human intervention.

References

1. McEnery, T., Wilson, A.: Corpus Linguistics: An Introduction, 2nd edn. Edinburgh University Press, Edinburgh (2001)
2. Wehr, H.: Arabic English dictionary. In: Cowan, M. (ed.) The Hans Wehr Dictionary of Modern Written Arabic, 4th edn. Spoken Language Services, Inc., Urbana (1994)
3. AlSaid. M.: A Corpus-based historical Arabic dictionary. Ph.D. thesis in linguistics, Cairo University (2010)
4. Kadri, Y., Nie, J.Y.: Effective stemming for Arabic information retrieval. In: Proceedings of the Challenge of Arabic for NLP/MT Conference, Londres, United Kingdom (2006)
5. Trench, R.: On some deficiencies in our English dictionaries: being the substance of two papers read before the philological society, London (1857)
6. Larkey, S.L., Ballesteros, L.: Light stemming for Arabic information retrieval. In: Arabic Computational Morphology, pp. 221–243 (2007)
7. Larkey, L., Ballesteros, S., Connell, E.M.: Improving stemming for Arabic information retrieval light stemming and co-occurrence analysis. In: Proceedings of the 25th Annual International Conference on Research and Development in Information Retrieval (SIGIR 2002), Tampere, Finland, August 11–15, pp. 275–282 (2005)
8. Larkey, S.L., Ballesteros, L., Connell, E.M.: Light stemming for Arabic information retrieval. In: Arabic Computational Morphology: Knowledge-Based and Empirical Methods (2005)
9. Souteh, Y., Bouzoubaa, K.: SAFAR platform and its morphological layer. In: Proceeding of the Eleventh Conference on Language Engineering ESOLEC'2011, 14–15 December, Cairo, Egypt (2011)

10. Habash, N.: Arabic natural language processing: words. In: Summer School on Human Language Technology Johns Hopkins University, Baltimore, 6 July 2005
11. Larkey, L., Ballesteros, S., Connell, E.M.: Light stemming for Arabic information retrieval, Arabic computational morphology: knowledge-based and empirical methods (2005)
12. Larkey, L., Ballesteros, S., Connell, E.M.: Improving stemming for Arabic information retrieval light stemming and co-occurrence analysis. In: Proceedings of the 25th Annual International Conference on Research and Development in Information Retrieval (SIGIR 2002), Tampere, Finland, pp. 275–282, 11–15 August 2002
13. Khalifa, I., Feki, Z., Farawila, A.: Arabic discourse segmentation based on rhetorical methods. Int. J. Electr. Comput. Sci. IJECS-IJENS **11**, 10–15 (2011)
14. Semmar, N., Elkateb-Gara, F., Fluhr, C.: Using a stemmer in a natural language processing system to treat Arabic for cross-language information retrieval. In: International Conference on Machine Intelligence, Tozeur, Tunisia (2005)
15. Jacquemin, C., Daille, B., Royante, J., Polanco, X.: In vitro evaluation of a program for machine-aided indexing. Inf. Process. Manag. **38**(6), 765–792 (2002)

Extracting Knowledge Using Wikipedia Semi-structured Resources

Nazanin Firoozeh[(⊠)]

Laboratoire d'Informatique de Paris-Nord, Université Paris 13,
Pixalione SAS, Paris, France
`nazanin.firoozeh@lipn.univ-paris13.fr`

Abstract. Automatic knowledge discovery has been an active research field for years. Knowledge can be extracted from source files with different data structures and using different types of resources. In this paper, we propose a pattern-based approach of extraction, which exploits Wikipedia semi-structured data in order to extract the implicit knowledge behind any unstructured text. The proposed approach first identifies concepts of the studied text and then extracts their corresponding common sense and basic knowledge. We explored the effectiveness of our knowledge extraction model on city domain textual sources. The initial evaluation of the approach shows its good performance.

Keywords: Wikipedia semi-structured resources · Knowledge discovery · Common sense knowledge

1 Introduction

Knowledge discovery is the nontrivial extraction of implicit, previously unknown, and potentially useful information from data [2]. Knowledge can be obtained from sources with different types of data structures: unstructured, structured and semi-structured. While structured and semi-structured sources have predefined data models, unstructured data has no organization to facilitate the task of extraction. Unstructured data files often contain a considerable amount of knowledge, which can be used in different applications of Artificial Intelligence.

In knowledge discovery, resources with different types of structures can be exploited [4]. The same as source data, resources are also in three types. Machine readable structured resources, such as thesauri, are easy to exploit but difficult to create and maintain. Due to these difficulties, they may not cover all domains and languages. Unstructured resources, on the other side, are collections of machine-unreadable multimedia content and extracting reliable knowledge from such resources is a very challenging task. Hence, structured resources extract knowledge with high accuracy but low coverage rate, while unstructured resources cover all the domains but the knowledge extracted from such resources is less reliable. In order to make use of the positive points of each type and reduce the limitations, in this paper we focus on semi-structured resources. Wikipedia is one

© Springer International Publishing Switzerland 2016
E. Métais et al. (Eds.): NLDB 2016, LNCS 9612, pp. 249–257, 2016.
DOI: 10.1007/978-3-319-41754-7_22

of the major resources, which is updated regularly and contains many statements in natural language. In this work, we exploit category names and infobox tables of Wikipedia as semi-structured resources in order to extract the implicit knowledge behind any given unstructured text.

Two kinds of knowledge are targeted in our work: basic and common sense (CS). By basic knowledge, we mean any kind of knowledge that provides basic information about the studied concept. Considering Paris as an example of a concept, information about its population, mayor, etc., can be considered as basic knowledge. Common sense knowledge is however defined as the background knowledge that human has. This type of knowledge is so obvious that is rarely expressed in documents. As an example, the fact that "Paris is a city" is a knowledge that majority of people know as a background knowledge.

The rest of the paper comprises the background (Sect. 2), the proposed model (Sect. 3), evaluation and results gained (Sect. 4), and conclusions and suggestions for future work (Sect. 5).

2   Background

Extracting knowledge from Wikipedia has been one of the interesting domains of research and different systems have been developed for extracting knowledge from Wikipedia. YAGO [9] is an ontology which makes use of Wikipedia. In this project, instead of using information extraction methods, categories of Wikipedia articles are used as resources of knowledge. This system makes use of both Wikipedia articles and WordNet database [5] in order to extract facts. One limitation of YAGO is that it extracts only 14 types of relations and some relations cannot be detected using this system. Authors also extended the system to YAGO2 [3], which is able to represent the facts along dimensions of time and space.

In another work, WordNet database is enriched using new relations extracted from Wikipedia [8]. One disadvantage of this system is that it needs to go through definition of each entity in Wikipedia to look for hyperlinks and to extract relations between different concepts. This increases the time complexity of the system. In [7], large scale taxonomy was built from Wikipedia. In that work, semantic relations between categories are found using a connectivity network and lexico-syntactic matching. Although this method extracts relation between category names, it cannot find relation between a specific concept and its corresponding category names. Wikipedia articles have been also used for extracting relations between different concepts of Wikipedia. In [10], an unsupervised approach is used along with linguistic analysis with web frequency information. The goal of using this analysis is to improve unsupervised classification performance. Unlike our approach, which focuses on semi-structured content of Wikipedia, here, unstructured parts of articles are taken into consideration.

DBpedia [1] is another knowledge extraction system, which is now available on the World Wide Web. The goal of DBpedia is first to extract structured information from Wikipedia and then to allow users to ask queries against it.

One drawback of DBpedia is that its data can be based on several months old data. DBpedia-live [6] solves this problem by providing a live synchronization method based on the update stream of Wikipedia.

Similar to the described systems, we also make use of Wikipedia articles as a rich resource of information. However, unlike YAGO, YAGO2 and the work done by Ruiz-Casado et al. [8], the system does not use WordNet database. Hence, the complexity of the system is reduced. In addition, while the work done by Ponzetto et al. [7] tries to find types of relations between different category names, our system finds the relations between any given concept and its related category names.

DBpedia can be considered as the closest work to our proposed approach. Comparing to DBpedia, our approach is simpler and less structured. This work is however an initial effort on knowledge extraction in order to propose a simple but efficient approach for extracting knowledge from Wikipedia. As will be seen later, as a future work, we aim to compare our extracted knowledge with DBpedia knowledge. Comparing the results, we will be then able to see if the proposed method can be considered as a complementary tool for DBpedia.

3   A Model of Extraction

Extracting a good amount of basic and common sense knowledge and using it in interactive applications is an important step in Artificial Intelligence field. In this work, we make use of Wikipedia as a resource of knowledge. Although Wikipedia has both unstructured and semi-structured content, we focus only on category names and infobox tables, which are semi-structured contents of Wikipedia.

In general, our model consists of four main steps: 1. HTML parsing, 2. concept extraction, 3. extracting infobox-based facts, 4. extracting category-based facts. It should be noted that by category-based and infobox-based facts, we respectively mean facts that are extracted from category names and infobox tables. Following is the detailed description of each step.

HTML Parsing. The input of the system is either a URL or a plain text. In the former case, the system directly starts from "Concept Extraction" step. Having a URL as the input of the system, the HTML source code is then parsed and the content of the page is extracted. In this step, some HTML tags which do not bring information about the content of the page are removed. Header and Footer are examples of these uninformative HTML tags.

Concept Extraction. Having the extracted text, the system then extracts its concepts using both part of speech information and Wikipedia titles. In fact, we assume that each Wikipedia title corresponds to one concept in the real world. Some concepts however have different representations. For these concepts, Wikipedia redirects users to the same article and shows the same page for different representations. Using the redirection feature of Wikipedia, we avoid generating concepts with the same meaning but different formulations. As an example, UK and United Kingdom are redirected to the same URL in Wikipedia. Hence, only one of them is taken into consideration for further processing.

Extracting Category-Based Facts. All Wikipedia articles have a category section, where navigational links to other Wikipedia pages are provided. In order to reveal the semantics encoded in category names of Wikipedia articles, first the Wikitext of the Wikipedia article is extracted. Wikitext is a markup language used for writing the content of wiki websites, such as Wikipedia. Category names of a Wikipedia article are extracted by parsing the Wikitext obtained as the answer of a proper query[1]. To avoid having uninformative category names, we ignore all the categories under Wikipedia administration.

In order to extract the knowledge behind each category name, we proposed 23 extraction patterns. To generate the patterns, different criteria such as type and position of the prepositions as well as occurrences of some keywords like Type, Establishment, Disestablishment, etc., are taken into consideration. Table 1 represents the proposed extraction patterns. These patterns encode relations between the studied concept and different concepts that appear in the category section of the article. For each category name, the system checks if it matches any of the patterns. If so, the associated fact(s) is returned. It is important to mention that in Table 1, X refers to the studied concept, and Y and Z refer to the concepts that appear in the category names. In addition, $YEAR$ simply indicates a year and Xs shows the plural form of X. $X1$, $X2$ and $Y1$, $Y2$ also show respectively the sub-terms of X and Y.

Category-based facts are in two types; "R-specific", which explicitly specifies types of relations and "R-generic", which just indicates that the concepts are related without explicitly showing type of the relation. In Table 1, patterns 1 to 22 generate R-specific facts, whereas R-generic facts are extracted using pattern 23.

Extracting Infobox-Based Facts. Infobox is another resource that we use for extracting knowledge. For each extracted concept, the system checks the existence of an infobox template in the extracted Wikitext. The template starts with "{{Infobox" and ends with corresponding "}}". This section is called infobox template and is used for further processing. If Wikitext of a concept does not contain this template, we stop extracting infobox-based facts for that concept.

As not all the infobox elements produce interesting facts, we first discard useless attributes including image, caption, logo, alt, coat, footnote, etc. We then define different extraction patterns as regular expressions for extracting attribute names and values. It should be noted that attributes with no informative value are discarded.

At the end of the extraction task, the extracted category-based and infobox-based facts are written as sentences in a text file.

---

[1] Example: http://en.Wikipedia.org/w/index.php?action=raw\&title=Paris.

Table 1. Extraction patterns and associated facts for category names of Wikipedia

	Extraction pattern	Extracted fact(s)
1	Y of X	Y is an attribute of X
2	Type(s) of Y	X has a type of Y
3	Y in X	X has Y
4	Y in Z (Y contains just letters)	X is a Y in Z
		X is a Y
		X is in Z
5	Y of Z	X is a Y of Z (Singular Y)
		X is one of the Y of Z (Plural Y)
6	Y in YEAR	In year YEAR, there was a Y of X
7	YEAR introductions	X was introduced in YEAR
8	X in YEAR	X was in YEAR
9	Y established in YEAR	X has been established in YEAR
		X is a Y (Y singular)
		X is one of the Y (Y plural)
10	YEAR establishment(s)	X has been established in YEAR
11	Y establishment(s) in Z	X has been established in Y (if Y contains digit)
		X is in Z
12	Y disestablished in YEAR	X has been disestablished in YEAR
		X was a Y (Y singular)
		X was one of the Y (Y plural)
13	YEAR disestablishment(s)	X has been disestablished in YEAR
14	Y disestablishment(s) in Z	X has been disestablished in Y (if Y contains digit)
		X was in Z
15	YX (one concept)	YX is one form of X
16	Y=(Y1 Y2) &X=(X1 X2) (if Y2 = X2)	X is a Y (Y is singular)
		X is one of the Y (Y is plural)
17	Y by type	X has a type of Y
18	Y invention	X is a Y invention
19	Y format(s)	Format of X is Y
20	YEAR birth	X was born in YEAR
21	YEAR death	X died in YEAR
22	Xs	X is a subset of Xs
23	If none of the above relations	X R Y (relates)

4   Evaluation and Results

Accuracy and reliability of the extracted knowledge is one of the main steps in knowledge discovery. Using incorrect knowledge in different applications decreases their performance. In this work, we evaluate our approach by defining three labels that indicate the quality of the extracted facts. The labels are correct, incorrect and ambiguous. A correct fact is a fact with a correct meaning and a proper formulation. Incorrect fact, on the other hand, refers to the fact with an incorrect meaning. Among the extracted facts, some have correct meanings but wrong formulations. These facts are labeled as ambiguous. By differently labeling these facts, we aim at differentiating them from the incorrect ones, since we believe that as a very first step of the future work, we can refine the system to get correct formulations for these cases.

The initial evaluation of the approach was performed manually by scanning all the facts and finding the ratio of the correct, incorrect and ambiguous facts. To do this, 10 students were asked to label the extracted facts. The facts were evaluated in terms of both meaning and formulation and the final label was assigned based on the majority vote. The students were also asked to label the facts as basic or common sense. Similarly, the final type of knowledge was determined based on the majority vote of the students.

To show the performance of the proposed approach, we took into account 10 input texts either from news websites or city-related web pages. From the ten input URLs, 313 concepts and 1443 facts were extracted. Having list of the extracted facts, each fact was evaluated by all the ten students. Table 2 shows examples of the extracted facts with different evaluation labels.

Table 2. Examples of the extracted facts

Fact	Label	Basic/CS
Paris is a capital in Europe	Correct	CS
Versailles is an art museum and gallery	Ambiguous	–
Eiffel Tower is a Michelin Guide	Incorrect	–
Latitude of Paris is 48.8567	Correct	Basic
Roof of Eiffel Tower is, abbr=on	Incorrect	–
Gini year of Spain is 2005	Ambiguous	–

Figure 1 compares the average rates of accuracy, error and ambiguity for both category-based and infobox-based facts and over all the evaluated examples. In the former case, the evaluation metrics are calculated over R-specific facts, since this type shows the performance of the extraction patterns. In calculations, accuracy, error, and ambiguity rates are obtained by respectively dividing the number of correct, incorrect, and ambiguous facts to the total number of the facts. It should be noted that in case of extracting semantically similar facts, only one of them is taken into consideration in the evaluation phase. According to Fig. 1, the average rate of accuracy for the extracted infobox-based facts is higher than the one for category-based facts. However, difference in their error rates is minor.

The ratio between R-specific and R-generic category-based facts is shown in Fig. 2. As expected, more facts are categorized as R-generic. The reason is that many of the category names have no specific structure. Figure 3 also compares the amount of common sense and basic knowledge in both category-based and infobox-based facts. According to the figure, infobox table is not a good resource for extracting common sense knowledge. This is due to the fact that Wikipedia has infobox tables for proper nouns such as Paris and not for general concepts such as shopping. As a result, the extracted facts in most cases cannot be considered as general knowledge. On the other hand, category names are better resources for extracting common sense knowledge.

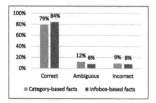

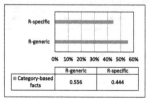

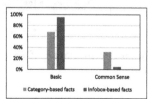

Fig. 1. Comparing the average rates of correct, incorrect and ambiguous facts. (Color figure online)

Fig. 2. Ratio of R-generic and R-Specific facts.

Fig. 3. Comparing types of the extracted facts. (Color figure online)

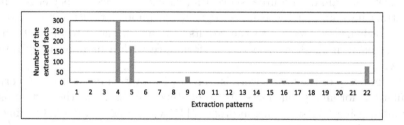

Fig. 4. Total number of the facts extracted by each pattern over five input texts.

We also evaluated the performance of the proposed extraction patterns. In this step, we studied only pattern 1 to 22, which generate R-specific facts. Figure 4 shows the total number of the facts extracted using each pattern.

Performance of each pattern is also represented in Fig. 5. According to the figure, patterns 3, 10, 17, 19, 20, and 21 have the highest accuracy rate. However, these patterns extract a few numbers of facts. Considering the patterns with high rates of extraction, i.e., patterns 4, 5 and 22, pattern 22 outperforms the other two patterns due to the high rate of accuracy (96.06 %). On the other side, pattern 15 cannot efficiently extract knowledge from the resources.

According to Fig. 4, no fact is extracted using patterns 11, 12, 13, and 14. In fact, these patterns extracted facts which were semantically similar to the previously extracted facts. However, it is important to keep these patterns, as due to their structure, in some cases they might extract new facts.

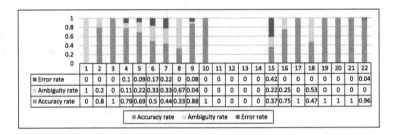

	1	2	3	4	5	6	7	8	9	10	11	12	13	14	15	16	17	18	19	20	21	22
Error rate	0	0	0	0.1	0.09	0.17	0.22	0	0.08	0	0	0	0	0	0.42	0	0	0	0	0	0	0.04
Ambiguity rate	1	0.2	0	0.11	0.22	0.33	0.33	0.67	0.04	0	0	0	0	0	0.22	0.25	0	0.53	0	0	0	0
Accuracy rate	0	0.8	1	0.79	0.69	0.5	0.44	0.33	0.88	1	0	0	0	0	0.37	0.75	1	0.47	1	1	1	0.96

■ Accuracy rate   ■ Ambiguity rate   ■ Error rate

Fig. 5. Performance of each extraction pattern over five input texts. (Color figure online)

5   Conclusion and Future Work

This paper addresses the problem of discovering knowledge behind an unstructured data using semi-structured resources. Information expressed in infobox tables and category names of Wikipedia are specifically used for this purpose. The proposed pattern-based approach aimed at extracting both basic and common sense knowledge behind any concepts of an unstructured text. Initial evaluation on city domain data sources shows that category names are better resources for extracting common sense knowledge, while infobox tables mostly extract basic knowledge. According to the results, among all the category extraction patterns, pattern 22 shows a better performance with a high accuracy rate of 96.06 %.

The first step of the future work is to refine the system in order to get correct formulations for ambiguous facts. As other steps, we can make the system more automated using bootstrapping approach [11], which makes use of a training set including pairs from infobox tables and the extracted facts and does a recursive self-improvement. After having an automatic evaluation phase, the next step is to compare our results with the ones obtained from the previous works, especially DBpedia. Finally, we should convert the extracted facts into RDF triples in order to make them machine-readable for further use in real systems.

References

1. Auer, S., Lehmann, J.: What have innsbruck and leipzig in common? Extracting semantics from wiki content. In: Franconi, E., Kifer, M., May, W. (eds.) ESWC 2007. LNCS, vol. 4519, pp. 503–517. Springer, Heidelberg (2007)
2. Frawley, W.J., Piatetsky-Shapiro, G., Matheus, C.J.: Knowledge discovery in databases: an overview. AI Mag. **13**(3), 57–70 (1992)
3. Hoffart, J., Suchanek, F.M., Berberich, K., Weikum, G.: Yago2: a spatially and temporally enhanced knowledge base from wikipedia. Artif. Intell. **194**, 28–61 (2010)
4. Hovy, E., Navigli, R., Ponzetto, S.P.: Collaboratively built semi-structured content and artificial intelligence: the story so far. Artif. Intell. **194**, 2–27 (2013)
5. Miller, G.A.: WordNet: a lexical database for english. Commun. ACM **38**, 39–41 (1995)

6. Morsey, M., Lehmann, J., Auer, S., Stadler, C., Hellmann, S.: DBpedia and the live extraction of structured data from wikipedia. Program Electron. Library Inform. Syst. **46**, 157–181 (2012)
7. Ponzetto, S.P., Strube, M.: Deriving a large scale taxonomy from Wikipedia. In: National Conference on Artificial Intelligence, vol. 2, pp. 1440–1445 (2007)
8. Ruiz-Casado, M., Alfonseca, E., Castells, P.: Automatic extraction of semantic relationships for wordnet by means of pattern learning from wikipedia. In: Montoyo, A., Muñoz, R., Métais, E. (eds.) NLDB 2005. LNCS, vol. 3513, pp. 67–79. Springer, Heidelberg (2005)
9. Suchanek, F.M., Kasneci, G., Weikum, G.: Yago: A core of semantic knowledge. In: International Conference on World Wide Web, USA, pp. 697–706 (2007)
10. Yan, Y., Okazaki, N., Matsuo, Y., Yang, Z., Ishizuka, M.: Unsupervised relation extraction by mining Wikipedia texts using information from the web. In: ACL/IJCNLP, USA, vol. 2, pp. 1021–1029 (2009)
11. Zhao, S.H., Betz, J.: Corroborate and learn facts from the web. In: ACM SIGKDD International Conference on Knowledge Discovery and Data Mining, pp. 995–1003 (2007)

Mixing Crowdsourcing and Graph Propagation to Build a Sentiment Lexicon: Feelings Are Contagious

Mathieu Lafourcade[1(✉)], Nathalie Le Brun[2], and Alain Joubert[1]

[1] Lirmm, Université Montpellier, Montpellier, France
{lafourcade, joubert}@lirmm.fr
[2] Imagin@T, 34400 Lunel, France
imaginat@imaginat.name

Abstract. This paper describes a method for building a sentiment lexicon. Its originality is to combine crowdsourcing via a Game With A Purpose (GWAP) with automated propagation of sentiments through a spreading algorithm, both using the lexical JeuxDeMots network as data source and substratum. We present the game designed to collect sentiment data, and the principles and assumptions underlying the action of the algorithm that propagates them within the network. Finally, we give a qualitative evaluation of the data obtained for both the game and the spreading done by the algorithm.

Keywords: Sentiments · Crowdsourcing · GWAP · Lexical network · Spreading

1 Introduction

The ability to automatically characterize the sentiments that emerge from a text has become a major issue for applications such as the analysis of political speeches, or the opinion survey in all kinds of areas, ranging from tourism to cultural offer, through the consumer goods, education and learning, health and human services, art and culture, sport, recreation and environment. Building a lexical resource of sentiments (in which each term is connected to a weighted set of sentiments) is a prerequisite for this type of study, whatever type of approach, statistics supervised or more linguistic one [1]. Sentiments can be expressed from a predetermined closed set [2] or defined in an open way, what is potentially richer with the risk of lower accuracy [3]. For example [4] used a set of seven emotions (trust, fear, sadness, anger, surprise, disgust and joy) to build Emolex, a polarity resource/sentiments for English obtained by crowdsourcing with Amazon Mechanical Turk (which can be problematic, see [5]). Each term of their lexicon (about 14,000 words) is bound to zero, one or more elements of this set of 7 feelings. The value "indifference" is assigned to terms that are not linked to any of these 7 feelings. This resource has been translated into French [6]. [7] produced a free lexical resource based on WorldNet, which is in the form of a lexicon connecting each *synset* with 3 polarity values (positive, negative, and neutral). This is not strictly speaking a lexicon of feelings, although the terminology often shows confusion

© Springer International Publishing Switzerland 2016
E. Métais et al. (Eds.): NLDB 2016, LNCS 9612, pp. 258–266, 2016.
DOI: 10.1007/978-3-319-41754-7_23

between sentiments and polarity. The detection of feelings and opinions is the subject of numerous works: data is usually identified and extracted automatically from some corpus, see [8–12]. We propose a different approach.

A lexical network, such as that obtained through the online game (GWAP) Jeux-DeMots (JDM) [13], includes a large set of terms linked by lexical-semantic relations. JDM project enabled not only the establishment of a lexical network in constant expansion whose data are freely accessible, but also the validation of relations between terms through many side games [14]. Thus, such a resource is particularly suitable for testing new data acquisition methods. Our hypothesis is as follows: to build a lexical resource of sentiments/emotions, it may be advantageous to combine a GWAP (where data are provided by the players, like in [24] for other kinds of NLP data) with a propagation algorithm that will automatically assign feelings data to many terms of the network. The data of sentiments provided by the gamers will thus be propagated towards the terms lacking of sentiment information. This approach is based on the implicit idea that gamers provide good quality data [15].

In this article, we first show how one can get associations between words and the emotions they evoke, through crowdsourcing, using the *Emot* GWAP. We then present *Botemot*, an algorithm inferring sentiments from the network. Results are quantified and qualitatively discussed on the basis of a comparison between the feelings proposed by the *Botemot* algorithm and those already present in the lexical network, which are thus considered as valid.

2   Crowdsourcing and Lexical Network

The JDM project [13, 14] combines a collection of games and contributory interfaces intended to build a large lexical-semantic network for the French. Initiated in 2007, from a set of 150,000 words without any relation between them, the network has now about 850,000 terms linked by more than 43 million lexical and semantic relations. Its development results from players activity combined with algorithms that continuously propose new relations inferred from existing ones (never ended learning). Such a structure is suitable for inferring new information by diffusion within the network. The JDM project demonstrated that in its application context, *i.e.* the creation of a lexical network of general knowledge and common sense, crowdsourcing via games was effective, either quantitatively or qualitatively.

2.1   Lexical Network: A Graph Structure

A lexical network is a graph in which nodes are lexical items, and arcs are lexical or semantic relations between them. Structures like Wordnet [16, 17] Babelnet [18] or HowNet [19], which are built according to this model, can also be considered as such networks. As far as we know, beside JDM, no other lexical network has involved a large community of volunteer gamers [14]. In the JDM network, about an hundred types of binary, oriented and weighted relations are defined. A relation with a negative weight is considered false, (for example: *fly: agent/-25: ostrich*) and can be considered inhibitory, because it will block the spreading of information in its vicinity during the propagation process. Similarly, a negative relationship is considered false by the inference mechanism, even if a generic term of the concerned one satisfies the relation in question (*fly : agent/>0: bird*). In JDM, over 12,000 polysemous terms are refined in 37,146 meanings and uses. This enlarges continuously, fed by many GWAPs and other crowdsourcing activities, coupled with processes of inference and consistency check. The resource is freely available (http://jeuxdemots.org).

In this article, a *working hypothesis* is that the values of feelings associated with terms related to the target word generally verify a form of transitivity. This allows us to automatically infer the feelings of a term if it is sufficiently provided with information itself and if its neighbors are reasonably provided too.

2.2 *Emot*, a Game to Capture Feelings

The *Emot* game (http://www.jeuxdemots.org/emot.php) offers the player to associate one or several feelings/emotions to a given word, either by choosing one among the displayed feelings (*e.g. love, joy, sadness, fear, …*), or by entering some other(s) feeling(s) via a text field, if none of the proposals suits him. The player can also "pass" if he doesn't know the word, or select "indifference" if it doesn't evoke any feeling. The main appeal of the game is the possibility for a player to compare his own answers with those of others. Moreover, regarding the proposed vocabulary, he can choose between two difficulty levels. Most players start with the easy level to get familiar with the game, and then go to the next level, which is more difficult and therefore more rewarding. Each sentiment proposed by a player for a given word affects the lexical network through the insertion of a specific sentiment relation between both terms (or a small augmentation of its weight on an open scale, if this relation already exists). The resulting data (the set of sentiment relations) were systematically and regularly manually evaluated by 4 native speakers of French to ensure their validity (so as to be sure that these data were valid as they were intended to be used by our algorithm). On about 1,500 contributions, 90 % were perfectly relevant, 9 % debatable, and only 1 % of the contributions were to be rejected. The questionable contributions are almost always related to minority opinions but possible (*e.g. disgust* associated with *salmon*). The cases of rejection clearly correspond to errors or malicious action of trollers (*shyness* associated with *potato*). Obtained data are freely accessible.

3   An Algorithm for Inferring Relations

Botemot is an algorithm designed to propagate some information through the network. For each selected word, *Botemot* proposes the feelings associated with its neighbors, *i.e.* those to which it is linked by certain relations. The negatively associated terms are ignored. A positively associated term with sentiment(s) negatively weighted is normally retained, the negative weight being subtracted from the sum of weights. The execution of *Botemot* is looped continuously together with the activity of players. To associate feelings with terms of the network, we apply the following procedure.

From a given term (T) as input, the algorithm infers a weighted list (L) of sentiments. At the beginning, L is initialized the empty set. if the term T is not associated with any polarity, positive or negative, then the empty set is returned It is the same if the number of words associated to T is too low (<5). Otherwise, for each t associated to T, we enumerate its associated weighted list of sentiments. We add this set to L (as a weighted union). Items of L are further filtered with a **tolerance factor**. We choose the T term among words that are names (common or proper), verbs, adjectives or adverbs. The knowledge base from which we extract associated terms is JDM, but this could be any other lexical network. A sentiment that appears n times with an average value of p will have a score of $n * p$. A sentiment term of the resulting list L might have a negative value and be kept as such. The 6 types of relation that seem useful for inferring sentiments are: *associated ideas, hypernyms, characteristics, synonyms, semantic refinements, consequences*. The assumption is that if two terms are connected by one of these relations, their associated feelings are globally "contagious" and can be transmitted. For example, the following general scheme can often be checked: *If* T→consequence→C *and* C→sentiments/emotions→S, *then* T→sentiments/emotions→S. A particular example could be: If *tumor*→consequence→*death* and *death*→sentiments/emotions→*fear*, then tumor→sentiments/emotions→*fear*.

The weighted union of lists is the set union of list elements in which weights of common elements are combined. *Botemot* is applied to each of the approximately 500,000 eligible terms of the network, and is part of the never ended learning loop. For a given T term, it may return an empty list, if: (1) the term is not characterized by a strong polarity (i.e.>25 %); (2) the term T has no related term for the 6 considered relation types; (3) the term T has some related terms but none of them has any associated feelings. The polarity, as defined in the JDM project, was presented in (Lafourcade et al., 2015a) and we use it as a filter in our algorithm. *Botemot* propagates sentiments depending on the network topology and it is done together with the activity of players, whose contributions serve as reference values.

The **threshold filtering** is used to determine the most relevant part of the list L to improve accuracy (to the detriment of the recall, see qualitative assessment). Terms of the list L being weighted, we calculate the average *mu* of their weights. Let *alpha* be the **tolerance factor**, defined on R+. We keep only the terms the weight of which is upper to the threshold *mu * alpha*. The higher is *alpha*, the stricter the filtering. For example, suppose the following list L :{*fear: 110, excitement*: 50, *joy*: 30, *anguish*:10}. The average is equal to (110 + 50 + 30 + 10)/4 = 200/4 = 50. If alpha = 2, we keep only the words whose weight is greater than or equal to 100, i.e. the set {fear: 110}. If alpha = 0.5, we keep the words the weight of which is greater than or equal to 25, i.e. the set {fear: 110, excitement: 50, joy: 30}. By lowering the acceptance threshold, a tolerant filtering increases the proportion of proposals with low weight. Increasing the recall also increases the error rate, and lowers the accuracy. Our second filter is based on the polarity [20] of the terms of the lexical network. In our experiments, we limited the action of the algorithm to the terms with positive or negative polarity higher than 25 %. For these words, we selected only the sentiments polarized in the same way. Note that due to polysemy or point of view, many terms may have a mixed polarity, both negative and positive. The neutral polarity is ignored: it typically does not arouse sentiments other than indifference. Upstream, we perform a filtering based on level of information (i.e. the sum of the weights of the 6 relations considered by *Botemot*), to avoid terms with too few associated relations. In practice, the terms with a level of information less than 1000 are ignored.

4   Evaluations

About information of sentiments, the JDM lexical network, modified by our experience, includes: 112,643 terms associated with at least one sentiment/emotion, including *indifference*; 110,671 terms associated with at least one sentiment/emotion, except *indifference*; 566,298 relations between these terms and sentiments/emotions, i.e. approximately 5 sentiments per term.

Botemot proposed a total of 972,467 sentiments for 154,099 words. Approximately 45,000 among them (30 %) had no associated feeling. About 620,000 are original proposals (not existing in the network) and 350,000 are proposals already present before our experience. Our hypothesis is the following: if a proposal of *Botemot* is already among the validated sentiments for a given term, then we consider that it is correct, and by extrapolation, we can have confidence in the proposals of the algorithm for the words that had no validated sentiments. One must keep in mind that *Botemot* does not know what are the feelings already validated for the term for which it tries to make proposals. Note that *Botemot* succeeded in selecting 19 words with associated sentiments for 1 which arouses only the indifference, what represents a noise about 5 %. Thus, we can consider that the algorithm is relatively effective in detecting words for which linkage with any sentiments (apart from indifference) is relevant.

4.1   Qualitative Assessment by Counting and Weighting

A good proposal is an already validated sentiment (positive weight). A bad proposal is an invalidated sentiment (negative weight). A *new proposal* is a sentiment that has never been validated or invalidated for the target word. For the evaluation, we consider the *amount of information* of each term, *i.e.* the sum of the weights of the 6 relations considered by *Botemot*. Terms with an amount of information between 5,000 and 100,000 are the ones to consider. Below 5,000, words have too little information to lead to correct inferences. And terms with a level of information higher to 100,000 are too few to lead to reliable statistics. The *evaluation by counting (c)* is only based on the criterion presence/absence of the words, regardless of their weight. However, it seems fair to consider that for the algorithm, regain a sentiment already present and strongly bound (high weight) is better than finding a more weakly bound sentiment (low weight): this is an *evaluation by weighting (w)*. So we also calculated precision (w) recall (w) and F1-score (w) by adding the weight of the words rather than their numbers, and we allocated to every good proposal its weight in the network, while we assigned an arbitrary weight of 25 to every new proposal. On average, for both methods, precision = 0.93, recall = 0.98, and F1-score = 0.95 (for about 350,000 proposals of sentiments).

Precision in weighting is always greater than precision in counting (it is the same for F1-score) of about 0.03. Indeed, taking weights into account allows a much finer modulation in the calculation than simply counting the number of words. The F1-score, which ranges between 94 and 98 %, validates our initial hypothesis: we can infer relevant feelings from existing ones, and, by extrapolation, consider valid the sentiments transmitted in this way to a word which had no validated sentiment, with a maximum error risk of 6 %.

4.2 Qualitative Assessment by Manual Validation of New Proposals

Manually checking for new feelings proposed by the algorithm was performed by 4 native speakers of French on more than 500 words, and shows that 93 to 97 % of the proposed feelings are valid, thus a rejection rate from 3 % to 7 %. According to both evaluation methods: (a) using weights in order to distinguish representative feelings and anecdotal ones seems relevant, and (b) evaluating the inference method by comparing the data produced with those supplied by the contributors seem to be a reliable approach. By changing the numerator in formulas of precision and recall, we may consider the new proposals made by the algorithm in two ways: the optimistic point of view would be to assume they are valid by assigning them, for counting, the value of 0.5. It would increase recall and precision of 2 %; the pessimistic view would be to suppose they are irrelevant by assigning them, for counting, the value of -0.5. It would decrease recall and precision of 2 % (on average, both with a tolerance of 1).

Data of sentiments provided by the players are very precise and diversified. This justifies that these data are the reference for the diffusion algorithm, which is very accurate. There are little new proposals (10 %), but by manual verification we estimated that 95 % are fair and 4.5 % debatable (0.5 % false). The algorithm doesn't propose any sentiment of negative weight, and doesn't propose again those invalidated, which avoids looping. The invalidated feelings are assigned a negative weight and this makes them inhibitors in the mechanism of inference.

The algorithm seems efficient to assign sentiments to the words of the network which had no information of this type, provided that these words have neighbors with sentiments data. The quality of inferred data is obviously highly dependent on the quality of data provided by the players. But as the crowdsourcing approach involves a very large number of contributors, the large majority of supplied data are correct.

4.3 Qualitative Assessment by Comparison with Emolex Lexicon

We compared the sentiments proposed by *Botemot* with the lexicon of [21], in the version translated by [6]. We remind that this lexicon allocates to about 14,000 words a variable number of sentiments selected from a predetermined set of seven. This number varies from 0 (indifference) to 6, each sentiment being activated (value 1) or not (value 0). For 99.9 % of the words, at least one of the feelings associated in the Emolex lexicon is also part of the proposals of our algorithm. For 69 % of the words to which no feeling is associated in Emolex our algorithm proposes at least "indifference". Total Inclusion: 92 % (For a given word, all the feelings associated in the Emolex lexicon are among the proposals of *Botemot*). Strict inclusion: 5 % (for a given word, the feelings associated in the Emolex lexicon are exactly identical to the proposals of *Botemot*). About twenty words from Emolex are not in the JeuxDeMots network but these are malformed words, probably mistranslated. Regarding the strict inclusion, the result (5 %) is due to the fact that our algorithm offers many synonyms of the same feeling (e.g. *fear, anxiety, apprehension* whereas only the word "*fear*" is listed in Emolex). When for a word *Botemot* proposes sentiments absent from Emolex, these words are in 70 % of cases synonymous with a sentiment of Emolex. In the remaining 30 %, there are only 0.5 % of inadequate sentiments (comparison carried out on a sample of 1500 words related to 8,000 sentiments). We think our approach is more adapted to reflect the richness and diversity of feelings and collect vocabulary to designate them as shown in [22, 23].

5 Conclusion

The use of GWAPs gives good results for the lexical crowdsourcing, but we can improve the effectiveness (in coverage) by combining the crowdsourcing with mechanisms of inference by diffusion. Note that such approach can be applied to any other type of information. Here, the inference is performed by an algorithm, which "contaminates" a term of the lexical network with information gleaned from its close neighbors. This mechanism enlarges the lexicon from a nucleus built by the gamers, which diversifies and is growing constantly under the combined action of the gamers and the algorithm (Never Ended Learning). Thus, one can predict that the inferred feelings will become increasingly accurate and relevant over time. *Botemot* uses as reference data the sentiments provided by the players, who provide novelty, richness and diversity of information to infer: the algorithm cannot infer a feeling that would not be associated with any of the neighbors of the target word. It can be noticed that thanks to the redundancy of the lexical network, polysemy does not compromise inference of sentiments. But the meanings that evoke the most pronounced feelings tend to contaminate the general meaning. The resource that we get reflects the diversity and richness of views, even opposite ones. The *Botemot* algorithm, very simply designed, produces very relevant proposals as shown by our different assessments, and the evaluation method of inferred data, by calculating the precision and the recall can be easily and effectively transposed to any field where a resource is in permanent construction.

References

1. Brun, C.: Detecting opinions using deep syntactic analysis. In: Proceedings of Recent Advances in Natural Language Processing, (RANLP 2011), Hissar, Bulgaria, pp. 392–398 (2011)
2. Taboada, M., Brooke, J., Tofiloski, M., Voll, K., Stede, M.: Lexicon-based methods for sentiment analysis. Comput. Linguist. **37**(2), 267–307 (2011)
3. Whissell, C.: The Dictionary of Affect in Language. Academic Press, Cambridge (1989)
4. Saif, M., Turney, P.: Crowdsourcing a word-emotion association Lexicon. Comput. Intell. **29**(3), 436–465 (2013)

5. Fort, K., Adda, G., Sagot, B., Mariani, J., Couillault, A.: Crowdsourcing for language resource development: criticisms about amazon mechanical turk overpowering use. In: Vetulani, Z., Mariani, J. (eds.) LTC 2011. LNCS, vol. 8387, pp. 303–314. Springer, Heidelberg (2014). ISBN 978-3-319-08957-7

6. Abdaoui, A., Azé, J., Bringay, S., Poncelet P.: FEEL: French Extended Emotional Lexicon (2014). ISLRN: 041-639-484-224-2

7. Esuli, A., Sebastiani, F.: SentiWordNet: a publicly available lexical resource for opinion mining. In: Proceedings of LREC-06, Gênes, Italie, p. 6 (2006)

8. Kim, S.-M. Hovy, E.: Determining the sentiment of opinions. In: Proceedings of the International Conference on Computational Linguistics (COLING), pp. 1367–1373 (2004)

9. Kim, S.-M., Hovy, E.: Extracting opinions, opinion holders, and topics expressed in online news media text. In: Proceedings of the Workshop on Sentiment and Subjectivity in Text, SST 2006, Stroudsburg, PA, USA, pp. 1–8. Association for Computational Linguistics. (2006)

10. Mudinas, A., Zhang, D., Levenem, M.: Combining lexicon and learning based approaches for concept-level sentiment analysis. In: Proceedings of the First International Workshop on Issues of Sentiment Discovery and Opinion Mining, WISDOM 2012, pp. 5:1–5:8. ACM, New York (2012)

11. Strapparava, C., Mihalcea, R.: Learning to identify emotions in text. In: Proceedings of the 2008 ACM Symposium on Applied Computing, SAC 2008, pp. 1556–1560. ACM, New York (2008)

12. Kiritchenko, S., Xiaodan, Z., Saif, M.: Sentiment analysis of short informal texts. J. Artif. Intell. Res. **50**, 723–762 (2014)

13. Lafourcade, M.: Making people play for lexical acquisition. In: Proceedings of SNLP 2007, 7th Symposium on Natural Language Processing, Pattaya, Thailande, p. 8, 13–15 December 2007

14. Lafourcade, M., Le Brun, N., Joubert, A.: Games with a Purpose (GWAPS), p. 158. Wiley-ISTE, New York (2015). ISBN 978-1-84821-803-1

15. Simperl, E., Acosta, M., Flöck, F.: Knowledge engineering via human computation. In: Michelucci, P. (ed.) Handbook of Human Computation, pp. 131–151. Springer, New York (2013)

16. Miller, G.A.: WordNet: a lexical database for English. Commun. ACM **38**(11), 39–41 (1995)

17. Fellbaum, C. (ed.): WordNet: An Electronic Lexical Database. MIT Press, Cambridge (1998)

18. Navigli, R., Ponzetto, S.P.: BabelNet: building a very large multilingual semantic network. In: Proceedings of the 48th Annual Meeting of the Association for Computational Linguistics, pp. 216–225 (2010)

19. Dong, Z.D., Dong, Q., Hao, C.L.: Theoretical findings of HowNet. J. Chin. Inf. Proc. **21**(4), 3–9 (2007)

20. Lafourcade, M., Le Brun, N., Joubert, A.: Collecting and evaluating lexical polarity with a game with a purpose. In: Proceedings of International Conference on Recent Advances in Natural Language Processing (RANLP 2015), Hissar, Bulgaria, p. 9, 5–11 September 2015

21. Saif, M., Turney, P.D.: Emotions evoked by common words and phrases: using mechanical turk to create an emotion Lexicon. In: Proceedings of the NAACL-HLT 2010 Workshop on Computational Approaches to Analysis and Generation of Emotion in Text, LA, California, pp. 26–34, June 2010

22. Ekman, P.: An argument for basic emotions. Cogn. Emot. **6**, 169–200 (1992)

23. Tausczik, Y.R., Pennebake, R.J.W.: The psychological meaning of words : LIWC and computerized text analysis methods. J. Lang. Soc. Psychol. **29**, 24–54 (2010)
24. Bos, J., Nissim, M.: Uncovering noun-noun compound relations by gamification. In: Megyesi, B., (Ed.) Proceedings of the 20th Nordic Conference of Computational Linguistics (NODALIDA 2015), pp. 251–255 (2015)

Comparing Two Strategies for Query Expansion in a News Monitoring System

Parvaz Mahdabi[(✉)] and Andrei Popescu-Belis

Idiap Research Institute, Martigny, Switzerland
{parvaz.mahdabi,andrei.popescu-belis}@idiap.ch

Abstract. In this paper, we study query expansion strategies that improve the relevance of retrieved documents in a news and social media monitoring system, which performs real-time searches based on complex queries. We propose a two-step retrieval strategy using textual features such as bi-gram word dependencies, proximity, and expansion terms. We compare two different methods for query expansion: (1) based on word co-occurrence information; (2) using semantically-related expansion terms. We evaluate our methods and compare them with the baseline version of the system by crowdsourcing user-centric tasks. The results show that word co-occurrence outperforms semantic query expansion, and improves over the baseline in terms of relevance and utility.

Keywords: Query analysis · Query expansion · Web IR and social media search

1 Introduction

The information discovery paradigm has recently gained attention in the industry as well as in the academia [1,7]. Unlike traditional web search results, displayed for instance as "ten blue links", an information discovery system finds insights that are otherwise hidden in various information sources such as Wikipedia, news feeds, or social media. These insights are an alternative way of answering information needs, and can be directly shown to users in addition to ranked lists of links to document results.

This paper studies two types of cues for query expansion, namely semantic vs. distributional ones, in order to enhance a commercial information discovery system. As our purpose is to compare the two types of cues, we will not advertise the name of the initial baseline system, but provide instead a sufficient description of it. The system facilitates access to evolving sources of information such as news or social media, thanks to complex queries formulated by users, who also receive assistance in formulating these queries. The queries are used to generate up-to-date "magazines" that are typically consulted daily by their users. The baseline version of the system implements a hybrid semantic search approach combining full-text search with entity-based search to overcome the vocabulary mismatch limitation of traditional keyword search.

© Springer International Publishing Switzerland 2016
E. Métais et al. (Eds.): NLDB 2016, LNCS 9612, pp. 267–275, 2016.
DOI: 10.1007/978-3-319-41754-7_24

To improve the list of documents retrieved by the system, we propose and compare two methods for query expansion with new words or concepts, within the existing hybrid semantic search framework. The first method finds candidate expansion terms from the systems's document repository using a global analysis of the word distributions in the corpus. We follow a pseudo-relevance feedback approach, assuming that the initial ranked list contains relevant documents, and drawing candidate expansion terms from this set. The second approach uses two external semantic resources, DBpedia and WordNet, to find candidate expansion terms. We analyze the initial ranked list and identify search terms from the query that did not occur in the retrieved documents. We expand these terms with synonyms from WordNet and titles of redirect pages from DBpedia.

To assess the merits of our proposals, we evaluate them over a set of queries obtained from users of the system. We design a comparative evaluation approach which relies on human subjects to compare two magazines in terms of relevance, diversity, and utility to potential users. First, we validate the results obtained from non-expert subjects (via crowdsourcing), by comparing them with evaluations of expert users. Then, using mostly non-expert users, we show that the proposed query expansion methods achieve promising results in terms of relevance and utility compared to the baseline, and that the distributional method outperforms the semantic one.

The paper is organized as follows. In Sect. 2 we present the framework and baseline system, describing its indexing, search and retrieval components. Section 3 presents our two-step retrieval strategy and the two query expansion methods, and compares them with the initial system. The experimental setup is presented in Sect. 4. Results are presented and discussed in Sect. 5.

2   A System for News and Social Media Monitoring

Query Representation. The information discovery engine underlying this study helps users find relevant information to satisfy a recurring information need, expressed as a complex query. The results, presented as a "magazine," are made of documents, news, or tweets. The system allows users to construct Boolean queries composed of keywords and phrases, which are represented in Conjunctive Normal Form (CNF). A CNF formula is an "AND of ORs", i.e. a Boolean conjunction of clauses, where a clause is a disjunction of literals. While CNF queries are mainly used by expert searchers such as librarians, lawyers, or patent searchers, our system assists general users with query construction through a user-friendly interface which provides an intuitive view of AND/OR connectives and recommends related concept names for each field.

For instance, one of the CNF complex queries considered in this paper is: "((FedEx OR Parcel Post OR Postage Stamp OR Royal Mail OR United Parcel Services OR United States Postal Service OR Universal Postal Union) AND (Privatization OR Private Sector OR Public Sector OR Postal Services))."

Data Repository. The document collection consists of semi-structured documents (obtained from Google, Yahoo News, and Twitter) with the following fields: title, excerpt, and text. These fields are stored in the system's index, created using the ElasticSearch toolkit (www.elastic.co). We perform minimal stemming on the words, by only removing plural and possessive forms. Upon indexing, each document also undergoes semantic annotation, which enriches the text documents by finding key phrases in them, then disambiguating these phrases, and linking them to DBpedia concepts (www.dbpedia.org). The linked concepts, identified by their DBpedia URIs, are added to the document representation and are stored in the system's index, on condition that the linking confidence value is higher than a threshold that was set experimentally. These semantic annotations are used both in the baseline and in the proposed approach.

3 Search, Retrieval and Re-ranking Methods

The architecture for hybrid semantic search is represented in Fig. 1. We first present below an overview of the innovations proposed in this paper, and then describe a new strategy for term weighting and two new query expansion methods that we study.

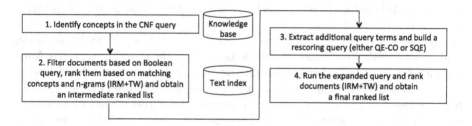

Fig. 1. Architecture for hybrid semantic search.

Overview of the System. In the first step, the CNF query is annotated using labels of concepts found in DBpedia. The goal of this step is to enhance the query by linking it to DBpedia URIs. In Step 2, the Boolean structure of the query is used to filter documents which lexically match either the query or the concepts associated to it. Then, this document set is ranked by calculating a relevance score. As a result, a unified ranked list is produced, which serves as an intermediate result set. In Step 3, additional expansion terms are selected, according to either of the two approaches proposed for query expansion, noted QE-CO and SQE. In Step 4, the results of Step 2 are re-ranked using the expansion terms selected in Step 3. Finally, the re-ranked results are presented to the user as a ranked list or "magazine".

Baseline. The baseline system only includes Step 2 from Fig. 1. In Step 2, to generate the magazine, the baseline system uses the Boolean query to filter documents and uses tf-idf scoring to rank documents that were found according to matching on concepts and unigrams.

Term and N-gram Weighting for Retrieval. The following indexing and retrieval strategies are used at Steps 2 and 4 of the proposed system. We consider both the unigram and the bigram statistics of these fields. Documents are ranked according to Language Modeling with Laplacian smoothing [6]. We hereafter refer to these enhancements as the 'IRM' option. Furthermore, we improve term weighting in several ways. We increase the weight of shorter fields (title and excerpt) and we boost matches of bigrams from queries with respect to unigrams, to increase precision. We also increase the weights of query terms that appear alone inside an AND clause, to compensate the fact that these terms are not accompanied by synonyms or semantically related terms in the query formulated by the user. We refer to these enhancements as the 'TW' option.

Query Expansion Using Word Co-occurrence (QE-CO). The first method for query expansion, noted 'QE-CO', relies on word co-occurrence information for finding related terms. To perform co-occurrence analysis at Step 3, we compare the foreground model (documents belonging to the ranked list generated at Step 2 for a query) with the background model (the entire dataset), and select terms that are indicators of the foreground model (more frequently appearing in the initial ranked list together with query terms, as opposed to the rest of the corpus). Various methods exist to perform this comparison and calculate term relatedness [5], such as chi-square, mutual information, and Google normalized distance [3]. We use here mutual information, following preliminary experiments in which we compared the terms obtained by each approach. We observed that the results of the mutual information approach contained fewer rare terms, and avoided the selection of misspellings.

Mutual information compares the joint probability that a term t appears in a document collection set with the probabilities of observing t and set independently [5]. The mutual information between U and C is calculated as follows:

$$I(U;C) = \sum_{e_t \in \{0,1\}} \sum_{e_{set} \in \{fg, \bar{fg}\}} P(U = e_t, C = e_{set}) \log \frac{P(U = e_t, C = e_{set})}{P(U = e_t) \, P(C = e_{set})}$$

where U is a random variable indicating the presence ($e_t = 1$) or absence ($e_t = 0$) of term t, and C is a random variable which represents the presence of the document in the foreground collection ($e_{set} = fg$) or its absence ($e_{set} = \bar{fg}$). These probabilities are estimated using Maximum Likelihood Estimate (MLE), using counts of documents that have the respective values of e_t and e_{set}.

All unigrams and bigrams appearing in the documents of the initial ranked list for a given query are sorted according to their mutual information scores and the top k terms with largest values are selected and used for query expansion. The expansion terms are only bigrams to increase the precision. Only the title fields of documents in the initial ranked list are examined, and the maximum number of expansion terms is set to 10, to keep the response time for a query acceptable to the users.

Selective Query Expansion with Semantic Cues (SQE). The second method for query expansion is based on a predictive analysis that selects search terms which correspond to vocabulary mismatches between the query and relevant documents (similar to [2]). In this approach, we perform the analysis over the initial set of results with the goal of finding terms from the initial query which are not present in top k retrieved documents in the ranked list. These terms are then marked as problematic query terms or indicators of vocabulary mismatch, generally because the actual use of a concept (surface form) in a document may differ from the query term chosen by the user. We check the top 9 documents, which is the number of documents generally visible in a magazine without scrolling. The presence of a query term is checked in the 'title', 'excerpt' and 'concept' fields and not in the 'text' field, as it appeared that all query terms were present in the 'text' fields of all retrieved documents.

For each AND clause composed of one or more disjunction (OR), if at least on disjunction element is present in the top k retrieved documents in the ranked list, then no disjunction element in the AND clause is marked as problematic query term. Then, to address vocabulary mismatch, we use two complementary types of expansion terms: semantically related terms and synonyms. To find the former, we use the DBpedia 'redirect' fields, while for the latter we use WordNet. The expansion terms extracted with either of our two methods are finally joined by the OR operator to the original query.

Examples. For the sample CNF query shown in Sect. 2, QE-CO generates expansion words such as: 'Royal Mail', 'Postal Service', 'U.S Postal', 'Post Office', 'Market Flash', or 'Parcel Post'. SQE generates, e.g.: 'Commercial Sector', 'Non-state Sector', 'British Mail', or 'British Post Office'.

Related Work. A frequently used approach for query expansion is Pseudo Relevance Feedback, which assumes that the top ranked documents from an initial ranked list are possibly related to the original query. This document set is then used to extract relevant terms to expand the original query [2]. Selective query expansion is a less studied approach, which resorts to a diagnosis to identify which parts of the query really need to be expanded [8], followed by automatic or manual query expansion or refinement.

4   Experimental Setup

Benchmark Query Set. To quantify the improvement over the baseline of our query expansion approaches, we gathered a benchmark set of nine queries from the system query log obtained from customers. These queries were created by experts, who also provided descriptions of their information needs. We gathered comparative evaluation judgments from two groups of evaluators: expert users of our system (9 users per query), and non-specific users (10 users per query) recruited via Amazon Mechanical Turk (AMT, www.mturk.com).

Design of Evaluation Tasks. The evaluation tasks were designed so as to compare two magazines (top nine documents of each) for a given query of the benchmark set. The description of the information need and the query are first shown to a user. Below the query, we show a pair of ranked lists. In order to avoid any bias regarding the presentation order of the two ranked lists, we randomly change the display order of the ranked lists. For each article in the ranked list, we display the title, the text, and a picture, as they actually appear in the ranked list (magazine). We then ask users to answer four multiple-choice comparison questions regarding the ranked lists, targeting relevance, diversity, and utility, plus a control question. AMT users had a time limit of two minutes to finish each task, but used on average about one minute.

We compare the performance of our query expansion approaches with the results of the baseline system. The comparison scores are computed using the *PCCH* method [4], which factors out unreliable workers and undecided tasks. *PCCH* generates two scores, one for each compared system, summing up to 100 %. The higher one score is above 50 %, the better the system is in comparison to the other one.

5   Experimental Results

In this section, we first report results that validate the evaluations collected using crowdsourcing by comparing them with the ones obtained from expert users, and then present the results of our proposed methods.

Comparing the Crowdsourced Judgments with Expert Ones. Table 1 compares the performance of the baseline system and of the two query expansion approaches through judgments from experts and AMT users. To assess the quality of the judgments obtained from the AMT users, we present side by side the two *PCCH* values summed over all 9 queries. The results show that the judgments obtained from non-specific users of the AMT crowdsourcing platform are correlated in terms of *PCCH* value with the judgments obtained from the experts for the QE-CO method. For instance, the difference of the *PCCH* values between the two user groups is on average smaller than 4 % absolute on all dimensions. Moreover, the values are always on the same side of the 50 % mark,

Table 1. Performance comparison of QE-CO and SQE approaches on the relevance, diversity, and utility dimensions according to AMT and expert users.

Dimension	$PCCH_{QE-CO}$		$PCCH_{SQE}$	
	AMT users	Expert users	AMT users	Expert users
Relevance	62 %	61 %	59 %	57 %
Diversity	45 %	40 %	50 %	27 %
Utility	56 %	63 %	57 %	51 %

i.e. either both above 50 % (meaning the QE-CO approach is better than the baseline) or both below 50 % (meaning the opposite).

Similarly, the results for SQE indicate that the AMT users and the experts agreed on relevance and utility (in similar proportions as above), though not on diversity. Here, SQE and the baseline are equally diverse from the viewpoint of AMT users, while expert users considered the baseline more diverse. Overall, these results validate to a large extent the quality of the judgments obtained from the users of the AMT crowdsourcing platform, thus allowing us to avoid soliciting experts for every comparison task.

Experimental Comparisons and Scores. The generic comparisons presented above, between the QE-CO/SQE approaches and the baseline, were mainly intended to validate the evaluation protocol and get a sense of the merits of QE-CO/SQE, which appeared to outperform the baseline. We now present more detailed comparative evaluations of the various enhancements of the baseline, assessed by AMT users through the same comparative task as above.

First, we assess the merits of the indexing improvements (IRM) with respect to the baseline. Then, we compare the QE-CO and the SQE approaches (combined with IRM) with the baseline. For each of these three comparisons, we also study the merits of term weighting (TW), by comparing each of the three systems (IRM, QE-CO and SQE) with or without TW against the baseline. All scores are given in Table 2.

The IRM approach obtains a higher $PCCH$ compared to the baseline (above 50 %) on the relevance dimension, i.e. the ranked lists obtained by IRM were found on average more relevant to the query. Applying TW on top of IRM achieves a positive effect in terms of $PCCH$. IRM+QE-CO also obtains higher $PCCH$ values compared to the baseline. However, applying the TW method over this query expansion method decreases the $PCCH$ value for relevance. The IRM+SQE method outperforms the baseline in terms of $PCCH$ but adding the TW component results in a lower $PCCH$ value. The conclusion is that all the proposed methods obtain improvements over the baseline in terms of relevance, and IRM+QE-CO performs better than IRM+SQE.

Table 2 also shows the performance comparison of the proposed methods against the baseline on the diversity dimension. IRM improves over the baseline in terms of $PCCH$ score. Adding TW to IRM preserves the improvements compared to the baseline. IRM+QE-CO scores below the baseline ($PCCH$ below 50 %) but IRM+SQE scores similarly to the baseline. Adding the TW method together with QE-CO and SQE approaches decreases the diversity of the ranked lists. A possible reason for this result could be the tendency of AMT users to judge relevance and diversity as inversely correlated.

Table 2. Comparison of methods for relevance, diversity, and utility.

Method	$PCCH$		
	Relevance	Diversity	Utility
IRM	58 %	58 %	63 %
IRM+TW	60 %	53 %	60 %
IRM+QE-CO	62 %	45 %	56 %
IRM+QE-CO+TW	55 %	43 %	52 %
IRM+SQE	59 %	50 %	57 %
IRM+SQE+TW	53 %	40 %	53 %

Table 2 further compares the performance of the proposed methods against the baseline on the utility dimension. IRM and both query expansion methods increase the perceived utility of the ranked list, as they achieve better $PCCH$ values compared to the baseline. However, applying the TW method to IRM, SQE, and QE-CO impacts negatively the perceived utility of the ranked lists. From the results of Table 2, we conclude that IRM and both query expansion approaches achieve a positive effect on improving the relevance and perceived utility of the ranked list with respect to the baseline. However, the diversity of the ranked lists generated by the two query expansion approaches is lower compared to the baseline, while IRM alone achieves improvements in diversity.

6   Conclusion

In this paper we considered the challenge of improving the ranking of relevant documents in a news and social media monitoring system. We presented a two stage process, where in the first stage an initial ranker retrieves a set of documents and in the second stage a model re-ranks the first list before presenting the results to the user. We studied two query expansion methods to improve the re-ranking stage: the first one employs corpus statistics to find terms related to the query, while the second one performs selective query expansion by predicting which query terms require expansion and then finding semantically related terms from external resources. The first method outperformed the second one based on crowdsourced comparative judgments, and both were above the baseline system in terms of relevance and utility.

Acknowledgments. This work was funded by the Swiss Commission for Technology and Innovation.

References

1. Arguello, J., Diaz, F., Callan, J., Crespo, J.: Sources of evidence for vertical selection. In: Proceedings of ACM SIGIR Conference on Research and Development in Information Retrieval, pp. 315–322 (2009)
2. Cao, G., Nie, J., Gao, J., Robertson, S.: Selecting good expansion terms for pseudo-relevance feedback. In: Proceedings of ACM SIGIR Conference on Research and Development in Information Retrieval (SIGIR), pp. 243–250 (2008)
3. Cilibrasi, R., Vitanyi, P.M.: The Google similarity distance. IEEE Trans. Knowl. Data Eng. **19**, 370–383 (2007)
4. Habibi, M., Popescu-Belis, A.: Using crowdsourcing to compare document recommendation strategies for conversations. In: Workshop on Recommendation Utility Evaluation, Held in Conjunction with ACM RecSys (2012)
5. Manning, C., Raghavan, P., Schütze, H.: Introduction to Information Retrieval. Cambridge University Press, New York (2008)
6. Ponte, J.M., Croft, B.: A language modeling approach to information retrieval.In: Proceedings of ACM SIGIR Conference on Research and Development in Information Retrieval (SIGIR), pp. 275–281 (1998)
7. Tablan, V., Bontcheva, K., Roberts, I.: Mímir: an open-source semantic search framework for interactive information seeking and discovery. Web Semant. Sci. Serv. Agents World Wide Web **30**, 52–68 (2015)
8. Zhao, L., Callan, J.: Term necessity prediction. In: Proceedings of the ACM International Conference on Information and Knowledge Management (CIKM), pp. 259–268 (2010)

An Active Ingredients Entity Recogniser System Based on Profiles

Isabel Moreno[1]([✉]), Paloma Moreda[1], and M.T. Romá-Ferri[2]

[1] Department of Software and Computing Systems,
University of Alicante, Alicante, Spain
{imoreno,moreda}@dlsi.ua.es
[2] Department of Nursing, University of Alicante, Alicante, Spain
mtr.ferri@ua.es

Abstract. This paper describes an active ingredients named entity recogniser. Our machine learning system, which is language and domain independent, employs unsupervised feature generation and weighting from the training data. The proposed automatic feature extraction process is based on generating a profile for the given entity without traditional knowledge resources (such as dictionaries). Our results (F1 87.3 % [95 %CI: 82.07–92.53]) proves that unsupervised feature generation can achieve a high performance for this task.

Keywords: Named Entity Recognition · Active ingredient · Profile · SVM · Voted Perceptron · Machine learning · Spanish

1   Introduction

There is a huge amount of information concerning health and healthcare, which is stored in many and different sources [8]. For example, patient's health records, discharge summaries, health forums; etc. Employing all this available information, most of which is in textual form, is critical for all healthcare aspects [8]. For instance, several studies [5,9] suggest that, during day-to-day care, the prescription of pharmacological treatments is a common critical situation. In a limited time, healthcare professionals need to compare efficacy and safety of several treatments before prescribing the adequate one. Nevertheless, it is unmanageable to physicians [10], due to their limited consultation time; and it is inaccessible for any automated processing system, since an intermediate step that transforms text to structured data is required [8].

Such transformation can be achieved applying Natural Language Processing (NLP). In particular, Named Entity Recognition (NER) systems convert key elements, locked in text, to an structured format and assign them a category [6]. In this manner, this structured data could be used by critical applications in the healthcare domain, which help physicians in their daily workload [8], such as: decision support systems, cohort identification, patient management, etc. There are many key concepts involved during pharmacological treatment prescription

© Springer International Publishing Switzerland 2016
E. Métais et al. (Eds.): NLDB 2016, LNCS 9612, pp. 276–284, 2016.
DOI: 10.1007/978-3-319-41754-7_25

that can be eligible for being a named entity (NE). However, prescription is mainly related with active ingredients, which is the name of a substance giving a drug its effect (e.g. "paracetamol", in English "acetaminophen").

Taking into consideration all these aspects, our aim is to present a profile-based NER system. It identifies entities and assigns a category in a single step. Although our methodology is domain and language-independent, it is focused on identifying whether candidates are mentions of active ingredients or not. Given that most efforts are concentrated in English [22], this work is evaluated over documents written in Spanish to enhance available resources in this language.

The remainder of the paper is structured as follows. Section 2 reviews existing approaches for NER in this narrow field. Then, Sect. 3 describes our approach. The evaluation performed is provided in Sect. 4. Finally, Sect. 5 concludes the paper and outlines future work.

2 Background

NER has been faced through two main approaches: hand-crafted rules or statistical models. Most systems are rule-based and make extensive use of lexicons with good results. These dictionaries were built combining several sources: (i) biomedical knowledge resources [3,4,12,16,20,21,23]; (ii) training data [12,23,24]; (iii) unlabelled data [12]; and (iv) the web [21,23]. Given their lexicons, these systems employ pattern matching and regular expressions [3,12,16,23,24], join a spell checker with an existing NER system (*MedEx*) [4], combine rules from experts with a concept recognition tool (*Mgrep*) [20] or integrate a dictionary-based NER (*Textalytics*) and GATE gazetteers [21]. A drawback shared by these dictionary rule-based systems is an expensive maintenance to make them updated for including new instances and their variations [16].

There is a trend to employ Machine Learning (ML) algorithms [1,2,11,13,18] for NER, which generates good results with or without biomedical knowledge resources. The algorithms more commonly used are: (i) J48 implementation of C4.5 decision tree to identify and recognise entities independently [2]; (ii) Conditional Random Fields (CRF) to found and classify entities [11,13,18]; and (iii) Support Vector Machines (SVM) to categorize entities [1]. The features employed vary from domain-independent ones based on context from different levels of analysis (lexical, syntactical and semantic) [1,2,11,13], as well as domain-specific tools (*MMTx* [1], TEES [1], *Chemspot* [13]) and domain-specific knowledge bases such as dictionaries [13,18], ontologies (*ChEBI* [11,13] and *PHARE* [13]) or databases (*DrugBank* [1]). These ML approaches often define their features by hand, which leads to manually redefine and adjust the features of the system. For instance, when new types of entities appear, but also if either domain or language changes.

As for languages tackled, only two systems were developed for Spanish [16,21], whereas English has received more attention [1–4,11–13,18,20,23]. Thus, there is a scarcity of Spanish NER tools in this domain.

Our hypothesis is that medical NER systems can be effective out of knowledge resources and can take advantage of an unsupervised feature generation process from a training corpus. To this end, we experiment with one entity (active ingredient), two automatic feature generation spaces (sentences and sections) and one approach for entity recognition based on ML.

3   Method: Named Entity Recognition Through Profiles

Our methodology performs active ingredients mentions extraction using profiles. The term profile refers to a collection of unique entity descriptors that are a set of relevant concept bearing words found in a corpus [14].

The proposed approach is an adaptation for NER from the method defined in [14]. There are two main differences between their work and ours. On the one hand, while Lopes et al. [14] obtain their descriptors from a concept extractor system; ours are derived from nouns, verbs, adjectives and adverbs. On the other hand, their categorization is performed by ranking possible entities using their own similarity measure, but ours calculate similarity between profiles and entities through ML.

Our method has two main stages: (i) *profile generation* step, whose main goal is to build a profile for a given NE type from which to train a ML model for recognising such NE; and (ii) *profile application* step, whose aim is to recognise entities from a text through a ML model. It computes similarity measures between a profile and the candidates' profile. Next, each phase is further described.

1. **Profile generation** for a NE type is an off-line process that follows these steps:
   (a) *Linguistic annotation*: a corpus previously annotated with this NE is tokenized, sentence-splitted, morphologically analysed, PoS-tagged, and shallow-parsed by using Freeling [17]. Finally, noun phrases are extracted.
   (b) *Split the training corpus*: the dataset is divided in two sets called target and contrasting. The former set represents a fragment of the corpus capable to characterize that a given NE belongs to a certain class (i.e. positive examples or sentences containing active ingredients). While the latter is composed by negative examples (i.e. sentences with named entities different from active ingredients). According to [14], the use of contrasting entities (or negative examples) discovers frequent terms in all types of entities, and thus it will allow to find common relevant descriptors between all entities (both positive and negative examples).
   (c) *Descriptors extraction*: For each set, we extract content bearing terms and their frequency, called top and common descriptors. The former contains descriptors from the target set, whereas the latter includes common descriptors in both target and contrasting set. As noted above, content bearing terms are lemmas of nouns, verbs, adjectives and adverbs that appear in a window (i.e. 10 descriptors before and 10 after).
   (d) *Relevance computation*: For both top and common descriptors lists, a relevance index is assigned to weight them. The *term frequency, disjoint corpora frequency* index [14] determines the relevance of each item from

a top list applying (1), where: d is a descriptor, t is the target set, c is the contrasting set, and $occ(d,t)$ is the occurrences of a term d in a set t. idx_{top} penalize descriptors that appear in the contrasting set, thus higher idx_{top} values estimate more relevant descriptors. Similarly, to weight descriptors from the commons list, the relevance common index is defined in (2).

$$idx_{top}(d,t,c) = \log(1 + \frac{occ(d,t)}{1 + \log(1 + occ(d,c))}) \qquad (1)$$

$$idx_{common}(d,t,c) = \log(1 + occ(d,t) - \frac{occ(d,t)}{1 + \log(1 + occ(d,c))}) \qquad (2)$$

This step produces a profile of a named entity category (such as active ingredient p_a), which is a top T_{p_a} and common C_{p_a} descriptors lists. Each item from these lists is a pair representing a term and its relevance index: $\{term(i), idx(term(i), t, c)\}$. Such index is used in next steps to determine the category of this NE and weights the relevance of each descriptor. These lists contains every extracted descriptor whose $occ(d) \geq 1$.

(e) *Model training*: Our proposal creates ML models for computing profile similarity between an NE class and its candidates. Thus, in this step, a model is generated for each entity profile. Two algorithms have been employed: (i) SVM [19], an algorithm commonly used for classification purposes [1]; and (ii) Voted Perceptron (VP) [7], a simplified version of CRF algorithm which is frequently used in NER [11,13,18]. The built classifiers have as features the profile of the entity, that is the T_p and C_p lists. The value of each feature (i.e. a descriptor) is its relevance index, idx_{top} (1) and idx_{common} (2), respectively.

2. **Profile application** process for recognising a NE type takes these steps:
   (a) *Linguistic annotation*: text is tokenized, sentence-splitted, morphologi-cally analyzed, PoS-tagged, and shallow-parsed by using Freeling [17]. Finally, noun phrases are annotated.
   (b) *Candidate detection*: an active ingredient is a noun (e.g. "amoxicilina", in English "amoxicilin") or a set of nouns (e.g. "ácido acetilsalicilico", in English "aspirin"). Thus, all extracted noun phrase are considered as a candidate, as long as they do not have numerals.
   (c) *Descriptors extraction*: For each candidate, we extract the top and com-mon descriptors that appear in a window (i.e. 10 terms before and 10 after) using the same restrictions as in the generation phase (nouns, verbs, adjectives and adverbs). Common descriptors are obtained using the con-trasting set employed for profile generation.
   (d) *Relevance computation*: Only the top relevance index idx_{top} (1) is com-puted, as [14] does.
   (e) *Similarity computation &classification*: Once an entity candidate has filled its profile, this one is compared against the previously generated one to compute their similarity. It was estimated using two ML algo-rithms (i.e. SVM and VP). The classification is based on a model whose features are the descriptors of the given entity (T_p and C_p).

4   Results Analysis

The aim of our experiments is to estimate similarity between the active ingre-dient profile and a candidate NE using SVM and VP algorithms. Each experi-ment preserved the remaining variables (i.e. maximum window size, linguistic annotations, relevance computation, candidate and descriptors extraction) and computed Recall (R), Precision (P) and F-measure (F1). This section describes: (i) baseline; (ii) evaluation set-up; (iii) evaluation results; and (iv) comparison with similar systems.

Baseline: Similarity between a profile and a candidate was computed setting an arbitrary threshold (i.e. 20) to the measure proposed by [14].

Evaluation Set-up: Our NER tool was evaluated against DrugSemantics cor-pus [15], a collection of 5 Spanish Summaries of Product Characteristics manu-ally annotated. It contains 670 sentences and 582 active ingredients.

This evaluation employed two spaces, sentences and sections, to investigate whether our estimation is robust given a maximum window size but potential bigger contexts. This implies that sentence-level tests have smaller windows. For example, first and final descriptors in a sentence only take into account half of the context; whereas a descriptor in the middle of a sentence considers the context completely. As opposed, section-level space has a bigger window, since first and final descriptors in a sentence have context descriptors from previous and following sentences until the maximum window size is reached. Given that each space is characterised differently, distinct assessment methodologies were used.

Evaluation at sentence-level used stratified sampling. It allows to randomly select each sentence based on its features to keep proportionality as in DrugSe-mantics corpus. Here, the features were: (i) lack of any NEs; (ii) contain exclu-sively other NEs; (iii) include only one active ingredient; and (iv) contain more than one active ingredient. Four samples were created from DrugSemantics. The size of each sample was decided considering a 95 % Confidence Interval (CI) and an error rate of 3 % so as to ensure the representativeness of every sample from DrugSemantics. Briefly, 2 samples were derived to provide target (125 sentences) and contrasting sets (128 sentences). The other two samples were for validation (261 sentences) and evaluation purposes (156 sentences), but only the latter was used. These experiments generated profiles at sentence-level, regardless of the window size.

Evaluation at section-level was performed through a 5-fold cross-validation (i.e. 4 SPCs for training and one for evaluation). Common cross-validation was discarded due to the profiles-dataset relation. Independently of the window size, this experiment generated profiles considering first and second level headings.

Evaluation Results: Results for both spaces are in Table 1. Regardless of the dimension, the more balanced results are achieved when VP estimates similar-ity between active ingredients profile and each candidate. Sentence-level tests,

Table 1. Active ingredient NER through profiles evaluation

Dim	Sample	Similarity	Recall (%)	Precision (%)	F-measure (%)
Sen	Test	BSL	24 [17.3–30.7]*	18 [11.97,24.03]*	21 [14.61,27.39]*
Sen	Test	SVM	84.7 [79.05–90.35]*	87 [81.72–92.28]*	85.7 [80.21–91.19]*
Sen	**Test**	**VP**	**89.1** [84.21–93.99]*	**86.1** [80.67–91.53]*	**87.3** [82.07–92.53]*
Sec	MAv	BSL	78.4	12.6	21.2
Sec	MAv	SVM	81.5	84.66	82.92
Sec	**MAv**	**VP**	**89.08**	**85.38**	**86.4**
Sec	Fold1	BSL	93	25	39
Sec	Fold1	SVM	68.4	76.8	71.9
Sec	Fold1	VP	82.4	78.5	80
Sec	Fold2	BSL	71	7	13
Sec	Fold2	SVM	83.2	88.3	85.6
Sec	Fold2	VP	92.6	87.5	90
Sec	Fold3	BSL	70	5	9
Sec	Fold3	SVM	94.3	94	94.2
Sec	Fold3	VP	95.3	94	94.7
Sec	Fold4	BSL	81	15	26
Sec	Fold4	SVM	84.6	85.6	85.1
Sec	Fold4	VP	86.4	81.8	82.8
Sec	Fold5	BSL	77	11	19
Sec	Fold5	SVM	77	78.6	77.8
Sec	Fold5	VP	88.7	85.1	86.7

Acronyms, by appearance: (i) Dim: Dimension; (ii) Sen: Sentence; (iii) BSL: baseline [14];(iv) *: 95 % confidence interval;(v) SVM: Support Vector Machine; (vi) VP: Voted Perceptron; and (vii) MAv: Macro average.

which provide a 95 %CI and 5 % error rate, produce the best result with VP and present the greater F1 (87.3 % [95 %CI: 82.07–92.53]).

State of the Art Comparison: Comparing our system results with others can not be done directly because the training and testing samples are different. Although, to prove that our results are in line with the state of the art, Table 2 lists systems described in Sect. 2 and our best result (called 'Profile'). Our system only detects active ingredients, but we can ensure its effectiveness margin. The best ML-system [18] reported R and F1 inside our CI. Examining the best dictionary-based systems [4,24], their results are within our 95 % CI. This positive comparison is repeated for Spanish NER systems [16,21]. It should be noted that [16,21] use lexicons, whereas wegot such results without them.

Table 2. Active ingredient NER by F1, 1st ML systems and 2nd dictionary-based

System	Lang	Recall(%)	Precision(%)	F1(%)	Corpus	AI entities
Sydney [18]	EN	87.77	**92.89**	**90.26**	i2b2	-
Profile	ES	**89.1**	86.1	87.3	DrugSemantics	117
WBI-NER [13]	EN	85.2	73.6	79	DDIExtraction	351
UTurku [1]	EN	-	-	78	DDIExtraction	351
LASIGE [11]	EN	70	75	73	DDIExtraction	351
UMCC-DLSI [2]	EN	56	20	30	DDIExtraction	351
OpenU [24]	EN	-	-	**89**	i2b2	-
Vanderbilt [4]	EN	**87**	**90**	88	i2b2	-
MaNER [16]	ES	86	87	88	DrugSemantics	**616**
BME-Humboldt [23]	EN	82	92	87	i2b2	-
SpanishADRTool [21]	ES	80	87	83	SpanishADR	188

Acronyms: (i) AI: Active Ingredient; (ii) ES: Spanish; (iii) EN: English; (iv) Lang:Language

5   Conclusions and Future Work

This paper presented a supervised ML NER system. This proposal, which is language and domain independent, employs unsupervised feature generation and weighting only from training data. It was evaluated on a Spanish corpus, DrugSemantics [15]. Our VP classifier results (F1 87.3 % [95 %CI: 82.07–92.53]) proves our hypothesis that an active ingredient NER system can perform this task successfully without knowledge bases. As future work, we plan to enhance our system with other NEs relevant for day-to-day care, as well as traditional NEs.

Acknowledgments. This paper has been partially supported by the Spanish Government (grant no. TIN2015-65100-R).

References

1. Björne, J., Kaewphan, S., Salakoski, T.: UTurku: drug named entity recognition and drug-drug interaction extraction using svm classification and domain knowledge. In: Proceedings of the Seventh International Workshop on Semantic Evaluation, vol. 2, pp. 651–659 (2013)
2. Collazo, A., Ceballo, A., Puig, D.D., Gutiérrez, Y., Abreu, J.I., Pérez, R., Fernández Orqín, A., Montoyo, A., Muñoz, R., Camara, F.: UMCC_DLSI: Semantic and Lexical features for detection and classification Drugs in biomedical texts. In: Proceedings of the Seventh International Workshop on Semantic Evaluation, vol. 2, pp. 636–643 (2013). http://www.aclweb.org/anthology/S13-2106
3. Deléger, L., Grouin, C., Zweigenbaum, P.: Extracting medical information from narrative patient records: the case of medication-related information. J. Am. Med. Inform. Assoc. **17**(5), 555–558 (2010)

4. Doan, S., Bastarache, L., Klimkowski, S., Denny, J.C., Xu, H.: Integrating existing natural language processing tools for medication extraction from discharge summaries. J. Am. Med. Inform. Assoc. **17**(5), 528–531 (2010)
5. Ely, J.W., Osheroff, J.A., Ebell, M.H., Bergus, G.R., Levy, B.T., Chambliss, M., Evans, E.R.: Analysis of questions asked by family doctors regarding patient care. Br. Med. J. **319**, 358–361 (1999)
6. Feldman, R., Sanger, J.: The Text Mining Handbook: Advanced Approaches in Analyzing Unstructured Data. Cambridge University Press, New York (2007)
7. Freund, Y., Schapire, R.E.: Large margin classification using the perceptron algorithm. Mach. Learn. **37**(3), 277–296 (1999)
8. Friedman, C., Rindflesch, T.C., Corn, M.: Natural language processing: state of the art and prospects for significant progress, a workshop sponsored by the National Library of Medicine. J. Biomed. Inform. **46**(5), 765–773 (2013)
9. Gonzalez-Gonzalez, A.I., Dawes, M., Sanchez-Mateos, J., Riesgo-Fuertes, R., Escortell-Mayor, E., Sanz-Cuesta, T., Hernandez-Fernandez, T.: Information needs and information-seeking behavior of primary care physicians. Ann. Family Med. **5**, 345–352 (2007)
10. González-González, A.I., Sánchez Mateos, J., Sanz Cuesta, T., Riesgo Fuertes, R., Escortell Mayor, E., Hernández Fernández, T.: Estudio de las necesidades de información generadas por los médicos de atención primaria (proyecto ENIGMA)*. Atención primaria, **38**(4), 219–224 (2006)
11. Grego, T., Couto, F.M.: LASIGE : using Conditional Random Fields and ChEBI ontology. In: Proceedings of the Seventh International Workshop on Semantic Evaluation, vol. 2, pp. 660–666 (2013). http://aclweb.org/anthology/S13-2109
12. Hamon, T., Grabar, N.: Linguistic approach for identification of medication names and related information in clinical narratives. J. Am. Med. Inform. Assoc. **17**(5), 549–554 (2010)
13. Huber, T., Linden, U.D., Rockt, T.: WBI-NER : The impact of domain-specific features on the performance of identifying and classifying mentions of drugs. In: Proceedings of the Seventh International Workshop on Semantic Evaluation, vol. 2, pp. 356–363 (2013). http://www.aclweb.org/anthology/S13-2058
14. Lopes, L., Vieira, R.: Building and applying profiles through term extraction. In: X Brazilian Symposium in Information and Human Language Technology, Natal, Brazil, pp. 91–100 (2015)
15. Moreno, I., Moreda, P., Romá-Ferri, M.: Reconocimiento de entidades nombradas en dominios restringidos. In: Actas del III Workshop en Tecnologías de la Informática, Alicante, Spain, pp. 41–57 (2012)
16. Moreno, I., Moreda, P., Romá-Ferri, M.T.: MaNER: a MedicAl named entity recogniser. In: Biemann, C., Handschuh, S., Freitas, A., Meziane, F., Métais, E. (eds.) NLDB 2015. LNCS, vol. 9103, pp. 418–423. Springer, Heidelberg (2015)
17. Padró, L., Stanilovsky, E.: FreeLing 3.0: towards wider multilinguality. In: Proceedings of the Language Resources and Evaluation Conference, Istanbul, Turkey (2012)
18. Patrick, J., Li, M.: High accuracy information extraction of medication information from clinical notes: 2009 i2b2 medication extraction challenge. J. Am. Med. Inform. Assoc. **17**(5), 524–527 (2010)
19. Platt, J.: Using analytic QP and sparseness to speed training of support vector machines. In: Proceedings of the Advances in Neural Information Processing Systems, vol. 11, pp. 557–563 (1999)

20. Sanchez-Cisneros, D., Aparicio Gali, F.: UEM-UC3M: An Ontology-based named entity recognition system for biomedical texts. In: Proceedings of the Seventh International Workshop on Semantic Evaluation, vol. 2, pp. 622–627 (2013)

21. Segura-Bedmar, I., Revert, R., Martínez, P.: Detecting drugs and adverse events from Spanish health social media streams. In: Proceedings of the 5th International Workshop on Health Text Mining and Information Analysis, pp. 106–115 (2014)

22. Singh, A.: Natural language processing for less privileged languages: Where do we come from? Where are we going? In: IJCNLP-08 Workshop on NLP for Less Privileged Languages, Hyderabad, India, pp. 7–12 (2008)

23. Tikk, D., Solt, I.: Improving textual medication extraction using combined conditional random fields and rule-based systems. J. Am. Med. Inform. Assoc. **17**(5), 540–544 (2010)

24. Yang, H.: Automatic extraction of medication information from medical discharge summaries. J. Am. Med. Inform. Assoc. **17**(5), 545–548 (2010). http://dx.doi.org/10.1136/jamia.2010.003863

Improving the Sentiment Analysis Process of Spanish Tweets with BM25

Juan Sixto[(⊠)], Aitor Almeida, and Diego López-de-Ipiña

DeustoTech-Deusto Institute of Technology, Universidad de Deusto,
Avenida de las Universidades 24, 48007 Bilbao, Spain
{jsixto,aitor.almeida,dipina}@deusto.es

Abstract. The enormous growth of user-generated information of social networks has caused the need for new algorithms and methods for their classification. The Sentiment Analysis (SA) methods attempt to identify the polarity of a text, using among other resources, the ranking algorithms. One of the most popular ranking algorithms is the Okapi BM25 ranking, designed to rank documents according to their relevance on a topic. In this paper, we present an approach of sentiment analysis for Spanish Tweets based combining the BM25 ranking function with a Linear Support Vector supervised model. We describe the implemented procedure to adapt BM25 to the peculiarities of SA in Twitter. The results confirm the potential of the BM25 algorithm to improve the sentiment analysis tasks.

Keywords: BM25 · Linear support vector · Sentiment analysis · Term frequency

1 Introduction

Sentiment analysis is the task of identifing and extracting the subjective information of a document or set of documents. Through the use of natural language processing (NLP) mechanisms, sentiment analysis attempts to determine the opinion expressed in a document. The final purpose is to establish the attitude of a user in a context or in relation with a specific topic. The heterogeneous sources of the documents studied by sentiment analysis require different approaches and customized techniques, according to the different features of the documents. In Twitter texts, the opinions are expressed in micro-blogging texts that may contain only one or two words with subjective information. This raises the need of implement new methods to address this problem.

A vast amount of techniques have been developed by researchers in the sentiment analysis field during the last years. However, most of these approaches are language dependent, and could not be applied to different languages due to the considerable differences between languages and the difficulty of establish standard linguistic rules between them [8]. The existence of different lexicons and data sets for each language increases the difficulty in the research of efficient solutions for sentiment analysis task.

© Springer International Publishing Switzerland 2016
E. Métais et al. (Eds.): NLDB 2016, LNCS 9612, pp. 285–291, 2016.
DOI: 10.1007/978-3-319-41754-7_26

In this paper, we present our adaptation of the BM25 ranking function for Sentiment Analysis and their latest results with a Linear Support Vector supervised model. Our work is based on the Spanish language and is evaluated using the corpus of TASS 2015 workshop, composed by tweets written is Spanish. The paper is organized as follows: In Sect. 2, the context of this work is presented, including the Sentiment Analysis at global level task of the TASS workshop. In Sect. 3, the Okapi BM25 ranking function is described, in conjuction with the contextual adaptation process. Section 4 covers the experimental procedures, and conclusions are in Sect. 5.

2 Related Work

BM25 Okapi is a ranking function developed by Robertson et al. [14,15] closely related to the standard TF-IDF (*frequency-inverse document frequency*) retrieval function. Even so, BM25 has two parametrizable variables that allow a superior optimization of the function. In addition, the asymptotic maximum of the function establishes a limit for high frequency terms [16], unlike the standard TF-IDF. A large number of comparisons has been conducted between BM25 and other ranking functions like TF-IDF, Dirichlet or Pivoted normalization [5]. Also, the performance of BM25 has been evaluated with several datasets formed by Twitter-like short texts about product reviews [4], with certain similarities with the work presented in this paper. In [21], BM25 Okapi score has been used to calculate the term similarity between queries and documents, similarly to [11,12].

In this paper, we use the datasets and results of the sentiment analysis task at TASS'15[1] workshop [20]. This is an evaluation workshop for sentiment analysis focused on Spanish language, organized as a satellite event of the annual conference of the Spanish Society for Natural Language Processing (SEPLN)[2]. This paper is focused on the first task of the workshop that consists in performing an automatic sentiment analysis to determine the global polarity of each message in the provided corpus. Tweets are divided into six different polarity labels: strong positive (P+), positive (P), neutral (NEU), negative (N), strong negative (N+) and no sentiment tag (NONE), and form two sets, the training dataset with 7.219 (11 %) items and the test dataset with 60.798 (89 %) items. Additionally, the task includes a 1.000 items dataset, a subset of the test dataset, for an alternative evaluation of the performance of systems.

Due to the sentiment analysis task consisting of a classification problem, a very large selection of machine learning classification algorithms has been used by researchers. The most frequent algorithms used in the recent years are Support Vector Machine (SVM) [9,17], Logistic Regression [18], Naïve Bayes [7] and Decision Tree [19]. Anta et al. [1], present a comparative analysis of classification techniques for sentiment analysis on spanish tweets. During our work these algorithms and some others were tested in order to achieve the highest accuracy results. This process are described in Sect. 4.

---

[1] Workshop on Sentiment Analysis at SEPLN Conference.

[2] http://www.sepln.org/.

3   Okapi BM25 Ranking Function

In order to use the BM25 function as a sentiment resource, it has been necessary to implement an adaptation of the algorithm to the concrete features of the task. Previously to the automatic training step, a terms dictionary must be generated through the BM25 algorithm. This dictionary allows the system to obtain the relevance weights of each term and polarity in the corpus texts. The formula used to implement BM25 in the system and the conceptual necessary adaptations are defined below. The software implementation of the function has been realized in Python Programming Language, in order to be combined with the *scikit-learn*[3] toolkit.

BM25 uses *term frequency* (TF) and *inverse document frequency* (IDF) to weight terms. The relevance of a term q_i in a polarity D is calculated acording to the Eq. 1, where $f(q_i, D)$ is the frequency of the term (TF) in the document (polarity) D. In the Okapi model, D concerns the documents used for similarity scoring. However, in our context the documents should be the six polarities existing in the corpus. To do this, the complete collection of tweets in the training dataset must be compiled by polarity, in order to generates a single document for each polarity. During this step, the use of a part-of-speech tagger and a text normalization algorithm reduces the length of the dictionary and increases their quality. This requires the use of a similar preprocessing for the texts before weighting them. The variable $|D|$ is the length of the document in words and *avgdl* is the average of the longitude of all documents in the collection. k_1 and $b \in [0, 1]$ are free parameters and their optimization is described in Sect. 4. Both for the D documents generation and for the ranking of tweets, a conversion process applies to the texts is necessary. This includes the removal of the Twitter special characters and the tokenization in order to be evaluated separately.

$$score(D, Q) = \sum_{i=1}^{n} IDF(q_i) \cdot \frac{f(q_i, D) \cdot (k_1 + 1)}{f(q_i, D) + k_1 \cdot (1 - b + b \cdot \frac{|D|}{avgdl})} \tag{1}$$

The *inverse document frequency* (IDF) equation used in BM25 is not similar to the used in standard TF-IDF algorithm, using the variation described in Eq. 2. The N value is the quantity of document that exists in the collection and $n(q_i)$ is the quantity of documents that contains the keyword (q_i).

$$IDF(q_i) = log \frac{N - n(q_i) + 0.5}{n(q_i) + 0.5} \tag{2}$$

We use this ranking function over the training dataset in order to create a lexical dictionary that is used to rank each tweet in the test datasets. To rank the tweets, the dictionary weights the text to each polarity, generating a six values array that it is used as feature in a automatic classifier. In this way, the algorithm presents better results than a direct election of the highest values. It should be taken into consideration that the six categories are not completely

---

[3] http://scikit-learn.org/.

different categories, while the pairs (P, P+) and (N, N+) comprehend the same sentiment with different intensity. Therefore they can share terms but differing in frequency.

4   Experimental Investigation

Our proposal is to assess the precision of our adaptation of BM-25 to the SA of spanish tweets task. In order to measure the efficiency of the algorithm, a set of experiments have been performed, with the TASS datasets and using the ranked weights as only lexical feature. For this work we use a LinearSVC Classifier as base classifier, due to its good performance in similar approaches [10,17]. The selected tests consist in classify both test datasets (1.000 and 60.798 items) in the six polarity categories (P+, P, NEU, N, N+, NONE). The measures selected for the experiments are accuracy and F-score, because they are usually used to order and compare classifier systems in state-of-art and represent well the success in the task.

4.1   Results

A set of experiments have been developed in order to evaluate the best values of parameters b and k_1 in the task. These experiments consist on an array of performance tests, alternating the values of both parameters and extracting the accuracy and F-score of each test. The optimal values are different to each dataset, having the highest accuracy values the parameters ($b = 0.6$ and $k_1 = 35$) for the 1 K dataset and ($b = 0.7$ and $k_1 = 40$) for the full dataset. The complete results of this experiments are resumed in Figs. 1 and 2. This values are clearly superior to the typical values, probably according to the particularities of the task. In [16], Robertson and Zaragoza asserts that values such as $0.5 < b < 0.8$ and $1.2 < k_1 < 2$ are reasonably good in many circumstances, according to their experiments, a very low values for K_1 in comparison with our optimal value.

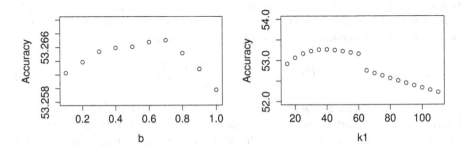

Fig. 1. Accuracy of $k_1 = 40$ (left) and $b = 0.7$ (right) using the full corpus.

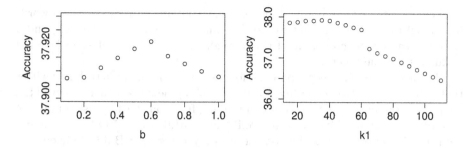

Fig. 2. Accuracy of $k_1 = 35$ (left) and $b = 0.6$ (right) using the 1.000 corpus.

Table 1. Accuracy and F1-measure of studied systems using both datasets.

System	Features	ML Methods	Accuracy		F1-Measure	
			1k	Full	1k	Full
S1	Monograms	Support Vector Machines	0.371	**0.533**	0.285	0.343
S2	Monograms	Logistic Regression	**0.384**	0.520	0.311	**0.369**
S3	Bigrams	Logistic Regression	0.234	0.266	0.203	0.192
S4	Trigrams	Logistic Regression	0.207	0.375	0.234	0.283
S5	Bigrams+Trigrams	Logistic Regression	0.354	0.517	0.314	0.333
S6	Bigrams+Trigrams	Gradient Boosting	0.335	0.459	0.278	0.316
S7	S5+S6	VoteClassifier	0.334	0.475	0.302	0.323
S8	S2+S7	MajorityVote	0.367	**0.533**	**0.383**	0.349

Once the free parameters have been optimized, we studied the complementary posibilities of the system in order to improve the results. Table 1 presents the most relevant results using monograms, bigrams or trigrams as input of the BM25 function. The studied classifiers are Support Vector Machines (SVM) [2], Logistic Regression (LR) [3] and Gradient Boosting Classifier (GBC) [6]. Also, we used a VoteClassifier in order to combine several systems and the majority voting scheme described in [13].

5   Conclusion and Further Works

In this paper, we have presented our approach to use the BM25 Okapi function in the sentiment analysis of spanish tweets. We have studied the influence of their multiple configurations of linguistic processing on the results. Also, its association with a set of automatic classifiers of contextual relevance has been exposed. During the study of the b and k_1 free parameters, has been confirmed that the optimal values of k_1 for this task is far from the typical values presented in other research fields. Our theory is that this happens because, in contrast with the application over large documents, tweets have a very short longitude

(140 characters maximum). A tweet text may contains only two or three words with a subjective content that would determine the polarity of the text, and a very low asymptotic maximum of BM25 might not generate a sufficient levels of granularity in the function.

We have presented a new approach for the sentiment analysis task in Twitter that we consider a strong baseline for future developments. We have achieved the adaptation of one of the most common ranking functions for the task, obtaining a competitive result (38,4 % of a maximum of 51,6 % and 53,3 % of a maximum of 67 %) considering that our study only covers the BM25 funcion. We think that this approach will be useful for future works in the sentiment analysis field. For further work, we would like to improve the present system including polarity lexicons and semantic resources, similarly to another approaches, and pre-processing tools to enhance the results. Our objective is to develop a complete system that can prove the real accuracy of the BM25 ranking function in the sentiment analysis task.

References

1. Anta, A.F., Chiroque, L.N., Morere, P., Santos, A.: Sentiment analysis and topic detection of Spanish tweets: a comparative study of NLP techniques. Procesamiento del lenguaje natural **50**, 45–52 (2013)
2. Cortes, C., Vapnik, V.: Support-vector networks. Mach. Learn. **20**(3), 273–297 (1995)
3. Cox, D.R.: The regression analysis of binary sequences. J. Roy. Stat. Soc. Series B (Methodological) **20**, 215–242 (1958)
4. Esparza, S.G., O'Mahony, M.P., Smyth, B.: Mining the real-time web: a novel approach to product recommendation. Knowl.-Based Syst. **29**, 3–11 (2012)
5. Fang, H., Tao, T., Zhai, C.: A formal study of information retrieval heuristics. In: Proceedings of the 27th Annual International ACM SIGIR (2004)
6. Friedman, J.H.: Greedy function approximation: a gradient boosting machine. Ann. Stat. **29**, 1189–1232 (2001)
7. Gamallo, P., Garcia, M., Fernández-Lanza, S.: A Naive-Bayes strategy for sentiment analysis on Spanish tweets. In: Workshop on Sentiment Analysis at SEPLN (TASS 2013), pp. 126–132 (2013)
8. Han, B., Cook, P., Baldwin, T.: Unimelb: Spanish text normalisation. In: Tweet-Norm@ SEPLN (2013)
9. Hurtado, L.F., Pla, F., Buscaldi, D.: ELiRF-UPV en TASS 2015: Anlisis de Sentimientos en Twitter. In: TASS 2015: Workshop on Sentiment Analysis at SEPLN (2015)
10. Hurtado, L.F., Pla, F.: ELiRF-UPV en TASS 2014: Analisis de sentimientos, deteccin de tpicos y anlisis de sentimientos de aspectos en twitter. Procesamiento del Lenguaje Natural (2014)
11. Sparck-Jones, K., Walker, S., Robertson, S.E.: A probabilistic model of information retrieval: development and comparative experiments: Part 2. Info. Process. Manage. **36**, 809–840 (2000)
12. Liu, S., Liu, F., Yu, C., Meng, W.: An effective approach to document retrieval via utilizing WordNet and recognizing phrases. In: Proceedings of the 27th Annual International ACM SIGIR (2004)

13. Pla, F., Hurtado, L.-F.: Sentiment analysis in twitter for Spanish. In: Métais, E., Roche, M., Teisseire, M. (eds.) NLDB 2014. LNCS, vol. 8455, pp. 208–213. Springer, Heidelberg (2014)
14. Robertson, S.E., Walker, S.: Some simple effective approximations to the 2-poisson model for probabilistic weighted retrieval. In: Proceedings of the 17th Annual International ACM SIGIR Conference (1994)
15. Robertson, S.E., Walker, S., Jones, S., Hancock-Beaulieu, M.M., Gatford, M.: Okapi at TREC-3. NIST SPECIAL PUBLICATION SP (1995)
16. Robertson, S., Zaragoza, H.: The Probabilistic Relevance Framework: BM25 and Beyond. Now Publishers Inc., Hanover (2009)
17. Sixto, J., Almeida, A., López-de-Ipiña, D.: DeustoTech Internet at TASS 2015: Sentiment analysis and polarity classification in spanish tweets. In: TASS 2015: Workshop on Sentiment Analysis at SEPLN (2015)
18. Thelwall, M., Buckley, K., Paltoglou, G.: Sentiment strength detection for the social web. J. Am. Soc. Inform. Sci. Technol. 63(1), 163–173 (2012)
19. Valverde, J., Tejada, J., Cuadros, E.: Comparing Supervised Learning Methods for Classifying Spanish Tweets. In: TASS 2015: Workshop on Sentiment Analysis at SEPLN, p. 87 (2015). Comit organizador
20. Villena-Román, J., García-Morera, J., García-Cumbreras, M., Martínez-Cámara, E., Martín-Valdivia, M., Ureña-López, L.: Overview of TASS 2015. In: TASS 2015: Workshop on Sentiment Analysis at SEPLN (2015)
21. Zhang, W., Yu, C., Meng, W.: Opinion retrieval from blogs. In: Proceedings of the Sixteenth ACM Conference (2007)

Validating Learning Outcomes of an E-Learning System Using NLP

Eiman Aeiad[✉] and Farid Meziane

School of Computing, Science and Engineering, University of Salford,
Salford M5 4WT, UK
E.Aeiad@edu.salford.ac.uk, F.Meziane@salford.ac.uk

Abstract. Despite the development and the wide use of E-Learning,
developing an adaptive personalised E-Learning system tailored to the
needs of individual learners remains a challenge. In an early work, the
authors proposed APELS that extracts freely available resources on the
web using an ontology to model the leaning topics and optimise the information extraction process. APELS takes into consideration the leaner's
needs and background. In this paper, we developed an approach to evaluate the topics' content extracted previously by APELS against a set
of learning outcomes as defined by standard curricula. Our validation
approach is based on finding patterns in part of speech and grammatical
dependencies using the Stanford English Parser. As a case study, we use
the computer science field with the IEEE/ACM Computing curriculum
as the standard curriculum.

Keywords: NLP · Keyword extraction · Key phrases · Dependency
relation

1 Introduction

E-learning is a modality of learning using Information and Communication Technologies (ICTs) and advanced digital media [14]. It offers education to those who
cannot access face to face learning. However, the needs of individual learners
have not been addressed properly [1]. A system to address this problem was
proposed by Aeiad and Meziane [1] in the form of an E-Learning environment
that is Adaptive and Personalised using freely available resources on the Web
(the APELS System). The aim of APELS is to enable users to design their own
learning material based on internationally recognised curricula and contents. It
is designed to first identify learners' requirements and learning style and based
on their profile, the system uses an ontology to help in extracting the required
domain knowledge from the Web in order to retrieve relevant information as per
users'requests. A number of modules were developed to support this process [1].
The contents of the retrieved websites are then evaluated against a set of learning
outcomes as defined by standard curricula and this constitutes to the main focus
of this paper. We used a set of action verbs based on Bloom's taxonomy [2] to

© Springer International Publishing Switzerland 2016
E. Métais et al. (Eds.): NLDB 2016, LNCS 9612, pp. 292–300, 2016.
DOI: 10.1007/978-3-319-41754-7_27

analyse the learning outcomes. Bloom's taxonomy classifies action verbs into six levels representing the following cognitive skills: Remembering, Understanding, Applying, Analysing, Evaluating and Creating. For example, action verbs such as define, describe and identify are used to measure basic levels of cognitive skills in understanding, while action verbs such as carry out, demonstrate, solve, illustrate, use, classify and execute are used to measure basic levels of the applying cognitive skills. In addition, we used the Stanford Parser, an implementation of a probabilistic parser in Java which comprises a set of libraries for Natural Language Processing (NLP) that together make a solid unit capable of processing input text in natural language and produces part-of-speech (PoS) tagged text, context-free phrase structure grammar representation and a typed dependency representation [6]. In this paper, a case study using the IEEE/ACM Computing Curriculum [12] will be used to illustrate the functionality of APELS.

The rest of the paper is structured as follows: the next section reviews some related work by outlining different approaches for extracting keywords and keyphrases. Section 3 presents a revised architecture of the APELS system and the knowledge extraction module in details. Section 4 illustrates the functionality of the APELS system using examples from the ACM/IEEE Computer Society Computer Science Curriculum. and the system evaluated. Finally, conclusion and future developments are given in Sect. 5.

2   Project Background

Personalised E-learning systems have attracted attention in the area of technology-based education, where their main aim is to offer to each individual learner the content that suits her/his learning style, background and needs. Previously, we designed APELS that extracted information from the WWW based on an ontology and tailored to individual learners' profile [1]. However, the suitability of the contents of the selected websites should be evaluated to ensure that they fit the learner's needs and this will be addressed in this paper. Matching the content to learning outcomes of curricula, is very important when assessing the suitability of the selected websites. Learning outcomes are statements of what a student is expected to know, understand and/or be able to demonstrate after the completion of the learning process [7]. Each learning outcome contains an action verb followed by usually a noun phrase that acts as the object of the verb. Together, the action verbs and noun phrases are referred to as Keywords or key phrases. These are used in academic publications to give an idea about the content of the article to the reader as they are a set of representative words, which express the meaning of an entire document. Various systems are available for keywords extraction such as automatic indexing, text summarization, information retrieval, classification, clustering, filtering, topic detection and tracking, information visualization, report generation, and web searches [3]. Automatic Keyword Extraction methods are divided into four categories: statistical methods [5], machine learning methods [13], linguistic methods [9] and hybrid methods [10].

The APELS architecture given in Fig. 1, is a revised version of the one proposed in [1]. It is modified based on our experience while developing this project. APELS is based on four modules: student profile, student requirement, knowledge extraction and content delivery. The student profile and the student requirement modules are similar to the ones presented in [1]. We updated the knowledge extraction module adding the learning outcomes validation in order to evaluate the topics' contents against a set of learning outcomes as defined by standard curricula and the details of this module are given in Sect. 3.

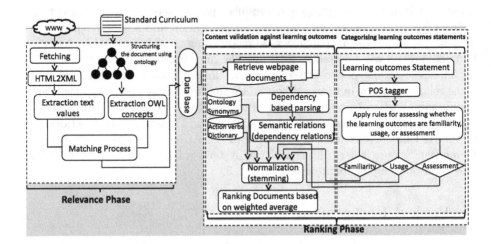

Fig. 1. Knowledge extraction module

3   The Knowledge Extraction Module

The Knowledge extraction module is responsible for the extraction of the learning resources from the Web that would satisfy the learners' needs and learning outcomes. The Module comprises two phases; the Relevance phase and the Ranking phase. The relevance phase uses an ontology to retrieve the relevant information as per users' needs. In addition, it transforms HTML documents to XML to provide the information in a friendly accessible format and easier for extraction and comparison. Moreover, we implemented a process called the matching process that computes the similarity measure between the subset of the ontology that models the learning domain and the values element extracted from the websites. The website with the highest similarity is selected as the best matching website that satisfies the learners' learning style.

After the matching process occurs, there is a further step that is required to evaluate that content adheres to a standard set of guidelines for studying the chosen subject. Hence, the learning outcome validation was added to ensure the selection of the most relevant websites that satisfy the learning outcomes set by standard curricula. This is the purpose of the ranking phase that is composed of two components (i) categorising learning outcomes statements and content validation against learning outcomes.

3.1 Categorising Learning Outcomes Statements

The learning outcomes statements are analysed by selecting a set of action verbs based on the Bloom's taxonomy [2]. Each learning outcome contains an action verb associated with the intended cognitive level of the bloom's Taxonomy, followed by the object of the verb (specific subject material). The Stanford parser is used as the pre-processor of the input statement, which is a learning outcome. It takes the learning outcome statement, written in natural language, and marks it with the PoS tagger, builds tree representation of the sentence from the sentence's context-free-phrase-structure-grammar parse, and eventually builds a list of typed dependencies [6]. Here we used only the PoS tagger to analyse the learning outcome and followed by Nouns and Verbs Extractor to classify the learning outcomes. The PoS tagger is used to identify the nouns and verbs by tagging each word in the text (e.g. 'drink': verb, 'car': noun, 'lucky': adjective, 'vastly': adverb, 'on': preposition etc.). It has been widely proposed by many authors [4,8] as the main task for analysing the text syntactically at the word level. After all the words in the learning outcomes statements are tagged, Nouns and Verbs Extractor is used to extract the nouns and verbs by selecting the pattern tags of the PoS. The current pattern tags of Stanford parser is defined as follow:

define/VB and/CC describe/VB variable/NN ./.

A set of rules are used to identify the learning outcomes statement by searching the pattern token in the tagged verb in the action verbs dictionary that have been manually defined based on the Bloom's Taxonomy. The rules that are used to assign learning outcomes based on action verbs in bloom's Taxonomy have the form:

```
if pattern token in tagged verb belongs to Level A,
    then learning outcome = "A"
```

The six levels representing of the cognitive skills defined in Sect. 1. We associate a set of action verbs with each level which will be used to identify the level. The actions verbs associated with the Applying level for example are also given in Sect. 1.

3.2 Content Validation Against Learning Outcomes

The evaluation of the topic's content will be against the identified learning outcomes statements. The Stanford typed dependencies representation is used to extract a topic name, an action verb and their relationship. Moreover, we adopted a rule based linguistic method to extract key phrases and keywords from text.

The Stanford typed dependencies representation provides a simple description of the grammatical relationships in a sentence, establishing relationships between "head" words and words which modify those heads ("refer"). Furthermore, the Stanford dependency parser consists of three variables namely: Type dependency name, governing of the dependency and Subordinate of the dependency.

To extract action verbs, topic names and their relationships, two types of dictionaries are used. The action verbs dictionary that contains the action verbs that have been manually defined based on the Bloom's Taxonomy and the topic name synonym dictionary whose terms are retrieved from the ontology. The system checks the output of a typed dependency pattern to check if the governor of the dependency is an action verb and its subordinate a topic name or if the governor of the dependency is a topic name and its subordinate an action verb. Moreover, we used Porter's stemming Algorithm [11] to produce the roots of the words. Once the Stanford parser produces the typed dependency between a pair of words, these are analyzed to get the root of the word that will be looked up in the action verb dictionary and the topic name synonyms from the ontology. The other distinctive feature of Normalisation (stemming) is to reduce the size of the action verbs dictionary and topic name synonym as they contain all the different forms of the word.

The typed dependency parsing approach is only used to analyze the text in order to identify the potential relationship between the action verbs and topic names. This is not enough to fully validate the content against the learning outcomes. Hence, we used rule based linguistic methods to filter out the key phrases and keywords by using the linguistic features of the word (i.e., PoS tags) to determine key phrases or keyword from the text. These rules are employed to identify familiarity, usage, and assessment levels which are illustrated in the case study section.

4   Case Study and Evaluation

4.1   Description of the Case Study

The ACM/IEEE Computer Science Society Curriculum [12] was used to illustrate the functionality of APELS. The IEEE/ACM Body of Knowledge (BoK) is organized into a set of 18 Knowledge Areas (KAs) corresponding to typical areas of study in computing such as Algorithms and Complexity and Software Engineering. Each Knowledge Area (KA) is broken down into Knowledge Units (KUs). Each KU is divided into a set of topics which are then classified into a tiered set of core topics (compulsory topics that must be taught) and elective topics (significant depth in many of the Elective topics should be covered). Core topics are further divided into Core-Tier1 topics and Core-Tier2 topics (Should almost be covered). The software development fundamentals area for example is divided into 4 KUs. The Algorithm and design KU is divided into 11 Core-Tier1 topics. Learning outcomes are then defined for each class of topics. We will specifically look at designing an advanced programming module in C++ from the IEEE/ACM fundamental Programming Concepts KU using APELS. Moreover, we used the learning outcomes that have an associated level of mastery in the Bloom's Taxonomy, which have been well explored within the Computer Science domain based on the IEEE/ACM Computing curriculum. The level of mastery is defined in the Familiarity, Usage and Assessment levels. Each level has a special set of action verbs. The linguistic rules used in APELS include:

- **Rule 1:** At the "Familiarity" level, the student would be expected to know the definition of the concept of the specific topic name in the content text. Thus this rule is utilized to extract the key phrases when the topic name is followed by verb "to be" expressed as "is" and "are" such as in the phrases "variable is" and "algorithms are". In these kind of key phrases the noun "variable" does not depended on the verb to be "is". The PoS tag is used to identify the grammatical categories for each word in the content of the text. Then, the system will extract a noun followed by the verb "to be" by selecting the pattern token in the tagged noun followed by the pattern's token in the tagged verb. We first identify the token with the noun tag in the topic name synonyms from the ontology and check if it is followed by the token with the verb tag ("is" or "are" in this case).
- **Rule 2:** At the "usage" level, the student is able to use or apply a concept in a concrete way. Using a concept may include expressions made up of two words such as "write program", "use program" and "execute program". In these expressions, where words such as "write", "use" and "execute" are dependent on "program", the system is able to recognize these expressions automatically from the text using dependency relations.
- **Rule 3:** Students who take courses in computer science domain will have to apply some techniques or use some programs. Therefore, the content may include examples to illustrate the use these concepts. To search whether the content has terms such as "example" or "for example". A PoS tagger is used to tag each word in the text. The system will then extract nouns by selecting the pattern token in the tagged noun. Finally, the system checks if the pattern token in the tagged noun matches with the word "example".
- **Rule 4:** at the "assessment" level, we have designed a special kind of rules because at this level there are two types of information that needs to be evaluated. First, the student is required to understand a specific topic and be able to use the topic in a problem solving scenario for example. In this case the system will apply rules 1 to 3 for the specific topic. Second, the student should be able to select the appropriate topic among different topics, hence the system apply again rules 1 to 3 for each topic.

4.2 Results and Evaluation

The APELS system produced a list of websites for learning the C++ language with the highest accuracy rating [1]. Now, the validity of these selected websites will be assessed by matching their content to the targeted learning outcomes as described by IEEE/ACM curriculum. We selected one of the outcome "Define and describe variable" to be tested by our system. First, the system identified the learning outcome in the Familiarity level because it contains the action verbs "define" and "describe". Hence in this case, three kinds of grammatical patterns are implemented for keywords and key phrases extraction as shown in Table 1. One is the potential relationship between the action verb and the topic name, second if the potential syntactic structure of the sentences includes a noun phrase followed by a verb phrase and the third pattern of the PoS includes a noun phrase. The system ranks the relevant documents using the weighted average method, which is used to calculate the average value of a particular set of occurrence of keywords and key phrases in a document with different weights. The weighted average formula is defined as follows:

$$WeightAvg(x) = w_1 x_1 + w_2 x_2 + w_3 x_3 + ... + w_n x_n$$

Where W = weight, x = occurrence of keywords and key phrases.

The weights are determined based on the importance of each mastery level of the learning outcomes. The weights are chosen manually where sentence definition of the topic name taken from Rule 1 is worth 0.70. This is because understanding the definition of the concept is most important for the students before they do any further learning around it. For example, the learning outcome "define and describe variables", if the student does not understand the variable concept he/she might not be able to implement it in their work. The dependency relationship between action verbs and topic name and Rule 3 are both given a weighted 0.15. This is because both criteria have the same importance. To calculate a weighted average, each value must first be multiplied by its weight. Then all of these new values must be added together. Thus, the overall calculation for the website (www.cplusplus.com/doc/tutorial/variables) would be (43 * 0.15) + (8 * 0.70) + (16 * 0.15) = 14.45. It is crucial that each criteria are given the correct weights based on their importance. If more weights were given to less important criteria compared to important one, it would give inaccurate ranking for the WebPages. For example, although (www.cplusplus.com/doc/tutorial/variables) has highest dependency relation,

Table 1. Ranking of WebPages based on Weighted average

WebPages	Occurences			Weighted average			Total
	Dependency Relation	Rule 1	Rule 3	*0.15	*0.70	*0.15	
fresh2refresh.com/c-programming/c-variables	16	15	15	2.4	10.5	2.25	15.15
www.cplusplus.com/doc/tutorial/variables	43	8	16	6.45	5.6	2.4	14.45
en.wikibooks.org/wiki/C_Programming/Variables	32	4	20	4.8	2.8	3	10.6
microchip.wikidot.com/tls2101:variables	37	2	3	5.55	1.4	0.45	7.4
http://www.doc.ic.ac.uk/~wjk/C++Intro /RobMillerL2.html	11	2	25	1.65	1.4	3.75	6.8
www.penguinprogrammer.co.uk/c-beginners -tutorial/variables	11	3	1	1.65	2.1	0.15	3.9

(fresh2refresh.com/c-programming/c-variables) ranked first because it has the highest score for Rule 1 which is most important.

5 Conclusion and Future Work

Validating the website content against the learning outcome would add a great value to the system making it more specific. In the future work, we need to finalise the system so that the leaner will be able to adapt and modify the content and learning style based on the interactions of the users with the system over a period of time. The information extracted by the system will be passed to a Planner module that will structure it into lectures/tutorials and workshops based on some predefined learning times.

References

1. Aeiad, E., Meziane, F.: An adaptable and personalised e-learning system based on free web resources. In: Biemann, C., Handschuh, S., Freitas, A., Meziane, F., Métais, E. (eds.) NLDB 2015. LNCS, vol. 9103, pp. 293–299. Springer, Heidelberg (2015)
2. Bloom, B.S., Engelhart, M.D., Furst, E.J., Hill, W.H., Krathwohl, D.R.: Taxonomy of Educational Objectives: Handbook 1, The Cognitive Domain. Allyn & Bacon, Boston (1956)
3. Bracewell, D.B., Ren, F., Kuriowa, S.: Multilingual single document keyword extraction for information retrieval. In: Proceedings of the International Conference on Natural Language Processing and Knowledge Engineering, pp. 517–522. IEEE (2005)
4. Brill, E.: A simple rule-based part of speech tagger. In: Proceedings of the 3rd Conference on Applied Natural Language Processing, pp. 152–155. ACL, Stroudsburg (1992)
5. Cohen, J.D.: Highlights: language- and domain-independent automatic indexing terms for abstracting. J. Assoc. Inf. Sci. Technol. **46**(3), 162–174 (1995)
6. De Marneffe, M.-C., Manning, C.D.: Stanford typed dependencies manual. Technical report, Stanford University (2008)
7. ECTS Users' Guide. Computer science curricula (2005). http://ec.europa.eu/education/programmes/socrates/ects/doc/guide_en.pdf
8. Hepple, M.: Independence, commitment: assumptions for rapid training and execution of rule-based pos taggers. In: Proceedings of the 38th Annual Meeting on Association for Computational Linguistics, ACL 2000, pp. 278–277. Association for Computational Linguistics, Stroudsburg (2000)
9. Hulth, A.: Improved automatic keyword extraction given more linguistic knowledge. In: Proceedings of the 2003 Conference on Empirical Methods in NLP, pp. 216–223 (2003)
10. Krulwich, B., Burkey, C.: Learning user information interests through extraction of semantically significant phrases. In: Proceedings of the AAAI Spring Symposium on Machine Learning in Information Access, pp. 100–112 (1996)
11. Porter, M.F.: An algorithm for suffix stripping. Program **14**(3), 130–137 (1980)
12. ACM/IEEE Societies. Computer science curricula (2013). http://www.acm.org/education/CS2013-final-report.pdf

13. Uzun, Y.: Keyword extraction using naive bayes. Technical report, Bilkent University, Computer Science Dept., Turkey (2005)
14. Wentling, T.L., Waight, C., Strazzo, D., File, J., La Fleur, J., Kanfer, A.: The future of e-learning: A corporate and an academic perspective (2000). http://learning.ncsa.uiuc.edu/papers/elearnfut.pdf

Aspects from Appraisals!! A Label Propagation with Prior Induction Approach

Nitin Ramrakhiyani[1]([⊠]), Sachin Pawar[1,2], Girish K. Palshikar[1], and Manoj Apte[1]

[1] TCS Research, Tata Consultancy Services, Pune 411013, India
{nitin.ramrakhiyani,sachin7.p,gk.palshikar,manoj.apte}@tcs.com
[2] Department of CSE, Indian Institute of Technology Bombay, Mumbai 400076, India

Abstract. Performance appraisal (PA) is an important Human Resources exercise conducted by most organizations. The text data generated during the PA process can be a source of valuable insights for management. As a new application area, analysis of a large PA dataset (100K sentences) of supervisor feedback text is carried out. As the first contribution, the paper redefines the notion of an aspect in the feedback text. Aspects in PA text are like activities characterized by verb-noun pairs. These activities vary dynamically from employee to employee (e.g. conduct training, improve coding) and can be challenging to identify than the static properties of products like a camera (e.g. price, battery life). Another important contribution of the paper is a novel enhancement to the Label Propagation (LP) algorithm to identify aspects from PA text. It involves induction of a prior distribution for each node and iterative identification of new aspects starting from a seed set. Evaluation using a manually labelled set of 500 verb-noun pairs suggests an improvement over multiple baselines.

1 Introduction

Performance Appraisals (PA) are carried out in various organizations to measure growth and productivity of employees. Apart from gauging employee performance, the PA process is used for employee promotions, re-numerations and rewards. In this work, PA process of a large Information Technology services organization is considered. The organization has more than 300 K employees, leading to large amount of appraisal text (∼10 million sentences) getting generated every year.

The organization's PA methodology consists of three steps - (i) supervisor setting goals for an employee, (ii) employee recording his self-appraisal for each goal and (iii) supervisor recording his feedback. It is important to summarize such appraisal dialogue. For summarization in the product review domain, "aspect" based analysis is carried out. Aspects are certain informative noun phrases useful in summarization and sentiment analysis. A novel contribution of this work is the redefinition of an aspect in PA domain which is more an activity than a single

© Springer International Publishing Switzerland 2016
E. Métais et al. (Eds.): NLDB 2016, LNCS 9612, pp. 301–309, 2016.
DOI: 10.1007/978-3-319-41754-7_28

noun phrase. As the second contribution, the Label Propagation algorithm [12] is supplemented to introduce an instance-wise prior which indicates an instance's belongingness to the target labels. The approach starts with a seed set and then learns new aspects iteratively using weakly supervised label propagation.

2   Notion of an Aspect in the PA Domain

Based on a perspective to regard a performance appraisal as a kind of "person review", the term "aspect" is borrowed from the domain of product review analysis. Observation of data from both review texts reveals a host of differences in terms of writing style, sentential structure, level of sentiment and noise content. In the product review domain, aspect based analysis is well researched [7] and is used widely to derive useful information. Majority of the work in the product review domain assumes noun phrases in the review sentences as candidates for aspects. In the PA domain, a noun phrase like training or unplanned leave alone doesn't characterize the facets of the dynamic pair of an employee and his work. As an example, the noun phrase training can be an important facet of an employee's work. However, for an experienced employee it may be related to conducting trainings and for an amateur it may be related to attending trainings. For proper summarization and analysis of PA text, it is necessary to differentiate these modifications to the noun phrase training. Based on these observations, it is proposed that aspects in PA text should be modelled like an activity rather than a noun phrase. Further, the activity should be characterized as a verb-noun pair, e.g. conduct training and attend training. Multiple examples in Table 1 emphasize the effectiveness of this verb-noun pair notion of an aspect. It is observed that many highly frequent verb-noun pairs carry marginal semantic content to be regarded as an aspect and in turn for summarization. As examples, consider verb-noun pairs like spend-time, help-project and similar which are regularly said by supervisors. One can see that these verb-noun pairs are too generic to bring an information gain to an appraisal summary and hence, are not accounted as aspects. It also establishes that pairs showing high co-occurrence are not necessarily aspects. Guidelines for tagging verb-noun pairs provided to the annotators were based on the above arguments. A verb-noun pair which describes a specific activity in a clear manner and is fit for inclusion in a summary should be tagged as an aspect and non-aspect otherwise.

Table 1. Verb-noun pairs as aspects

Verb-Noun pairs
develop--tool, create--tool, learn--tool
deliver--code, develop--code, review--code
propose--idea, suggest--idea, implement--idea, share--idea

3   Prior Induction in the Label Propagation Algorithm

Given the above definition of an aspect in PA text, the task of identifying meaningful aspects is posed as a classification of candidate verb-noun pairs into aspects and non-aspects. The semi-supervised graph based Label Propagation Algorithm [12] (LPA), is employed. It represents labelled and unlabelled instances as nodes in a graph with edges reflecting the similarity between nodes. The label information of a node is propagated to its nearby nodes iteratively through the weighted edges and labels of unlabelled examples are inferred when the propagation converges.

Now there can be scenarios when prior domain knowledge about nodes is available and can be a crucial aid in classification of the nodes. Providing a skewed label distribution at initiation won't work as Zhu and Gheramani [12] proved that the final labels which get assigned to unlabelled nodes are independent of their label distribution at initialization. As a major contribution of the paper, a novel technique to induce this prior knowledge is proposed which involves addition of dummy nodes, one per label, to the graph. The initial distribution of a dummy node is skewed to hold full probability for the label it represents and zero for other labels. Now for a dummy node representing label l_j, all unlabelled nodes and nodes labelled l_j should be connected to it. Weights of the edges connecting unlabelled nodes to dummy nodes can be used to add the desired prior knowledge. To keep this edge weight annotation generic in nature it can be constrained to behave as a prior probability distribution. So it can be ensured that the sum of weights of edges connecting a node to the dummy nodes should be 1. For instance, in a graph with M dummy nodes if the prior inclination to the ith label is expressed as μ_i, the corresponding edge weight should be the normalized value $\frac{\mu_i}{\sum_{j=1}^{M} \mu_j}$. The paper generalizes this notion further so that the edge weight sum can also be a fixed constant k instead of 1. This would change the corresponding edge weight in the example to $\frac{k*\mu_i}{\sum_{j=1}^{M} \mu_j}$. Using such a constant, helps in controlling the influence of the prior during label propagation iterations. It is also used in the ensuing discussion on fixed and variable prior induction.

The proposed graph setup is accompanied by an assumption that a well defined procedure (domain specific or independent) to express the prior information as a real number can always be devised. Based on the proposed premises, there can be two ways to induce the prior - fixed percentage prior induction and node-wise variable prior induction. For a node in the modified graph, with M dummy nodes and N other nodes, it can be inferred that the prior information from all dummy nodes to a node will be weighted by a factor $\frac{k}{(\sum_{i=1}^{N} w_i)+k}$. Now under fixed percentage prior induction, it is intended to infuse the prior information by a fixed percentage (say $q\%$). The constant k_r for a node r with n_r non-dummy neighbours, can then be expressed as a function of the total non-dummy edge weight incident at node r and the percentage q through the following formalism.

$$q = \frac{k_r}{(\sum_{i=1}^{n_r} w_{ri}) + k_r} \quad \Rightarrow \quad k_r = \frac{q * (\sum_{i=1}^{n_r} w_{ri})}{1 - q}$$

Under node-wise variable induction, it is intended that the prior infusion gets controlled by the connectivity at a node which is dynamic and dependent on the graph. For densely connected nodes the prior infusion should be minimal allowing the label propagation to play a major role and vice versa. A simple variable prior can be defined based on the mean of sum of non-dummy edge weights acting at all non-dummy nodes. For a graph with N non-dummy nodes and the pth node having n_p non-dummy neighbours, the value for constant k_r for a node r can be devised as follows.

$$k_r = \frac{\frac{1}{N} \sum_{p=1}^{N} \sum_{i=1}^{n_p} w_{pi}}{\sum_{i=1}^{n_r} w_{ri}}$$

In the above formulation, the numerator is the mean of sum of edge weights incident at all nodes in the graph and the denominator captures the density at a node through the edge weight sum. Hence, the ratio represents the necessary variability with a higher k_r indicating sparse nodes and lower k_r indicating dense nodes.

The label propagation algorithm can be executed iteratively on this new graph which now includes M dummy nodes along with N nodes of the graph resulting into labels for unlabelled nodes. It is also important to clamp the labels of the dummy nodes like labelled nodes, across iterations.

4   Experimental Analysis

4.1   Corpus

Performance appraisal text is a confidential dataset present with an organization. For experiments in this paper, text generated during the PA process of a large Information Technology services organization was considered. A set of 100 K supervisor feedback sentences recorded for average performing employees working in a programming role was compiled for the experiments.

For developing a test dataset, 500 most frequently occurring verb-noun pairs were collected from the feedback sentences and a manual annotation of the pairs as aspect or otherwise, was carried out. Three annotators were employed for the task. Cohen's Kappa score [3] for measuring inter-annotator agreement was computed pair wise for the three annotations. The maximum Kappa score was observed to be less than 0.6 making it less satisfactory. To get best of the three annotations, a majority vote among them for each verb-noun pair was taken and a majority vote annotation was devised. The Cohen's Kappa scores between the individual annotations and the majority vote annotation were observed to be 0.783, 0.7485 and 0.7853 respectively. Consequently, the majority vote annotation was considered the ground truth for evaluation. The test dataset was fairly balanced (269 aspects against 231 non-aspects).

4.2 Experiments

To obtain candidate verb-noun pairs, dependency parsing of each sentence was performed using the Stanford CoreNLP pipeline [8]. Next, the verbs and nouns connected with relations - *nsubj, dobj, iobj, nsubjpass, rcmod* and *xcomp*[1] were extracted. These relations ensured that the verbs and nouns formed a meaningful activity pair. The extracted verbs and nouns are organized in the form of a bipartite graph with one set of vertices consisting of the verbs and the other the nouns. The edges carry co-occurrence frequency of the enclosing nodes.

Discussion on Baselines: Three different co-occurrence based baselines are used for comparison. The first baseline is Point-wise Mutual Information (PMI) [2] of the verb-noun pair. The test set pairs are ordered according to largest PMI first. A classification is achieved by demarcating a threshold PMI value above which all pairs are considered as aspects and the rest as non-aspects. The best threshold is determined empirically. As another baseline the Lexicographer's Mutual Information (LMI) [6] is computed. It multiplies the PMI with the co-occurrence frequency, hence giving priority to pairs with high co-occurrence. A product baseline which computes a score by multiplying the conditional probabilities $P(v/n)$ and $P(n/v)$ is also employed. $P(v/n)$ is the conditional probability of observing verb v in a verb-noun pair given that noun is n. A maximum likelihood estimation is used which is the ratio of the number of times v and n occur together as a verb-noun pair to the number of times n occurs in any verb-noun pair. A similar argument holds for $P(n/v)$.

Experiments with Label Propagation: For experiments with the Label propagation algorithm, an open implementation available at [9] is used. The verb/noun bipartite graph developed earlier is processed further. For each noun and verb, three information measures are computed. The first measure is Resnik's Information Content (IC) [10] which is calculated as the negative log of probability of a word's occurrence in a corpus. The second measure is the entropy of a noun/verb. The entropy is computed using probabilities of co-occurrence of a noun (verb) with each verb (noun) in the opposite set of vertices. The third measure, named Specificity, is introduced to combine both the measures - IC and entropy. As shown in Eq. 1, Specificity of a verb v is formulated as a modified entropy over all nouns, incorporating a weighted probability where the weights are IC values of the nouns. Specificity of a noun can be visualized on similar lines. Through these measures, a relative importance of the word can be established and hence its belongingness to the aspect class can be characterized. These measures are regarded as verb/noun properties.

$$S(V_i) = \sum_{j=1}^{|N|} -\acute{p}_{ij} * log\,\acute{p}_{ij} \quad where \quad \acute{p}_{ij} = \frac{|(v_i, n_j)| * IC(n_j)}{\sum_{k=1}^{|N|} |(v_i, n_k)| * IC(n_k)} \qquad (1)$$

Considering the verb-noun pair notion of an aspect, a bipartite graph separating the verbs and nouns is conducive to extract candidate aspects. As the

---

[1] http://nlp.stanford.edu/software/dependencies_manual.pdf.

next step, an edge-to-node transformation is performed to convert the bipartite graph to a line graph. In the new graph, each node is a verb-noun pair and an edge exists between nodes with a common verb or noun. This transformation allows setting weight of an edge using different combinations of properties of the connector verb/noun and the adjacent (connected) verbs/nouns. Three different combinations of verb/noun properties are used to put weights on edges in the line graph:

- Connector's verb/noun property. E.g., verb properties of the verb `fix` connecting nodes `fix-bugs` and `fix-queries`.
- Linear combination of prop erties of connector and connected verbs/nouns. E.g., for `fix-bugs` and `fix-queries`, $2 * property(verb = fix) + property (noun1 = bugs) + property(noun2 = queries)$.
- Representing each node's verb and noun properties in two-dimensional space, the so obtained Euclidean distance between the connected verb-noun pairs.

Along with the three properties IC, Entropy and Specificity and the above combinations, a total of 9 configurations for the line graph are generated. Another such 9 configurations of the line graph are generated now with two dummy nodes corresponding to the two labels **aspect** and **non-aspect**. Two different types of priors are tried. For the first type, the prior depends on the property (IC or entropy or specificity) being used in the configuration. To compute the prior values, an average of the verb property and noun property for the verb-noun pair in the node is computed. All these average values are then sorted and ranked. A percentile based on the rank is assigned to each verb-noun pair and is used as the prior information.

For the second type of prior, Latent Dirichlet Allocation (LDA) based values are computed for each word and used. LDA [1] is a probabilistic generative model which can be used for uncovering the underlying semantic structure of a set of documents in the form of "topics". In simple words, a "topic" consists of a cluster of words that frequently occur together and are semantically similar. Considering number of topics to be T, the following topic-based specificity score (τ) was defined for each word which is part of at least one topic representation:

$$\tau(w) = \log \left(\frac{T}{\sum_{t_i \in T} \mathbb{I}(w, t_i, k)} \right) \ where$$

$$\mathbb{I}(w, t_i, k) \quad = \quad 1 \ (if \ w \in top \ k \ words \ in \ topic \ t_i); \ 0 \ (otherwise)$$

The Mallet software package[2] was used for learning 100 topics from the corpus of supervisor comments. Each topic was represented by top 20 most probable words. The score for all other words which are not part of any topic representation is set to 0. The intuition behind this score definition is that the words which are part of many topics, are not "specific" enough and hence get a lower score. Whereas the words which are part of a few topics, are quite "specific" and get a

---

[2] http://mallet.cs.umass.edu.

higher score. An average of these scores for the verb and noun in the verb-noun pair is computed and a rank based percentile score is used as the prior.

To form the seed set, 11 verb noun pairs comprising of 6 aspects and 5 non-aspects were chosen randomly (ensuring agreement of all three annotators) from the annotated set of 500 verb-noun pairs. Experiments are conducted for fixed percentage prior induction by varying the percentage from 1 to 100. Additionally, results based only on the prior information are also computed for each configuration.

4.3 Evaluation

For evaluation, accuracy values from baseline methods and the runs of LP algorithm are compiled. Accuracy values are preferred over F1-measures because in this case of two-class classification, the accuracy turns out to be the micro-averaged F1-measure for the two classes. Table 2 shows the various accuracy values.

It can be observed that co-occurrence baselines perform better than the normal label propagation and only prior techniques. However, it is use of prior information in label propagation that improves the results and helps outperform the baselines. The combined approach also proves to be better than the LPA and prior individually barring a few cases. Use of LDA information as prior boosts the accuracy further taking it as high as 76.2 % which approaches the average inter-annotator agreement score. Considering the case when LDA priors are used, in most configurations the fixed percentage prior marginally outperforms node-wise variable prior. However, it is preferable to employ the node-wise variable prior as that doesn't require learning/tuning of any external parameters (like q) and is completely dependent on node connections.

Table 2. Experimentation Results (% Accuracy values)

Baseline	PMI		LMI		Product	
Accuracy	0.68		0.668		0.672	
Property	LPA	Property	LPA + Property Prior		LPA + LDA Prior	
+ Combination		prior	Fixed (q%)	Variable	Fixed (q%)	Variable
IC + Connector	60.48	59.04	64.11 (42)	70.28	75 (21)	75.1
IC + Linear Comb.	55.44	59.04	64.31 (36)	70.88	75.4 (2)	75.1
IC + Euclidean Dist.	60.48	59.04	64.71 (25)	**71.08**	76 (27)	75.7
E + Connector	52.82	40.16	68.15 (3)	54.21	75.6 (23)	74.1
E + Linear Comb.	53.63	40.16	67.94 (3)	54.82	75.4 (25)	74.7
E + Euclidean Dist.	59.48	40.16	68.75 (2)	52.41	75.6 (10)	74.9
S + Connector	52.62	39.76	67.14 (3)	53.82	75.4 (24)	73.7
S + Linear Comb.	53.62	39.76	67.34 (2)	53.61	75.4 (25)	74.7
S + Euclidean Dist.	60.08	39.76	69.15 (2)	54.21	**76.21** (10)	76.1

LDA Prior performs at 72.69 %; E: Entropy; S: Specificity

Analysis of Results: The strong LDA prior baseline misses several aspects which are correctly identified by the LPA + LDA prior setting, e.g. `improve-competency`, `follow--ethic`, `conduct--kss`[3], `complete--documentation`. Further, the LDA prior incorrectly labels certain non-aspect verb-noun pairs as aspects, which the LPA + LDA Prior setting corrects, e.g. `accept--comment`, `reduce--count`, `rework--effort`, `code--standard`. It can be observed that the former two verb-noun pairs do not provide much information but more importantly, the latter two are not even proper verb-noun pairs as they are result of incorrect POS-tagging (`rework` and `code` incorrectly tagged as verbs).

5   Related Work

Liu [7] provides a good survey of various types of aspect extraction techniques. But none of these techniques deals with "aspects" in the form of verb-noun pairs. On the contrary, Widdows [11] and Cohen et al. [4] have experimented with verb-noun pairs but not in the context of aspect extraction or summarization.

There have been some attempts to introduce "dummy" nodes in graph based semi-supervised learning algorithms. Zhu et al. [13] proposed an approach that is based on a Gaussian random fields with methods for incorporating class priors and output of some external classifier. Here, the general class priors at the mass level are incorporated, whereas the proposed approach inducts the specific class priors for each unlabelled node in the graph. They also used "dongle" nodes, connected to each unlabelled node for incorporating output of the external supervised classifier. Whereas the proposed approach only introduces limited number of "dummy" nodes, one for each class label and all such "dummy" nodes are connected to all the unlabelled nodes representing the prior knowledge about their class assignment. Kajdanowicz et al. [5] also use "dummy" nodes, one for each labelled node in order to preserve labels of the labelled nodes.

6   Conclusion and Future Work

The process of performance appraisals in large organizations is considered and the problem of identifying "aspects" from supervisor feedback is explored. This is the first attempt to automatically extract insights from PA text sentences. Unlike the usual notion of noun phrase based aspects used in literature, a novel definition of aspect in PA domain in the form of verb-noun pairs is presented. As a major contribution, a novel way of inducing prior knowledge in the label propagation algorithm is proposed. This is achieved by introducing "dummy" nodes (one for each class label) and connecting them to all the nodes in the graph along with the edge-weights commensurate with the prior knowledge. A better performance of the proposed approach is reported in comparison to various baseline methods, basic label propagation and only prior.

In future, focus will be on exploring the effectiveness of the approach on a dataset from other domains. Realizing more complex priors is another area to be tried.

---

[3] KSS stands for Knowledge Sharing Session.

References

1. Blei, D.M., Ng, A.Y., Jordan, M.I.: Latent dirichlet allocation. J. Mach. Learn. Res. **3**, 993–1022 (2003)
2. Church, K.W., Hanks, P.: Word association norms, mutual information, and lexicography. Comput. Linguist. **16**(1), 22–29 (1990)
3. Cohen, J.: A coefficient of agreement for nominal scales. Educ. Psychol. Measur. **20**(1), 37–46 (1960)
4. Cohen, W.W., Carvalho, V.R., Mitchell, T.M.: Learning to classify email into "speech acts". In: EMNLP, pp. 309–316 (2004)
5. Kajdanowicz, T., Indyk, W., Kazienko, P.: Mapreduce approach to relational influence propagation in complex networks. Pattern Anal. Appl. **17**(4), 739–746 (2014)
6. Kilgarriff, A., Rychly, P., Smrz, P., Tugwell, D.: Itri-04-08 the sketch engine. Inform. Technol. **105**, 116 (2004)
7. Liu, B.: Sentiment analysis and opinion mining. Synth. Lect. Hum. Lang. Technol. **5**(1), 1–167 (2012)
8. Manning, C.D., Surdeanu, M., Bauer, J., Finkel, J., Bethard, S.J., McClosky, D.: The stanford corenlp natural language processing toolkit. In: ACL: System Demos, pp. 55–60 (2014)
9. Ozaki, K.: (2011). https://github.com/smly/java-labelpropagation
10. Resik, P.: Using information content to evaluate semantic similarity. In: Proceedings of the 14th IJCAI, pp. 448–453 (1995)
11. Widdows, D.: Semantic vector products: Some initial investigations. In: Second AAAI Symposium on Quantum Interaction, vol. 26, p. 28 (2008)
12. Zhu, X., Ghahramani, Z.: Learning from labeled and unlabeled data with label propagation. Technical report, Citeseer (2002)
13. Zhu, X., Ghahramani, Z., Lafferty, J., et al.: Semi-supervised learning using gaussian fields and harmonic functions. ICML **3**, 912–919 (2003)

Mining Hidden Knowledge
from the Counterterrorism Dataset Using
Graph-Based Approach

Kishlay Jha$^{(\boxtimes)}$ and Wei Jin

North Dakota State University, Fargo, ND, USA
{kishlay.jha,wei.jin}@ndsu.edu

Abstract. Information overloaded is now a matter of fact. These enormous stack of information poses huge potential to discover previously uncharted knowledge. In this paper, we propose a graph based approach integrated with statistical correlation measure to discover latent but valuable information buried under huge corpora. For given two concepts, C_i and C_j (e.g. bush and bin ladin), we find the best set of intermediate concepts interlinking them by gleaning across multiple documents. We perform query enrichment on input concepts using Longest Common Substring (LCSubstr) algorithm to enhance the level of granularity. Moreover, we use *Kulczynski* correlation measure to determine the strength of interdependence between concepts and demote associations with relatively meager statistical significance. Finally, we present our users with ranked paths, along with sentence level evidence to facilitate better interpretation of underlying context. Counterterrorism dataset is used to demonstrate the effectiveness and applicability of our technique.

1 Introduction

It is known that the rise in text data is on unprecedented scale. These huge corpora of literature necessitate the intelligent text mining techniques which can explore non-apparent but meaningful information buried under them. "The wealth of recorded knowledge is greater than sum of its parts"- a hypotheses proposed by Davies in 1989 [1] underneaths the essence of this paper. Swanson and Smalheiser [2], the pioneers of Literature based discovery (LBD) (viz. a methodology of finding new knowledge from complementary and non-interactive set of articles) used aforementioned hypothesis as a basis to propound several novel hypothesis in the area of biomedicine, which were later validated by bioscientists. They postulated a simple ABC model, where AB and BC refer to direct relationships in the literature reported explicitly wherein the goal is to find any plausible relationship between A and C via intermediate B. Meanwhile, the pioneering contributions of Swanson and Smalheiser [3] made major strides in LBD, it also stimulated researchers to make notable contribution in practical areas of biomedicine, healthcare, clinical study and counterterrorism. In context of counterterrorism, [4] found hidden connections between concepts across multiple documents using concept chain queries.

© Springer International Publishing Switzerland 2016
E. Métais et al. (Eds.): NLDB 2016, LNCS 9612, pp. 310–317, 2016.
DOI: 10.1007/978-3-319-41754-7_29

Although the meticulous research in intelligent text mining algorithms have immensely advanced the domain of LBD, there are two questions which are largely pondered upon by information scientist working in this area of study (a) How to find prudent and scalable correlation measures which help to determine the strength of interdependence between concepts while taking into account the sheer size of documents? (b) How to lessen the need for manual intervention or domain knowledge required during the discovery phase? In this paper, we intend to investigate into these problems by using Kulczynski correlation measure. The motivation behind is - the use correlation measure propels us atleast one step towards our goal of presenting users with paths which are statistically meaningful and which in turn provides a rationale for discovery of hitherto unknown knowledge. To advance with our second question, we present paths with sentence level evidence to facilitate in understanding the scenario from multiple perspectives.

The unfortunate incident of 9/11, lead to exponential growth in documentation related to terrorism. Research works in this area could heavily benefit intelligence analyst to discover covert association between entities (e.g. person, organization, events) and also help them to monitor criminal activities. Relevant to this area of study, [5] combined text mining with link analysis techniques to discover new anti-terror knowledge (i.e. they found Osama bin laden's organization was responsible for assassination attempt of John Paul II). In our approach, we model the problem of finding hidden but precious knowledge from huge corpus using graph-based approach. We capture both unexpressed and precise knowledge in paths between query terms (e.g. bush and bin ladin) by eradicating the need for any user intervention. We also perform query enrichment [Sect. 3.3.1] to add finer granularity and rank retrieved paths based on correlation measure.

The reminder of this paper is structured as follows. Section 2 discusses related work. In Sect. 3, we present an overview of our approach in detail. In Sect. 4 we present experiments and evaluation results. And finally Sect. 5 brings conclusion and gives directions for future work.

2   Related Work

Fish Oil-Raynauds disease(FO-RD) discovery by Swanson [6] initiated the work in this area of study and was successful in attracting researchers from diverse domain. Though, Swanson's work entrenched a role model for many text mining researchers, it had a few setbacks. One of the major setback was the requirement of strong domain knowledge and manual settings during various stages of Knowledge Discovery (KD). To overcome this limitation, many LBD researchers proposed distinct techniques. [7] used a Natural Language Processing (NLP) component to extract biomedically relevant concepts and Unified Medical Language (UML) provided semantic types to eradicate spurious connections. Gordon and Dumais applied [8] their popular technique of Latent semantic indexing (LSI) to solve this problem and several other distributional approaches such as "term frequency-inverse document frequency (tf-idf)", "token frequency", "record frequency" [9,10] were used to find bridge concepts. In retrospect, although traditional techniques have heavily relied upon frequency of concept co-occurrence, they seem to have ignored the nature of correlation that exists between them. In our approach, after collecting corpus level statistical information, an appropriate correlation measure is further used to capture this nature and estimate the strength of associations between concepts.

In the years followed, Srinivasan [11] proposed open and closed text mining algorithm to discover new biomedical knowledge from MEDLINE[1] using MESH[2] terms. Her approach was a major improvement in LBD which successfully replicated several Swanson's discoveries. While this model may seem lucrative, there are nuances which should be carefully thought upon. One of them is the number of $A \rightarrow B$ and $B \rightarrow C$ combinations which could grow exponentially large in massive text data. To deal with such a large search space, Pratt and Yetisgen-Yildiz [12] incorporated several ordering and filtering capabilities. More recently, Xiaohua Hu [13,14] proposed semantic based association rule mining algorithm to discover new knowledge from biomedical literature. They utilized semantic knowledge (e.g. semantic types, semantic relations and semantic hierarchy) to filter irrelevant association rules.

LBD in counterterrorism was limited till recent past. Research work in this domain include work of [15,16], who proposed an Unapparent Information retrieval (UIR) framework using concept chain graph (CCG) to find interesting associations. Our current work is motivated by [17], which in itself is based on idea of concept graph. In this work, we have incorporated the following specific points in the algorithm design (a) we perform query enrichment on input query terms to incorporate the comprehensive detail of available literature (b) we use Kulczynski correlation measure to promote paths with high statistical correlation. To the best of our knowledge, both of the these aforementioned points are nascent in their application to LBD.

3   Graph Based Text Mining Model

We model the problem of finding novel knowledge from existing Knowledge Base (KB) as one of ranking paths generated between query terms from the concept co-occurrence graph. We use an open source tool *Open Calais*[3] and an existing IE engine Sementax [16] to extract meaningful concepts from documents. Concepts extracted are used to create a KB, wherein, we issue queries (e.g. "bush" and "bin ladin") to generate ranked paths utilizing correlation measure. A basic architecture of our method is shown in Fig. 1.

3.1   Concept Extraction

In the Concept Extraction (CE) phase, we extract meaningful information from the available text. Concepts or instances refer to the name of a person, place, organization, event etc. Each concept belong to one or more category. We process

---

[1] https://www.nlm.nih.gov/pubs/factsheets/medline.html.

[2] https://www.nlm.nih.gov/mesh.

[3] http://www.opencalais.com/opencalais-demo/.

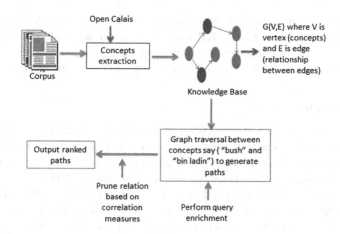

Fig. 1. Basic architecture proposed method

our entire dataset with Open Calais/Semantex [16] and map the extracted concepts to their respective category or semantic type. An example of concept extraction for our dataset is shown in Table 1.

Table 1. Example of concept extraction from documents.

Category or Semantic type	Concepts
Person	Bin ladin, George W bush
Organization	Al-qaeda, Muslim brotherhood
Position	President, Emir, Messenger
Industry term	Bank accounts, Oil wealth
City	New York, Kabul, Boston

3.2   Knowledge Base Creation

The concepts extracted in CE phase are used to build a KB. This KB forms the central component of our system. It is modeled as graph G (V, E), where V refers to a set of vertices and E is a set of edges. Each unique concept $c_i \in V$ is considered a vertex and each $e_{ij} \in E$ refers to a relationship between c_i and c_j. The relationship between concepts can be defined in several ways. One way is to consider the frequency of co-occurrence between concepts in a specific window unit (e.g. sentence, paragraph, document). In our approach, we define the relation between concepts at sentence level. In other words, for a given document, we build a concept-by-concept co-occurrence frequency matrix at sentence level (viz., all concepts occurring within this window unit is considered co-occurring with each other). By shifting the window unit across the documents, an overall co-occurrence matrix for all the concepts $c_i \in V$ is produced. We use this co-occurrence value as an edge property (cooccurCount) for e_{ij} between concepts. Altogether, we have a KB modeled in form of graph, where concepts are the vertices and concept-concept co-occurrence frequency is the weight of relationship between them. We use *Neo4j*[4] graph database to store our KB.

3.3   Algorithm

Having built the Knowledge base (KB), now our goal is to find interesting intermediate terms which connect query pair in a coherent manner. For the input query terms, we perform query enrichment [Sect. 3.3.1] to capture the available knowledge in comprehensive manner. We intend to highlight that while sentence level relation in itself is considered granular [4,5], we further add to it by performing query enrichment over input terms. It is also important to note that as our graph is a concept cooccurrence graph, a path of length K = 2 between query pair will result in finding one intermediate concept.

3.3.1   Query Enrichment

A concept name may have several lexicographic variants. For example, a concept "bin ladin" may have variants as "bin laden", "osama bin ladin", "usama bin ladin". Though, all these variants refer to the same person semantically, they vary lexicographically. We may risk missing important interlinking connections without incorporating them into queries pairs. Therefore, we perform *Longest common substring(LCSubstr)*[5] matching to find variants of input query. Using LCSubstr, we generate a small list of variants which are closest to the query terms and present them to our user. The user makes a final decision on which possible variants may be added to the query. The reason behind not fully automating this process is to prevent false positives from being added to the query.

3.4   Graph Traversal

In this phase, we perform graph traversal between query terms in our created graph (i.e. KB). The user inputs start concept (C_i), end concept (C_j), and desired path length (K). Our system performs query enrichment over input terms followed by graph traversal. In graph traversal, the algorithm enumerates all possible paths of specified length. The paths generated are ranked using Kulc correlation measure. Finally, the user is presented ranked paths with sentence level evidence.

3.5   Ranking

Once we obtain the paths between enriched query pairs, we rank them using *Kulczynski* correlation measure. *Kulczynski* correlation measure for concepts C_i and C_j is defined as:

---

[4] http://neo4j.com.
[5] http://en.wikipedia.org/wiki/Longest_common_substring_problem.

$$Kulczynski(C_i, C_j) = \frac{1}{2}(P(C_i|C_j) + P(C_j|C_i)) \tag{1}$$

4 Experiments

For experiments, we use publicly available 9/11 commission report[7] as our corpus. It includes executive summary, preface, thirteen chapters, appendix and notes. Each of them is treated as a separate document. An open source tool Open Calais and an existing IE engine Semantex is used to extract relevant concepts from documents as described in Sect. 3.1.

4.1 Evaluation Set

Unfortunately, there is no standard goldset available for performing quantitative evaluation in this domain. Therefore, a major task in this step was to construct our own evaluation dataset. Our goal was to calculate the precision and recall of paths generated by our system. Precision was calculated by manual inspection (i.e., manually reading the sentences of generated paths to check if there is a logical connection) of top N chains. For recall, we synthesize an evaluation set by selecting paths of length ranging from 1 to 4. To choose paths, we ran queries on corpus with start and end concepts as named entities. For the selected pairs, we manually inspected the relevant paragraphs and selected those where there exists a logical connection. Finally, we generated paths for these pairs as evaluation data. The parameters used to select the target paths are as follows:-

- **Entity-to-Entity link:** A correlation between entities.
- **Entity-to-Event link:** If two entities participated in the same event.
- **Entity-to-person link:** If two persons are connected by the same entity.

4.2 Evaluation Result

The above mentioned process resulted in approximately 35 query pairs as evaluation dataset. In our experiment, we executed each of the query pairs and selected top 5 paths for various path lengths. The evaluation looked at (a) Weather the target path was found in top 5 paths returned by system? (b) the overall precision of top 5 paths for various path length. Our system obtained an overall recall of **73.50 %**. Tables 2 and 3 show recall and precision, respectively for top 5 paths of various lengths.

The recall parameter shows the strength and applicability of our technique. Using this methodology, we obtain a sound coverage of intermediate links with higher precision. Also, as expected, the precision decreases with increase in path length (K), as paths with greater K generate more generic intermediates. The reason for missing some truth chains (Paths) is - our technique does not take

---

[7] http://govinfo.library.unt.edu/911/report/911Report.pdf.

Table 2. Recall on test data

Path Length (K)	Paths found in top 5	Total Paths
1	10	13
2	5	7
3	8	11
4	2	3

Table 3. Precision on test data

Path Length (K)	Precision
1	1
2	0.88
3	0.80
4	0.78

into account the anaphors (he, she, his, her) of persons names. Additionally, although our query enrichment technique finds lexicographic variants of input terms, it does not capture the semantically related terms. We are looking further into these issues.

5   Conclusions

In this paper, we propose a graph based approach amalgamated with correlation measure to discover new knowledge from existing knowledge base by sifting across multiple documents. We conclude that, query enrichment over input terms along with sentence level association between concepts helped us achieve finer granularity required for plausible hypothesis generation. The incorporated correlation measure successfully identifies significant relationships between concepts and ranks them high. In addition to generation of ranked paths, we present users with relevant sentence extracted from multiple documents to assist for further investigation.

In further research, in addition to the specific points raised in the paper, we are researching on alternative correlation measures which could incorporate more context in the discovery process to produce more meaningful correlations. We also intend to extend concept chain queries to concept graph queries where three or more concepts are involved. This will enable users to quickly generate hypotheses graph specific to a corpus.

Acknowledgments. This work was supported by National Science Foundation grant IIS-1452898.

References

1. Davies, R.: The creation of new knowledge by information retrieval and classification. J. Documentation **45**(4), 273–301 (1989)
2. Swanson, D.R., Smalheiser, N.R.: Implicit text linkage between medline records; using arrowsmith as an aid to scientific discovery. Libr. Trends **48**, 48–59 (1999)
3. Swanson, D.R., Smalheiser, N.R.: An interactive system for finding complementary literatures. Artif. Intell. **91**, 183–203 (1997)
4. Jin, W., Srihari, R.K.: Knowledge discovery across documents through concept chain queries. In: Proceedings of the 6th IEEE International Conference on Data Mining Workshop on Foundation of Data Mining and Novel Techniques in High Dimensional Structural and Unstructred Data, pp. 448–452 (2006)
5. Ben-Dov, M., Wu, W., Feldman, R., Cairns, P.A., House, R.: Improving knowledge discovery by combining text-mining and link analysis techniques. In: Proceedings of the SIAM International Conference on Data Mining (2004)
6. Swanson, D.: Fish oil, raynauds syndrome, and undiscovered public knowledge. Perspect. Biol. Med. **30**, 7–18 (1986)
7. Weeber, M., Klein, H., Berg, L., Vos, R.: Using concepts in literature-based discovery: simulating swansons raynaud-fish oil and migraine-magnesium discoveries. J. Am. Soc. Inf. Sci. **52**(7), 548–557 (2001)
8. Gordon, M., Dumais, S.: Using latent semantic indexing for literature based discovery. JASIS **49**(8), 674–685 (1998)
9. Lindsay, R., Gordon, M.: Literature-based discovery by lexical statistics. JASIS **50**(7), 574–587 (1999)
10. Gordon, M., Lindsay, R.: Toward discovery support systems: a replication, re-examination, and extension of swansons work on literature based discovery of a connection between raynauds and fish oil. JASIS **47**(2), 116–128 (1996)
11. Srinivasan, P.: Text mining: generating hypotheses from medline. JASIS **55**(5), 396–413 (2004)
12. Yetisgen-Yildiz, M., Pratt, W.: Using statistical and knowledge-based approaches for literature-based discovery. J. Biomed. Inf. **39**(6), 600–611 (2006)
13. Hu, X., Zhang, X., Yoo, I., Wang, X., Feng, J.: Mining hidden connections among biomedical concepts from disjoint biomedical literature sets through semantic-based association rule. Int. J. Intell. Syst. **25**, 207–223 (2010)
14. Hu, X., Zhang, X., Yoo, I., Zhang, Y.: A semantic approach for mining hidden links from complementary and non-interactive biomedical literature. In: SDM, pp. 200–209 (2006)
15. Srihari, R., Lamkhede, S., Bhasin, A.: Unapparent information revelation: a concept chain graph approach. In: Proceedings of the ACM Conference on Information and Knowledge Management, pp. 200–209 (2005a)
16. Srihari, R.K., Li, W., Niu, C., Cornell, T.: infoxtract: a customizable intermediate level information extraction engine. J. Nat. Lang. Eng. **14**(01), 33–69 (2008)
17. Jin, W., Srihari, R.K., Ho, H.H.: A text mining model for hypothesis generation. In: 19th IEEE International Conference on Tools with Artificial Intelligence 2007, ICTAI 2007, vol. 2, pp. 156–162. IEEE (2007)

An Empirical Assessment of Citation Information in Scientific Summarization

Francesco Ronzano$^{(\boxtimes)}$ and Horacio Saggion

Natural Language Processing Group, Universitat Pompeu Fabra,
Barcelona, Spain
{francesco.ronzano,horacio.saggion}@upf.edu

Abstract. Considering the recent substantial growth of the publication rate of scientific results, nowadays the availability of effective and automated techniques to summarize scientific articles is of utmost importance. In this paper we investigate if and how we can exploit the citations of an article in order to better identify its relevant excerpts. By relying on the BioSumm2014 dataset, we evaluate the variation in performance of extractive summarization approaches when we consider the citations to extend or select the contents of an article to summarize. We compute the maximum ROUGE-2 scores that can be obtained when we summarize a paper by considering its contents together with its citations. We show that the inclusion of citation-related information brings to the generation of better summaries.

Keywords: Citation-based summarization · Scientific text mining · Summary evaluation

1 Introduction: Citation-Based Summarization

A new scientific article is published every 20 s and such publication rate is going to increase during then next decade [1]. Considering this trend, everyone can agree that any person who has to deal with scientific publications, including researchers, editors, reviewers, is flooded by a huge amount of information. Keeping track of recent advances has become an extremely difficult and time consuming activity.

New and improved approaches to select and organize contents of interest from the huge amount of scientific papers that are accessible on-line could help to solve or at least reduce this problem. In this context, automated summarization techniques play a central role. By exploiting these techniques, it is possible to identify the set of excerpts of a text that better recap its contents, thus

This work is (partly) supported by the Spanish Ministry of Economy and Competitiveness under the Maria de Maeztu Units of Excellence Programme (MDM-2015-0502) and by the European Project Dr. Inventor (FP7-ICT-2013.8.1 - Grant no: 611383).

E. Métais et al. (Eds.): NLDB 2016, LNCS 9612, pp. 318–325, 2016.
DOI: 10.1007/978-3-319-41754-7_30

substantially reducing the effort needed to organize and evaluate scientific information. Current approaches to summarize scientific literature usually extend existing document summarization techniques [2] by taking advantage of distinctive traits of scientific publications. Citations represent by far the feature of scientific articles that has been more often exploited in order to improve the quality of automatically generated summaries.

When a paper is cited, the part of the citing article that explains the reason of the citation is usually referred to as *citation context* or *citance* [3]. Each citance includes the sentence where the citation occurs (citing sentence) or part of it and possibly one or more surrounding sentences. To automatically identify the *citation context* different approaches have been proposed relying on the dependency tree of the citing sentence [6], on classifiers like Support Vector Machines or on sequence taggers, including Conditional Random Fields [7,8].

Several analyses aim at characterizing what type of information about a paper is provided by its citances. [5] analyze a corpus of open access articles from PubMed Central[1] and notice that there is a small but quantifiable difference between the informative content of the abstract of a paper and the information provided by its set of citances. [9] prove the utility of citances to improve the quality of multi-document summaries: graph-based summarization methodologies are evaluated over two topic-homogeneous collections of papers from the ACL Anthology[2].

Distinct approaches have been proposed to exploit citances so as to select the set of clues or sentences that better summarize a paper. Starting from an article, [10] build a similarity graph to clusters sentences concerning the same topic and explore different selection strategies to choose from each cluster the ones to include in the summary [11]. Exploit the citances of a paper to build a language model that, in turn, is used to select the most influential sentences. [13] show that the citation information improves sensibly the quality of a summary by relying on different summarization approaches and exploiting the BioSumm2014 dataset. In [4] the contents of a paper are summarized by extracting relevant keyphrases (n-grams) from the collection of citances of the considered paper. A background language model is built by gathering several papers of the same domain of the one to summarize. The summary includes the sentences with the lowest keyphrase overlap with the background language model. [6] rely on the citances of an article to improve the diversity and readability of its automatically generated summary.

We investigate to what extent the citances of a paper are useful to create an improved summary of its contents. In particular we analyze how the contents of different parts of a paper, including abstract, body and citances, contribute to maximize a widespread summary evaluation metric, ROUGE-2. To this purpose we exploit the Biomedical Summarization Dataset (BioSumm2014), released in the context of the Biomedical Summarization Track of the Text

---

[1] http://www.ncbi.nlm.nih.gov/pubmed.
[2] https://aclweb.org/anthology/.

Analysis Conference 2014[3]. The BioSumm2014 dataset consists of 20 collections of annotated papers, each one including a reference article and 10 citing articles. Section 2 provides a detailed description of the BioSumm2014 dataset and the citation-related manually annotated information that it includes. In Sect. 2 we also briefly describe the pre-processing steps we perform over the contents of the BioSumm2014 dataset to support further data analyses. In Sect. 3 we present and discuss how the maximum ROUGE-2 of a summary of a paper varies if we consider its citances to automatically build its summary. In Sect. 4 we provide our conclusions, outlining future venues of research.

2   The BioSumm2014 Dataset

The BioSumm2014 dataset has been released in the context of the Biomedical Summarization Trak (Text Analysis Conference 2014) to provide a manually annotated corpus of scientific publications useful to investigate several aspects of citations. All the experiments presented in the following Sections of this paper rely on this dataset.

The BioSumm2014 dataset includes 20 collections of scientific publications dealing with the bio-molecular, pharmaceutical and medical domains, indexed by PubMed Central. Each collection consists of one cited paper (also referred to as Reference Paper) and 10 Citing Papers. Every Citing Paper of each collection cites one or more times the corresponding Reference Paper (see Fig. 1). Each citation of each Citing Paper has been manually annotated by four annotators who were asked to identify:

- the *context of the citation* (i.e. citance), consisting of one to three text spans and including the related in-line citation marker (*[1]*, *(Rossi et al., 2010)*, etc.);
- the *citing spans*, that are one to three text spans in the Reference Paper that most accurately reflect the citance, thus representing the part of the Reference Paper that contains the reason why such paper has been cited by the Citing Paper.

In each Reference Paper we can identify the following parts (see Fig. 1): the *abstract*, the *body* and, for each citation of a Citing Paper from the same collection, *four citing spans*, each one identified by a different annotator. On average, one third of the sentences the body of each Reference Paper intersects one or more citing spans. In addition, for each collection of papers, each annotator provided a summary of the Reference Paper having an approximate length of 250 words and including the opinions expressed by the related citations. In the BioSumm2014 dataset, for each citation of a Reference Paper, different citances have been identified by distinct annotators in the related Citing Paper. We associate to each citation, a *global citance span*. The global citance span represents the union of the citance text spans identified by the four human annotators for that particular citation. All the experiment we discuss in the next Section are based

---

[3] http://www.nist.gov/tac/2014/BiomedSumm/.

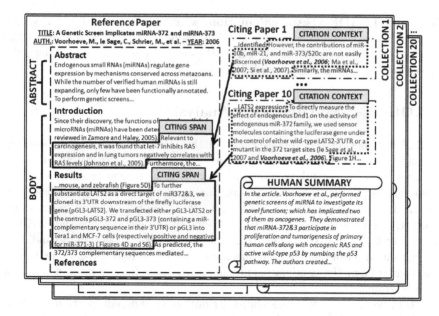

Fig. 1. Structure of the BioSumm2014 dataset.

on the manual annotations of the citation contexts and the citing spans. The automatic identification of these text spans is out of the scope of the experiments presented in this paper.

2.1   Pre-processing BioSumm2014 Papers

To enable the execution of the experiments described in this paper, we imported the articles of the BioSumm2014 dataset together with their stand-off annotations into the text engineering framework GATE[4]. Then, we performed the following set of pre-processing steps on the textual contents of each article (both Reference and Citing Papers):

- **Custom sentence splitting:** we customized GATE regular-expression sentence-splitter in order to properly identify sentence boundaries in the scientific texts included the in the BioSumm2014 dataset. We added rules to correctly deal with abbreviations that are characteristic of scientific literature (Fig., Sect., etc.);

- **Tokenization and POS-tagging:** we exploited GATE English Tokenizer and POS-tagger to perform these actions on the texts;

- **Sentence sanitization:** we filtered out incorrectly annotated sentences, relying on a set of rules and heuristics. For instance, we excluded sentences with no verbal tokens or short sentences, with less than 4 tokens (excluding punctuation). These sentences often represent headers or captions;

---

[4] https://gate.ac.uk/.

- **Sentence TF-IDF vector calculation:** we determined the TF-IDF vector associated to each sentence, where the IDF values are computed over all the papers of the related collection (10 Citing Papers and one Reference Paper). We removed stop-words to compute TF-IDF vectors.

3   ROUGE-2 Maximization Analysis

The ROUGE metrics [12] represent one of the most adopted sets of measures that aim at evaluating the quality of an automatically generated summary (also referred to as candidate summary) with respect to one or more reference summaries, usually created by humans. Summary evaluation results based on ROUGE metrics are proven to have a high correlation with human judgements. In our experiments we rely on ROUGE-2, one of the measures of the ROUGE family of metrics. ROUGE-2 quantifies the bi-gram recall between a candidate summary and a set of reference summaries: higher is the ROUGE-2 value, higher is the similarity of the candidate summary with the reference ones in terms of shared bi-grams.

In this Section we evaluate the contribution of different portions of Bio-Summ2014 papers (abstracts, body, citing spans, etc.) with respect to the variations of ROUGE-2. Specifically, we compute the maximum ROUGE-2 that can be obtained when we produce a summary of 250 words by picking sentences that belong to distinct portions of the papers. We truncate the contents of the summaries to evaluate to their 250th word.

Relying on the sentence detection output (see pre-processing Subsect. 2.1), we are able to divide the sentences of the papers of each collection into the following groups:

- **ABSTRACT:** sentences that belong to the abstract of the Reference Paper;
- **BODY:** sentences of the body of the Reference Paper;
- **CITING SPANS:** sentences of the body of the Reference Paper that intersect totally or partially a citing span (abstract sentences intersecting a citing span are not included in this group);
- **CITATION CONTEXT:** sentences that intersect at least one global citance span in a Citing Paper (i.e. sentences that belong to the context of the citations of the Reference Paper).

In each collection of papers, for each combination of the previous groups of sentences, we determine the subset of sentences (up to 250 words) that maximizes the ROUGE-2, by considering two types of reference summaries:

- **ABST_SUMM:** the abstract of the Reference Paper;
- **GS_SUMM:** the four 250-words summaries written by BioSumm2014 annotators.

In Tables 1 and 2 we show the average value (across the 20 collections) of the maximized ROUGE-2, when we generate a 250-words summary by choosing sentences from different groups.

Table 1. Average max ROUGE-2 across collections with respect to GS_SUMM when the sentences to include in the 250-words generated summary are chosen by combining the following groups: abstract (ABST.), body (BODY), citing spans (CIT. SP.) and citation context (CIT. CTX). The best ROUGE-2 is highlighted in bold.

ABST.	ABST. BODY	ABST. CIT. SP.	ABST. CIT. CTX	ABST. BODY CIT. CTX	ABST. CIT. SP. CIT. CTX
0.126486	0.229957	0.228127	0.224505	**0.245432**	0.242601

BODY	BODY CIT. CTX	CIT. SP. CIT. CTX	CIT. SP.	CIT. CTX
0.211558	0.234545	0.230904	0.204874	0.167024

Table 2. Average max ROUGE-2 across collections with respect to ABST_SUMM (abstract as reference summary) when the sentences to include in the 250-words generated summary are chosen by combining the following groups: body (BODY), citing spans (CIT. SP.) and citation context (CIT. CTX). The best ROUGE-2 is highlighted in bold.

BODY	BODY CIT. CTX	CIT. SP CIT. CTX	CIT. SP	CIT. CTX
0.305349	**0.317036**	0.290046	0.275215	0.141528

3.1 ROUGE-2 Maximization Discussion

In both Tables 1 and 2, as expected, the best ROUGE-2 result is obtained when we consider all the sentences available (from the abstract, body and citation context in Table 1 and from the body and citation context in Table 2).

When we refer to human generated reference summaries (Table 1), we can notice that if we consider only the sentences from the abstract or the sentences from the citation context, lower quality extractive summaries are generated with respect to other combinations of groups of sentences. As a consequence the contents of the abstract or the contents of the citation context alone are not enough for a good quality summarization.

If we add the sentences of the citation contexts to the sentences of the Reference Paper (abstract and body), we can observe a sensible improvement of the average maximum ROUGE-2, from 0.229957 to 0.245432. We can observe the same trend if we add the sentences of the citation context to the sentences of the body (from 0.211558 to 0.234545). Both trends show that by considering the sentences of the citation context of a paper, we can potentially build a summary with a higher ROUGE-2 score.

The sentences of the body of each Reference Paper that belong to a citing span represent about one third of the whole set of sentences of each Reference Paper body. The summary generated thanks to these citing span sentences has a quality comparable to the summary generated by considering all the sentences of the body. As a consequence, we can state that without a considerable loss of summarization quality, we can select from the sentences of the body of a Reference

Paper, the ones that most accurately reflect the contents of its citation contexts and then generate an extractive summary from this subset of sentences. This selection would act as a filter, thus considerably reducing the number of candidate sentence to evaluate when generating a summary of the Reference Paper.

When our reference summary is the abstract of the Reference Paper (Table 2), we can note that the sentences of the citation contexts alone do not constitute the best choice to summarize the paper (citation context column, avg. max ROUGE-2 equal to 0.141528). Moreover, as previously observed with human generated reference summaries, if we add to the sentences of the body of the paper, the sentences of the citation context, the average maximum ROUGE-2 improves (from 0.305349 - only body - to 0.317036 - body and citation context). These fact confirms the usefulness of citation context sentences to generate better summaries of a paper.

4   Conclusions and Future Work

Nowadays, the growing publication rate of scientific results is more and more stressing the importance of effective approaches to easily structure and summarize the contents of articles. We have investigated how the citations of an article can contribute to generate a better summary. To this purpose we relied on the BioSumm2014 dataset that includes 20 manually annotated collections, each one made of one cited paper and 10 other papers that cite the first one. We evaluated how different parts of each cited paper (abstract, body, citing spans) together with its citation context contribute to the generation of a summary. We adopted the following approach: by considering any possible combination of parts of the cited paper and its citation context, we computed the maximum ROUGE-2 score achievable. As a result, we noted that the maximum ROUGE-2 score achievable sensibly increases when we exploit the sentences from the citation context to generate a summary by considering as reference summaries both human generated summaries and the abstract.

In general, we have to notice that when we want to generate a citation-based summary of a paper, we need to gather and process the set of articles that cite the paper. Currently this process requires a considerable effort and the contents of all the citing articles are not always available. As a consequence, besides the improvement of the quality of the summary, when we consider citation-based summarization we have also to take into account the efforts needed in order to collect the citation related information concerning a paper.

As venues for future research we would like to validate the set of experiments presented in this paper with scientific publications from different domains. It would be interesting to perform a detailed evaluation of the consistency and diversity of the information included in automatically generated summaries, since the contents of a paper are often repeated or paraphrased by its citation contexts. Moreover, we would like to investigate into details to what extent different summarization approaches in different domains generate better summaries when we consider citation contexts as well as other kinds of information from citing papers. We would also perform more extensive analyses by contemplating, besides the ROUGE-2 score, alternative summary evaluation metrics.

References

1. AAAS Science Magazine: Communication in Science Pressures and Predators. Science 342(6154), 58–59 (2013). doi:10.1126/science.342.6154.58
2. Gupta, V., Lehal, G.S.: A survey of text summarization extractive techniques. J. Emerg. Technol. Web Intell. 2(3), 258–268 (2010)
3. Nakov, P.I., Schwartz, A.S., Hearst, M.: Citances: citation sentences for semantic analysis of bioscience text. In: Proceedings of the SIGIR 2004 Workshop on Search and Discovery in Bioinformatics, pp. 81–88 (2004)
4. Qazvinian, V., Radev, D.R., Özgür, A.: Citation summarization through keyphrase extraction. In: Proceedings of the 23rd International Conference on Computational Linguistics, pp. 895–903. Association for Computational Linguistics (2010)
5. Elkiss, A., Shen, S., Fader, A., Erkan, G., States, D., Radev, D.: Blind men and elephants: what do citation summaries tell us about a research article? J. Am. Soc. Inf. Sci. Technol. 59(1), 51–62 (2008)
6. Abu-Jbara, A., Radev, D.: Coherent citation-based summarization of scientific papers. In: Proceedings of the 49th Annual Meeting of the Association for Computational Linguistics: Human Language Technologies, vol. 1, pp. 500–509. Association for Computational Linguistics (2011)
7. Abu-Jbara, A., Radev, D.: Reference scope identification in citing sentences. In: Proceedings of the 2012 Conference of the North American Chapter of the Association for Computational Linguistics: Human Language Technologies, pp. 80–90. Association for Computational Linguistics (2012)
8. Qazvinian, V., Radev, D.R.: Identifying non-explicit citing sentences for citation-based summarization. In: Proceedings of the 48th Annual Meeting of the Association for Computational Linguistics, pp. 555–564. Association for Computational Linguistics (2010)
9. Mohammad, S., Dorr, B., Egan, M., Hassan, A., Muthukrishan, P., Qazvinian, V., Radev, D., Zajic, D.: Using citations to generate surveys of scientific paradigms. In: Proceedings of Human Language Technologies: The 2009 Annual Conference of the North American Chapter of the Association for Computational Linguistics, pp. 584–592. Association for Computational Linguistics (2009)
10. Qazvinian, V., Radev, D.R., Mohammad, S., Dorr, B.J., Zajic, D.M., Whidby, M., Moon, T.: Generating extractive summaries of scientific paradigms. J. Artif. Intell. Res. (JAIR) 46, 165–201 (2013)
11. Mei, Q., Zhai, C.: Generating impact-based summaries for scientific literature. In: ACL 2008, pp. 816–824 (2008)
12. Lin, C.Y.: Rouge: a package for automatic evaluation of summaries. In: Text Summarization Branches Out: Proceedings of the ACL 2004 Workshop, vol. 8 (2004)
13. Cohan, A., Goharian, N.: Scientific article summarization using citation-context and articles discourse structure. In: Empirical Methods in Natural Language Processing (2015)

PIRAT: A Personalized Information Retrieval System in Arabic Texts Based on a Hybrid Representation of a User Profile

Houssem Safi[✉], Maher Jaoua[✉], and Lamia Hadrich Belguith[✉]

MIRACL Laboratory, Faculty of Economics and Management of Sfax,
ANLP Research Group, Sfax, Tunisia
safi.houssem@gmail.com,
{Maher.Jaoua,l.beguith}@fsegs.rnu.tn

Abstract. The work presented in this paper aims at developing a Personalized Information Retrieval system in Arabic Texts ("PIRAT") based on the user's preferences/interests. For this reason, we proposed a user's modeling and a personalized matching method document-query. The proposed user's modeling is based on a hybrid representation of the user profile. In this approach, we introduce an algorithm which automatically builds a hierarchical user profile that represents his implicit personal interests and domain. It is to represent the interests and the domain with a conceptual network of nodes linked together through relationships respecting the linking topology defined in the domain of hierarchies and ontologies (hyperonymy, hyponymy, and synonymy). Then, we address the problem of unavailable language resources by building (i) a large Arabic text corpus entitled "WCAT" and (ii) Building our own Arabic queries corpus entitled "AQC2" in order to evaluate the suggested PIRAT system and AXON system. The results of this evaluation are promising.

Keywords: Information retrieval · Personalization · User profile · Ontology

1 Introduction

Personalization is an effort to uncover most relevant documents using information about the user's target, domain of interest, browsing history, query context etc., that gives higher value to the user from the large set of results [1]. However, in certain languages, the linguistic tools, such as the case of Arabic language are not developed enough. In fact, there is a lack of resources and automated tools to deal with this language. In addition, the Arabic language is distinguishable from the Western ones as it is a flexible and derivative language. First, unlike the Indo-European languages such as English and French, stemming in Arabic is more difficult, due mainly to the fact that Arabic is an agglutinative and derivational language. Indeed, in Arabic, there are many more affixes than in English, which leads to a large number of word forms. Besides, as a property of words in Semitic language, Arabic stem has an additional internal structure consisting of two parts, namely a "root" and a "pattern". The root is usually a sequence of three consonants and has some abstract meanings, while the pattern is a template that defines the placement of each root letter among vowels and possibly other

© Springer International Publishing Switzerland 2016
E. Métais et al. (Eds.): NLDB 2016, LNCS 9612, pp. 326–334, 2016.
DOI: 10.1007/978-3-319-41754-7_31

consonants. For example, the word (الكتابان) "AlkitabAn", meaning the two books, has a root composed of the letters k, t, and b; and its pattern is "Root1+i+Root2+a+ Root3" [2].

It is emphasized that the present work is an extension of previous studies [4, 5]. Indeed, we integrated the stemmer MADAMIRA instead of the analyzer Alkhalil. Then, we proposed a personalized matching method document query. Moreover, we have implemented these methods to produce a system entitled "PIRAT" based on dynamic ontology profile as an extension of the AXON system [5]. Finally, we performed an evaluation of PIRAT system on a large and closed corpus of documents, about 30550 items. The results are very encouraging.

In the second section, we present a brief overview of the Personalized Information Retrieval (PIR). In the third section, we present in detail the proposed user's modeling. In the forth section, we will see in detail the integration of the user profile in the proposed method of PIR. In the last section, we provide a description of the PIRAT system as well as an evaluation of our own evaluation corpus.

2 Personalized Retrieval Information

Any information about the user or his *profile*, his preferences, his main interests, his information needs and his research environment are consequently supposed to be relevant and usable by the system of personalization [6]. There are mainly tree types of representation of user's profile [5]: Set, Semantics and Multidimensional. The upgrading of the user profile means the adaptation to changes in interest centers that the users describe, and therefore their information needs over time. There are two types of the user's needs: short-term profile and long-term profile. For more details of the overview on the user's modeling, it is worth referring to the work of [5].

A personalized web search is usually achieved in practice by incorporating the user's profile in one of the three main retrieval steps: query refinement, document retrieval and document re-ranking.

In this context, the works on IRS and query expansion for Arabic texts are not much known. First, the system of [8] adopts the notion of schema as a basis for lemmatize words and substitutes it with its lemmas in the operations of indexing and search. Then, the studies carried out done in [9] showed that manual reformulation reweighting of query terms leads to improved performance (recall and precision) of SRI Arabic. For their experiment, [9] used a corpora of 242 documents and a set of nine queries. Additionally, the work of [10] on the query expansion by terms from a thesaurus shows an improvement in the recall of the Arab SRI. We note that the corpus used is the Koran. On the other hand, the works of [11] show that the use of a thesaurus (i.e. Wikipedia) improves significantly (18 %) the performance of an Arab SRI. [11] also showed that the use of an indexation based on the roots is more efficient than the use of schemes for Arabic texts. Moreover, the works of [3] on the query expansion using ontology of the legal field and WordNet provided significant improvements in the performance of IRS. The works of [7] used a reformulation based on an external resource (Arabic WordNet and DSA). The results of comparison between the use of

Arabic WordNet and DSA are in favor of the first. The evaluation corpora for this work consists of a set of 50 keyword queries.

It should be noted that the works already mentioned, and despite the fact that some of them have proven their performance, they also have some limitations. Indeed, some works used an ontology or semantic resource for a specific field such as the case of the works of [3, 10]. In addition, in other works, we see the non-exploitation of conceptual relationships ontology. Finally, there is a lack of studies about the contribution of each semantic relationship used in the enrichment process for some query expansion systems for the Arabic language as the case of the works of [3, 7].

According to the state of the art of IRS and QE, we notice that the contribution of this work is distinguished from that of the work cited above by (i) the coupling between the user profile information and the one contained in ontology is an interesting approach that should be developed (ii) integrating the user profile in all levels of the RIP process.

3 Suggested Modeling User

The proposed user's modeling is based on a hybrid representation of the user profile, an explicit and implicit construction of the profile as well as an updating the short and the long term user profile. In this approach, we introduce an algorithm which automatically builds a hierarchical user profile that represents his implicit personal interests and domain. It is to represent the interests and the domain with a conceptual network of nodes linked together through relationships respecting the link topology defined in domain of hierarchies and ontologies. The model proposed is based on the notion of hyperonymy, hyponymy, and synonymy for a more realistic representation of the user. In what follows, we present the proposed approach.

3.1 Hybrid Representation of the User Profile

The representation of the user profile in our method is based on a hybrid approach (combination of the set-approach and conceptual approach). Thus, the proposed modeling approach is based on two models: static Model and dynamic model. The proposed static model of the user profile (User Profile set-representation) is based on dimensions that can be the user's demographic attributes (identity, personal data), professional attributes (employer, address and type), behavioral attributes and centers of interest attributes. The proposed approach is to consider a multidimensional way of modeling the interest in the user profile by integrating a technical term weighted by the formula in TF * IDF [3, 4]. However, the dynamic model (Profile semantic representation) also presents the centers of interest and the domain of user in the form of semantic representation. Indeed, the profile is presented as a hierarchy of personal concepts based on the knowledge contained in the ontologies Arabic Wordnet[1] (AWN) and Amine Arabic Wordnet[2](Amine AWN). The construction of user's domain

---

[1] http://wwww.globalwordnet.org/AWN/AWNBrowser.html.

[2] http://amine-platform.sourceforge.net.

and centers of interest tree is based on the extraction of semantic relationships found in ontologies (AWN and Amine AWN).

3.2 Initial Building of Profile

This stage occurs in order to avoid cold start and primarily to prompt the user to fill out personal information form related to him. Then, the system builds a hierarchy of concepts based on the user's domain already indicated and his/her interests. The treatment is to build, from Arabic WordNet Amine and Arabic WordNet ontologies, a tree containing all the information on the domain and interests of the concerned person. This step includes the user's registration and the building of the domain tree that will be the subject of the following two sections.

The Registration. The user is invited to fill in a registration form containing his personal information, such as his/her domain and interests. This is an explicit acquisition approach of the user's data. Indeed, the profile is composed of three categories (i) personal data (the name, first name, login, password, etc.), (ii) professional Data (the user's profession and the field) and (iii) interest centers (the preferences).

Construction of the Profile as a Hierarchy of Concept. Building a hierarchy of concepts is based on the semantic relationships found in the ontology in question. This step is based on the extraction of concepts from Amine AWN ontology and AWN ontology that are in semantic relationships with the field in question and with each center of interest. For a given field it is to extract for each concept the synonym (s), the hyperonymy (s) and hyponymy (s) if they exist in the ontologies. The hierarchy of concepts below corresponds to a sub-tree of the field (economy) "إقتصاد" (Fig. 1).

Fig. 1. Example of a hierarchy extracted from the AWN ontology for (economy) "إقتصاد" field

3.3 Updating the User Profile

The proposed approach manages to update the short-term, long-term profile or both at the same time in terms of the time dimension. The capture of changes in the interest centers is visualized by adding a search history (query and search results) in the short-term profile. Indeed, the proposed approach defines a personalization score based on building and updating of a user profile from its relevance judgments. Besides, each user's request will be added to his short-term profile. In addition, we calculate the formula tf * idf of each term of each document is considered to be relevant or very relevant by the user. Then, we add the first three terms with the highest tf * idf in the short-term user profile.

Thus, updating is considered by the recent information classification visualized or created by the user in the ontology concepts. This contributes to the accumulation of the weight of the initial concepts of the profile or activation of new concepts to be considered in the representation of the profile.

In fact, the approach proposed attributes an initial weight to each concept of the hierarchy that is going to be updated based on the user's relevance judgments. Thus, the user profile is nodes set which compound is (i) interest score $S(C)$ to the concept c initialized to 1, (ii) when updating the profile from the user's relevance judgments, each interest score $S(C)$ is incremented in relation to a score of updating which depends on the similarity $Sim(d_i, c)$ of the document selected by the user and the concept c. The score c. updating c.evolution of concept c is estimated as follows:

$$c.evolution = IS(C).Sim(d_i, c) \qquad (1)$$

The highest score of concept c has spread to the entire nearest neighbor of concepts c based on the weight of the relationship between these two concepts. The operation is performed iteratively.

The proposed approach of an updating long-term profile is to add or modify a context implicitly formed of a pair of concepts associated with a user's query. The identification of a similar context to the query profile implies its combination with the latter and the possible updating of the final representation in the long-term profile. If no context of all the already learned contexts matches the context of the query being evaluated, a new context is added to the long term profile. Similarly, the long-term profile can also be updated by the user by explicitly modifying the field.

4   Proposed Methods for AXON and PIRAT

The proposed reformulation method distinguishes five implemented steps. For more details about this method, one could refer to the work of [3, 5]. In the stemming phase, the Alkhalil stemmer was used. However, the latter does not provide a good result since its lexical database is poor and it does not give the lemma of a given term in its current version. To our knowledge, MADAMIRA[3] is the only available light stemmer which performs morphological analysis and disambiguation of Arabic. For this reason, we used it to apply a light stemming on each query. Light stemming aims at reducing words to their lemma forms. For example, if the query is " الأمراض الوراثية" (the genetic diseases) then the pretreated query is " وراثي مَرَض" (genetic disease).

The personalized matching method consists of two steps, namely the document-query matching step by integrating the user profile in calculating the relevance score, and the user's relevance feedback step which consists in updating the short-term as well as long-term profile.

---

[3] http://nlp.ldeo.columbia.edu/madamira/#locale=ar.

4.1 Document-Query Matching Step

By exploiting the body of indexed documents and the enriched query according to the profile, matching will take place. In this step, matching is composed of two phases, namely the classical matching and personalized matching. The classic matching is done by computing a similarity function between document "d" and query "q" RSV (d, q). Indeed, this matching based on the search tool offered by Lucene for indexing a document corpus and research documents and their correspondence with the query. The correspondence between the query and the documents is made by operating the index and display of the search results.

In personalized matching step, the user remains an inescapable component thereby obtaining a new relevance function RSV (D, Q, U) where D, Q and U represent a document, a user and a query. It is therefore appropriate to replace the RSV function of Lucene with another function which takes the user into account. Thus, the personalization technique is to include the user profile in computing the document relevance. This function is calculated as follows:

$$RSV(D, Q, U) = RSV(D, Q) + RSV(D, U) \tag{2}$$

4.2 User's Relevence Feedback

This phase is considered to be an explicit user's feedback. It is very helpful in the proposed evaluation process. Furthermore, its output serves as an input to the proposed reranking method using the learning technique (this method will be described in other studies). Thus, the user is asked to evaluate the documents deemed relevant by an indexing tool (Lucene) to the query entered by the user. According to his/her needs and interests, the user can choose the level of importance of each document by assigning a percentage of 20 % (Not relevant) 40 % (a little relevant), 60 % (Medium relevant), 80 % (relevant) or 100 % (very relevant). The chosen documents and their relevance levels will be saved in the short-term user profile.

4.3 Updating the Short Term and Long Term Profile

Updating the short-term user profile is to enrich its representation on the basis of correlated queries. Likewise, updating the long-term profile is based on the assumption of recurrence of needed information during research sessions. Indeed, the most consulted profile in the short term is regarded as a long-term profile (see Sect. 3.3).

5 Experimentation and Discussion of Results

We implemented a personalized matching Arabic text query system entitled "PIRAT". To evaluate this system, we used the search engine "Lucene[4]". The later is capable of processing large volumes of documents with its power and speed due to indexing.

---

[4] https://lucene.apache.org/.

In our system, we used the 4.0.0 version of lucene to index a corpus of documents, analyze the queries, search for the documents and present document results.

5.1   Arabic Corpora: WCAT and AQC

Given that the Arabic language resources are very rare or even unavailable, we have created a large Arabic text corpus entitled WCAT (Wikipedia Corpus for Arabic Text). This corpus is segmented into 30550 text article, extracted from Wikipedia[5], in order to evaluate PIRAT system. This corpus contains texts dealing with topics related to the "natural sciences" domain. Moreover, each article has one or more categories related to the root category of "natural sciences". We generated 7200 sub-categories from the "natural sciences" category. The study and the treatment of Wikipedia categories can only be performed by a manual procedure. In what follows, we will give some features of the WCAT (Table 1).

Table 1. Features of WCAT

Size of corpus	Average size of an item	Size of items	Num. of words	Language
100 MB	5 KB	5 KB	21492628	Arabic

We have used Lucene tool to index our corpus. In this phase, the indexing step of the corpus consists in word stemming, removing stop words, indexing and extracting key words of each document in the corpus. A list of keywords thus allows defining the themes represented in a document, that is to say, we will calculate the TF*IDF of each term of the document to extract first five keywords with the highest TF*IDF for each document. These terms will be used later to update the short-term user profile.

We built also our own Arabic queries corpus entitled "AQC". This corpus is composed of 100 queries submitted by 5 different users. The AQC dealing with topics related to the "natural sciences" domain. An Arabic query corpus consists of 20,301 words or 93,021 characters and 1.16 megabyte size. Thus, the evaluation corpus of our system contains different types of queries suggested by various users.

5.2   PIRAT System and the Discussion of the Evaluation Results

Given that there are no custom access standard assessment frameworks to PIR, particularly suitable for short-term personalization, we proposed context-oriented assessment frameworks based on simulation collections of TREC campaign by simulated user profiles and search sessions. We have exploited these evaluation frameworks to validate our contribution. We have also experimentally compared our "PIRAT" contribution to "AXON" method.

---

[5] The Wikipedia account 236,019 items in 02/12/2015.

Calculation of Precision Average. The results for the PIRAT system (with hybrid profile) are better than those of the AXON system with set-profile and the Lucene system (without profile) (Table 2). Indeed, the precisions P10, P20, P30 and P50 of the AXON are better than the one in the Lucene system, however, they are below P10, P20, P30 and P50 of those of the PIRAT system. However, the PIRAT system shows all these performances for the first 10 documents by P10 = 16 and its precision is better than the AXON system. As a conclusion, we have demonstrated that personalizing the IR showed better results with a hybrid profile than IR with a set profile.

Table 2. Performance gain of personalized search (precision and MAP measures)

Precision	AXON	Lucene	PIRAT	MAP	AXON	Lucene	PIRAT
%P10	10.2	9	16	%MAP5	8	7	10
%P20	8	10.1	14.3	%MAP10	6.9	5	11
%P30	4.1	3.2	11	%MAP15	4.1	6	13
%P50	8	6	11	%MAP	6.33	6	11.33

Calculation of MAP (Mean Average Precision). We notice that the results obtained with the PIRAT system are better than those obtained with the AXON system (with a set profile) and the Lucene system (without a profile) (Table 2). Indeed, the MAP5, MAP10 and MAP15 for AXON are better than those of the Lucene system and are respectively lower than MAP5, MAP10 and MAP15 of the PIRAT system with hybrid profile. However, PIRAT system shows all these performances for the first 15 documents by MAP15 = 13 and its MAP is better than the AXON system and Lucene system. Similarly, we can see that the personalization of IR showed better results with hybrid profile than with set profile.

6 Conclusion and Prospects

In this work, we have suggested a PRI method for Arabic texts based on the user's preferences/interests. Therefore, we proposed a user's modeling and a personalized matching method document query. The representation of the user profile in our method is based on a hybrid approach. Thus, the proposed modeling approach is based on two models: a static Model and a dynamic one. It should be noted that the construction of user's domain and centers of interest tree is based on the extraction of semantic relationships found in ontologies (AWN and Amine AWN) i.e. synonymy, hyperonymy and hyponymy if they exist. Moreover, in personalized matching method, we replaced the RSV function of Lucene by a new relevance function RSV. Then, we experimentally compared the "PIRAT" system to "AXON" system and showed that the PIRAT approach is causing a significant performance gain.

In future studies, we plan to evaluate the quality of the profile graphs based on the query collection and the collection of WCAT documents.

References

1. Jenifer, K., Yogesh Prabhu, M., Gunasekaran, N.: A survey on web personalization web approaches for efficient information retrieval on user interests. J. Recent Res. Eng. Technol. **2**(6) (2015). ISSN: 2349–2260
2. Heintz, I.: Arabic language modeling with stem-derived morphemes for automatic speech recognition. Ph.D. dissertation, The Ohio State University, USA (2010)
3. Zaidi, S., Laskri, M.: Expansion de la requête Arabe sur le réseau internet. In: Barmajiat CSLA, Alger (2007)
4. Safi, H., Jaoua, M., Belguith, L.: AXON: a personalized retrieval information system in Arabic texts based on linguistic features. In: 6th International Conference on SIIE 2015, Hammamet, Tunisia, 12–14 February 2015
5. Safi, H., Jaoua, M., Belguith, L.: Intégration du profil utilisateur basé sur les ontologies dans la reformulation des requêtes Arabes. In: 2ème Colloque CEC-TAL, 23–25 Mars, Sousse, Tunisie (2015)
6. Akhila, G.S., Prasanth, R.S.: A survey on personalized web search. Int. J. Adv. Res. Trends Eng. Technol. (IJARTET) **II**(Special Issue VI), February 2015
7. Abderrahim, M.-A.: Utilisation des ressources externes pour la reformulation des requêtes dans un système de recherche d'information. Prague Bull. Math. Linguist. **99**, 87–99 (2013)
8. Bessou, S., Saadi, A., Touahria, M.: Vers une recherche d'information plus intelligente application à la langue arabe. In: Actes of SIIE 2008, Hammamet, Tunisia (2008)
9. Kanaan, G.R., Al-Shalabi, R., Abu-Alrub, M., Rawashdeh, M.: Relevance feedback: experimenting with a simple Arabic information retrieval system with evaluation. Int. J. Appl. Sci. Comput. **12**(2) (2005)
10. Hammo, B., Azzam, S., Mahmoud, E.-H.: Effectiveness of query expansion in searching the Holy Quran. In: Proceedings of CITALA 2007, vol. 7, pp. 18–19, Morroco (2007)
11. Xu, Y., Jones, G.J.F., Wang, B.: Query dependent pseudorelevance feedback based on wikipedia. In: Proceedings of the Annual International SIGIR Conference on Research and Development in Information Retrieval, SIGIR 2009, pp. 59–66. ACM (2009)

YATS: Yet Another Text Simplifier

Daniel Ferrés, Montserrat Marimon, Horacio Saggion[✉],
and Ahmed AbuRa'ed

TALN - DTIC, Universitat Pompeu Fabra, Barcelona, Spain
{daniel.ferres,montserrat.marimon,horacio.saggion,
ahmed.aburaed}@upf.edu

Abstract. We present a text simplifier for English that has been built with open source software and has both lexical and syntactic simplification capabilities. The lexical simplifier uses a vector space model approach to obtain the most appropriate sense of a given word in a given context and word frequency simplicity measures to rank synonyms. The syntactic simplifier uses linguistically-motivated rule-based syntactic analysis and generation techniques that rely on part-of-speech tags and syntactic dependency information. Experimental results show good performance of the lexical simplification component when compared to a hard-to-beat baseline, good syntactic simplification accuracy, and according to human assessment, improvements over the best reported results in the literature for a system with same architecture as YATS.

1 Introduction

Automatic text simplification is a research field which studies methods to simplify textual content. Text simplification methods should facilitate or at least speed up the adaptation of textual material, making accessible information *for all* a reality. The growing interest in text simplification is evidenced by the number of languages and users which are targeted by researchers around the globe. In automatic text simplification, the algorithms can involve either or both lexical and syntactic simplifications. Lexical simplification is the task of identifying and substituting complex words for simpler ones in given contexts. Syntactic simplification is the task of reducing the grammatical complexity of a sentence while keeping the original information and meaning. An example involving both lexical and syntactic simplifications is given in (1), where the passive voice is changed into active and complex words are substituted for simpler synonyms.

(1) a. The poem was **composed** by the **renowned** artist.

  b. The **famous** artist **wrote** the poem.

In this paper we present YATS, a state-of-the-art text simplifier for English that has been built with the aim to improve text readability and understandability in order to help people with intellectual disabilities. The system, however, is highly adaptable and the resources used can easily be changed to meet the needs of people with special demands.[1]

---

[1] http://taln.upf.edu/pages/yats.

E. Métais et al. (Eds.): NLDB 2016, LNCS 9612, pp. 335–342, 2016.
DOI: 10.1007/978-3-319-41754-7_32

2 Related Work

Early work on text simplification relied on hand-crafted rules to perform syntactic simplifications. Chandrasekar et al. [4] developed a linear pattern-matching hand-crafted rule system that simplified a few specific constructions, namely relative clauses, appositions, and coordinated clauses. Then, Siddharthan [11] described a rule-based system that performed syntactic simplifications in three phases: (1) analysis, which used chunking and PoS tagging, followed by pattern-matching, (2) transformation, which applied a few rules for dis-embedding relative clauses, splitting conjoined clauses and making new sentences out of appositives, and (3) regeneration, which fixed mistakes by generating referring expressions, selecting determiners, and preserving discourse structure to improve text cohesion. More recent hand-crafted systems made use of transfer rules that operated on the output of a parser: Siddharthan [12] applied transformation rules to a typed dependency representation produced by the Stanford parser. This is the approach that has been mostly followed in non-English projects [9].

Coster and Kauchak [5] and Wubben et al. [15] used monolingual translation from English Wikipedia (EW) to Simple English Wikipedia (SEW) and they applied phrase based Machine Translation to the text simplification augmented with a phrasal deletion model [5] and a post-hoc reranking procedure that ranked the output [15]. Siddharthan and Angrosh [13] presented a hybrid system that combined manually written synchronous grammars for syntactic simplifications with an automatically acquired synchronous grammar for lexicalized constructs.

Work on lexical simplification began in the PSET project [6]. The authors used WordNet to identify synonyms and calculated their relative difficulty using Kucera-Francis frequencies. Biran et al. [1] used word frequencies in EW and SEW to calculate their difficulty, and Yatskar et al. [16] used SEW edit histories to identify the simplify operations.

3 Lexical Simplification in YATS

The lexical simplifier based on [2,8] is composed of the following phases (executed sequentially): (i) Document analysis, (ii) Complex words detection, (iii) Word Sense Disambiguation (WSD), (iv) Synonyms ranking, and (v) Language realization. The document analysis phase extracts the linguistic features from the documents. It uses the GATE[2] NLP API and its ANNIE pipeline to perform: tokenization, sentence splitting, part-of-speech (PoS) tagging, lemmatization, named entity recognition and classification, and co-reference resolution.

The complex word detection phase identifies a word as complex when the frequency count of the word in a given psycholinguistic database is in a range determined by two threshold values (i.e. w is complex if $min \leq w_{frequency} \leq max$). The two psycholinguistic resources that can be used separately in our lexical simplification system are: Age-of-Acquisition norms[3] and Kucera-Francis[4] frequency counts.

---

[2] http://gate.ac.uk.
[3] http://crr.ugent.be/archives/806.
[4] http://www.psych.rl.ac.uk/kf.wds.

The WSD phase selects the most appropriate word replacement out of a list of "synonyms". The WSD algorithm used is based on the Vector Space Model [14] approach for lexical semantics which has been previously used in Lexical Simplification [1]. The WSD algorithm uses a word vector model derived from a large text collection from which a word vector for each word in WordNet-3.1[5] is created by collecting co-occurring word lemmas of the word in N-window contexts (only nouns, verbs, adjectives, and adverbs) together with their frequencies. Then, a common vector is computed for each of the word senses of a given target word (lemma and PoS). These word sense vectors are created by adding the vectors of all synonyms in each sense in WordNet. When a complex word is detected the WSD algorithm computes the cosine distance between the context vector computed from the words of the complex word context (at sentence or document level) and the word vectors of each sense from the model. The word sense selected is the one with the lowest cosine distance (i.e. greater cosine value) between its word vector in the model and the context vector of the complex word in the sentence or document to simplify. The data structure produced following this procedure has 63,649 target words and 87,792 entries. The plain text of the Simple English Wikipedia (99,943 documents[6]) has been extracted using the WikiExtractor[7] tool. The FreeLing 3.1[8] NLP API has been used to analyze and extract the lemmas and the PoS tags. These lemmas were extracted from a 11-word window (5 lemmas are extracted to each side of the target words). We rank synonyms in the selected sense by their simplicity and find the simplest and most appropriate synonym word for the given context. The simplicity measure implemented is word frequency (i.e. more frequent is simpler) [3]. Several frequency lists were compiled for YATS, however for the experiments to be described here the Simple English Wikipedia frequency list was used.

The language realization phase generates the correct inflected forms of the selected synonym word substitutes in the contexts. The SimpleNLG Java API[9] is used with its default lexicon to perform this task considering the context and the PoS tag of the complex word.

3.1 Intrinsic Evaluation of the YATS Lexical Simplifier

Horn et al. [7] have produced a dataset for evaluation of lexical simplification systems containing 500 examples. Each example contains a sentence and a target word, randomly sampled from alignments between sentences in pairs of articles from English Wikipedia and Simple English Wikipedia produced by [5]. Fifty Mechanical Turkers[10] provided simplifications (i.e. lexical substitutes) for each sentence in the dataset. Moreover, counts for the lexical substitutes proposed

---

[5] http://wordnet.princeton.edu/.

[6] simplewiki-20140204 dump version.

[7] http://github.com/attardi/wikiextractor.

[8] http://nlp.cs.upc.edu/freeling/.

[9] http://github.com/simplenlg/simplenlg.

[10] https://www.mturk.com.

were obtained so as to produce a frequency-based rank for the set of replacements. One example of the evaluation dataset is shown below:

> **Sentence:** A haunted house is defined as a house that is believed to be a center for supernatural <u>occurrences</u> or paranormal phenomena.
> **Replacements:** events (24); happenings (12); activities (2); things (2); accidents (1); activity (1); acts (1); beings (1); event (1); happening (1); instances (1); times (1); situations (1)

The example shows: (i) a sentence where the *target* word to simplify is underlined (i.e. occurrences) and (ii) its possible replacements, together with the number of annotators selecting the replacement (e.g. the word *events* was chosen 24 times as simpler synonym for *occurrences*). We have carried out a hard intrinsic evaluation of the lexical simplification system using the above dataset. We have used the YATS lexical simplifier to select the most appropriate and simpler synonym of each target word in the dataset. Note that this is not a real application scenario of our lexical simplifier, since we already know which target word to simplify and therefore the task is somehow simpler. As a baseline we used two approaches: (i) the system proposed by [3] and replicated in [10] which simply selects the most frequent synonym of the target word ignoring the possible polysemy of the target word (i.e. no WSD), and (ii) the rules induced by the system proposed by [16] which are freely available (see Sect. 2). The frequency-based baseline uses the same resources as YATS, that is, WordNet for finding synonyms, and same file for lemma frequencies used in YATS.

In order to carry out the evaluation some transformations have to be applied to the dataset: (i) all replacements have to be lemmatized and counts merged for replacements with the same lemma (e.g. in the example above *activities* and *activity* have to be collapsed under *activity* with count $1 + 2 = 3$); (ii) for evaluating both YATS and the frequency-based baseline only replacements (i.e. lemmas) appearing as synonyms of the target word are considered (e.g. in the example above *occurrence* has only 3 possible synonyms in our lexical resource: *happening*, *presence*, and *event*, therefore all other listed replacements will not be considered since they can not be produced by the considered system). We use *lexical simplification accuracy* as a metric defined for a sentence S, target word T, a set of weighted replacements $Replacements(S, T)$ of T in S, and $Syno$: the synonym chosen by the system, as follows:

$$Lex\_Simp\_Acc(S, T, Syno) = \frac{\sum_{R \in Replacements(S,T)} Match(R, Syno, R_n)}{\sum_{R \in Replacements(S,T)} R_n}$$

where R_n is the number of annotators who have chosen R as replacement, $Match(R, Syno, R_n)$ is R_n if $R = Syno$ and 0 otherwise. That is, the system wins as many points as annotators have chosen $Syno$ as replacement. The denominator of the formula is a normalization factor which indicates how many annotators have provided replacements for the target word. As an example, if for the instance above a system selects "event" as a substitute of "occurrences", then it will obtain 24 (=23+1)points. Given k pairs of sentences and target words $< S, T >$, the overall lexical simplification accuracy of a system will be its average lexical simplification accuracy. The results of evaluating the three systems: YATS, frequency-based baseline, and rules are shown in Table 1. We also include the maximum possible lexical simplification accuracy a system could obtain (i.e. an oracle).

Table 1. Average lexical simplification accuracy of three systems and an oracle on the Horn et al.'s lexical simplification dataset [7].

System	Oracle	YATS	Frequency	Rules
Lex.Simp.Acc.	0.81	**0.21**	0.19	0.11

Results indicate a good performance of the YATS system, overtaking both a hard to beat baseline and rule-based system which rules were induced from corpora. Moreover, YATS provided valid replacements for 146 sentences while the frequency-based baseline provided 141 replacements and the rule-based system only 71.

4 Syntactic Simplification in YATS

The syntactic simplification is organized as a two-phase process: *document analysis* and *sentence generation*. The document analysis phase uses three main resources to identify complex syntactic structures: (i) the GATE/ANNIE analysis pipeline used for lexical simplification; this step performs tokenization, sentence splitting and NE recognition, (ii) the Mate Tools dependency parser[11,12] which adds a dependency labels to sentence tokens, and (iii) a set of GATE JAPE (Java Annotation Patterns Engine) grammars which detect and label the different kind of syntactic phenomena appearing in the sentences.

JAPE Rules were manually developed in an iterative process by using dependency-parsed sentences from Wikipedia. The process resulted in a set of rules able to recognize and analyze the different kinds of syntactic phenomena appearing in the sentences (described below). These rules rely on dependency trees, which allow a broad coverage of common syntactic simplifications with a small hand-crafted rules. Given the complex problem at hand, it is not enough to perform pattern matching and annotation of the matched elements, also the different annotations matched instantiating the pattern have to be properly annotated and related to each other. Each syntactic phenomena dealt with in the system has a dedicated grammar (i.e. a set of rules). The complete rule-based system is composed of the following sets of rules (applied in priority order when several syntactic phenomena are detected): (i) appositives (1 rule), (ii) relative

---

[11] http://code.google.com/archive/p/mate-tools/.
[12] It uses the Mate Tools' PoS tagger and lemmatizer before parsing.

clauses (17 rules), (iii) coordination of sentences and verb phrases (10 rules), (iv) coordinated correlatives (4 rules), (v) passive constructions (14 rules), (vi) adverbial clauses (12 rules), and (vii) subordination (i.e. concession, cause, etc.) (8 rules). The system recursively simplifies sentences until no more simplifications can be applied.

The *sentence generation phase* uses the information provided by the analysis stage to generate simple structures. It applies a set of rules which are specific for each phenomenon. These rules perform the common simplification operations, namely sentence splitting, reordering of words or phrases, word substitutions, verbal tense adaptions, personal pronouns transformations, and capitalization and de-capitalization of some words.

4.1   Intrinsic Evaluation of the YATS Syntactic Simplifier

We have carried out an intrinsic evaluation of the rule-based systems in terms of precision. We have collected 100 sentence examples (not used for system development) per syntactic phenomena we target. Each grammar was then applied to the set of sentences the grammar was covering so as to analyze the performance of the rules on unseen examples. Results are presented on Table 2.

Most rules are rather precise, except perhaps those dealing with coordination which is a very difficult phenomena to recognize given its ambiguity. An analysis of the errors showed us that the JAPE rules produced errors due to the lack of coverage of certain structures, e.g. coordination of anchors, coordination of antecedents, or coordination of main verbs taking a subordinated clause. Some of the errors produced by the dependency parser were: the parser PoS tagger assigned a wrong PoS tag, the parser assigned a wrong function (e.g. the anchor was identified with a wrong token, the relative pronoun got a wrong function) or a wrong dependency (e.g. the subordinated clause depended on a wrong head, the parser did not identify all the dependents of a head or it assigned a wrong token dependency), and the parser analyzed a syntactic structure wrongly.

Table 2. Evaluation of the JAPE grammars. The first column lists the syntactic phenomena, the second column indicates the number of sentences used which contained the sought syntactic phenomena, the third column indicates the percent of the grammar fired, the fourth column indicates the precision of the grammar, the fifth column is the percent of wrong grammar applications, and finally the last column is the percent of times the grammar did not fire.

Grammar	#sents	%fired	%right	%wrong	%ignored
Appositions	100	100%	79%	21%	0%
Relative Clauses	100	93%	79%	14%	7%
Coordination	100	62%	56%	6%	38%
Subordination	100	97%	72%	25%	3%
Passives	100	91%	85%	6%	9%
Total	500	89%	74.2%	14.4%	11.4%

5 Human Evaluation of YATS

We performed manual evaluation relying on eight English proficient human judges,[13] who assessed our system w.r.t. fluency, adequacy, and simplicity, using the evaluation set used by Siddharthan and Angrosh [13], from which we randomly selected 25 sentences. Participants were presented with the source sentence from the English Wikipedia (EW) followed by three simplified versions from Simple English Wikipedia (SEW) (i.e. a sentence from SEW aligned to the sentence in EW), the system developed by Siddharthan and Angrosh [13], and YATS, in a randomized order, and they were asked to rate each of the simplified version w.r.t. the extend to which it was grammatical (fluency), the extend to which it had the same meaning as the complex sentence (adequacy), and the extend to which it was simpler than the complex and thus easier to understand (simplicity). We used a five point rating scale (high numbers indicate better performance). Table 3 shows the results for the complete data set. Our system achieved the same mean score for simplicity as Siddharthan and Angrosh's [13], and is slightly better at fluency and adequacy, though not statistically significant.

Table 3. Average results of the human evaluation with eight human judges.

System	Fluency	Adequacy	Simplicity
Simple English Wikipedia (Human)	4.58	3.76	3.93
Siddharthan and Angrosh [13] (Automatic)	3.73	3.70	**2.86**
YATS (Automatic)	**3.98**	**4.02**	**2.86**

6 Conclusion

In this paper we have presented YATS, a text simplifier for English with lexical and syntactic simplification capabilities. The lexical simplifier uses a vector space model approach to obtain the most appropriate sense of a given word in a given context and word frequency simplicity measures to rank synonyms. We have shown that our lexical simplifier outperforms a hard-to-beat baseline procedure based on frequency and a rule-based system highly cited in the literature. The syntactic simplifier, which is linguistically motivated and based on peer-reviewed work, uses rule-based syntactic analysis and generation techniques that rely on part-of-speech tags and dependency trees. Experimental results of human assessment of the system output showed improvements over the best reported results in the literature. Future research includes: (a) experiments to better assess the performance of the system (e.g. lexical simplification in other available datasets), (b) improve the coverage of the syntactic simplifier by retraining the parser, (c) extending the scope of the lexical simplifier relying on more advanced vector representations (e.g. embeddings), and (d) porting the system to other languages.

---

[13] None of them developed the simplifier.

Acknowledgments. We are grateful to three anonymous reviewers for their useful comments, to the participants in our human evaluation experiments, and to A. Siddharthan for sharing his dataset. This work was funded by the ABLE-TO-INCLUDE project (European Commission CIP Grant No. 621055). Horacio Saggion is (partly) supported by the Spanish MINECO Ministry (MDM-2015-0502).

References

1. Biran, O., Brody, S., Elhadad, N.: Putting it simply: a context-aware approach to lexical simplification. In: Proceedings of the ACL 2011, pp. 496–501 (2011)
2. Bott, S., Rello, L., Drndarevic, B., Saggion, H.: Can Spanish be simpler? LexSiS: lexical simplification for Spanish. In: Proceedings of the COLING 2012, Mumbai, India, pp. 357–374 (2012)
3. Carroll, J., Minnen, G., Canning, Y., Devlin, S., Tait, J.: Practical simplification of English newspaper text to assist aphasic readers. In: Proceedings of the AAAI 1998 Workshop on Integrating AI and Assistive Technology, pp. 7–10 (1998)
4. Chandrasekar, R., Doran, C., Srinivas, B.: Motivations and methods for text simplification. In: Proceedings of the COLING 1996, pp. 1041–1044 (1996)
5. Coster, W., Kauchak, D.: Learning to simplify sentences using wikipedia. In: Proceedings of ACL 2011 Workshop on Monolingual Text-To-Text Generation, Portland, Oregon, USA, pp. 1–9 (2011)
6. Devlin, S., Tait, J.: The use of a psycholinguistic database in the simplification of text for aphasic readers. In: Linguistic Databases, pp. 161–173 (1998)
7. Horn, C., Manduca, C., Kauchak, D.: Learning a lexical simplifier using Wikipedia. In: Proceedings of ACL 2014, pp. 458–463 (2014)
8. Saggion, H., Bott, S., Rello, L.: Simplifying words in context. Experiments with two lexical resources in Spanish. Comput. Speech Lang. **35**, 200–218 (2016)
9. Saggion, H., Stajner, S., Bott, S., Mille, S., Rello, L., Drndarevic, B.: Making it simplext: implementation and evaluation of a text simplification system for spanish. TACCESS **6**(4), 14 (2015)
10. Shardlow, M.: Out in the open: finding and categorising errors in the lexical simplification pipeline. In: Proceedings of LREC 2014, Reykjavik, Iceland (2014)
11. Siddharthan, A.: Syntactic simplification and text cohesion. In: Proceedings of the LEC 2002, pp. 64–71 (2002)
12. Siddharthan, A.: Text simplification using typed dependencies: a comparision of the robustness of different generation strategies. In: Proceedings of the 13th European Workshop on Natural Language Generation, Nancy, France (2011)
13. Siddharthan, A., Angrosh, M.: Hybrid text simplification using synchronous dependency grammars with hand-written and automatically harvested rules. In: Proceedings of the EACL 2014, Gothenburg, Sweden (2014)
14. Turney, P.D., Pantel, P.: From frequency to meaning: vector space models of semantics. J. Artif. Int. Res. **37**(1), 141–188 (2010)
15. Wubben, S., Bosch, A., Krahmer, E.: Sentence simplification by monolingual machine translation. In: Proceedings of ACL 2012, pp. 1015–1024 (2012)
16. Yatskar, M., Pang, B., Danescu-Niculescu-Mizil, C., Lee, L.: For the sake of simplicity: unsupervised extraction of lexical simplifications from Wikipedia. In: Proceedings of HLT-NAACL 2010 (2010)

Are Deep Learning Approaches Suitable for Natural Language Processing?

S. Alshahrani$^{(\boxtimes)}$ and E. Kapetanios

Cognitive Computing Research Group, Computer Science Department,
University of Westminster, London, UK
w1484137@my.westminster.ac.uk,
E.Kapetanios@westminster.ac.uk

Abstract. In recent years, Deep Learning (DL) techniques have gained much attention from Artificial Intelligence (AI) and Natural Language Processing (NLP) research communities because these approaches can often learn features from data without the need for human design or engineering interventions. In addition, DL approaches have achieved some remarkable results. In this paper, we have surveyed major recent contributions that use DL techniques for NLP tasks. All these reviewed topics have been limited to show contributions to text understanding, such as sentence modelling, sentiment classification, semantic role labelling, question answering, etc. We provide an overview of deep learning architectures based on Artificial Neural Networks (ANNs), Convolutional Neural Networks (CNNs), Long Short-Term Memory (LSTM), and Recursive Neural Networks (RNNs).

Keywords: Deep learning · Natural language processing · Artificial neural networks · Convolutional neural networks · Long short-term memory · Recursive neural networks

1 Introduction

Machine Learning (ML) is a robust AI tool, which has shown its usefulness in our daily lives, for example, with technologies used in search engines, image understanding, predictive analytics, transforming speech to text and matching relevant text. All ML approaches can be roughly classified as supervised, unsupervised and semi-supervised.

Building a machine-learning system with features extraction requires specific domain expertise in order to design a classifier model for transforming the raw data into internal representation inputs or vectors. These methods are called representation learning (RL), in which the model automatically feeds in raw data to detect the needed representation.

In particular, the ability to precisely represent words, phrases, sentences (statement or question) or paragraphs, and the relational classifications between them, is essential to language understanding. Deep learning approaches are similar to RL methods using multiple levels of representational processing [1].

Deep Learning (DL) involves multiple data processing layers, which allow the machine to learn from data, through various levels of abstraction, for a specific task

© Springer International Publishing Switzerland 2016
E. Métais et al. (Eds.): NLDB 2016, LNCS 9612, pp. 343–349, 2016.
DOI: 10.1007/978-3-319-41754-7_33

without human interference or previously captured knowledge. Therefore, one could classify DL as unsupervised ML approach. Investigating the suitability of DL approaches for NLP tasks has gained much attention from the ML and NLP research communities, as they have achieved good results in solving bottleneck problems [1]. These techniques have had great success in different NLP tasks, from low level (character level) to high level (sentence level) analysis, for instance, sentence modelling [12], Semantic Role Labelling [4], Named Entity Recognition [19], Question Answering [15], text categorization [11], opinion expression [8], and Machine Translation [9].

The focus of this paper is on DL approaches that are used for NLP tasks. ANNs are discussed in Sect. 2 below, while CNNs are considered in Sect. 3. Section 4 discusses the suitability of these techniques for NLP and the implications for future research. We then conclude with a brief overview in Sect. 5.

2   Artificial Neural Networks Approaches

A standard neural network consists of many connected units called neurons, each generating a sequence of real-valued activations. Neurons are activated by previous neurons in the circuit, via weighted connections. Each link transforms chains of computational sequences between neurons in a non-linear way. ANNs are robust learning models that are about precisely assigning weights across many levels. They are broadly divided into two types of ANN architectures that can be feed-forward networks (FF-NNs), Recurrent Neural Networks (RNNs) and Recursive Neural Networks [10]. FF-NNs architecture consists of fully connected network layers. The RNNs model, on the other hand, consist of a fully linked circle of neurons connected for the purpose of back-propagation algorithm implementation.

FF-NNs applied to NLP tasks consider syntax features as part of semantic analysis [26]. ANN learning models have been proposed that can be applied to different natural language tasks, such as semantic role labelling and Named Entity Recognition [23]. The advantage of these approaches is to avoid the need for prior knowledge and task specific engineering interventions. FF-NNs models have achieved an efficient performance in tagging systems with low computational requirements [4]. A Neural-Image-QA approach has been proposed, which combines NLP and image representation for a textual question answering system regarding images [16]. However, this system has shown incorrect answers regarding the size of the objects in an image, and also with spatial reasoning questions, such as "which is the largest object?". In another study [28], a novel approach was proposed to match queries with candidate answers (sentences) depending on their semantic meaning, by combing distributed representations and deep learning without the need for features engineering or linguistic expertise. State-of-the-art neural network-based distributional models have achieved a high performance in various natural NLP tasks; for instance, document classification [13] and Entity Disambiguation and Recognition [23]. Another FF-NNS approach is the ANNABEL model [7], which aspires to simulate the human brain and to become able to process verbal and non-verbal natural language without prior knowledge. This work is based on a large-scale neural architecture which acts as the central executive and interacts with humans through a conversational interface. The lack of control over information flow through memory inspired the ANNABEL team to build their mental action to control the Short-Term Memory system.

2.1 Recurrent Neural Network Approaches

The RNNs model fully links neurons in a circle of connections for the purpose of back-propagation algorithm implementation. Recurrent Neural Networks have been recommended for processing sequences [10], while Recursive Neural Networks are collections of recurrent networks that can address trees [6]. Another application uses Recurrent Neural Networks for question answering systems about paragraphs [15], and a Neural Responding Machine (NRM) has been proposed for a Short-Text Conversation generator, which is based on neural networks [21]. In addition, Recurrent Neural Networks models offered state-of-the-art performance for sentiment classification [13], target-dependent sentiment classification [25] and question answering [15]. Adaptive Recurrent Neural Network (AdaRNN) is introduced for sentiment classification in Twitter promotions based on the context and syntactic relationships between words [2]. Furthermore, Recurrent Neural Networks are used for the prediction of opinion expression [8].

3 Convolutional Neural Networks Architecture

CNNs evolved in the field of vision research, where the first use was for performing image classification learning to detect the edges of an image from its raw pixels, in the first layer, and then using the edges to identify simple shapes located in the second layer. These shapes are then used to identify higher-level features such as facial shapes in the higher layers of the network [27].

Applying a non-linear function over a sequence of words, by sliding a window over the sentences, is the key advantage of using CNNs architecture for NLP tasks [57]. This function, which is also called a 'filter', mutates the input (k-word window) into a d-dimensional vector that consists of the significant characteristic of the words in the window. Then, a pooling operation is applied to integrate the vectors, resulting from the different channels, into a single n-dimensional vector. This is done by considering the maximum value or the average value for each level across the different windows to capture the important features, or at least the positions of these features. For example, Fig. 1 gives an illustration of the CNNs' structure where each filter executes convolution on the input, in this case a sentence matrix, and then produces feature maps, hence it showing two possible outputs. This example is used in the sentence classification model.

Most NLP classification tasks use CNN models and pioneering work has been done using these methods for semantic-role labelling [3], sentiment and question-type classification [14], text understanding from the basic character level [57] and text categorization [11]. A further use of CNNs is in the visual question-answering model, known as mQA model, which is able to answer a question about image contents [5]. This model comprises of four parts: a Long Short-Term Memory (LSTM) to extract the query representation, a CNN, another LSTM for processing the context of the answers,

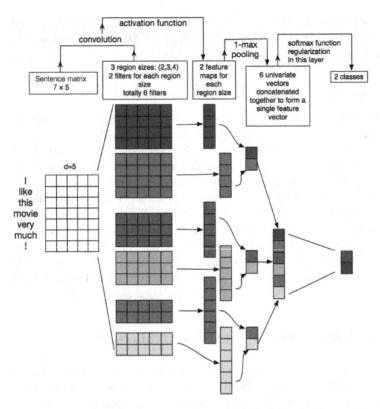

Fig. 1. Model of three filter division sizes (2, 3 and 4) of CNNs architecture for sentence classification. (Source: Zhang and Wallace (2015)). (Color figure online)

and a process part for generating the answer, which can collect the information from the first three parts in order to provide answers. This can be in the form of a word, a phrase or a sentence. A new convolutional latent semantic approach for vector representation learning [22] uses CNNs to deal with ambiguity problems in semantic clustering for short text. However, this model can work appropriately for long text as well [25]. CNNs are proposed for sentiment analysis of short texts that learn features of the text from low levels (characters) to high levels (sentences) to classify sentences in positive or negative prediction analysis. However, this approach can be used for different sentence sizes [18]. The Dynamic CNN (DCNN) is embraced for semantic sentence modelling which includes extracting features from sentences. DCNN is trained by selectively weighting functions between linked network layers [12].

4   Discussion and Future Research

DL approaches are used for a variety of NLP tasks, as shown in Table 1. This gives an overview of different NLP tasks using NN, CNN and RNN (Recurrent Neural Network) approaches for NLP tasks related to semantic and context analysis. The RNNs

Table 1. Gives an overview of different NLP tasks using DL approaches for for NLP tasks related to semantic and context analysis.

Text understanding task	ANNs	CNNs	RNNs
Sentence classification		✓	
Named entity recognition (NER)	✓		
Text categorization		✓	
Semantic role labelling		✓	
Semantic clustering		✓	
Short-text conversation	✓		✓
Question answering	✓	✓	✓
Sentiment analysis	✓		✓
Paraphrase detection	✓		
Document classification	✓		
Topic categorization		✓	
Opinion expression prediction			✓

model can achieve a competitive performance in NLP tasks with sequence input; but, if the problem deals with sparsity in its input, this can be solved using CNNs. The CNN implementation is most commonly used in NLP classification tasks, such as sentiment classification, question answering, and sentence selection. CNNs have achieved state-of-the-art performance in solving data sparsity problems in large structures of NLP tasks.

In addition, DL learning techniques provide an enhanced performance for NLP tasks due to their ability to its distinguishing features. Firstly, they use powerful performance, provided by advanced CPUs, to improve training processes which are implemented in current deep learning techniques, for NLP tasks. Secondly, they provide functional evidence in terms of representation, as convolutional filters can automatically learn the term representations without requiring prior knowledge. Furthermore, RNNs deal with sequence inputs, and we therefore expect models which use these methods to have a large impact on natural language understanding over the next couple of years.

Conversely, conventional deep learning approaches that may be very appropriate for processing of raw pixels in images cannot work properly for text processing, due to the need for more features and data in the hidden layer parameters [24].

Popular word-embedding models are GloVe (Global Vectors for word representation) [17], Word2vec [20], and embedding algorithms, such as dynamically sized context windows [4]. Word2vec is an open-source Google text processing tool published in 2013. This approach is used for word representation. Word2vec relies upon two algorithms, skip-grams and continuous bag of words (CBOW). The skip-gram model process the current input word using a linear classifier to predict surrounding words in a specific scope, whereas CBOW is a trained model for understanding ambiguous words, based on their context [20]. GloVe is a vector representation word-learning method.

GloVe works by computing the large word co-occurrence matrix in memory, and is dependent on matrix factorization algorithms, making it a good model for optimization. In contrast, Word2Vec goes through sentences directly, treating each co-occurrence separately. The advantage of these approaches (word-embedding) can be combined with other deep learning models to enhance performance for NLP tasks. In addition, a gradient-based method can be used in NN training, which helps to reduce error over a training set.

In addition, further research may explore the suitability of DL for Conversational User Interfaces and/or for dynamic Question Answering, particularly for graph based pattern search and recognition, over Linked Open Data.

5   Conclusion

In this paper, we have presented an overview of the deep learning based approach to natural language processing tasks. Different studies are included in this review, covering various NLP tasks that implemented ANNs, CNNs, RNNs and LSTM. These approaches can be combined with other deep learning models to develop improved performance for NLP tasks. The advantage of using these approaches is to avoid the need for prior knowledge and human engineering interventions. The key conclusion from this overview is that deep learning approaches are ideal for solving data sparsity problems in large structures, while CNNs have the advantage of fast performance and also providing functional evidence for representational learning and feature extraction.

We have not discussed statistical surveys or performance comparisons relating to deep learning techniques used for NLP tasks. This should be explored comprehensively as part of future research.

References

1. Ba, L., Caurana, R.: Do Deep Nets Really Need to be Deep? **521**(7553) 1–6 (2013). arXiv preprint arXiv:1312.6184
2. Dong, L., Wei, F., Tan, C., Tang, D., Zhou, M., Xu, K.: Adaptive recursive neural network for target-dependent Twitter sentiment classification. In: ACL-2014, pp. 49–54 (2014)
3. Collobert, R., Weston, J.: Fast semantic extraction using a novel neural network architecture. In: Proceedings of the 45th Annual Meeting of the ACL, pp. 560–567 (2007)
4. Collobert, R., Weston, J., Bottou, L., Karlen, M., Kavukcuglu, K., Kuksa, P.: Natural language processing (almost) from scratch. J. Mach. Learn. Res. **12**, 2493–2537 (2011)
5. Gao, H., Mao, J., Zhou, J., Huang, Z., Wang, L., Xu, W.: Are you talking to a machine? Dataset and methods for multilingual image question answering. In: Arxiv, pp. 1–10 (2015)
6. Goller, C., Kuchler, A.: Learning task-dependent distributed representations by backpropagation through structure. In: Proceedings of the ICNN 1996, pp. 347–352. IEEE (1996)
7. Golosio, B., Cangelosi, A., Gamotina, O., Masala, G.L.: A cognitive neural architecture able to learn and communicate through natural language. PLoS ONE **10**(11), e0140866 (2015)
8. Irsoy, O., Cardie, C.: Opinion mining with deep recurrent neural networks. In: EMNLP-2014, pp. 720–728 (2014)

9. Jean, S., Cho, K., Memisevic, R., Bengio, Y.: On using very large target vocabulary for neural machine translation. In: Proceedings of the ACL-IJCNLP (2015)
10. Elman, J.L.: Finding structure in time. Cogn. Sci. **14**(2), 179–211 (1990)
11. Johnson, R., Zhang, T.: Semi-supervised Convolutional Neural Networks for Text Categorization via Region Embedding, pp. 1–12 (2015)
12. Kalchbrenner, N., Grefenstette, E., Blunsom, P.: A convolutional neural network for modelling sentences. In: Proceedings of the 52nd Annual Meeting of the Association for Computational Linguistics, ACL 2014, 22–27 June 2014, Baltimore, MD, USA, vol. 1, pp. 655–665 (2014). Long Papers
13. Hermann, K.M., Blunsom, P.: Multilingual models for compositional distributional semantics. In Proceedings of ACL (2014)
14. Kim, Y.: Convolutional neural networks for sentence classification. In: Proceedings of the 2014 Conference on Empirical Methods in Natural Language Processing (EMNLP 2014), pp. 1746–1751 (2014)
15. Iyyer, M., Boyd-Graber, J., Claudino, L., Socher, R., Daumé III, H.: A neural network for factoid question answering over paragraphs. In: EMNLP (2014)
16. Malinowski, M., Rohrbach, M., Fritz, M.: Ask your neurons: a neural-based approach to answering questions about images. In: IEEE International Conference on Computer Vision, pp. 1–9 (2015)
17. Pennington, J., Socher, R., Manning, C.D.: GloVe: global vectors for word representation. In: EMNLP (2014)
18. dos Santos, C.N., Gatti, M.: Deep convolutional neural networks for sentiment analysis of short texts. In: COLING-2014, pp. 69–78 (2014)
19. dos Santos, C.N., Guimarães, V.: Boosting named entity recognition with neural character embeddings. In: ACL 2014, pp. 25–33 (2015)
20. Schmidhuber, J.: Deep learning in neural networks: an overview. Neural Netw. **61**, 85–117 (2015)
21. Shang, L., Lu, Z., Li, H.: Neural responding machine for short-text conversation. In: ACL-2015, pp. 1577–1586 (2015)
22. Shen, Y., He, X., Gao, J., Deng, L., Mesnil, G.: A latent semantic model with convolutional-pooling structure for information retrieval. In: Proceedings of the 23rd ACM International Conference on Conference on Information and Knowledge Management - CIKM 2014, pp. 101–110 (2014)
23. Sun, Y., Lin, L., Tang, D., Yang, N., Ji, Z., Wang, X.: Modelling mention, context and entity with neural networks for entity disambiguation. In: IJCAI, pp. 1333–1339 (2015)
24. Mikolov, T., Corrado, G., Chen, K., Dean, J.: Efficient estimation of word representations in vector space. In: Proceedings of the International Conference on Learning Representations (ICLR 2013) (2013)
25. Wang, P., Xu, J., Xu, B., Liu, C., Zhang, H., Wang, F., Hao, H.: Semantic clustering and convolutional neural network for short text categorization. In: Proceedings of the ACL 2015, pp. 352–357 (2015)
26. Weston, J., America, N.E.C.L., Way, I.: A unified architecture for natural language processing: deep neural networks with multitask learning. In: ICML 2008, pp. 160–167 (2008)
27. LeCun, Y., Bengio, Y.: Convolutional networks for images, speech, and time-series. pp. 255–258. MIT Press (1995)
28. Yu, L., Hermann, K.M., Blunsom, P., Pulman, S.: Deep learning for answer sentence selection. In: NIPS Deep Learning Workshop, 9 p. (2014)
29. Zhang, X., LeCun, Y.: Text Understanding from Scratch (2015)

An Approach to Analyse a Hashtag-Based Topic Thread in Twitter

Ekaterina Shabunina[✉], Stefania Marrara, and Gabriella Pasi

Dipartimento di Informatica Sistemistica e Comunicazione,
Università degli Studi di Milano-Bicocca, Viale Sarca 336, 20126 Milan, Italy
{ekaterina.shabunina,stefania.marrara,pasi}@disco.unimib.it

Abstract. In last years, the spread of social Web has promoted a strong interest in analyzing how information related to a given topic diffuses. Nevertheless, this is still quite an unexplored field in the literature. In this paper we propose a general approach that makes use of a set of Natural Language Processing (NLP) techniques to analyse some of the most important features of information related to a topic. The domain of this study is Twitter, since here topics are easily identified by means of hashtags. In particular, our aim is to analyse the possible change over time of the content sub-topicality and sentiment in the tracked tweets, and bring out their relationships with the users' demographic features.

Keywords: Corpus analysis · Natural language processing techniques · Twitter · Information demographic · User generated content analysis

1 Introduction

In last years, with the spread of social media the analysis of how content related to a given topic is spread through the Web constitutes an interesting research issue. In particular, Twitter is a microblogging platform where the discussion around a given topic can be easily identified and tracked by means of the so-called *hashtags* (e.g., #HeForShe). The assumption at the basis of this paper is that a hashtag is representative of an *eccentric* topic, as outlined in [4]. This allows to easily crawl a thread of tweets, which can offer a good repository to perform various kinds of analysis, related to different characteristics of both the content and the users generating it. Analyses of this kind can give several interesting insights on how a certain topic diffuses among users also dependeing on users' characteristics.

The objective of this paper is to propose a tool that, based on a set of standard NLP techniques, allows to analyse some of the most important characteristics of the Twitter topics and their contributing users. In particular, the aim of the proposed approach is to monitor over time the content sub-topicality and sentiment in the tracked tweets, and its relation with demographics such as gender and age. Differently to the well studied phenomenon of information diffusion, in our work we aim to study the evolution of content on a given Twitter topic in a limited timeframe.

© Springer International Publishing Switzerland 2016
E. Métais et al. (Eds.): NLDB 2016, LNCS 9612, pp. 350–358, 2016.
DOI: 10.1007/978-3-319-41754-7_34

The paper is organised as follows: in Sect. 2 we present the context of the current study. In Sect. 3 we introduce the approach proposed in the current work, followed by the application of our approach on a specific topic in Sect. 4. Finally, Sect. 5 concludes the paper.

2   Related Work

In this section we shortly review the literature that shares some common background with the approach proposed in this paper.

Several research works focus on the study of the demographics of the social network users who generate the information spread. In [5] the gender of the Twitter users is detected through a comparison of the first word of the self-reported name in the user's profile to the list of the most popular names reported by the USA Social Security Administration. In the research conducted in [1] the tweets from reporters of 51 US newspapers have been analyzed for the gender ratio in the quotes. The result reported in [1] is in line with most previous works on this topic, which present that women are less likely to be used as quoted sources overall. Additionally, in [7] it has been discovered that blogs are more likely to quote the source without introducing any changes, in contrast to professional journalists.

Another research line analyzes hashtags, their diffusion and the characteristics of their spread [4]. In [6,8] the authors outline that different topics have distinct patterns of information spread.

The spread of units of information or concepts on the Web, known as Information Propagation, has been widely studied as a modelling and prediction problem [2] or as function of the influence among nodes of a network [8]. Unlike the aim of our work, these approaches do not focus on the evolution of the content itself but rather on the evolution of the network of users. However, in [3] the authors point out the importance of the content of a piece of information when predicting its propagation.

In this paper we claim that the analysis of the user generated content over time is an important research issue. To this aim we propose a simple approach with a related tool that can easily support a few basic kinds of analysis related to both users and tweets. The proposed approach is presented in Sect. 3.

3   The Proposed Approach

In this section we describe the proposed approach for analysing some characteristics of the evolution of a topic (identified by a hashtag in Twitter) in a given time interval.

3.1   The Twitter Topic Analysis Questions

In this section we propose a set of questions that can be used as guidelines for the analysis of any stream of tweets associated with a topic:

1. What is the main sentiment along the tweets stream? Does it change over time and w.r.t. user gender or age?
2. How are the users who contributed to the analysed topic divided w.r.t. their topics of interests?
3. Who are the main contributors to the tweets stream? Are they private users or corporates?
4. Does the topic have a goal? If so, are the statistics of the analysis in line with the main purpose of the topic creation?
5. What are the characteristics (e.g. gender, age, etc.) of the leading quoted sources?

The above questions can be more specifically tailored to the topic under consideration. To answer the above questions, we propose a combined usage of open source NLP tools and a module that we have developed to perform some specific statistical analyses; we refer to this combination as the Analysis Tool.

The tool, presented in Sect. 3.2, works with a set of sentiment and demographic dimensions aimed at gathering the analysis statistics. The considered dimensions are:

- the **gender**, male or female, of the user;
- the **age** of the user, to identify if the topic touched the diverse groups of the Twitter population;
- the **user sentiment**;
- the **tweet sentiment**;
- the main **topic of interest** of each user (e.g., Arts, Business...), to identify the interests of the users who posted in the analysed topic;
- the **quoted sources** (via the analysis of *retweets*), to detect the trusted voice of the crowd and the dominant contributors to the spread of information;
- **sub-topicality** (via terms frequency and jointly used hashtags), to analyse changes in the information carried by the given hashtag.

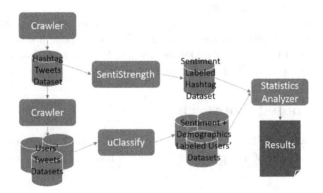

Fig. 1. The tool for analysing the content evolution of a stream of tweets associated with a given hashtag.

3.2   The Analysis Tool

In this section we describe the Analysis Tool we have developed. The tool has been created to provide all the necessary means to answer the questions reported in Sect. 3.1. The analysis is performed by computing various statistics on the stream of tweets based on the dimensions identified in Sect. 3.1 in the time interval of observation.

The tool is sketched in Fig. 1 and it is constituted of four components: the **Crawler** that we have created, the open source software **uClassify**[1] and **SentiStrength**[2], and the **Statistics Analyzer** that we have generated. As shown in Fig. 1, an important phase is the creation of a target tweets collection that shares the considered hashtag (*Hashtag Tweets Dataset*); this is done by means of the *Crawler* over a considered time span.

Another important analysis pre-requisite is to gather a dataset for each user who has been identified as a contributor to the hashtag related tweets in the explored time span (the *Users' Tweets Datasets* in Fig. 1).

To the purpose of text classification we have selected the open source machine learning web service *uClassify*[1]; in our opinion this web service constitutes a good choice as it allows to easily train a custom classifier on any dataset. Based on each *Users' Tweets Dataset*, the selected text classifiers provide labels for the users on the categories that have been identified as interesting dimensions. In Fig. 1 the outcome of this classification phase is the named *Sentiment + Demographics Labeled Users' Datasets*.

For the single tweet sentiment classifier a popular option is the open source software *SentiStrength*[2] that is specifically tailored to evaluate the sentiment carried by short texts. When applied to the *"Hashtag Tweets Dataset"* SentiStrength produces the *Sentiment Labeled Hashtag Dataset* as depicted in Fig. 1.

All tweets and their associated labels constitute the input to the *Statistics Analyzer*, which has the aim of performing various statistical, demographic and sentiment analyses of the information associated with a given hashtag as it changes over the time. We have developed this software component to define some specific analysis, as it will be illustrated in Sect. 4.

4   Application of the Analysis Tool to a Case Study

As a demonstrative example of our tool, we have analysed the tweets related to a solidarity movement initiated by Emma Watson. In the microblogging platform Twitter this discussion topic is identified by the hashtag #HeForShe.

We have tailored the analysis questions, presented in Sect. 3.1, w.r.t. the topic of this case study. In particular, we were interested to discover if the majority gender of the hashtag contributors was masculine as planned by the campaign. Another ensuing question concerned the dominant gender of the quoted sources. Moreover, we were interested to test the hypothesis if the most popular quoted

---

[1] http://www.uclassify.com/.

[2] http://sentistrength.wlv.ac.uk/.

sources belonged to celebrities rather than common people. The next question was related to the dominant age groups of the contributors to the promotion of gender equality. Lastly we explored the expectation that the dominant sentiment of the tweets stream was positive, based upon the noble intentions of the campaign.

4.1 Dataset Description

The tweets collection was generated using the Twitter Search API 1.1 on a weekly basis. The tweets associated with the considered hashtag, i.e., #HeForShe, were crawled, limited to those in English. Our dataset covers one month of Twitter microblogs, containing the hashtag #HeForShe, crawled from March 08 to April 08, 2015. In total it consists of 72.932 tweets, posted by 53.700 distinct users. The retweets were included in the dataset since they comprise a Twitter mechanism to favor the propagation of information found interesting by the users; this constitutes an important information to the purpose of our analyses. Only tweets from users with protected accounts were eliminated from this dataset (about 1,5% of the initial dataset). These users were deleted since their tweets are protected, thus not allowing their classification. In addition to the main dataset, we have gathered 50.445 collections of the last tweets belonging to the users who contributed to the discussion under the #HeForShe hashtag.

All tweets datasets were fed to the text classifiers of *uClassify* and to the sentiment classifier *SentiStrength* to generate labels on five categories: Gender, Age Groups, User Sentiment, Tweet Sentiment and Topic. The geographic distribution of the tweets was not considered in the analysis due to the extreme sparseness of the geographic information in the tweets and in the users' accounts.

4.2 Results of the Performed Analyses

In this section we present the results of the performed analysis with our tool.

Demographic. A first issue we addressed was to verify if the peculiarity of the hashtag (campaign) served its purpose. Although this hashtag was created for men to participate in the gender equality fight, only 18 % of all tweets and only 17 % of all retweets were spread by men. This gender domination trend was consistent throughout the whole time period under study at a daily granularity.

Additionally, it was interesting to compare the retweets and tweets amounts on a daily scale. Over the investigated time period, 71.829 tweets were posted on the gender equality campaign in Twitter. Out of these, 57.520 tweets were actually "retweets", and consequently only 14.309 were "original tweets" containing new content. This explains the identical oscillation pattern on a daily basis of the retweets number and the "all posts" one, observable in Fig. 2.

Looking at the age distribution per each gender group in Fig. 3, we found that for women the strong majority of users (57,8 %) was classified to the "13–17 years old" group persistently along the 32 studied days. While, interestingly,

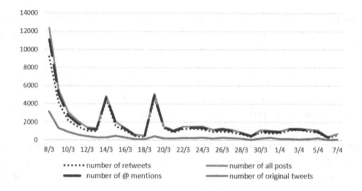

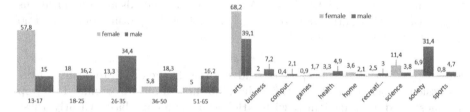

Fig. 2. Number of all posts, original tweets, retweets and user mentions.

a constant majority of men (34, 4 %) was identified in the young adults ("26–35 years old") group, presenting an almost perfect platykurtic age distribution in the examined time span.

Fig. 3. The median percentage age distribution for both genders.

Fig. 4. The median percentage topic distribution for both genders.

Sentiment. The sentiment analysis of the collected tweets and of the users who posted them showed that mainly positive people participated in this campaign, by posting mainly positive tweets. In fact, 61 % of tweets were classified as positive, 7, 9 % as negative and 30, 6 % as neutral. By analysing the sentiments of the users, 79 % of them were found positive and 20, 8 % negative. Concerning the gender ratio: 66, 5 % of women and 53, 6 % of men posted positive tweets.

Analysis of Topical Interests. The analysis of the users' topical interests (Fig. 4) showed that a strong majority of users (65, 7 %) were classified as interested in the "Arts" topic. Followed by 9, 7 % of users interested in Society and 9, 2 % in Science.

Quoted Sources. Since our study was carried out on Twitter, where quoting is performed via retweeting, we analyzed the rate at which new popular retweet sources were introduced, as well as the gender of the users who were in the top five quoted per day.

First of all, we found that for one third of the days under study two new popular retweet sources were introduced per day. Another one third of the days, only one new popularly quoted account per day was introduced. And for 28 % of the days there was no new quoted sources.

Interestingly, 41, 5 % of the most popular retweet sources were organizations, such as "NATO" and "United Nations". Followed by 36, 6 % quoted celebrities: singers, actors, etc. News media sources were popular quoted sources in 7, 3 % of cases. And non famous people in 14, 6 % of cases.

As for the gender ratio of the popular retweet sources: 39 % of accounts were identified as male, 26, 8 % as female and 34, 2 % were without gender information. 92, 8 % of the Twitter account with unknown gender were organizations.

The geographical spread of the popular quoting sources was diverse. The majority, namely 34, 1 % of retweet sources were from USA and 14, 6 % from UK. These two locations could be a biased result due to the choice to analyse only tweets in English. 12, 2 % of the retweet sources were from Asia, 9, 8 % from European countries, and 4, 9 % from Africa. 7, 3 % were funds and organizations stating that their location is "worldwide". 17, 1 % of the quoting sources had no geolocation information in their Twitter account.

Sub-topicality. To the purpose of studying the change in sub-topicality of the discussion associated with the target hashtag, first of all, we examined the top five most used hashtags jointly with #HeForShe on a daily basis. Clearly, these additional popular hashtags are related to subtopics ranging from those corresponding to the #HeForShe topic (#equalpay, #feminism, etc.) and closely related (such as International Women's Day: #iwd2015, #womensday), to those not really in the gender equality topic: #UncleOfTheYear, #facebook, #urbandictionary, etc. These additional hashtags, which are not apparently related to the topic of #HeForShe, are a demonstration of a daily shift of attention and misuse of hashtags in Twitter.

Regarding the duration of the identified hashtags popularity, only the #genderequality has been constantly used over 30 out of 32 days examined. On the second position with 16 days duration of popularity is #feminism. And on the third is the #yesallwomen with 7 days of popularity duration.

Next we looked at the words frequencies for each day under study. For the visualization we have employed the "Voyant"[3] tool, which is an open source web-based environment for texts analysis. Figure 5 presents the word clouds of the dataset for all 32 days studied with and without all hashtags and all user mentions, respectively. Although the additional hashtags analysis revealed general unrelated subtopics, these word clouds clearly display that the collected tweets were truly focused on promoting "men" "supporting" "women" for "equality".

---

[3] http://voyant-tools.org/.

Fig. 5. The #HeForShe word clouds for all the tweets for the considered time period. Left: Dataset untouched. Right: Dataset without all hashtags and user mentions.

4.3 Discussion of the Case Study Analysis Results

First of all, let we notice that the 80 % average daily retweet rate is a positive result due to the overall goal of the campaign to create awareness and to spread information about gender equality. This goal is further supported by over 90 % of all tweets including mentions of other Twitter users, which strongly helps the spread. Surprisingly, even though the campaign purpose was to activate men to fight for gender equality, the figures clearly outline that this goal was not achieved, with only 18 % of tweets belonging to male users. Regarding the involved age groups, the majority of men (34, 4 %) was identified in the young adults "26–35 years old" group. While for women the strong majority of users (57, 8 %) were classified to the "13–17 years old"group. This is quite surprising since the topic of the campaign probably was expecting a more active participation by adult women. On the opposite side, the interest w.r.t this campagin shown by teenage girls indicates that the gender equility theme has a good level of awareness in the environments in which these girls are involved.

Even though there are plenty of emotionally coloured debates on the Web around the gender equality subject, the month of tweets with #HeForShe presented a strongly positive sentiment by both female and male users. The findings of [7] were confirmed also in our study: the majority of the popular quoted Twitter users by gender are male. But the most popular quoted sources were found to be actually the Twitter accounts of organizations, funds and media sources, closely followed by celebrities. Similar to [1], we discovered that in Twitter quoting is performed without introducing any additional information such as personal opinion on the quoted subject.

5 Conclusions

In this paper we have presented a general approach aimed to perform a qualitative and quantitative analysis of information carried by tweets gathered w.r.t a given topic. The application of this approach on the #HeForShe hashtag has proved to gather interesting and not explicitly visible insights about the spread of information on a Twitter topic, and has shown that information can follow unpredictable paths. Often the purpose of a hashtag may not be served as expected by the promoter. This study offers an interesting auxiliary tool for other fields of research, such as sociology.

References

1. Artwick, C.G.: News sourcing and gender on twitter. Journalism **15**(8), 1111–1127 (2014)
2. Cheng, J., Adamic, L., Dow, P.A., Kleinberg, J.M., Leskovec, J.: Can cascades be predicted? In: Proceedings of the 23rd International Conference on World Wide Web, WWW 2014, pp. 925–936. ACM, New York (2014)
3. Lagnier, C., Denoyer, L., Gaussier, E., Gallinari, P.: Predicting information diffusion in social networks using content and user's profiles. In: Serdyukov, P., Braslavski, P., Kuznetsov, S.O., Kamps, J., Rüger, S., Agichtein, E., Segalovich, I., Yilmaz, E. (eds.) ECIR 2013. LNCS, vol. 7814, pp. 74–85. Springer, Heidelberg (2013)
4. Lin, Y.-R., Margolin, D., Keegan, B., Baronchelli, A., Lazer, D.: #bigbirds never die: Understanding social dynamics of emergent hashtags (2013)
5. Mislove, A., Lehmann, S., Ahn, Y., Onnela, J., Rosenquist, J.N.: Understanding the demographics of twitter users. In: ICWSM (2011)
6. Romero, D.M., Meeder, B., Kleinberg, J.: Differences in the mechanics of information diffusion across topics: Idioms, political hashtags, and complex contagion on twitter. In: Proceedings of the 20th International Conference on World Wide Web, WWW 2011, pp. 695–704. ACM, New York (2011)
7. Simmons, M.P., Adamic, L.A., Adar, E.: Memes online: Extracted, subtracted, injected, and recollected. In: Proceedings of the Fifth International AAAI Conference on Weblogs and Social Media (2011)
8. Yang, J., Leskovec, J.: Modeling information diffusion in implicit networks. In: Proceedings of the IEEE International Conference on Data Mining, ICDM 2010, pp. 599–608. IEEE Computer Society, Washington (2010)

Identification of Occupation Mentions
in Clinical Narratives

Azad Dehghan[1,2,3(✉)], Tom Liptrot[1], Daniel Tibble[1],
Matthew Barker-Hewitt[1], and Goran Nenadic[2,3]

[1] The Christie NHS Foundation Trust, Manchester, UK
azad.dehghan@christie.nhs.uk
[2] School of Computer Science, The University of Manchester, Manchester, UK
gnenadic@manchester.ac.uk
[3] Health e-Research Centre (HeRC), Manchester, UK

Abstract. A patient's occupation is an important variable used for disease
surveillance and modeling, but such information is often only available in
free-text clinical narratives. We have developed a large occupation dictionary
that is used as part of both knowledge- (dictionary and rules) and data-driven
(machine-learning) methods for the identification of occupation mentions. We
have evaluated the approaches on both public and non-public clinical datasets.
A machine-learning method using linear chain conditional random fields trained
on minimalistic set of features achieved up to 88 % F_1-measure (token-level),
with the occupation feature derived from the knowledge-driven method showing
a notable positive impact across the datasets (up to additional 32 % F_1-measure).

Keywords: Information extraction · Natural language processing · Named
entity recognition · Lexical resource · Occupation · Profession

1 Introduction

The identification of occupation mentions in free-text has been highlighted as important
named entity recognition (NER) task for many applications [1]. This is particularly
important in the clinical domain, where patient's profession is important for disease
modeling and surveillance, but yet only available in free-text narratives (clinical notes,
discharge letters, etc.). Occupation information is also considered as personal identi-
fiable information and therefore important to capture in practical applications such as
automated de-identification of clinical documents. However, identification of occupation mentions is a challenging task – a recent community challenge demonstrated that
state-of-the-art NER approaches could merely achieve up to 69 % F_1-measure [2, 3].

In this paper we evaluate two approaches (knowledge- and data-driven) to occupation NER, which rely on a large lexical *Occupation* dictionary that we have engi-
neered semi-automatically. This novel and freely available lexical resource is one of the
main contributions of this paper.

© Springer International Publishing Switzerland 2016
E. Métais et al. (Eds.): NLDB 2016, LNCS 9612, pp. 359–365, 2016.
DOI: 10.1007/978-3-319-41754-7_35

2   Methods

2.1   Data

Two datasets (see Table 1) from different institutions and countries were used:

Table 1. Datasets summary.

Dataset	Document count		Label count (occupation)	
	Training	Held-out	Training	Held-out
Christie	780	520	486	234
i2b2	790	514	265	179

(1) clinical narrative notes and correspondence from the Christie NHS Foundation Trust, United Kingdom (henceforth referred to as the *Christie* dataset); and (2) clinical narratives from the Partners HealthCare obtained via the Informatics for Integrated Biology and the Bedside (i2b2)[1], United States (henceforth referred to as the *i2b2* dataset) [2].

The i2b2 and Christie datasets were annotated using multiple annotators with an entity level Inter Annotation Agreement (IAA) of 71.4 % and 76.6 % F_1-measure respectively. The training datasets were used for development sets by splitting them into validation-training and validation-test sets, and the held-out sets were used for evaluation.

2.2   Natural Language Processing Methods

We have developed both knowledge- (dictionary and rules) and data-driven (machine-learning) methods. Figure 1 provides an outline of the workflow and its components in our methodology, which are explained below.

Pre-processing. The data was pre-processed with GATE [4] and OpenNLP[2] to provide tokenization, sentence splitting, part-of-speech (POS), and shallow parsing or chunking.

Knowledge-Driven Method. This component is made up of a dictionary and a rule tagger, that are used in combination to identify occupation mentions in text.

Dictionary Tagger. This component uses a large-scale dictionary of *Occupations* that contains 19,148 lexical entries. This dictionary has been semi-automatically generated and curated from electronic patient records (The Christie NHS Foundation Trust) and other data sources retrieved from the Office for National Statistics (ONS) in the United

---

[1] https://www.i2b2.org.

[2] https://opennlp.apache.org/.

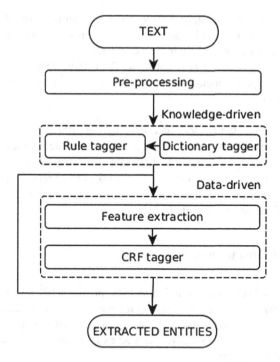

Fig. 1. Occupation NER architecture

Kingdom [5]. This component uses the dictionary with case insensitive and longest match to tag occupation mentions in text.

Rule Tagger. This component includes a set of zoning rules to restrict or reinforce the dictionary tagger to relevant sub-sections and specific contexts. For example, a considerable number of occupation mentions tend to be preceded by specific contextual clues such as "social history", "works as", "job as" or "occupation".

Data-Driven Method. This method uses the preceding components (pre-processing and knowledge-driven) to extract features that are subsequently used by the CRF tagger to tag token sequences that most probable contain occupation mentions.

Feature Extraction. This step extracts the relevant features for the CRF tagger. We derived 31 features from previous work in clinical NER [6, 7] and by using forward/backward feature selection strategies. Specifically, the following token level features were used:

- *Lexical features* (a total of 4 features) include (a) the word/token itself, (b) its stem (using Porter's stemmer [8]), (c) POS, and (d) chunking information.
- *Orthographic features* (2 features) include tokens' characteristics such as (e) word type (i.e., word, number, punctuation, or symbol) and (f) casing (i.e., upper initial, all capital, all lower case, or mixed capitalization).

- *Contextual (lexical and orthographic) features* (24 features) for all aforementioned feature types (a, b, c, d, e, and f) are used: for each token, a context window of two tokens before and two tokens after are considered.
- *Semantic feature* (1 feature) is a binary attribute indicating if a given token was tagged (as an occupation mention) by the knowledge-driven component (dictionary- and rule-taggers).

CRF Tagger. A state-of-the-art sequence labeling algorithm, linear chain conditional random field (CRF, [9]), was trained for each dataset separately with the above feature set to predict occupation mentions in textual data. The data was modeled at the token level using the Beginning, Inside, Outside (BIO) token level representation schema. In addition, the L2 regularization algorithm was adopted together with the default parameters (C = 1.00 and ETA = 0.001).[3]

3   Results and Discussion

Evaluation. We used standard evaluation metrics (precision (P), recall (R), F_1-measure (F)) and two approaches defined by different matching strategies: *token level matching* that requires at least one token of the gold standard span and the predicted/tagged span to match; and *entity level matching* that requires an exact match of the gold standard span with the predicted/tagged span.

The results on the held-out test datasets are given in Table 2. The best results were achieved using the data-driven methodology. Specifically, on the i2b2 dataset, our method achieved 70 % and 76 %, and on the Christie 81 % and 88 % F (entity and token level respectively). Our experiments indicate that the data-driven method seems to be particularly useful for entity level extraction. Overall, better results were achieved on the Christie dataset. This was expected given how the lexical resource was sourced and a difference in label distribution across datasets (more mentions in the Christie dataset). Yet, the results on the i2b2 data are better than previously published work [2, 3].

The dictionary tagger without zoning rules performed worst (Table 2). These results highlight the importance of the rules to optimize the performance of the dictionary tagger. However, more interestingly, the dictionary tagger on its own achieved the highest recall: 96.09 % and 88.68 % (token level) on the i2b2 and Christie datasets respectively. The high recalls achieved across the two different datasets indicate that the developed dictionary has an extensive coverage. The lower recall on the Christie dataset may partly be explained by the fact that some occupation mentions such as *retirement* and *unemployment* were captured by the rule tagger and were not included in the dictionary.

When comparing data- versus knowledge-driven methods, we note that the token level evaluation shows small difference (e.g. 1.10 % on the i2b2 data), while the differences in entity level evaluation, as expected, are larger (e.g. 6.56 % for the i2b2

---

[3] CRF implementation used: https://taku910.github.io/crfpp/.

Table 2. Results on the held-out datasets.

Dataset	Method	Token level matching			Entity level matching		
		P%	R%	F%	P%	R%	F%
Christie	Data-driven	95.96	80.75	87.70	88.34	74.34	80.74
Christie	Knowledge-driven	93.39	80.00	86.18	83.26	71.32	76.83
Christie	Dictionary tagger	29.27	88.68	44.01	20.05	60.75	30.15
i2b2	Data-driven	96.55	62.57	75.93	88.79	57.54	69.83
i2b2	Knowledge-driven	95.65	61.45	74.83	80.87	51.96	63.27
i2b2	Dictionary tagger	28.76	96.09	44.27	28.09	93.85	43.24

dataset). Given the high recall provided by the dictionary tagger, we hypothesize that the entity level results can be further improved using post-processing strategies explored in e.g., [6, 7].

Table 3 shows the evaluation on the held-out datasets when the semantic feature (generated by the knowledge-driven method) is removed from the CRF taggers. The gains of using the knowledge-driven component are 22.63 % and 32.39 % (entity level) and 14.28 % and 20.95 % (token level) for the Christie and i2b2 data respectively, indicating a notable impact and illustrating the potential of hybrid approaches.

Table 3. Results on held-out dataset with no knowledge-driven features.

Dataset	Method	Token level matching			Entity level matching		
		P%	R%	F%	P%	R%	F%
Christie	DD no KD feature	91.06	61.51	73.42	72.07	48.68	58.11
i2b2	DD no KD feature	95.83	38.55	54.98	65.28	26.26	37.45

(Data-driven (DD), Knowledge-driven (KD)).

Moreover, the results on the validation-test datasets (Table 4) show that our methods generalized well as there are no significant differences between the results on the validation-test and held-out test sets. Notably, there are small differences (considering F) between data- and knowledge-driven methods when comparing token level

Table 4. Results on the validation-test datasets.

Dataset	Method	Token level matching			Entity level matching		
		P%	R%	F%	P%	R%	F%
Christie	Data-driven	96.84	82.26	88.95	91.77	77.96	84.30
Christie	Knowledge-driven	96.25	82.80	89.02	87.50	75.27	80.92
i2b2	Data-driven	98.21	64.71	78.01	83.93	55.29	66.67
i2b2	Knowledge-driven	100.0	64.71	78.57	81.82	52.94	64.29

evaluation for the Christie (0.07 %) and i2b2 (0.56 %) datasets with advantage to knowledge-driven methods. This is largely consistent with the held-out test sets (see Table 2) where the smallest differences between methods are also observed in the token level results: the Christie data (1.52 %) and i2b2 data (1.10 %) with advantage to the data-driven methods. The small differences observed and the variable method advantage indicate that the results are inconclusive and further experiments using statistical significance testing would be appropriate to determine whether there is a real difference observed between the methods.

4 Conclusion

In this paper we have evaluated knowledge- and data-driven approaches to identify occupation mentions in clinical narratives. The results show that using a large dictionary as part of a data-driven pipeline enables state-of-the-art performance in identifying occupation mentions in free-text clinical narratives. Notably, the automated methods presented in this paper were comparable to human benchmarks. We propose that future work focuses on boundary adjustment and post-processing strategies in order to improve entity level performance (e.g., [6, 7, 10]). In addition, using two-pass recognition [6] may also improve results by capturing occupation mentions that lack the specific contextual clues modeled.

The methods and data sources presented herein are available online at: https://github.com/christie-nhs-data-science/occupation-ner.

Acknowledgement. We would like to thank Bob Seymour from ONS for his help with finding relevant resources.

References

1. Stubbs, A., Kotfila, C., Uzuner, O.: Automated systems for the de-identification of longitudinal clinical narratives: overview of 2014 i2b2/UTHealth shared task track 1. J. Biomed. Inform. **58**(Suppl.), S11–S19 (2015). ISSN: 1532-0464, http://dx.doi.org/10.1016/j.jbi.2015.06.007

2. Stubbs, A., Uzuner, O.: Annotating longitudinal clinical narratives for de-identification: the 2014 i2b2/UTHealth corpus. J. Biomed. Inform. **58**(Suppl.), S20–S29 (2015). ISSN: 1532-0464, http://dx.doi.org/10.1016/j.jbi.2015.07.020

3. Yang, H., Garibaldi, J.M.: Automatic detection of protected health information from clinic narratives. J. Biomed. Inform. **58**(Suppl.), S30–S38 (2015). ISSN: 1532-0464, http://dx.doi.org/10.1016/j.jbi.2015.06.015

4. Cunningham, H., Tablan, V., Roberts, A., Bontcheva, K.: Getting more out of biomedical documents with GATE's full lifecycle open source text analytics. Comput. Biol. **9**(2), e1002854 (2013). doi:10.1371/journal.pcbi.1002854

5. Office for National Statistics (ONS), the Standard Occupational Classification (SOC) 2010 Index. http://www.ons.gov.uk/ons/search/index.html?newquery=soc2010. Accessed 7 Dec 2015

6. Dehghan, A., Kovacevic, A., Karystianis, G., Keane, J.A., Nenadic, G.: Combining knowledge- and data-driven methods for de-identification of clinical narratives. J. Biomed. Inform. **58**(Suppl.), S53–S59 (2015). ISSN: 1532-0464, http://dx.doi.org/10.1016/j.jbi.2015.06.029
7. Dehghan, A.: Temporal ordering of clinical events (2015). arXiv:1504.03659
8. Porter, F.M.: An algorithm for suffix stripping. Program **14**(3), 130–137 (1980)
9. Lafferty, J.D., McCallum, A., Pereira, F.C.N.: Conditional random fields: probabilistic models for segmenting and labeling sequence data. In: Proceedings of the Eighteenth International Conference on Machine Learning, pp. 282–289 (2001)
10. Dehghan, A., Keane, J.A., Nenadic, G.: Challenges in clinical named entity recognition for decision support. In: 2013 IEEE International Conference on Systems, Man, and Cybernetics, Manchester, pp. 947–951 (2013). doi:10.1109/SMC.2013.166

Multi-objective Word Sense Induction
Using Content and Interlink Connections

Sudipta Acharya[1(✉)], Asif Ekbal[1], Sriparna Saha[1],
Prabhakaran Santhanam[1], Jose G. Moreno[2], and Gaël Dias[3]

[1] Indian Institute of Technology Patna, Patna, India
sudiptaacharya.2012@gmail.com
[2] LIMSI, CNRS, Université Paris-Saclay, 91405 Orsay, France
[3] University of Caen Lower Normandy, Caen, France

Abstract. In this paper, we propose a multi-objective optimization based clustering approach to address the word sense induction problem by leveraging the advantages of document-content and their structures in the Web. Recent works attempt to tackle this problem from the perspective of content analysis framework. However, in this paper, we show that contents and hyperlinks existing in the Web are important and complementary sources of information. Our strategy is based on the adaptation of a simulated annealing algorithm to take into account second-order similarity measures as well as structural information obtained with a pageRank based similarity kernel. Exhaustive results on the benchmark datasets show that our proposed approach attains better accuracy compared to the content based or hyperlink strategy encouraging the combination of these sources.

1 Introduction

Word Sense Induction (WSI) is a crucial problem in Natural Language Processing (NLP), which has drawn significant attention to the researchers during the past few years. WSI concerns the automatic identification of the senses of a known word, which is an expected capability of modern information retrieval systems. In recent times, there are few research works that address this problem by analyzing Web contents and exploring interesting ideas to extract knowledge from external resources. One important work is proposed in [6], which shows that increased performance may be obtained for Web Search Results Clustering (SRC) when word similarities are calculated over the Google Web1T corpus. The authors propose a comparative evaluation of WSI systems by the use of an end-user application such as SRC. The key idea behind SRC system is to return some meaningful labeled clusters from a set of snippets retrieved from a search engine for a given query. So far, most of the works have focused on the discovery of relevant and informative clusters [4] in which the results are organized by topics in such a way as WSI.

In this paper, we present a strategy for WSI based on content and link analyses over Web collections through the use of a Multi-Objective Optimization

E. Métais et al. (Eds.): NLDB 2016, LNCS 9612, pp. 366–375, 2016.
DOI: 10.1007/978-3-319-41754-7_36

(MOO) technique [5]. The underlying idea is grounded on the hypothesis that word senses are related to the distribution of words on Web pages and the way that they are linked together. In other words, Web pages containing the same word meaning should share some similar content-based and link-based values. This study supposes that (1) word distribution is different for each sense, (2) links over the Web express knowledge complementary to content and (3) web domains provide an unique meaning of a word, thus extrapolating the "one sense per discourse" paradigm [7].

Our proposal addresses the WSI problem using a MOO framework that has a different perspective compared to Single Objective Optimization (SOO). In SOO, we concentrate in optimizing only a single objective function, whereas MOO deals with the issue of simultaneously optimizing more than one objective function. Here, we first pose the problem of WSI within the framework of MOO, and thereafter solve this using a simulated annealing based MOO technique called AMOSA [3]. Specifically, we are interested in optimizing the similarity of a cluster of documents in terms of both their content and interlink similarity. The combination of these sources is not straightforward, and this calls for the use of MOO techniques.

The main contributions of this paper are summarized as follows. First, we propose a new MOO algorithm for WSI which automatically determines the number of senses. Next, we propose an evaluation of alternative sources as a combination to improve existing solutions to the WSI problem. Evaluation results show that our MOO based clustering algorithm performs better compared to the hyperlink-based techniques, and very closely compared to approaches with strong content-based solutions when evaluated using WSI measures.

2 Related Works

As far as our knowledge goes, no existing methodology for WSI uses the hyperlink information and content information in an unified setting. The popular techniques either use one or the other sources of information. *Content-based* techniques such as [6] are based on the use of word distributions over the huge collections of n-grams mapped to a graph where the nodes are the words (sense candidates) and the arcs are calculated based on the frequencies in which two words are found together. Once the graph is built the words are grouped together based on simple graph patterns such as curvature clustering or balanced maximum spanning tree clustering, where each obtained cluster represents a sense. Similarly, [8] proposes the use of extra frequency information extracted from Wikipedia and forms groups using a variation of the well known latent dirichlet allocation (LDA) algorithm. Each topic obtained by LDA is considered as a cluster. In contrast, *hyperlink-based* techniques are quite rare. In [11] authors have proposed a technique exclusively based on hyperlink information. There are works where similarity between documents is calculated using a Jensen-Shannon kernel [9] and then clustering is performed by the use of a classical spectral clustering algorithm. Each cluster represents a sense in the solution.

Both approaches manage to adequately discover the word senses in the documents. However, reported results show a superiority of content-based techniques over hyperlink-based techniques.

In the field of application of MOO, as far as we know, within text applications, [12] is the first work, which formulates text clustering as a MOO problem. In particular, they express desired properties of frequent itemset clustering in terms of multiple conflicting objective functions. The optimization is solved by a genetic algorithm and the result is a set of Pareto-optimal solutions. But, towards solution of WSI problem, according to best of our knowledge, this is the very first attempt that utilises MOO based clustering approach.

3   Combining Content and Hyperlink Approaches with a MOO

Most of the existing SRC techniques are based on a single criterion which reflects a single measure of goodness of a partitioning. However, one single source of information may not fit all problems. In particular, [2] have shown the utility of hyperlink information to cluster Web documents, whereas [11] have shown their applicability to the WSI problem. Manifestly, both SRC and WSI have been addressed by the analysis of document contents. Moreover, an effective combination of these two has not been yet proposed. Hence, it may become necessary to simultaneously optimize several cluster quality measures which can capture different content-based or hyperlink-based characteristics. In order to achieve this, MOO can be an ideal platform and therefore we pose the problem of finding word senses within this framework. Therefore, the application of sophisticated MOO techniques seems appropriate and natural.

Content Compactness based on SCP/PMI Measure: This type of indices measures the proximity among the various elements of the cluster. One of the commonly used measures for compactness is the variance. Documents are kept together if the distribution of words is similar. This measurement is based on either Symmetric Conditional Probability (SCP) or Pointwise Mutual Information (PMI), defined on Sect. 4.2.

Content Separability based on SCP/PMI Measure: This particular type of indices is used in order to differentiate between two clusters. Distance between two cluster centroids is a commonly used measure of separability. This measure is easy to compute and can detect hyperspherical-shaped clusters well. Word-distribution between senses must be as different as possible. This measurement is also based on computing either SCP or PMI.

Hyperlink Compactness: Well-connected Web documents must belong to the same sense cluster.

Hyperlink Separability: The number of interlinks between senses must be as small as possible.

4 MOO Based Clustering for WSI

In order to perform MOO based clustering we adapt Archived multi-objective simulated annealing (AMOSA) [3] as the underlying optimization strategy. It incorporates the concept of an archive where the non-dominated solutions seen so far are stored. Steps of the proposed approach are described in the following sections. For better understanding we also show the various steps of our proposed algorithm in Fig. 1.

4.1 Archive Initialization

As we follow an endogenous approach, only the information returned by a search engine is used. In particular, we only deal with web snippets and each one is represented as a word feature vector. So, our proposed clustering technique starts its execution after initializing the archive with some random solutions as archive members. Here, a particular solution refers to a complete assignment of web snippets (or data points) in several clusters. So, the first step is to represent a solution compatible with AMOSA, which represents each individual solution as a string. In order to encode the clustering problem in the form of a string, a center-based representation is used. Note that the use of a string representation facilitates the definition of individuals and mutation functions [3].

Let us assume that the archive member i represents the centroids of K_i clusters and the number of tokens in a centroid is p^1, then the archive member has length l_i where $l_i = p \times K_i$. To initialize the number of centroids K_i encoded in the string i, a random value between 2 and K_{max} is chosen and each of the K_i centroids is initialized by randomly generated token from the vocabulary.

4.2 Content-Based Similarity Measure

In order to compute the similarity between two Web documents (word vectors) we have used two well known content-based similarity metrics, the
$SCP(W_1, W_2) = \frac{P(W_1, W_2)^2}{P(W_1) \times P(W_2)}$ and $PMI(W_1, W_2) = \log(\frac{P(W_1, W_2)}{P(W_1) \times P(W_2)})$.

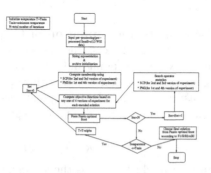

Fig. 1. Flowchart of proposed methodology

---

[1] A centroid is represented by a p word feature vector $(w_k^1, w_k^2, w_k^3, \ldots, w_k^p)$.

We have computed the SCP and PMI values between each pair of words in the global vocabulary (V), i.e., the set of different tokens in the list of all Web documents.

4.3 Hyperlink-Based Similarity Measure

Given a collection of Web documents relevant to a sense, we calculate their corresponding pagerank values pr_{d_i}. Similarity between documents is calculated through a kernel function between pagerank values. Specifically, we use the Jensen-Shannon kernel proposed by [9]: $k_{JS}(d_i, d_j) = ln2 - JS(d_q^i, d_q^j)$, where the $JS(d_i, d_j)$ value is defined under the hypothesis that each hypertext document has a probability distribution with two states: whether or not they are selected by a random walk. Following the proposal for pagerank similarities defined by [11], we calculate the similarity values between two Web pages as shown in Eq. 1

$$JS(d_i, d_j) = \frac{1}{2} \left[p_{d_i}^r ln \left(\frac{2 * p_{d_i}^r}{p_{d_i}^r + p_{d_j}^r} \right) + p_{d_j}^r ln \left(\frac{2 * p_{d_j}^r}{p_{d_i}^r + p_{d_j}^r} \right) \right]. \tag{1}$$

These similarity measures (two content-based and one hyperlink-based) are used in our proposed MOO based clustering framework.

4.4 Assignment of Web-Snippets and Objective Function Calculations

After initializing the archive the first step concerns the assignment of n word vectors or points (where n is the total number of Web snippets in a particular query) to different clusters. This assignment can be done using any one of content-based similarity measurement techniques (SCP or PMI). In the second step, we compute two cluster quality measures, cluster compactness and separation, by varying similarity computation (either content-based or hyperlink-based) and use them as objective functions of the string. Thereafter we simultaneously optimize these objective functions using the search capability of AMOSA.

Assignment of Web-Snippets and Updation of Centroids. In this technique, the assignment of points to different clusters is done based on the content similarity measurement between that point and different cluster centroids. Each Web document is assigned to that cluster center with respect to which it has the maximum similarity measure. In particular, any point j is assigned to a cluster t whose centroid has the maximum similarity to j using:

$$t = argmax_{k=1,...K} S(\overline{x}_j, \overline{m}_{\pi_k}). \tag{2}$$

K denotes the total number of clusters, $\overline{x}_j$ is the jth point (or Web document), $\overline{m}_{\pi_k}$ is the centroid of the kth cluster π_k and $S(\overline{x}_j, \overline{m}_{\pi_k})$ denotes similarity measurement between the point $\overline{x}_j$ and cluster centroid $\overline{m}_{\pi_k}$.

One possible way to compute the similarity between two word vectors is defined by Eq. 3, which is inspired by [1][2].

$$S(d_i, d_j) = \sum_{k=1}^{\|d_i\|} \sum_{l=1}^{\|d_j\|} SCP(W_{ik}, W_{jl}) \tag{3}$$

Here $\|d_i\|$ and $\|d_j\|$ respectively denote, total number of words in word vectors d_i and d_j. After assigning all Web snippets to different clusters, the cluster centroids encoded in that string are updated. For each cluster, p number of words from global vocabulary which are most similar to other documents of that particular cluster are chosen to form new centroid of that cluster.

Objective Functions. In order to compute the goodness of the partitioning encoded in a particular string, cluster compactness and separability are usually used as the objective functions in MOO clustering. The objective functions quantify two intrinsic properties of the partitioning. First, compactness is defined in Eq. 4 and it is maximized.

$$COM = \sum_{k=1}^{K} \sum_{x_i \in \pi_k} S(x_i, m_{\pi_k}) \tag{4}$$

Here m_{π_k} is the cluster centroid of the kth cluster consisting of p words $(w_1^{\pi_k}, ..., w_p^{\pi_k})$, K is the number of clusters encoded in that particular string and $S(x_i, m_{\pi_k})$ value is computed using Eq. 3. In Eq. 3, SCP can be replaced by PMI. Also cluster compactness can be measured using hyperlink similarity as given in Eq. 1. The compactness using hyperlink similarity is computed by the following equation:

$$COM = \sum_{k=1}^{K} \frac{\sum_{x_i, x_j \in \pi_k} JS(x_i, x_j)}{|\pi_k|} \tag{5}$$

Hence, total three different versions of cluster compactness can be computed by varying the similarity measures. Note that if words in a particular cluster are very similar to the cluster centroid and documents are highly interconnected then the corresponding COM value would be maximized. Also for hyperlink similarity based compactness if all the documents of any particular cluster are highly interconnected to each other then also corresponding COM gets maximized. Here our target is to form good clusters whose compactness in terms of similarity should be maximum.

The second objective function is the cluster separation which measures the dissimilarity between two given clusters. Purpose of any clustering algorithm is to obtain compact similar typed clusters which are dissimilar to each other. Here we have computed the summation of similarities between different pairs

---

[2] SCP could be replaced by PMI.

of cluster centers and then minimized this value just to produce well-separated clusters. The separation is defined in Eq. 6.

$$SEP = \sum_{k=1}^{K} \sum_{o=k+1}^{K} S(m_{\pi_k}, m_{\pi_o}) \tag{6}$$

Here m_{π_k} and m_{π_o} are the centroids of the clusters π_k and π_o, respectively. $S(m_{\pi_k}, m_{\pi_o})$ value is computed using Eq. 3.

Similar to compactness, SCP or PMI based similarity measure can be used to compute separatibility. The process to compute SEP value using hyperlink-based similarity measure for a string is given in Eq. 7.

$$SEP = \sum_{k=1}^{K} \sum_{o=k+1}^{K} \forall_{x_i \in k, x_j \in o} min(JS(x_i, x_j)) \tag{7}$$

It shows that the sum of maximum distance (i.e., minimum similarity) between the documents of all possible pairs of clusters in a string is represented as the separability measure. Minimizing this value represents well separated clusters.

Therefore, similar to compactness, separation SEP can be calculated in three different ways by varying the similarity measures. Out of total six compactness and separation based objectives any combination of them can be used. These objective functions are maximized using the search capability of AMOSA.

4.5   Search Operators

As mentioned earlier, the proposed clustering technique uses a multi-objective simulated annealing based approach as the underlying optimization strategy. As a simulated annealing step, we have introduced three types of mutation operations as used in [1]. These mutation operations can update, increase or decrease the size of a string. During smilarity measurement in mutation operations either SCP or PMI similarity matrix is used. In order to generate a new string any one of the above-mentioned mutation types is applied to each string with equal probability.

5   Experimental Setup

Dataset: In our experiments the SemEval13 Word Sense Induction dataset [13] was used. In brief, it is composed of 100 queries extracted from AOL query log dataset which has a corresponding Wikipedia disambiguation page. Each query has 64 web results classified in one of the senses proposed in the Wikipedia article. However, the Web results do not include the PageRank values. For that, we use the Hyperlink Graph publicly available in [10]. Each Web result is reduced to a Pay-Level-Domain (PLD) Graph and a PageRank value is assigned after calculating all of them for the entire PLD Graph. The HyperLink Graph is composed of more than 43 million PLD values and less than 1.3 % of the URLs of the SemEval13 dataset were not found. For these cases, the lowest PageRank value was assigned to avoid zero values. To evaluate the cluster quality, we selected the same SemEval13 metrics: F_1-measure (F1), RandIndex (RI), Adjusted RandIndex (ARI) and Jaccard coefficient (JI).

Baselines: As baselines, we use the well-known Latent Dirichlet Allocation (LDA) technique over the documents. This technique has been reported as suitable for this task [13]. All parameters were selected to guarantee the best performance of the algorithm. As a non-content baseline, we use the results reported by [11].

6 Results and Discussions

We execute our proposed MOO clustering technique on the SemEval2013 dataset [13]. The parameters of the proposed clustering technique are as follows: $T_{min} = 0.001$, $T_{max} = 100$, $\alpha = 0.9$, $HL = 30$, $SL = 50$ and $iter = 15$. They were determined after conducting a thorough sensitivity study. We perform experiments in four different ways. In the first version, we consider total four objectives: (i) SCP based compactness, (ii) SCP based separability, (iii) hyperlink or pagerank (PR) based compactness and (iv) PR based separability. For assigning points to different clusters and also to calculate similarity values during objective function calculation, SCP matrix is used. In the second version of our experiments, we use four objective functions: (i) PMI based compactness, (ii) PMI based separability, (iii) PR based compactness and (iv) PR based separability. In this version PMI based similarity measure is used for computing the membership matrix and

Table 1. Results over the SemEval13 WSI dataset.

Algorithm	Parameter	F1	JI	RI	ARI
MOO(SCP,PR)	5	0.618	**0.347**	0.604	0.096
	10	0.679	0.332	0.605	0.128
MOO(PMI,PR)	5	0.646	0.352	0.604	0.128
	10	0.668	0.339	**0.628**	0.118
MOO(SCP)	5	0.613	0.334	0.569	0.095
	10	0.644	0.329	0.609	0.120
MOO(PMI)	5	0.628	0.343	0.540	0.048
	10	0.630	0.330	0.552	0.059
Hyperlink baseline	5	0.609	0.210	0.605	0.079
	10	0.646	0.159	0.626	0.082
Content baseline	LDA-5	0.657	0.234	0.621	**0.151**
	LDA-10	**0.716**	0.168	0.626	0.131

objective functions. In the third version, we use two objective functions: (i) SCP based compactness and (ii) SCP based separability. In the fourth version we use two objective functions: (i) PMI based compactness (ii) PMI based separability.

Results are reported in Table 1. In the table second column (parameter) represents the number of clusters in corresponding version of our experiments. From the results it is evident that for all the validity metrices the first version performs better compared to the third version. It implies that inclusion of hyperlink information makes the clustering algorithm more efficient. Similarly, second version performs better than the fourth version in all aspects. Results also show that both the first two versions of the proposed algorithm (using SCP and PMI based similarity measures, respectively) perform better compared to the approach reported in [11], where only hyperlink information was used. However, the similar situation was not observed when the results of the proposed approach are compared with the content-based baseline. It is important to note that LDA is a strong baseline, and our algorithm shows slight under-performance for F1 (5 %) and RI (0.01 %). It is more significant for ARI (15 %) and on the other hand, MOO outperforms LDA by 30 % in terms of JI. Clearly, the use of content has helped in the sense identification, but fails to contribute to their maximum as it is obtained by the use of LDA. Moreover as mentioned in [3], selecting appropriate combination of parameters is very important for the good performance of a particular MOO based approach. Thus a proper sensitivity study is required to conduct to choose the correct values of parameters. In the current approach the same set of parameters as used in [1] is used. But as the targeted task is more complex compared to [1], it will be more interesting to conduct the sensibility analysis further.

7   Conclusion

In this paper, we have formulated the problem of WSI within the framework of MOO that combines atypical mixed sources of information. Our proposed approach differs from related work as clustering is performed over multiple objective functions that take into account document content and hyperlink connections. As far as we know, this is the first attempt towards this research direction in WSI studies.

In particular, we proposed the use of similarity metrics based on the frequencies of words in documents (SCP and PMI) to evaluate the content similarity and the use of a Jensen-Shannon kernel function based on PageRank to compute the Web pages interconnectivity. Four cluster indices are proposed to guide the optimization process. Results show that the combination of these two different sources outperforms clustering techniques that relay on just one.

References

1. Acharya, S., Saha, S., Moreno, J.G., Dias, G.: Multi-objective search results clustering. In: Proceedings of COLING 2014, the 25th International Conference on Computational Linguistics, pp. 99–108 (2014)

2. Avrachenkov, K., Dobrynin, V., Nemirovsky, D., Pham, S., Smirnova, E.: Pagerank based clustering of hypertext document collections. In: Proceedings of the 31st Annual International Conference on Research and Development in Information Retrieval (SIGIR), pp. 873–874 (2008)
3. Bandyopadhyay, S., Saha, S., Maulik, U., Deb, K.: A simulated annealing-based multiobjective optimization algorithm: amosa. IEEE Trans. Evol. Comput. **12**, 269–283 (2008)
4. Carpineto, C., Osinski, S., Romano, G., Weiss, D.: A survey of web clustering engines. ACM Comput. Surv. **41**(3), 1–38 (2009)
5. Deb, K.: Multi-objective Optimization Using Evolutionary Algorithms. Wiley, New York (2009)
6. Di Marco, A., Navigli, R.: Clustering and diversifying web search results with graph-based word sense induction. Comput. Linguist. **39**(4), 1–43 (2013)
7. Gale, W., Church, K., Yarowsky, D.: One sense per discourse. In: Proceedings of the Workshop on Speech and Natural Language (HLT), pp. 233–237 (1992)
8. Lau, J.H., Cook, P., Baldwin, T.: Unimelb: topic modelling-based word sense induction for web snippet clustering. In: Proceedings of the 7th International Workshop on Semantic Evaluation (SemEval 2013), June 2013
9. Martins, A., Smith, N., Xing, E., Aguiar, P., Figueiredo, M.: Nonextensive information theoretic kernels on measures. J. Mach. Learn. Res. **10**, 935–975 (2009)
10. Meusel, R., Vigna, S., Lehmberg, O., Bizer, C.: Graph structure in the web - revisited. In: Proceedings of the International World Wide Web Conference (WWW), pp. 1–8 (2014)
11. Moreno, J.G., Dias, G.: Pagerank-based word sense induction within web search results clustering. In: Proceedings of the 14th ACM/IEEE-CS Joint Conference on Digital Libraries, pp. 465–466 (2014)
12. Morik, K., Kaspari, A., Wurst, M., Skirzynsk, M.: Multi-objective frequent termset clustering. Knowl. Inf. Syst. **30**(3), 715–738 (2012)
13. Navigli, R., Vannella, D.: Semeval-2013 task 11: word sense induction & disambiguation within an end-user application. In: Proceedings of the International Workshop on Semantic Evaluation (SEMEVAL), pp. 1–9 (2013)

Re-expressing Business Processes Information from Corporate Documents into Controlled Language

Bell Manrique-Losada[1], Carlos M. Zapata-Jaramillo[2],
and Diego A. Burgos[3(⊠)]

[1] Universidad de Medellín, Medellin, Colombia
bmanrique@udem.edu.co
[2] Universidad Nacional de Colombia, Medellin, Colombia
cmzapata@unal.edu.co
[3] Wake Forest University, Winston-Salem, USA
burgosda@wfu.edu

Abstract. In this paper, we propose a top-down approach for converting business processes information from corporate documents into controlled language. This proposal is achieved with a multi-level methodology. We first characterize document structure by using rhetorical analysis to determine relevant sections for information extraction. Then, a verb-centered event analysis is performed to start defining the typical patterns featured by business processes information. Lastly, morpho-syntactic and dependency parsing is carried out for extracting this information. This multi-level knowledge is used to define rules for converting the extracted sentences into a controlled language, which is intended to be used in software requirements elicitation.

Keywords: Business processes · Controlled language · Rhetorical analysis · Information extraction

1 Introduction

Business knowledge and stakeholder needs regarding the development of a software product are analyzed and specified by means of requirements elicitation[1]. Traditionally, requirements elicitation has been carried out by human analysts by using interviews, observations, questionnaires, etc. Sometimes, the information obtained is converted by the analyst to a controlled language, which is used in further stages of the software implementation. This approach to requirements analysis, however, increase costs and imply a certain degree of subjectivity [1]. Alternatively, human analysts elicit requirements by hand from corporate documents. In this paper, we propose a multi-level methodology for automatically obtaining information about business

---

[1] This work has been partly funded by the University of Medellin's Research Vice-provost's Office, Wake Forest University, and National University of Colombia, under the project: *"Defining a Specific-Domain Controlled Language: Linguistic and Transformational Bases from Corporate Documents in Natural Language"*.

E. Métais et al. (Eds.): NLDB 2016, LNCS 9612, pp. 376–383, 2016.
DOI: 10.1007/978-3-319-41754-7_37

processes from corporate documents in English. A business process can be defined as an action in a corporate environment. This action—represented in a text by verbs or deverbal nouns—generally involves an agent and an object. The way process components relate to each other will be described in Sect. 3.

The methodology proposed here spans from rhetorical analysis to semantic and morpho-syntactic parsing. The rhetorical analysis determines the sections were information about business processes prevail. Then, a verb-centered event analysis defines the typical patterns of such information. Lastly, a morpho-syntactic and dependency parsing is combined with previous levels for extracting information satisfying such patterns. This multi-level knowledge is used to define rules for converting the extracted sentences into a controlled language[2] for the requirements elicitation process.

This paper is structured as follows. In Sect. 2 we review related key concepts and work on knowledge extraction and acquisition. In Sect. 3 we present our proposal. In Sect. 4 we describe how evaluation was carried out, and Sect. 5 draws some conclusions on the present work.

2 Related Work

Related to semi-structured document processing, we found the following contributions: the RARE project [3] is focused on parsing texts based on a semantic network assisted by a thesaurus. Concerning requirement texts, Cleland-Huang *et al.* [4] work on detection of viewpoints. Bajwa *et al.* [5] propose mapping business rules to semantic vocabulary. Meth *et al.* [6] focus on providing automated and knowledge based support for the elicitation process. These approaches are based on structured or semi-structured documents, but no experiments were developed for technical documents. Young and Antón [7] propose the identification of requirements by analyzing the commitments, privileges, and rights conveyed within online policy documents. The usage of corpora is a suitable means for describing and analyzing texts, as Wang [8] promotes for using classification rules based on set theory and corpus. In knowledge engineering, Dinessh *et al.* [9] propose the validation of regulations from organizational procedure texts by using formalizations. These authors depart from our approach in that they define the stakeholder behavior by means of phrases previously created. Several approaches for identifying domain knowledge from documents have been proposed including the following: techniques for formalizing business process from organizational domains [10], software design [11], and knowledge mapping [12]. Likewise, some techniques and methods for knowledge representation in requirement analysis and elicitation based on work processes specification are presented [13]. Most of the aforementioned studies report techniques to represent domain knowledge, and more particularly, for designing scenario maps and objective diagrams. From the knowledge engineering perspective, we find useful techniques to be used in the requirements elicitation process—such as business process modeling, semantic nets, and knowledge diagrams—to facilitate domain understanding.

---

[2] A controlled natural language is a sub-language of the corresponding natural language [2].

3   Re-expressing Business Processes Information

3.1   Rhetorical Analysis

Rhetorical analysis (RA) is concerned with discourse construction, prioritizing on the communicative purpose of each genre [14]. Rhetoric supports discourse from its intentional and instrumental perspective. RA is a discursive approach to carry out the structural analysis of a document. From RA, document analysis is discussed in terms of rhetorical moves, which refer to the functional parts or sections of a document corresponding to a specific genre. This approach for studying a particular genre comprises the analysis of a text and its description in terms of rhetorical structure sections (*i.e.*, moves). The particular configuration of the text surface is defined in terms of levels of text organization, also known as rhetorical discourse organization. The macro units identified by genre analysis can be characterized as moves (*i.e.*, document macro-sections) and steps (*i.e.*, sections), which are larger than clauses, complex clauses, and sentences. Also, macro units possess some unity grounded in a common function or meaning.

We define the rhetorical organizational model (ROM) in this study as follows. The corpus construction starts by collecting possible technical documents on the macro-genre circulating on the web. We broadly explore four types of technical documents, and collect and analyze a corpus of Standard Operating Procedures—SOP— for this case study. A SOP is a constitutive document of a quality system describing a set of recurring operations for illustrating how corporate policies are effectively implemented. We selected a sample of 32 documents from the corpus corresponding to 64 % of the total population. This is the minimum percentage statistically randomized, calculated with the Z-test of proportions. Then, we conducted the rhetorical analysis based on this sample. The preprocessing carried out on the corpus consisted in tokenization, keyword and stop-word identification, and creation of word frequency lists, among others. The corpus contains 9,252 word types and 167,905 tokens. No references were found in the literature regarding SOP models, so we follow Burdiles [15] and use an inductive method for defining a preliminary model of this kind of document: (i) We randomly select four sample documents from the corpus; (ii) We develop an incremental construction of a preliminary model from a by-hand review of the document structure and superstructure, in order to identify the common organization units (*i.e.*, moves) in the sample; (iii) Incrementally, we define rhetorical moves as functional sections of a genre [16], and according to the macro-move concept from Parodi [17], so that each macro-move serves a communicative purpose and all macro-moves shape the overall organization of the text; (iv) By analyzing the functional organization of the document, we identify macro-purposes, which comprise a set of more specific moves; (v) We define a preliminary model as a set of functional and structural features, which results from the identification of recurrent moves. In the reference model, we consider the most frequent moves, which comprise three macro-moves containing in turn 19 moves—showing more specific functional units—as we show in Table 1 (detailed moves are described by Manrique [18]).

Table 1. Rhetorical organization model

Code	Rhetorical unit type	Rhetorical unit name
1	Macromove	**Presenting the SOP:** preliminary statement presenting an introduction of the document, describing the document purpose, conventions, revision schedule, approval authority, document organization, etc.
2	Macromove	**Developing procedures:** Presents the procedures associated with each organizational process in detail. A series of specific purposes, responsibilities and functions, and procedural descriptions are defined
3	Macromove	**Ending the SOP:** Related to the moves I and II. It is optional, but it is intended to supplement the development macro-move

3.2 Event Identification

According to Pivovarova *et al.* [19], in the context of information extraction, events represent real-world facts, which should be extracted from plain text. Since events are unique, they receive in-depth attention in current research by trying to identify what events are mentioned within texts and how they are semantically related [20].

We orient our approach to event identification on text expressions referring to real-world events—also called event mentions—[21] from a set of clusters. Documents are analyzed on the basis of lexical chains defined by a set of semantically related words of given sentences. WordNet was used for constructing lexical chains with event mentions. A set of features and properties for each event is identified for characterizing the specific genre linked to the technical document. The preliminary results, in the form of events features and properties, are the basis of a processing module in an automated system for text processing. Then, this event-centered processing supports the identification of organizational domain knowledge and of business information, which constitute the first instances of the requirements elicitation process.

3.3 Morpho-Syntactic and Dependency Parsing

As described above, the most frequent verbs are classified by categories according to Vossen [22]. Then, we identify patterns for each verb type. Such patterns—informed by rhetorical, morphological, and lexical features—are the basis of rules for inferring and extracting business processes information from corporate documents. This way, we focus on the analysis of verbs regarding their usage in the SOPs. Based on the feature identification of verbs, we propose a set of rules for transforming each feature into a controlled language. We intend to take these sentences to the UN-Lencep[3] language, which is an intermediate representation between natural language and conceptual models for software engineering.

---

[3] UN-Lencep is the Spanish acronym for '*National University of Colombia—Controlled language for the specification of pre-conceptual models*.

In Table 2 we present the rules defined for such mapping. Each mapping rule is assigned to one category (first column), expressed in terms of its pattern in the SOP [If Pattern] and the generated expression in UN-Lencep [then (→) Expression]. Additionally, in the second row of each attribute (features conditioning each element of the pattern) we related the tags (*e.g.*, syntactic or semantic tag—*synt*—, function tag—*func*—, etc.) assigned by the parser[4]. In the third row of each rule, we included an example of a phrase matching the pattern and the resulting expression in UN-Lencep.

Table 2. Rules for sentence re-expression in controlled language

Rule Name	Description of Mapping Rules	
	Pattern in the SOP	Expression in UN-Lencep
i. Transitive verbs	If *Obj₁ + VB + Obj₂* →	*Obj₁ + VB + Obj*
	Obj₁ = func:ncsubj, tag:NN *VB* = synt:sv, tag:VB|VBZ|VBP	*Obj₂* = func:dobj, tag:NN
	Manager performs data analysis	*Manager+perform+analysis*
ii. Transitive verbs without inmediate subject	If *VB + Obj₁* →	*X + VB + Obj₁*[a]
	Obj₁ = func:dobj, tag:NN	VB = synt:sv, ag:VB|VBZ| VBD|VBP
	Assess the request	*X+assess+request*
iii:Passive voice	If *Obj₁ + [to be+VBN] +by+ Obj₂* →	*Obj₂ + VB + Obj₁*
	Else If *Obj₁ + [to be+VBN]* →	*X + VB + Obj₁*
	Obj₁,Obj₂ = func:dobj, tag:NN| NNS	[.] = tag:VBZ+VBN
	the *procedure* is *applied* by *Sector*	*Sector+apply+procedure*
iv.Construction of the form 'is a'	If *Obj₁ +[is (a|an)] + Obj₂* →	*Obj₁ + is + Obj₂*
	Obj₁,Obj₂ = func:dobj|ncsubj, tag:NN| NNS	[.] = tag:VBZ+VBN
	...*Internal Controls* is a *document*...	*Internal_control+is+document*
v. Noun phrase with post-modification	If *Obj₁ + [of the | of a] + Obj₂* →	*Obj₂ + has + Obj₁*
	Obj₁ & Obj₂ = func:dobj	[.] = tag:IN
	...is the *responsibility* of the *Sector*...	*Sector+has+responsibility*
vi. Noun phrase with pre-modification	If *[JJ | VBN] + Obj* →	*Obj + has + Attribute [value: JJ | VBN]*
	Obj = func:dobj|ncsubj	tag:NN|NNS, synt: sn-chunk|n-chunk
	JJ,VBN = func:ncmod, synt:attrib,	tag:JJ|VBN
	...Agency to manage the *electronic data*...	*data+has+Attribute [value:electronic]*
	...the *requested product* belongs...	*Product+has+Attribute value:requested]*
vii. Main clause + infinitive phrase	If *VB₁ + [to | for] + VB₂ + Obj* →	*VB₁ => VB₂ + Obj*
	Else If *VB₁ + [to] + VB₂* →	*VB₁ => VB + Y*
	VB₁ = synt:sv, tag:VBN| VB|VBZ	[.] = tag:TO|IN, func:iobj
	VB = func:cmod, synt:inf, tag:VB|VBG|VBZ|VBP	Obj = func:dobj, tag:NN|NNS
	...it is *used to classify event*...	*Use => classify+event*

[a]This rule corresponds to document related features, *e.g.*, when an agent is stated early in the document and a list of duties for him/her is presented later with a verb in the infinitive form.

The output expression in the rules shows what the parser and the semantic processor generate in the application of each rule (where the pattern matches the sentence). The parser applies multiple rules to the same sentence, so the output can be complemented by successive rules. Since most of the relevant verbs are either copulative or transitive, we initialize the information extraction process with the macro-rules below, which summarize the rules in Table 2.

---

[4] We used Freeling (http://nlp.lsi.upc.edu/freeling/) for dependency parsing.

- Copulative verbs: *Subject + copulative verb + attribute*
- Transitive verbs: (including those that take a prepositional complement): *Subject + verb chunk + noun chunk or prepositional chunk*
- Transitive verbs: *Subject + verb chunk + THAT clause*

4 Evaluation

4.1 Rhetorical Organization Model

We determined that rhetorical moves are crucial for writing/analyzing a SOP, as a part of the methodology based on corpus linguistics. We pursue the evaluation activities as follows: (i) Selecting the experts for the evaluation of the reference model; (ii) Designing a template and an instruction guide for the evaluation; (iii) Filling in the evaluation by selected experts; (iv) Analyzing and filtering the results of the evaluation, based on an inter-rater reliability analysis. Thus, we generate a new version of the model for the subsequent analysis, comprising only the moves considered mandatory by the experts, and the adjustments and changes by move in the necessary cases.

4.2 Sentences in Controlled Language

We compared the sentences in controlled language that an expert generates on the one hand, and the ones generated by our prototype on the other hand. We perform a preliminary assessment in terms of the potential business process sentences being extracted, the number of relevant business process sentences extracted by the expert and by the prototype, and the number of irrelevant business process sentences extracted. We measured precision (P) and recall (R) [23]. When comparing the general results between an expert and our prototype (85 vs. 77.5) compared to the potential sentences (n = 127), we obtained: R = 0.9118 and P = 0.939.

We computed the harmonic mean of P and R (*i.e.*, the F-measure) to determine efficiency. Usually beta is 0.5, by weighting precision and recall as equally important. In this way, the F-measure obtained by the prototype performance—regarding the ideal or potential relations—is 0.739, but regarding the expert mapping it is 0.925. The evaluation results in terms of precision, recall, and F-measure are shown in Table 3.

When the number of retrieved documents grows, recall increases since it is sorted according to relevance. The human output variable 'Mapping Time' directly depends on the expertise level and the familiarity with the mapping rules. The first run of the mapping took 45 min and the second one 30 min. It can be expected that the mapping

Table 3. Results of extracted controlled sentences (expert vs. prototype vs. potential)

	Exp. vs. pot.	Prot. vs. exp.	Prot. vs. pot.
Recall	0,66929134	0,91176471	0,61023622
Precision	0,94444444	0,93939394	0,93939394
F-measure	0,78341014	**0,92537313**	0,7398568

time decreases with periodical training of experts. However, this will not get close to the prototype mapping time (between 40 and 100 s).

5 Conclusions

In this paper we proposed a methodology for re-expressing business processes information from corporate documents into a controlled language. This approach aims at extracting knowledge from business-related technical documents. Our approach is based on a rhetorical analysis of discourse and natural language processing (NLP) methods, which take each document as input. We are promoting the relevant role of elicitation techniques based on document analysis as sources of domain knowledge and business information within the requirements elicitation process.

We describe how business process information from corporate documents is potentially useful for identifying domain knowledge following an elicitation technique based on document analysis. We follow a multi-level processing methodology comprising: rhetorical analysis, verb-centered event analysis, and morpho-syntactic and dependency parsing. All together, these processes identify and extract sentences from a natural language document to be translated into a controlled language.

By analyzing the processing results yielded by the prototype, some important findings arise: the linguistic quality of the input document affects processing performance; a refinement of the regular expressions for pre-processing is needed (regarding the treatment of lists, bullets, and non-textual elements in the source text, among others); design of new complex processing rules should be considered, *e.g.*, compound adjectives, complex noun phrases, past participles with gerund, modal structures, and phrasal verbs with long predicates. The evaluation of the prototype was presented in the form of percentage of correctness, *i.e.*, precision and recall.

As future work, we are interested in: (i) increasing the kind and the number of documents in the corpus and refining the study of lexical and semantic features; (ii) considering statistical association measures for reinforcing term identification and pattern extraction in the context of knowledge acquisition; (iii) validating extensively the processing methodology; and (iv) systematically addressing the challenges and limitations posed by the task such as text ambiguities and implicit agents.

References

1. Manrique-Losada, B., Burgos, D.A., Zapata-Jaramillo, C.M.: Exploring MWEs for knowledge acquisition from corporate technical documents. In: 9th Workshop on Multiword Expressions -MWE 2013, NAACL 2013, Atlanta, July 2013
2. Fuchs, N.E., Schwitter, R.: Specifying logic programs in controlled natural language. Technical report IFI 95.17, University of Zurich (1995)
3. Cybulski, J.L., Reed, K.: Requirements classification and reuse: crossing domain boundaries. In: Frakes, W.B. (ed.) ICSR 2000. LNCS, vol. 1844, pp. 190–210. Springer, Heidelberg (2000)

4. Cleland-Huang, J., Marrero, W., Berenbach, B.: Goal-centric traceability: using virtual plumblines to maintain critical systemic qualities. Trans. Soft. Eng. **34**, 685–699 (2008)
5. Bajwa, I.S., Lee, M., Bordbar, B.: SBVR business rules generation from natural language specification. In: AAAI Spring Symposium, pp. 2–8. AAAI, San Francisco (2011)
6. Meth, H., Li, Y., Maedche, A. Mueller, B.: Advancing task elicitation systems–an experimental evaluation of design principles. In: Proceedings of 33rd International Conference on Information Systems, pp. 54–68. AISEL, Florida (2012)
7. Young, J.D., Antón, A.I.: A method for identifying software requirements based on policy commitments. In: 18th International Requirements engineering Conference, pp. 47–56. IEEE, Sydney (2010)
8. Wang, F.H.: On acquiring classification knowledge from noisy data based on rough set. Expert Syst. Appl. **29**(1), 49–64 (2005)
9. Dinesh, N., Joshi, A., Lee, I. Webber, B.: Extracting formal specifications from natural language regulatory documents. In: ICoS-5, Buxton (2006)
10. Vegega, C., Amatriain, H., Pytel, P., Pollo, F., Britos, P., García, R.: Formalización de Dominios de Negocio basada en Técnicas de Ingeniería del Conocimiento para Proyectos de Explotación de Información. In: Proceedings of IX JIISIC, pp. 79–86. PUCP, Lima (2012)
11. Aysolmaz, B., Demirors, O.: Modeling business processes to generate artifacts for software development: a methodology. In: Proceedings of the 6th International Workshop on Modeling in Software Engineering, pp. 7–12. ACM, New York (2014)
12. Hao, J., Yan, Y., Gong, L., Wang, G., Lin, J.: Knowledge map-based method for domain knowledge browsing. Decis. Support Syst. **61**, 106–114 (2014)
13. Tavares, V., Santoro, F.M., Borges, M.R.S.: A context-based model for knowledge management embodied in work processes. Inf. Sci. **179**, 2538–2554 (2009)
14. Azaustre, A., Casas, J.: Manual de retórica española. Ariel, Barcelona (1997)
15. Burdiles, G.A.: Descripción de la organización retórica del género caso clínico de la medicina a partir del corpus CCM-2009. Ph.D. thesis in Applied Linguistics. Pontificia Universidad Católica de Valparaíso, Chile (2011)
16. Swales, J.M.: Research Genres: Explorations and Applications. Univ. Press, Cambridge (2004)
17. Parodi, G.: Lingüística de corpus: una introducción al ámbito. Revista de Lingüística Teórica y Aplicada **46**(1), 93–119 (2008)
18. Manrique-Losada, B.: A formalization for mapping discourses from business-based technical documents into controlled language texts for requirements elicitation. Ph.D. thesis, Universidad Nacional de Colombia (2014)
19. Pivovarova, L., Huttunen, S., Yangarber, R.: Event representation across genre. In: Proceedings of 1st Workshop on EVENTS: Definition, Detection, Coreference, and Representation, pp. 29–37 (2013)
20. Do, Q.X., Chan, Y.S., Roth, D.: Minimally supervised event causality identification. In: EMNLP 2011 (2011)
21. Bejan, C.A., Harabagiu, S.: Unsupervised Event Coreference Resolution. Computational Linguistics **40**(2) (2013)
22. Vossen, P. (ed.): EuroWordNet General Document. Version 3. University of Amsterdam, Amsterdam (2002)
23. Chaowicharat, E., Naruedomkul, K.: Co-ocurrence-based error correction approach to word segmentation. In: Boonthum-Denecke, C., McCarthy, P.M., Lamkin, T. (eds.) Cross-Disciplinary Advances in Applied Natural Language Processing. Issues and Approaches, pp. 354–364. Information Sciences Reference Publishers (2012)

Using Semantic Frames for Automatic Annotation of Regulatory Texts

Kartik Asooja[✉], Georgeta Bordea, and Paul Buitelaar

Insight Centre, National University of Ireland, Galway, Ireland
{kartik.asooja,georgeta.bordea,paul.buitelaar}@insight-centre.org

Abstract. The global legislation system is being actively updated, especially after the financial crisis of 2007–2008. This results in a significant amount of work load for the different industries, in order to cope up with the volume, velocity, variety, and complexity of the regulations in order to be compliant. So far, this is mainly being handled manually by the regulatory experts in the industries. In this paper, we explore the space of providing automatic assistance to experts in compliance verification pipeline. This work specifically focuses on performing automatic semantic annotations of the regulatory documents with a set of predefined categories. This is achieved by using text classification approaches using linguistically motivated features.

1   Introduction

There are fundamental challenges in terms of Governance, Risk and Compliance that the financial industry is facing nowadays. This is mainly due to the constantly changing regulations, which leads to an increased need for experts to perform compliance verification for regulatory legislation. This further increases the already high costs in terms of time and manual effort required in this process. In recent years, the automation of some of the tasks associated with compliance verification using recent advances in artificial intelligence is more and more seen as a solution to this problem [1,6].

An important stage in the compliance verification pipeline discussed by Asooja et al. [1] is the semantic annotation of the text sections in the regulatory documents. This requires a semantic framework consisting of generic and domain specific regulatory semantics. This work focuses on a text classification approach for automatic semantic annotation of the regulation documents. Such annotations would enable intelligent querying over the regulatory documents. In our context, we define these semantic annotations as pre-defined concepts from Subject Matter Experts (SMEs), like *Enforcement* or *Obligation*. This paper describes an approach for semantic annotation of regulations in the financial domain, which uses multi-label and multi-class text classification for tagging domain specific concepts and generic concepts respectively. Our main contributions in this work can be summarized as: (1) Gathering a larger dataset of annotated regulations compared to previous work, (2) Using semantic frames as features for automatic annotation.

© Springer International Publishing Switzerland 2016
E. Métais et al. (Eds.): NLDB 2016, LNCS 9612, pp. 384–391, 2016.
DOI: 10.1007/978-3-319-41754-7_38

2 Related Work

Most of the works targeting related text classification problems around regulatory texts revolve around the classification of generic concepts or modalities like Prohibition and Obligation [3,7,8]. First steps towards addressing the problem of annotating regulations with domain specific concept types are taken in Asooja et al. [1]. They focus mostly on shallow features such as n-grams and manually identified cue-phrases. We follow the problem definition as defined by them, and further propose additional semantics based features including semantic features from FrameNet [2] for the categorization of regulatory text. In addition to this, we also extend the dataset with more instances and modality classes, thus showing also an evaluation over modality classification.

3 Task Description

We perform automatic classification of regulatory text into different domain specific concepts like Customer Due Diligence, as well as generic concepts e.g. modalities like Prohibition or Obligation [1]. We utilize a multi-class classification for classifying a text into generic concepts (3 classes as shown in Table 2). We further classify each text into domain specific concepts using multi-label classification, since each text unit can represent more than one domain specific categories. The decision of using a multi-class classifier for generic concepts and a multi-label classifier for domain specific concepts is taken by the SMEs considering the size of the text unit, which can be aligned to a particular XML tag "P2" in the UK regulation document [1]. In our case, the domain is Anti Money Laundering (AML), and the identified 9 domain specific class labels are listed in Table 1.

Dataset: We extend the available dataset consisting of only domain specific concepts related to Anti Money Laundering (AML), with more domain specific class labels and generic class labels e.g. Prohibition, Obligation. These annotations are performed by the SMEs. We annotated 10 UK acts and 1 US act that relate to AML. Only the sections concerned with AML have been annotated. Following list enumerates the exact regulation documents that were annotated:

- **Anti Money Laundering (AML):** UK AML (1993, 2001, 2003, 2007), US Bank Secrecy Act Chapter X.
- **Crime:** Crime and Courts Act 2013, The Proceeds of Crime Act 2002 - Business in the Regulated Sector and Supervisory Authorities Order (2003, 2007), Serious Organised Crime and Police Act 2005
- **Terrorism:** Terrorism Act 2000, The Terrorism Act 2000 - Business in the Regulated Sector and Supervisory Authorities Order (2003, 2007)

Table 1 shows the different AML class labels and their distribution over the different regulation types in the domain specific multi-label dataset. In the new annotations, as compared to the existing ones [1], we merged the classes Customer Identification and Verification (CIV) and Customer Due Diligence (CDD),

as CDD, since SMEs proposed that it is a super class of CIV. The total number
of multi-label instances in the new dataset is 781 with a label cardinality of
1.625 in comparison to 171 instances with a label cardinality of 1.27 in the previous dataset. Regulation documents are segmented into smaller textual units
using a particular XML tag e.g. P2 tags in the case of UK XML format[1] for
regulatory documents. We relied on the SMEs for making the decisions on the
granularity level of the text. The number of data instances varies considerably
between different class labels. Table 2 gives the details of the generic concepts
multi-class dataset. There were no instances of modalities in the previous dataset.

Table 1. AML domain specific concepts distribution over different regulation types,
and comparison of this dataset to the existing dataset.

Class	AML	Crime	Terrorism	Current dataset total	Prev. dataset total
Customer Due Diligence	190	1	1	192	87
Defence	21	18	14	53	0
Enforcement	49	33	21	103	99
Internal Programme	51	0	0	51	0
Interpretation	455	46	16	517	0
Monitoring	12	0	0	12	12
Record-keeping	107	0	0	107	0
Reporting	247	10	7	264	14
Supervision	144	19	5	168	0
Total multi-label instances	648	83	50	781	172

Table 2. Generic concepts distribution over different regulation types.

Class	AML	Crime	Terrorism	Current dataset total	Prev. dataset total
Prohibition	57	4	1	62	0
Obligation	277	14	17	308	0
Others	531	68	31	630	0
Total multi-class instances	865	86	49	1000	0

The **regulations** **come** into **force** on 15th **December** 2007 .
Law   Arriving   Being_in_effect   Calendric_unit

Fig. 1. Frame-Semantic parsing output on an example of domain specific concept
"Enforcement"

---

[1] http://www.legislation.gov.uk/uksi/2007/2157/made/data.xml.

4   Experiments

We evaluate the proposed features in both cross-fold validation and train-test settings. In the cross-fold setting, the dataset comprised of all the instances from AML, Crime and Terrorism regulations. In the train-test setting, we use the data instances from the AML regulation documents for training, and the testing was performed on a dataset which combines the Crime and Terrorism instances. We report different metrics for evaluating multi-label classification, and F-measure for multi-class classification. We used Weka[2] toolkit for implementing SVM based multi-class classifier. For multi-label classification, we used Meka[3], which is a multi-label extension of Weka. We used Classifier Chains algorithm and J48 Tree as the base classifier for multi-label classification [9].

Table 3. Performance for different features in domain specific concept (multi-label) classification using 10 fold cross validation.

	Baseline	Baseline + Frames	Baseline + POS	Baseline + Frames + POS	Frames + POS
Class	F_1	F_1	F_1	F_1	F_1
Customer Due Diligence	0.572	0.620	0.589	**0.627**	0.496
Defence	**0.490**	0.425	0.428	0.444	0.456
Enforcement	0.526	0.567	0.550	0.551	**0.572**
Internal Programme	**0.660**	0.541	0.612	0.568	0.489
Interpretation	0.688	0.672	0.680	0.677	**0.692**
Monitoring	0.105	0.0	0.125	0.0	**0.211**
Record-keeping	**0.559**	0.500	0.507	0.513	0.423
Reporting	0.695	**0.713**	0.692	0.693	0.583
Supervision	0.403	0.411	**0.420**	0.394	0.377
Overall Metrics					
Exact Match	0.279	0.271	**0.283**	0.282	0.242
Hamming Score	0.860	0.863	**0.865**	0.863	0.844
Hamming Loss	0.140	0.137	**0.135**	0.137	0.156
F_1 (Micro avg.)	0.597	**0.598**	0.597	0.595	0.560
F_1 (Macro avg. by example)	**0.568**	0.563	0.556	0.557	0.531
F_1 (Macro avg. by label)	**0.522**	0.494	0.511	0.496	0.478

The customer	**should**	**refrain**	from	**registering**	**two**	**accounts**	.
	REQUIRED_EVENT	FORGOING		BECOMING_AWARE	CARDINAL_NUMBERS	TEXT	

Fig. 2. Frame-Semantic parsing output on an example of generic concept text

---

[2] http://www.cs.waikato.ac.nz/ml/weka/.
[3] http://meka.sourceforge.net/.

4.1 Features

We used multiple features for the classification including the ones described by Asooja et al. [1]. They use n-grams (uni/bi/tri grams), manually identified cue phrases, modal verbs (must, should etc.) and contextual features based on the succeeding and preceding n-grams extracted from the surrounding text units. However, we do not use the contextual features in this work, as we annotate only the sections related to AML in different regulation acts.

In this paper, we extend these with features based on FrameNet and part of speech tags/patterns. Below are the lists of different features used for classifying modalities and AML specific classes.

Modality Classification (Generic)

- Baseline: N-grams, presence of modal verbs, presence of negators like not, never
- Frames: Uni, bi-grams of FrameNet frames like Forgoing, Being_Obligated, Required_Event
- POS: Uni, bi-grams of POS tags

AML Concept Classification (Domain Specific)

- Baseline: N-grams
- Frames: Uni, bi-grams of FrameNet frames like Reporting, Being_into_effect
- POS: Uni, bi-grams of POS tags

FrameNet Based Features. FrameNet provides a lexical database of English based on examples of how words are used in actual texts [2]. It is formed around semantic frame, which is a conceptual structure describing an event, relation, or object and the participants in it. FrameNet lexical database contains around 1,200 semantic frames. In order to use it for our classification task, we first run a frame-semantic parser Semafor[4] over the text to perform an automatic analysis of the frame-semantic structure of it [4,5]. Essentially, we get in-context mapping of the vocabulary in our text to the semantic frames in FrameNet. Figure 2 and 1 shows examples of frame-semantic parsing output on texts of generic and domain specific classes respectively. Here, we can see that frames like "Required Event" and "Forgoing" in Fig. 2, and frame "Being_In_Effect" in Fig. 1 can be useful for classifying the corresponding sentences into generic class "Prohibition" and domain specific class "Enforcement" respectively. For our classification tasks, we use n-grams of the semantic frame sequence.

4.2 Results and Discussion

Tables 4 and 5 show results on cross fold validation and train-test results over the generic concept dataset respectively, where the proposed features show

---

[4] http://www.ark.cs.cmu.edu/SEMAFOR.

Table 4. F_1 for different features in generic concept (multi-class) classification using 10 fold cross validation.

Class	Baseline (n-grams + modal verbs + negators) F_1	Baseline + Frames F_1	Baseline + POS F_1	Baseline + Frames + POS F_1	Frames + POS F_1
Obligation	0.796	**0.813**	0.804	0.806	0.780
Prohibition	0.494	0.483	0.447	0.442	**0.522**
Others	0.903	**0.911**	0.906	0.910	0.891
Weighted Avg. F_1	0.844	**0.854**	0.846	0.849	0.834

Table 5. F_1 of different features in generic concept (multi-class) classification in train test setting, where train dataset is AML, and test dataset is Crime + Terrorism

Class	Baseline (n-grams + modal verbs + negators) F_1	Baseline + Frames F_1	Baseline + POS F_1	Baseline + Frames + POS F_1	Frames + POS F_1
Obligation	0.535	0.541	0.562	**0.571**	0.523
Prohibition	0.0	0.0	0.0	0.0	0.0
Others	0.760	0.767	0.761	**0.791**	0.739
Overall Metrics					
Weighted Avg. F_1	0.680	0.686	0.687	**0.712**	0.662

considerable improvement over baseline. Tables 3 and 6 show results on cross fold validation and train-test results only over the domain specific (AML) concept dataset respectively, where no clear trend is visible. Baseline performs better in some of the metrics. In general, cross fold setting results in both the cases produce higher scores than the train-test settings. This can be due to the possibility of data bias, as the dataset also contains different year versions of the same regulations, which brings in the possibility of repetition of some of the text sections in the dataset.

Also, the evaluations here may be very sensitive towards the change in the data samples, because of their very small size. Change of predictions for a few instances in the datasets can result in high deviations in the results. In short, the size and distribution of dataset is the main concern presently. Since the annotations are performed by SMEs, the data preparation process is slow and expensive.

Table 6. Performance of different features in domain specific concept (multi-label) classification in train test setting, where train dataset is AML, and test dataset is Crime + Terrorism. Here, NI denotes no instances of that class in the test instances.

	Baseline	Baseline + Frames	Baseline + POS	Baseline + Frames + POS	Frames + POS
Class	F_1	F_1	F_1	F_1	F_1
Customer Due Diligence	**0.057**	0.049	0.0	0.055	NaN
Defence	0.0	0.0	0.0	0.0	0.0
Enforcement	0.089	**0.213**	0.057	0.160	0.164
Internal Programme	NI	NI	NI	NI	NI
Interpretation	**0.562**	0.520	0.504	0.496	0.547
Monitoring	NI	NI	NI	NI	NI
Record-keeping	NI	NI	NI	NI	NI
Reporting	0.185	0.196	0.154	**0.211**	0.182
Supervision	0.047	0.091	0.140	**0.182**	0.043
Overall Metrics					
Exact Match	**0.090**	**0.090**	0.060	0.068	0.068
Hamming Score	**0.752**	0.742	0.736	0.744	0.743
Hamming Loss	**0.248**	0.258	0.264	0.256	0.257
F_1 (Micro avg.)	0.237	**0.244**	0.202	0.234	0.234
F_1 (Macro avg. by example)	0.230	**0.232**	0.190	0.213	0.221
F_1 (Macro avg. by label)	0.105	0.119	0.095	**0.122**	0.104

5   Conclusion

This paper discusses the use of semantic features to complement shallow features, providing an increase in performance especially in the generic classes. We plan to further extend and improve the dataset by using multiple annotators, and applying the approach to different domains than AML.

Acknowledgement. This work has been funded in part by a research grant from Science Foundation Ireland (SFI) under Grant Number SFI/12/RC/2289 (INSIGHT) and by Enterprise Ireland (EI) as part of the project Financial Services Governance, Risk and Compliance Technology Centre (GRCTC), University College Cork, Ireland.

References

1. Asooja, K., Bordea, G., Vulcu, G., O'Brien, L., Espinoza, A., Abi-Lahoud, E., Buitelaar, P., Butler, T.: Semantic annotation of finance regulatory text using multilabel classification (2015)
2. Baker, C.F., Fillmore, C.J., Lowe, J.B.: The berkeley framenet project. In: Proceedings of the 36th Annual Meeting of the Association for Computational Linguistics and 17th International Conference on Computational Linguistics - vol. 1, ACL 1998, Stroudsburg, PA, USA, pp. 86–90. Association for Computational Linguistics (1998)
3. Buabuchachart, A., Metcalf, K., Charness, N., Morgenstern, L.: Classification of regulatory paragraphs by discourse structure, reference structure, and regulation type (2013)
4. Chen, D., Schneider, N., Das, D., Smith, N.A.: Semafor: Frame argument resolution with log-linear models. In: Proceedings of the 5th International Workshop on Semantic Evaluation, pp. 264–267. Association for Computational Linguistics (2010)
5. Das, D., Chen, D., Martins, A.F., Schneider, N., Smith, N.A.: Frame-semantic parsing. Comput. Linguist. **40**(1), 9–56 (2014)
6. Elgammal, A., Butler, T.: Towards a framework for semantically-enabled compliance management in fiancial services. In: Toumani, F., et al. (eds.) ICSOC 2014. LNCS, vol. 8954, pp. 171–184. Springer, Heidelberg (2015)
7. Francesconi, E., Passerini, A.: Automatic classification of provisions in legislative texts. Artif. Intell. Law **15**(1), 1–17 (2007)
8. Morgenstern, L.: Toward automated international law compliance monitoring (tailcm). Technical report, Intelligence Advanced Research Projects Activity (IARPA) (2014)
9. Read, J., Pfahringer, B., Holmes, G., Frank, E.: Classifier chains for multi-label classification. Mach. Learn. **85**(3), 333–359 (2011)

Automatic Evaluation of a Summary's Linguistic Quality

Samira Ellouze(✉), Maher Jaoua, and Lamia Hadrich Belguith

ANLP Research Group, MIRACL Laboratory, University of Sfax, Sfax, Tunisia
ellouze.samira@gmail.com, {maher.jaoua,l.belguith}@fsegs.rnu.tn

Abstract. The Evaluation of a summary's linguistic quality is a difficult task because several linguistic aspects (e.g. grammaticality, coherence, etc.) must be verified to ensure the well formedness of a text's summary. In this paper, we report the result of combining "Adapted ROUGE" scores and linguistic quality features to assess linguistic quality. We build and evaluate models for predicting the manual linguistic quality score using linear regression. We construct models for evaluating the quality of each text summary (summary level evaluation) and of each summarizing system (system level evaluation). We assess the performance of a summarizing system using the quality of a set of summaries generated by the system. All models are evaluated using the Pearson correlation and the Root mean squared error.

Keywords: Automatic summary evaluation · Linguistic quality · Linear regression model · Machine learning

1 Introduction

The evaluation of a summary is an important and necessary task to improve the results of automatic summarization systems. It quantifies the informative and linguistic quality of a summary. There have been many studies in the field of automatic evaluation of summaries. Most of those studies focus on the evaluation of the content of a summary such as ROUGE [17], AutoSummENG [9], SIMetrix [16]. But, the fact that an automatic summary preserves the important ideas of the input text, is not enough; it should also have a good linguistic quality.

In the Text Analysis Conference (TAC, successor of Document Understanding Conference [DUC]), the linguistic quality (i.e. readability) of a summary was evaluated manually over the years. It should be noted that manual evaluation is a hard and expensive task. So there is a particular need to establish an automatic method for evaluating the linguistic quality of a summary. To urge researchers to automatically evaluate the linguistic quality of a summary, the TAC conference proposed, in the 2011 session, a new goal to the task of automatic evaluation of text summaries consisting in evaluating the participated metrics for their ability to measure the linguistic quality of a summary. In this conference linguistic quality evaluation involves five properties (cited in Sect. 2) which should all be

© Springer International Publishing Switzerland 2016
E. Métais et al. (Eds.): NLDB 2016, LNCS 9612, pp. 392–400, 2016.
DOI: 10.1007/978-3-319-41754-7_39

included in one automatic metric. For this reason, we suggest in this paper a new evaluation method based on a combination of relevant features from various feature classes that cover the maximum of the five aspects.

This paper is organized as follows: in Sect. 2, we give a brief historical overview of the linguistic quality evaluation methods used to assess automatic text summaries and other types of texts; Sect. 3 describes the proposed method, which operates by the linear combination of "Adapted ROUGE" scores and a variety of linguistic features. We describe the machine learning phase in Sect. 4. In Sect. 5, we experiment our method at the system and summary evaluation level.

2 Related Works

DUC and TAC conferences have measured readability manually using five linguistic quality properties: grammaticality, non-redundancy, structure and coherence, focus and referential clarity. Because of the time required to evaluate summaries with manual metrics, some studies have focused on automatic evaluation of the readability of a summary. In this context, Barzilay and Lapata [1] have evaluated the local coherence of a summary using an entity grid model which captures entity transitions between two adjacent sentences. Many other works like [18] have evaluated the readability of a summary or/and a text based on an entity grid model. Besides, Pitler et al. [19] have evaluated the five linguistic properties used in DUC to assess the readability of text summary by combining different types of features such as entity grid, modeling language, cosine similarity, Coh-Metrix [10], etc. Furthermore, [14] have predicted the readability of general documents using diverse features based on syntax and language models.

Apart from evaluating text summary readability, there are many works that have focused on the evaluation of text readability, as the school grade level of the reader. Early works like those of [8,11], combine superficial characteristics of a text like the average sentence length, the percentage of complex words, etc., to predict its readability. With the progress of research on natural language processing, several complex text properties have been experimented at many linguistic levels (e.g. the lexical level, the syntactic level). Some works, such as [3,5,6], use modeling language features. These features have proved to be useful in readability evaluation. Works like those of [6,13] used also syntactic features. [6,20] explored several part-of-speech features. Recently, Pitler and Nenkova [20] and Pitler et al. [19] have started to assess readability using discourse features like, lexical chains, entity grid features,...

3 Proposed Method

Our work differs from those previous studies in two points. First, we adapt ROUGE scores to evaluate the structure and the grammaticality of a text summary. Second, some similarity measures are used for the first time to evaluate local coherence. In [4], we have used a combination of content and some linguistic quality features to predict the PYRAMID score. To evaluate the linguistic quality, we have combined adapted content scores and many linguistic features that have an impact on the linguistic quality into one model. This model predicts the manual linguistic quality score which is given by human judges in the TAC conference. To build this model, we need to perform a machine learning phase.

4   Machine Learning Phase

4.1   Features' Extraction

The goal of this phase is to transform the text summary input into numeric features matrix. The literature explores various linguistic indicators of the linguistic quality of a text. We divided those indicators into several classes of features.

Adapted ROUGE scores: ROUGE scores measure the content of a text summary based on the calculation of the recall of words' N-grams between a candidate summary (CS) and one or more reference summaries (RSs). According to [2], ROUGE variants, which take into account large contexts, may capture the linguistic qualities of the summary such as some grammatical phenomena. In addition, a candidate summary is built by selecting a subset of existing words, phrases, or sentences from source documents. So, to evaluate linguistic quality with ROUGE, it is more suitable to compare CS with source documents, than with RSs. For this reason, we adapt ROUGE scores by replacing RSs with one document that contains all source documents. Then, instead of calculating recall for the original ROUGE, we calculating the precision of words overlap between a candidate summary and all source documents. So, for instance, "Adapted ROUGE-N" scores have the following formula:

$$Ad\_R - N = \frac{\sum_{gram_n \in CandidatSummary} Count_{match}(gram_n)}{\sum_{gram_n \in CandidatSummary} Count(gram_n)} \tag{1}$$

where n represents the length of n-grams, $Coun_{match}(gram_n)$ represents the maximum number of n-grams co-occurring in a candidate summary and a set of source documents. We have adapted the following ROUGE variants: Ad-R-2, Ad-R-3, Ad-R-4, Ad-R-5, Ad-R-L, Ad-R-S4 and Ad-R-W.

Traditional readability measures use some simple characteristics of documents like the average sentence length and the average number of syllables per word. In this study, we use as features the following set of most used traditional readability measures: FOG [11], Flesch Reading Ease [8], Flesch-Kincaid [15] and Automated Readability Index [21].

Shallow features: In fact, a text containing difficult words and long sentences will be more difficult to read and to understand. For this reason, we used as features the average number of syllables per word, the average number (NB) of characters per word, the average NB of words per sentence. Moreover, we noticed that many system summaries include very long sentences with a few sentences. This is why we calculate, for each summary, the NB of sentences, the NB of characters and the NB of words. These features are respectively equal to $-\log_2(NbSentences)$, $-\log_2(NbCharacters)$ and $-\log_2(NbWords)$.

Language modeling features: We use Language modeling features under the assumption that a text containing difficult words that are not frequently used by people will be less readable and vice versa. In our work, we have trained three language models over the Open American National Corpus. We used the SRI language modeling toolkit [22] to calculate for each summary the log probability and two measures of perplexity for unigrams, bigrams and trigrams models.

Part-of-speech Features: We divide Part-of-speech features into function words and content words. Features extracted from function words can tell us about the cohesion of a text. Indeed, according to Halliday and Hasan [12], the cohesion concept includes phenomena which allow a link between sentences or phrases. Since many functional words include link tools (e.g. while, and), we have decided to calculate for each summary the following features: the total NB, the average NB per sentence and the density of four categories of function words: determiners, conjunctions, prepositions and subordinating conjunctions, and personal pronouns. We have also calculated the same features as function words to four categories of content words: adjectives, nouns, verbs and adverbs.

Syntactic Features: In general, a summary can be unreadable due to ungrammatical language or unusual linguistic structures that may be expressed in the syntactic properties of the summary. We implement the following syntactic features: total number and average number per sentence of four phrases (i.e. noun phrases, verbal phrases, prepositional phrases and clauses). Furthermore, we add two other features: the average tree parse height of a summary and the average number of dependency relation per sentence in a summary.

Entity-based features: Feng et al. [6] hypothesize that the comprehension of a text is related to the number of entities presented in it. To extract entities we use the Stanford NER [7]. Based on the extracted named entities, we implement for three features: NB of named entities, density of named entities and average NB of named entities per sentence.

Local coherence features: We assess the local coherence of a text by measuring the topic continuity between two adjacent sentences in a summary. To measure this continuity, we determine the similarity between adjacent sentences using Levenshtein distance, Cosine similarity, Jaccard Coefficient, Pearson's correlation, Kullback-Leibler divergence, Jensen Shannon divergence, Dice coefficient and words overlap. For each measure, local coherence is defined as:

$$Coh_r = \frac{\sum_{i=1}^{(n-1)} SIM_{(i,i+1)}}{n-1} \qquad (2)$$

where n is the NB of sentences in a summary and $SIM_{(i,i+1)}$ is a similarity measure between sentence i and sentence i+1. This equation can be an indicator for continuity as well as an indicator for redundancy. As mentioned in [19], some repetition between two adjacent sentences is a good indicator of continuity while much repetition leads to redundant sentences.

4.2 Training and Validation

Before training our data, we select the relevant features to build our learning model using the "Wrapper subset evaluator" method implemented in the Weka platform. The input of the training phase is a feature Matrix, which contains selected features, and a manual readability score vector. In this phase, we build a predictive model (called "ranker") of the manual linguistic quality (readability) score. This manual score is given by human judges in the TAC conference on a scale ranging from 1(very poor) to 5(very good). We have tried to predict linguistic quality using several regression algorithms that are integrated in Weka environment. We notice that the best result is obtained by the linear regression function that estimates readability using the following equation:

$$\hat{y} = w_0 + w_1 x_1 + w_2 x_2 + ... + w_n x_n \qquad (3)$$

where $\hat{y}$ is the predictive value, n is the number of features, $x_1...x_n$ are the feature values, $w_1...w_n$ are the feature weights and w_0 is a constant. We used linear regression to find the linear combination that maximizes the correlation between the used features and the linguistic quality score. So the problem of linear regression is expressed as a set of features and their corresponding readability scores. Then, we determine a vector w of length n+1 maximizing the correlation as:

$$w = argmax\rho(w_0 + \sum_{j=1}^{n} a_{ij}w_j, b_i) \qquad (4)$$

where a_{ij} is the value of the jth feature for entry i; b_i is the readability score for entry i; and ρ is the Pearson correlation. We used the least squares method to minimize the sum of squared deviations between the readability score and the predicted readability score. Then, the equation of minimization is:

$$min \sum_{i=1}^{m} (y_i - \hat{y}_i)^2 \qquad (5)$$

To validate our model, we use the cross-validation method (with 10 folds).

5   Experiments

We Experiment our method in system level evaluation as well as in summary level evaluation. In our experiment we employ TAC2008's corpus, which contains 48 collections and 58 systems. For each collection, there are two sets of 10 documents: set (A) and set (B). Each system produces an initial summary constructed using only the first set of documents which chronologically precedes the second document set and an update summary built from the second set. In order to compare the baselines, the ranker and each set of features to the manual readability score, the Pearson correlation coefficient is computed. Since correlation ignores the size of errors made by wrong predictions, we decide to recite the "Root Mean Squared Error" (RMSE) which can be interpreted as the average deviation in readability scores between the predicted and the actual values.

5.1 System Level Evaluation

In this level, we measure the overall performance of a summarizing system by computing the average score for a system over the entire set of its produced summaries. Table 1 shows the correlation and the RMSE for each baseline, each class of features and ranker in both tasks on the system level evaluation. In initial summary task, the ranker out performs all the other experiments with a correlation of 0.8603 and an RMSE of 0.1994. This model has a strong correlation with manual readability score and a low RMSE. Added to that, we notice that the correlation of the "Adapted ROUGE" scores features class outperforms all other classes. In general, the majority of baselines has a weak correlation with linguistic quality.

In the update summary, as seen in Table 1, the best correlation and the lowest RMSE are between the ranker and the manual readability score. The modeling

Table 1. Pearson Correlation with Readability Score and RMSE (between brackets) in Both Tasks of System Level Evaluation ($p - value < 2.2e - 16$)

Features	Initial summary	Update summary
Baseline		
Gunning Fog Index	−0.4815(0.4026)	−0.3407(0.4907)
Flesch Reading Ease	−0.1584(0.399)	−0.0723(0.4850)
Flesch-Kincaid Index	−0.1679(0.4000)	−0.4058(0.4910)
Automated Readability Index	−0.1340(0.4005)	−0.2402(0.4878)
Our experiments		
Ad-ROUGE scores class	0.6481(0.2981)	0.6221(0.3727)
Traditional readability measures class	−0.2876(0.4113)	0.2830(0.4625)
Shallow features class	0.2051(0.3844)	0.1119(0.4814)
Language modeling features class	0.4642(0.3466)	0.7025(0.3382)
Part-of-speech class	0.2482(0.3908)	0.3987(0.4799)
Syntactic Features class	0.5008(0.3401)	0.4745(0.4261)
Named entity based features class	0.2782(0.3793)	−0.1271(0.5102)
local coherence features class	0.2906(0.4101)	0.3819(0.4514)
Ranker	0.8603(0.1994)	0.8485(0.2121)

language class is the best class in predicting the readability score. The "dapted ROUGE" scores class has also a good correlation with the manual readability score.

5.2 Summary Level Evaluation

Now, we investigate the predictive power of the used features at summary level evaluation in which we take, for each Summarizer system, each produced summary in a separate entry. The summary level evaluation is more difficult and the results of the correlation coefficient are lower compared to those in the system level evaluation. At this level, Table 2 shows that ranker gives the best correlation and the lowest RMSE on both tasks. Besides, the "Adapted ROUGE" scores class is the best class in predicting the manual readability score, in both tasks. Furthermore, the use of selected features in the ranker gives a better correlation compared to a single class of features, but this correlation remains low compared with the system level evaluation.

Table 2. Pearson Correlation with Readability Score and RMSE (between brackets) in both Tasks of Summary Level Evaluation ($p - value < 2.2e - 16$)

Features	Initial summary	Update summary
Baseline		
Gunning Fog Index	−0.0729(1.0607)	0.0284(1.0910)
Flesch Reading Ease	−0.0614(1.0614)	0.0284(1.0910)
Flesch-Kincaid Index	−0.0729(1.0607)	0.0291(1.0909)
Automated Readability Index	−0.0234(1.0609)	0.0496(1.0900)
Our experiments		
Ad-ROUGE scores class	0.2771(1.0186)	0.3124(1.0365)
Traditional readability measures class	0.0150(1.0606)	0.1195(1.0835)
Shallow features class	0.1719(1.0444)	0.1961(1.0701)
LM features class	0.2093(1.0368)	0.1821(1.0740)
Part-of-speech class	0.2402(1.0295)	0.1478(1.0807)
Syntactic Features class	0.1927(1.0405)	0.1837(1.0731)
Named entity based features class	0.0635(1.0581)	0,0700(1.0887)
Local coherence features class	0.0996(1.0551)	0.0406(1.0907)
Ranker	0.4323(0.9562)	0.4209(0.9908)

6 Conclusion

We have presented a method for summary readability evaluation. To predict readability score, we have combined "Adapted ROUGE" scores and a variety of linguistic features. The combination of features is performed using a linear regression method.

To evaluate our method, we have compared the correlation of ranker, of each class of features and of baselines, with the manual readability scores. We have evaluated our method in two levels of granularity: the system level and the summary level and in two evaluation tasks: the initial summary task and the update summary task. On both levels and in both tasks, our method has provided good performance compared to baselines. For both levels and both tasks, all built models use Adapted content features such as Ad-ROUGE-3, Ad-ROUGE-4,etc. This confirms the hypothesis of [2] which indicates that the integration of content scores which take into account large context may captivate some grammatical phenomena.

References

1. Barzilay, R., Lapata, M.: Modeling local coherence: an entity-based approach. Comput. Linguist. **34**(1), 1–34 (2008)
2. Conroy, J.M., Dang, H.T.: Mind the gap: dangers of divorcing evaluations of summary content from linguistic quality. In: Proceedings of the International Conference on Computational Linguistics, pp. 145–152 (2008)
3. Dell'Orletta, F., Wieling, M., Cimino, A., Venturi, G., Montemagni, S.: Assessing the readability of sentences: which corpora and features? In: Workshop on Innovative Use of NLP for Building Educational Applications, pp. 163–173 (2014)
4. Ellouze, S., Jaoua, M., Belguith, L.H.: An evaluation summary method based on a combination of content and linguistic metrics. In: Proceedings of RANLP Conference, pp. 245–251 (2013)
5. Falkenjack, J., Mühlenbock, K.H., Jönsson, A.: Features indicating readability in Swedish text. In: Proceedings of NoDaLiDa Conference, pp. 27–40 (2013)
6. Feng, L., Jansche, M., Huenerfauth, M., Elhadad, N.: A comparison of features for automatic readability assessment. In: Proceedings of COLING, pp. 276–284 (2010)
7. Finkel, J.R., Grenager, T., Manning, C.: Incorporating non-local information into information extraction systems by gibbs sampling. In: Proceedings of the Association for Computational Linguistics Conference, pp. 363–370 (2005)
8. Flesch, R.F.: How to Test Readability. Harper & Brothers, New York (1951)
9. Giannakopoulos, G., Karkaletsis, V., Vouros, G., Stamatopoulos, P.: Summarization system evaluation revisited: N-gram graphs. ACM Trans. Speech Lang. Process. J. **5**(3), 1–39 (2008)
10. Graesser, A.C., Mcnamara, D.S., Louwerse, M.M., Cai, Z.: Coh-Metrix: Analysis of text on cohesion and language. Behav. Res. Methods Instrum. Comput. J. **36**(2), 193–202 (2004)
11. Gunning, R.: The techniques of clear writing. McGraw-Hill, New York (1952)
12. Halliday, M.A.K., Hasan, R.: Cohesion in English. Longman, London (1976)
13. Islam, Z., Mehler, A.: Automatic readability classification of crowd-sourced data based on linguistic and information-theoretic features. Computacin Sistemas J. **17**(2), 113–123 (2013)
14. Kate, R.J., Luo, X., Patwardhan, S., Franz, M., Florian, R., Mooney, R.J., Roukos, S., Welty, C.: Learning to predict readability using diverse linguistic features. In: Proceedings of Coling Conference, pp. 546–554 (2010)

15. Kincaid, J.P., Fishburne, Jr., R.P., Rogers, R.L., Chissom, B.S.: Derivation of new readability formulas for Navy enlisted personnel. Research Branch Report 8–75, U.S. Naval Air Station, Memphis (1975)

16. Louis, A., Nenkova, A.: Automatically assessing machine summary contentwithout a gold standard. Comput. Linguist. J. **39**(2), 267–300 (2013)

17. Lin, C.Y.: Rouge: a package for automatic evaluation of summaries, text summarization branches out. In: Proceedings of the ACL-04 Workshop, pp. 74–81 (2004)

18. Lin, Z., Liu, C., Ng, H.T., Kan, M.Y.: Combining coherence models and machine translation evaluation metrics for summarization evaluation. In: Proceedings of Association for Computational Linguistics, pp. 1006–1014 (2012)

19. Pitler, E., Louis, A., Nenkova, A.: Automatic evaluation of linguistic quality in multi-document summarization. In: Proceedings of ACL, pp. 544–554 (2010)

20. Pitler, E., Nenkova, A.: Revisiting readability: a unified framework for predicting text quality. In: Proceedings of the EMNLP Conference, pp. 186–195 (2008)

21. Smith, E., Senter, R.: Automated readability index. AMRL-TR. Aerospace Medical Research Laboratories (6570th), 1 (1967)

22. Stolcke, A.: SRILM - an extensible language modeling toolkit. In: Proceedings of International Conference on Spoken Language Processing, vol. 2, pp. 901–904 (2002)

Poster Papers

Extracting Formal Models from Normative Texts

John J. Camilleri[1](✉), Normunds Gruzitis[2](✉), and Gerardo Schneider[1]

[1] CSE, Chalmers University of Technology and University of Gothenburg,
Gothenburg, Sweden
{john.j.camilleri,gerardo}@cse.gu.se
[2] IMCS, University of Latvia, Riga, Latvia
normunds.gruzitis@lu.lv

Abstract. *Normative texts* are documents based on the deontic notions of obligation, permission, and prohibition. Our goal is model such texts using the *C-O Diagram* formalism, making them amenable to formal analysis, in particular verifying that a text satisfies properties concerning causality of actions and timing constraints. We present an experimental, semi-automatic aid to bridge the gap between a normative text and its formal representation. Our approach uses dependency trees combined with our own rules and heuristics for extracting the relevant components. The resulting tabular data can then be converted into a *C-O Diagram*.

Keywords: Information extraction · Normative texts · C-O diagrams

1 Introduction

Normative texts are concerned with what must be done, may be done, or should not be done (*deontic norms*). This class of documents includes contracts, terms of services and regulations. Our aim is to be able to query such documents, by first modelling them in the deontic-based *C-O Diagram* [4] formal language. Models in this formalism can be automatically converted into networks of timed automata [1], which are amenable to verification. There is, however, a large gap between the natural language texts as written by humans, and the formal representation used for automated analysis. The task of modelling a text is completely manual, requiring a good knowledge of both the domain and the formalism. In this paper we present a method which helps to bridge this gap, by automatically extracting a partial model using NLP techniques.

We present here our technique for processing normative texts written in natural language and building partial models from them by analysing their syntactic structure and extracting relevant information. Our method uses dependency structures obtained from a general-purpose statistical parser, namely the Stanford parser [3], which are then processed using custom rules and heuristics that we have specified based on a small development corpus in order to produce a table of predicate candidates. This can be seen as a specific information extraction task. While this method may only produce a *partial* model which requires further post-editing by the user, we aim to save the most tedious work so that the user (knowledge engineer) can focus better on formalisation details.

© Springer International Publishing Switzerland 2016
E. Métais et al. (Eds.): NLDB 2016, LNCS 9612, pp. 403–408, 2016.
DOI: 10.1007/978-3-319-41754-7_40

2 Extracting Predicate Candidates

The proposed approach is application-specific but domain-independent, assuming that normative texts tend to follow a certain specialised style of natural language, even though there are variations across and within domains. We do not impose any grammatical or lexical restrictions on the input texts, therefore we first apply the general-purpose Stanford parser acquiring a syntactic dependency tree representation for each sentence. Provided that the syntactic analysis does not contain significant errors, we then apply a number of interpretation rules and heuristics on top of the dependency structures. If the extraction is successful, one or more predicate candidates are acquired for each input sentence as shown in Table 1. More than one candidate is extracted in case of explicit or implicit coordination of subjects, verbs, objects or main clauses. The dependency representation allows for a more straightforward predicate extraction based on syntactic relations, as compared to a phrase-structure representation.

Expected Input and Output. The basic requirement for pre-processing the input text is that it is split by sentence and that only relevant sentences are

Table 1. Sample input and partial output.

Refin.	Mod.	Subject (S)	Verb (V)	Object (O)	Modifiers
1. *You must not, in the use of the Service, violate any laws in your jurisdiction (including but not limited to copyright or trademark laws).*					
	F	User	violate	law	V: in User's jurisdiction V: in the use of the Service
2. *You will not post unauthorised commercial communication (such as spam) on Facebook.*					
	F	User	post	unauthorised commercial communication	O: such as spam O: on Facebook
3. *You will not upload viruses or other malicious code.*					
	F	User	upload	virus	
OR	F	User	upload	other malicious code	
4. *Your login may only be used by one person - a single login shared by multiple people is not permitted.*					
	P	person	use	login of User	S: one
5. *The renter shall pay all reasonable attorney and other fees, the expenses and costs incurred by owner in protection its rights under this rental agreement and for any action taken owner to collect any amounts due the owner under this rental agreement.*					
	O	renter	pay	reasonable attorney	V: under this rental agreement
AND	O	renter	pay	other fee	V: under this rental agreement
6. *The equipment shall be delivered to renter and returned to owner at the renter's risk.*					
	O	equipment	[is] delivered [to]	renter	V: at renter's risk
AND	O	equipment	[is] returned [to]	owner	V: at renter's risk

included. In this experiment, we have manually selected the relevant sentences, ignoring (sub)titles, introductory notes etc. Automatic analysis of the document structure is a separate issue. We also expect that sentences do not contain grammatical errors that would considerably affect the syntactic analysis and thus the output of our tool.

The output is a table where each row corresponds to a *C-O Diagram* box (clause), containing fields for: **Subject:** the agent of the clause; **Verb:** the verbal component of an action; **Object:** the object component of an action; **Modality:** obligation (O), permission (P), prohibition (F), or declaration (D) for clauses which only state facts; **Refinement:** whether a clause should be attached to the preceding clause by conjunction (AND), choice (OR) or sequence (SEQ); **Time:** adverbial modifiers indicating temporality; **Adverbials:** other adverbial phrases that modify the action; **Conditions:** phrases indicating conditions on agents, actions or objects; **Notes:** other phrases providing additional information (e.g. relative clauses), indicating the head word they attach to.

Values of the Subject, Verb and Object fields undergo certain normalisation and formatting: head words are lemmatised; Saxon genitives are converted to of-constructions if contextually possible; the preposition "to" is explicitly added to indirect objects; prepositions of prepositional objects are included in the Verb field as part of the predicate name, as well as the copula if the predicate is expressed by a participle, adjective or noun; articles are omitted.

A complete document in this format can be converted automatically into a *C-O Diagram* model. Our tool however does not necessarily produce a *complete* table, in that fields may be left blank when we cannot determine what to use. There is also the question of what is considered *correct* output. It may also be the case that certain clauses can be encoded in multiple ways, and, while all fields may be filled, the user may find it more desirable to change the encoding.

Rules. We make a distinction between rules and heuristics that are applied on top of the Stanford dependencies. Rules are everything that explicitly follow from the dependency relations and part-of-speech tags. For example, the head of the subject noun phrase (NP) is labelled by `nsubj`, and the head of the direct object NP—by `dobj`; fields Subject and Object of the output table can be straightforwardly populated by the respective phrases (as in Table 1).

We also count as lexicalised rules cases when the decision can be obviously made by considering both the dependency label and the head word. For example, modal verbs and other auxiliaries of the main verb are labelled as `aux` but words like "may" and "must" clearly indicate the respective modality (P and O). Auxiliaries can be combined with other modifiers, for example, the modifier "not" (`neg`) which indicates prohibition. In such cases, the rule is that obligation overrides permission, and prohibition overrides both obligation and permission.

In order to provide concise values for the Subject and Object fields, relative clauses (`rcmod`), verbal modifiers (`vmod`) and prepositional modifiers (`prep`) that modify heads of the subject and object NPs are separated in the Notes field. Adverbial modifiers (`advmod`), prepositional modifiers and adverbial clauses (`advcl`) that modify the main verb are separated, by default, in the Adverbials field.

If the main clause is expressed in the passive voice, and the agent is mentioned (expressed by the preposition "by"), the resulting predicate is converted to the active voice (as shown by the fourth example in Table 1).

Heuristics. In addition to the obvious extraction rules, we apply a number of heuristic rules based on the development examples and our intuition about the application domains and the language of normative texts.

First of all, auxiliaries are compared and classified against extended lists of keywords. For example, the modal verb "can" most likely indicates permission while "shall" and "will" indicate obligation. In addition to auxiliaries, we consider the predicate itself (expressed by a verb, adjective or noun). For example, words like "responsible" and "require" most likely express obligation.

For prepositional phrases (PP) which are direct dependants of Verb, we first check if they reliably indicate a temporal modifier and thus should be put in the Time field. The list of such prepositions include "after", "before", "during" etc. If the preposition is ambiguous, the head of the NP is checked if it bears a meaning of time. There is a relatively open list of such keywords, including "day", "week", "month" etc. Due to PP-attachment errors that syntactic parsers often make, if a PP is attached to Object, and it has the above mentioned indicators of a temporal meaning, the phrase is put in the Verb-dependent Time field.

Similarly, we check the markers (`mark`) of adverbial clauses if they indicate time ("while", "when" etc.) or a condition (e.g. "if"), as well as values of simple adverbial modifiers, looking for "always", "immediately", "before" etc. Adverbial modifiers are also checked against a list of irrelevant adverbs used for emphasis (e.g. "very") or as gluing words (e.g. "however", "also").

Subject and Object are checked for attributes: if it is modified by a number, the modifier is treated as a condition and is separated in the respective field.

If there is no direct object in the sentence, or, in the case of the passive voice, no agent expressed by a prepositional phrase (using the preposition "by"), the first PP governed by Verb is treated as a prepositional object and thus is included in the Object field.

Additionally, anaphoric references by personal pronouns are detected, normalised and tagged (e.g. "we", "our" and "us" are all rewritten as "<we>"). In the case of terms of services, for instance, pronouns "we" and "you" are often used to refer to the service and the user respectively. The tool can be customised to do such a simple but effective anaphora resolution (see Table 1).

3   Experiments

In order to test the potential and feasibility of the proposed approach, we have selected four normative texts from three different domains: (1) PhD regulations from Chalmers University; (2) Rental agreement from RSO, Inc.; (3) Terms of service for GitHub; and (4) Terms of service for Facebook. In the development stage, we considered first 10 sentences of each document, based on which the rules and heuristics were defined. For the evaluation, we used the next 10 sentences of each document.

We use a simple precision-recall metric over the following fields: Subject, Verb, Object and Modality. The other fields of our table structure are not included in the evaluation criteria as they are intrinsically too unstructured and will always require some post-editing in order to be formalised. The local scores for precision and recall are often identical, because a sentence in the original text would correspond to one row (clause) in the table. This is not the case when unnecessary refinements are added by the tool or, conversely, when co-ordinations in the text are not correctly added as refinements.

Table 2. Evaluation results based on a small set of test sentences (10 per document).

Document	Rules only			Rules & heuristics		
	Precision	*Recall*	F_1	*Precision*	*Recall*	F_1
PhD	0.66	0.73	0.69	0.82	0.90	0.86
Rental	0.75	0.67	0.71	0.71	0.66	0.69
GitHub	0.46	0.53	0.49	0.48	0.55	0.51
Facebook	0.43	0.54	0.48	0.43	0.57	0.49

The first observation from the results (see Table 2) is that the F_1 score varies quite a lot between documents; from 0.49 to 0.86. This is mainly due to the variations in language style present in the documents. Overall the application of heuristics together with the rules does improve the scores obtained.

On the one hand, many of the sentence patterns which we handle in the heuristics appear only in the development set and not in the test set. On the other hand, there are few cases which occur relatively frequently among the test examples but are not covered by the development set. For instance, the introductory part of a sentence, the syntactic main clause, is sometimes pointless for our formalism, and it should be ignored, taking instead the sub-clause as the semantic main clause, e.g. "User understands that [..]".

The small corpus size is of course an issue, and we cannot make any strong statements about the coverage of the development and test sets. Analysing the modal verb *shall* is particularly difficult to get right. It may either be an indication of an obligation when concerning an action, or it may be used as a prescriptive construct as in *shall be* which is more indicative of a declaration. The task of extracting the correct fields from each sentence can be seen as paraphrasing the given sentence into one of the known patterns, which can be handled by rules. The required paraphrasing, however, is often non-trivial.

4   Related Work

Our work can be seen as similar to that of Wyner and Peters [6], who present a system for identifying and extracting rules from legal texts using the Stanford parser and other NLP tools within the GATE system. Their approach is some-what more general, producing as output an annotated version of the original text. Ours is a more specific application of such techniques, in that we have a well-defined output format which guided the design of our extraction tool, which includes in particular the ability to define clauses using refinement.

Mercatali et al. [5] tackle the automatic translation of textual representations of laws to a formal model, in their case UML. This underlying formalism is of course different, where they are mainly interested in the hierarchical structure of the documents rather than the norms themselves. Their method does not use dependency or phrase-structure trees but shallow syntactic chunks.

Cheng et al. [2] also describe a system for extracting structured information for texts in a specific legal domain. Their method combines surface-level methods like tagging and named entity recognition (NER) with semantic analysis rules which were hand-crafted for their domain and output data format.

5   Conclusion

Our main goal is to perform formal analyses of normative texts through model checking. In this paper we have briefly described how we can help to bridge the gap between natural language texts and their formal representations. Though the results reported here are indicative at best (due to the small test corpus), the application of our technique to the case studies we have considered has definitely helped increase the efficiency of their "encoding" into *C-O Diagrams*. Future plans include extending the heuristics, comparing the use of other parsers, and applying our technique to larger case studies.

Acknowledgements. This research has been supported by the Swedish Research Council under Grant No. 2012-5746 and partially supported by the Latvian State Research Programme NexIT.

References

1. Alur, R., Dill, D.L.: A theory of timed automata. Theoret. Comput. Sci. **126**(2), 183–235 (1994)
2. Cheng, T.T., Cua, J., Tan, M., Yao, K., Roxas, R.: Information extraction from legal documents. In: SNLP 2009, pp. 157–162, October 2009
3. Klein, D., Manning, C.D.: Accurate unlexicalized parsing. In: ACL 2003, pp. 423–430 (2003)
4. Martinez, E., Cambronero, E., Diaz, G., Schneider, G.: A model for visual specification of e-Contracts. In: SCC 2010, pp. 1–8. IEEE Comp. Soc. (2010)
5. Mercatali, P., Romano, F., Boschi, L., Spinicci, E.: Automatic translation from textual representations of laws to formal models through UML. In: JURIX 2005, pp. 71–80. IOS Press (2005)
6. Wyner, A., Peters, W.: On rule extraction from regulations. In: Atkinson, K. (ed.) JURIX 2011, pp. 113–122. IOS Press, Amsterdam (2011)

Sentiment Analysis in Arabic

Sanjeera Siddiqui[1(✉)], Azza Abdel Monem[2], and Khaled Shaalan[1,3]

[1] British University in Dubai, Block 11, 1st and 2nd Floor,
Dubai International Academic City, Dubai, UAE
faizan.sanjeera@gmail.com, khaled.shaalan@buid.ac.ae
[2] Faculty of Computer and Information Sciences, Ain Shams University,
Abbassia, 11566 Cairo, Egypt
azza_monem@hotmail.com
[3] School of Informatics, University of Edinburgh, Edinburgh, UK

Abstract. The tasks that falls under the errands that takes after Natural Language Processing approaches includes Named Entity Recognition, Information Retrieval, Machine Translation, and so on. Wherein Sentiment Analysis utilizes Natural Language Processing as one of the way to locate the subjective content showing negative, positive or impartial (neutral) extremity (polarity). Due to the expanded utilization of online networking sites like Facebook, Instagram, Twitter, Sentiment Analysis has increased colossal statures. Examination of sentiments helps organizations, government and other association to extemporize their items and administration in view of the audits or remarks. This paper introduces an Innovative methodology that investigates the part of lexicalization for Arabic Sentiment examination. The system was put in place with two principles rules– "equivalent to" and "within the text" rules. The outcomes subsequently accomplished with these rules methodology gave 89.6 % accuracy when tried on baseline dataset, and 50.1 % exactness on OCA, the second dataset. A further examination shows 19.5 % in system1 increase in accuracy when compared with baseline dataset.

Keywords: Sentiment analysis · Opinion mining · Rule-based approach · Arabic natural language processing

1 Introduction

Web, additionally termed as World Wide Web, contains heaps of data. Web furnishes individuals with an open space to impart their insights or assumptions, their encounters, and their inclinations on a substance or product. The aim of Sentiment Analysis is to perceive the content with assessments and mastermind them in a way adjusting to the extremity (polarity), which incorporates: negative, positive or unbiased (neutral). Sentiments takes the organizations to tremendous statures [6, 7]. Dialects talked by individuals identifies with their way of life and what they talk, thus distinctive dialects are talked or learnt in better places, which contrast in components also in qualities. Arabic Natural dialect handling is moving a large portion of the scientist's outlook to Arabic, because of the expansion utilization of Arabic dialect by people and the expanded web Arabic clients. Arabic dialect holds one of the main ten position in the

© Springer International Publishing Switzerland 2016
E. Métais et al. (Eds.): NLDB 2016, LNCS 9612, pp. 409–414, 2016.
DOI: 10.1007/978-3-319-41754-7_41

overall utilized dialects. Arabic Natural dialect handling in sentiment investigation is taking gigantic consideration because of the inaccessibility of assets. This requires a need to develop the work in Arabic Sentiment Analysis.

The rest of this paper is organized as follows. Related work is covered in Sect. 2, Data collection is covered in Sect. 3 followed by system implementation in Sect. 4. Section 5 covers results and lastly Sect. 6 depicts conclusion.

2 Related Work

Shoukry and Refea [7] took after a corpus-based methodology, accomplished an accuracy of 72.6 %. Positive, negative and unbiased polarity characterization done by Abdullah et al. [1] displayed a dictionary and sentiment examination tool with an accuracy of 70.05 % on tweeter dataset and 63.75 % on Yahoo Maktoob dataset.

In order to take a shot at Sentiment Analysis, the key parameter is the dataset. Late endeavors by Shaalan [5] outlined the significance of crowdsourcing as an extremely effective system for clarifying dataset. SVM classifier accomplished 72.6 % accuracy on twitter dataset of 1000 tweets [2, 7].

Feldman [4] worked at record level opinion investigation utilizing a consolidated methodology comprising of a vocabulary and Machine Learning approach with K-Nearest Neighbors and Maximum Entropy on a blended area corpus including instruction, legislative issues and games achieved an F-measure of 80.29 %.

Aldayel and Azmi [2] utilizing a half and half approach that is Lexical and Support Vector Machine classifier created 84.01 % precision. El-Halees [3] accomplished 79.90 % precision with Hybrid methodology containing lexical, entropy and K-closest neighbor.

3 Data Collection

Collection of data is vital to perform sentiment analysis. In the field of sentiment analysis, the key underlying data used is the opinion or sentiment data for checking the polarity and lexicon for building the rules or for machine learning classifiers. Lexicons used in this paper is an extension of lexicon shared by [1]. These lexicons contains named entities, adjectives, randomly paced and most importantly words which appeared to be common in both positive and negative reviews. New addition to the lexicon created in this paper includes one word tweet or review in the appropriate lexicon list.

Hence we add Abdullah et al.'s [1] dictionaries with the expansion of words from the dataset which were found to seem more than once. Taking into account the attentiveness of reiteration of these words and their situation, they were incorporated into both the rundown that is sure and negative records. For instance, "كاتب" (Writer) was rehashed in both negative and positive surveys. Henceforth, it was incorporated into positive and negative dictionary.

4 Implementation of Arabic Sentiment Analysis

The spreadsheet guideline (rule) based framework proposed in this section included three key stages to show the outcomes in the wake of handling. The principal stage is to enter the tweet. The second stage incorporates rules centered check wherein the entered subjective statement is gone through an arrangement of standards rules which includes "equal to" and "within the text rules", examining for the tweet extremity either negative or positive. The third stage is to transform the content into "Red" color demonstrating content is positive or "Green" showing content is negative. Figure 1 delineates the review of the framework proposed in this paper.

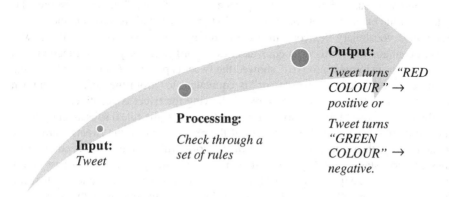

Output:

Tweet turns "RED COLOUR" →
positive or

Tweet turns "GREEN COLOUR" →
negative.

Processing:

Check through a set of rules

Input:
Tweet

Fig. 1. Overview of the system proposed

The methodology followed in this paper is guideline (rule) based, we allude (Shaalan 2010) for an audit about the importance of the principle based methodology and how it is utilized to apply distinctive NLP assignments or tasks. Parcel of examination till date included pre-handling or pre-processing as the real steps. In this exposition, with the very utilization of fitting standards the pre-preparing step was completely wiped out for the dataset, there were no changes done to the dataset. The tenet (rule) based methodology incorporate examples to gaze the entered tweet in the dictionary which in this case is the lexicon.

Methodology for the Formation of Rules. The very presence of the strategy that we have proposed in this paper, has enlivened a completely new methods, in order to address the basic and undealt issues with the lexicon based methodology. Both the extremity (polarity) dictionaries and extremity (polarity) opinions are scrutinized. This paper only focusses on positive and negative polarity examination.

The methodology in this research covers many crucial areas found to be missing in the literature review that we conducted. Firstly, the rules are not confined to search within text. Secondly, rules are not excluding common words found in both positive and negative lexicon and lastly, the proposed rule-based system covers all the directions where the word has been found to have a huge impact.

The rules incorporates two key types one rule checks for the word in the sentence, we framed it as "within the text" and for only one word tweet phrased as "equivalent to the text". The key hidden base ground components which offered us some assistance with formulating fitting standards incorporates examination of the tweets and the augmentation of positive and negative dictionaries. The examination of tweets brought about distinguishing relations relating to words which were either disjoint or coincided.

The words which were disjoint that is totally showing either positive or negative extremity were incorporated into their separate vocabularies. The words which coincided that is the ones which were observed to be normal in both negative and positive reviews were incorporated into positive and also negative dictionaries.

The system introduced in this paper encompasses two sorts of key principles which were composed taking into account "equivalent to" or "within" the text passages. For instance, if the entered tweet contains "أصدق" (I trust), which is recorded in the positive assumption vocabulary (lexicon), then the extremity is positive. All in all, if the tweet content contains the words or is equivalent to a word from the positive rundown then the content transformed into "red" showed the tweet is positive. On the off chance that the tweet content "contain words" or "is equivalent to" from the negative rundown, then the content transformed into "Green" showing the tweet is negative.

Be that as it may, the real turnover was in the tenets (the rules) subsequently made. The words observed to be basic in light of the examination of the tweets were incorporated into both positive and negative dictionaries which were legitimized amid the rules creation stage. One key expansion to the vocabulary utilized as a part of this exposition was the expansion of words in the rundown which were single words that is the tweet which included one and only word. In view of the extremity the words were consequently set in the fitting rundown.

5 Results

The outcomes incorporate the examination of the considerable number of investigations led in this paper. In order to do the examination the accuracy of the considerable number of investigations are utilized. The system was tried on [1] and OCA dataset. Table 1 delineates the trials results with respect to System proposed in this paper.

Table 1. Results of applying system on datasets

Datasets/Accuracy	System tested on [1] dataset	System tested on OCA dataset
Precision	87.4	50.4
Recall	93.3	97.6
Accuracy	89.6	50.1

Table 1 Unmistakably follows the outperformance of standards made in System with enormous accuracy for [1] when contrasted with the outcomes on OCA dataset. The system proposed in this paper gave to be fruitful 39.5 % more accuracy than [1] than OCA dataset. The outcome variety in both the datasets requires an extension to the current rules proposed in this paper, which could effectively enhance the exactness.

This inquiry is exceptionally very much replied with a basic correlation of our system with guideline (rule) fastening approach results with Abdullah et al.'s [1] lexicon based methodology results. As the dataset utilized as a part of this paper is taken from Abdullah et al. [1], we have looked at our outcome with this dataset. Our proposed framework (rule based) beat their outcomes with 17.35 % increment in exactness when tried on System 1.

On contrasting our technique and Abdullah et al. [1] presented some key increments. Abdullah et al. [1] concentrated on augmenting the dictionaries which was only expansion of new words to the rundown even irrelevant to the test set, has abandoned them with no change in accuracy. Consequently, the analyses outflanked when contrasted with the outcomes reported in [1]. Table 2 delineates the correlation of the tests directed in this paper with [1].

This section exhibited the system developed. System incorporated the rules which secured "within the text" and "equivalent to" standards.

Table 2. Results correlation with [1]

Rule-based vs lexicon-based approaches	Accuracy
System proposed	89.6 %
Abdullah et al. [1]	70.05 %

6 Conclusion

This paper delineated how Sentiment Analysis has discovered its presence with the extremely propelled developments in online data. The key fundamental parameter seen is the sharing of audits on any setting and how this effect the clients to take choices on numerous things right from purchasing a motion picture ticket to purchasing a property to numerous propelled operations.

Revealing insight into how a word in one assessment represents positive extremity and how the same word could bring about to make the sentence negative when utilized as a part of an alternate setting. Sentiment Analysis is observed to be exceptionally helpful in measuring the effect of an item or administration, through the surveys or reviews that the general population have shared on it. This paper beats the vocabulary (lexicon) building process through the proper arrangement of words too not barring the basic words found in both the tweets for the dictionaries. Sentiment Analysis, through the organized set guidelines (rules) and through the right utilization of various rules including "contains content" and "equivalent to".

The reasonable noteworthiness in results in a manner acquired through the standards made makes the rules based methodology the most alluring methodology. The greater part of the scientists have concentrated just on Lexicon based yield. Our yield is new to the sentiment investigation period. The yield was exhibited through the adjustment in shade of entered tweet to "Red" for positive tweets and "Green" for negative tweets with the use of two rules, which obtained 19.5 % increase in results when compared to Abdullah et al. [1] but resulted in only 50 % accuracy for OCA dataset. This calls for a need to extend this rule based system to be improvised and enhanced to be fitted into the context of any dataset. In spite of the fact that this framework (system) was just cantered around positive and negative tweets.

As a key errand for further creating and upgrading this proposed framework, we would be anticipating enhance the current proposed framework to cover more rules, subjectivity order and in this way show impartial(neutral) extremity also.

References

1. Abdulla, N.A., Ahmed, N.A., Shehab, M.A., Al-Ayyoub, M., Al-Kabi, M.N., Al-rifai, S.: Towards improving the lexicon-based approach for arabic sentiment analysis. Int. J. Inf. Technol. Web Eng. (IJITWE) 9(3), 55–71 (2014)
2. Aldayel, H.K., Azmi, A.M.: Arabic tweets sentiment analysis – a hybrid scheme. J. Inf. Sci. 0165551515610513 (2015)
3. El-Halees, A.: Arabic opinion mining using combined classification approach (2011)
4. Feldman, R.: Techniques and applications for sentiment analysis. Commun. ACM 56(4), 82–89 (2013)
5. Shaalan, K.: A survey of arabic named entity recognition and classification. Comput. Linguist. 40(2), 469–510 (2014). MIT Press, USA
6. Shoukry, A., Rafea, A.: Preprocessing Egyptian dialect tweets for sentiment mining. In: 4th Workshop on Computational Approaches to Arabic Script-Based Languages, pp. 47–56 (2012)
7. Shoukry, A., Rafea, A.: Sentence-level Arabic sentiment analysis. In: 2012 International Conference on Collaboration Technologies and Systems (CTS), pp. 546–550. IEEE, May 2012

Automatic Creation of Ontology Using a Lexical Database: An Application for the Energy Sector

Alexandra Moreira[1], Jugurta Lisboa Filho[1],
and Alcione de Paiva Oliveira[1,2(✉)]

[1] Departamento de Informática, Universidade Federal de Viçosa, Viçosa, Brazil
xandramoreira@yahoo.com.br, jugurta@ufv.br, alcione@gmail.com
[2] The University of Sheffield, Western Bank, Sheffield S10 2TN, UK

Abstract. The development of ontologies is a task that demands understanding of a domain and use of computational tools to develop them. One of the main difficulties is to establish the appropriate meaning of each term. There are several lexical bases that can help in this task, but the problem is how doing this automatically. This paper presents a technique for the automatic ontology development based on lexical databases available on the Web. The intention is that the whole process is to take place with a minimum of human intervention. The technique was able to generate the correct hierarchy for 64 % of the terms in a case study for the electricity sector. The tool was applied in the electrical power domain, but the goal is that the tool can be used for any domain.

Keywords: Ontology creation · Lexical database · Information extraction

1 Introduction

Ontologies, in the computational sense, are resources for the representation and retrieval of information. It is a resource often used when one wants to add semantics to a syntactic construct. However, as stated by Sanchez and Moreno [9], the construction of ontologies is a task that demands a lot of time and a thorough knowledge of the domain. The attribution of meaning to the terms of the domain for the composition of the ontology can be difficult due to the ambiguity inherent in natural language. One possible way that can facilitate the assignment of meaning to the terms of a domain is to make use of lexical databases and ontologies available on the web. Many researchers have already put much effort into the construction of these resources and they are mature enough to be useful for semantic attribution tasks. Among the lexical resources and ontologies available on the Web we can mention the FrameNet [1], the WordNet [4], and sumo ontology [8]. As those resources are available in digital format, accessible through a computer network, it would be interesting if a computer system could

A. de Paiva Oliveira—The author receives a grant from capes, process n.0449/15-6.

E. Métais et al. (Eds.): NLDB 2016, LNCS 9612, pp. 415–420, 2016.
DOI: 10.1007/978-3-319-41754-7_42

be built to make use of those resources. In this paper, we describe a computer system named AutOnGen (AUTomatic ONtology GENnerator), that makes use of these resources to attribute meaning to terms of a domain and to build up an ontology. The system generates the ontology fully automatically, however, the ontology generated should be later examined by an expert to carry out adjustments. The system was applied to the construction of an ontology for the sector of production and supply of electric power. This paper is organized as follows: the next section presents the work previously developed that are related to this research; Sect. 3 describes the proposed system; Sect. 4 presents the results obtained; and Sect. 5 presents the final remarks.

2   Related Works

Igor et al. [5] created a glossary automatically for the field of Geology. They extracted terms from a corpus composed of 140 specific texts in the field of Geology. Each term receives a definition obtained from Wikipedia sites and from glossaries for the relevant terms. In addition to the difference of the application domain, a distinction of our work in relation to the proposal of [5] is that we propose to build an ontology and not just a glossary.

Dahab et al. [2] have developed a system named TextOntoEx that constructs ontology from natural domain text using a semantic pattern-based approach. According to the authors, TextOntoEx analyses natural domain text to extract candidate relations and then maps them into a meaning representation to facilitate the construction of the ontology. They applied it in a case study of agricultural domain. The method is not fully automatic, and they do not use a top level ontology.

Ruiz-Casado et al. [7] presented a procedure for automatically enriching the WordNet with relationships extracted from Wikipedia. In this case, the point that most differentiates this work of ours is the fact they didn't create an ontology for a particular domain but rather enriched a general ontology.

3   The Proposed System

The AutOnGen (AUTomatic ONtology GENnerator) was developed in Python language and uses the framework NLTK[1]. It also relies heavily on tools used to process natural language as POS tagger and machine translation and lexical and semantic bases, in this case the Wordnet 3.0 [4] and the SUMO ontology [8]. We will now explain each module of the system, as well as the decisions adopted. The Fig. 1 illustrates the modules of the system and their interactions. The system entry is a list of selected terms of the domain. The selection of terms is a key step in the construction of ontology, however, this is not the focus of this paper. But it is essential that all the terms are relate to a specific domain. In the case of our project the list was generated automatically by the Exterm[2] software and formed by the Brazilian Portuguese language terms related to the electricity sector.

---

[1] http://www.nltk.org/.

[2] The article describing the software is being evaluated for publication.

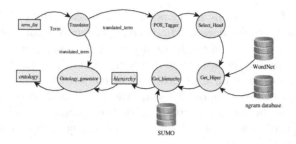

Fig. 1. Diagram showing the AutOnGen system modules. The rectangles are the input and output data and the ellipses denote the functions performed.

3.1   The Translation Module

At this point one can note the first critical decision in the system design. As the list of the terms is in Brazilian Portuguese then it would be expected the use of Portuguese lexical resources. However, in the light of scarce lexical resources for the Portuguese language, it was chosen to use lexical resources in English. For example, the current Brazilian WordNet version contains only verbs (3713 synsets) and the present work deals with nouns [3]. On the other hand, the American Word-Net has a much broader lexical coverage, with 82115 synsets for nouns and 13767 synsets for verbs, not to mention the synsets classes for other classes of words[3]. For this reason, the first step in the system is the translation of the terms of the list for the English language. To perform the translation it was used the libtranslate[4] library that works as a façade to web-based translation services such as Babel Fish, Google Language Tools and SYSTRAN. We decided to use only the Google Language Tools once it produced better results in our preliminary tests. As the Google tool is based on statistics generated by an immense ngrams database, in order to increase the quality of the translation, each term submitted for translation was attached to another term, called *domain anchor*. *Domain anchor* is an *ad hoc attempt* to establish the context of translation. It ensure, for instance, that the term "acordo" be translated as "agreement" and not "wake up" by simply adding the term "negócios" (business) as *domain anchor*. The choice of the term that will serve as the *domain anchor* must be done carefully in order to help characterize the domain. Despite the employment of this feature we have not yet measured the effectiveness of such a technique. After translation the term is analyzed by a POS tagger.

3.2   POS Tagging and Hypernym Finding

The translated terms are annotated according to their syntactic class with the goal to determine the core lexeme of the noun phrase. This is done because the next step in the process is to obtain the head of the term and, after that,

---

[3] http://wordnet.princeton.edu/wordnet/man/wnstats.7WN.html.

[4] http://www.nongnu.org/libtranslate/.

the hypernym of the term by querying the Wordnet. In the case of compound terms, when the WordNet does not find the hypernym of a compound term, the system uses the hypernym of the head lexeme of the compound term. The rule used for the selection of the *head* word of a compound term is the following: If the composite term does not have prepositions or conjunctions, it will be selected the last nominal of the compound term, otherwise it will be selected the last nominal occurring before the preposition or the conjunction. For instance, in the case of the compound term "the distribution system usage charge" that has the following POS annotation

('the','DT'), ('distribution','NN'), ('system','NN'), ('usage','NN'), ('charge','NN')

it would be selected as head term the word "charge". This rule for head selection of a compound word in Germanic languages, was proposed by [10], and is called The Right-hand Head Rule (RHHR): the head of a compound word is the right-hand member of that compound.

The biggest challenge in this step is to get the correct hypernym. Wordnet can return several senses to the lexeme and it's hard for the system to choose the correct one automatically. The choice of generic sense is made by the *get_hypernym* module. This module works as follows: when the WordNet returns more than one term candidate for being the hypernym, the module queries the google ngrams database[5] to verify which term has the highest probability of being the hypernym term. The query is done building for each term a bigram formed by concatenation of the hypernym term with hyponym term. Thus, for the term "business" and the hypernym candidate term "group" is formed the compound term "group business". The underlying hypothesis of this technique is that the construction $<noum\ noum>$, where the second term modifies or qualifies the first, it is a common linguistic construction in the English language. This type of compound was called subsumptive compound by Marchand [6]. Once obtained the hypernym the system moves to the step of obtaining the class hierarchy.

3.3   Getting the Hierarchy and Generating the Ontology

After obtaining the hypernym, obtaining the remainder of the class hierarchy is a relatively simple step, since we rely on SUMO ontology-related technologies. This is done by means of successive queries to SUMO's website, where in each query is retrieved the superclass of the current class, until the system returns to top class (Entity). Thus a query with the term "power company" would have to return the following class hierarchy:

Business→ Organization→ Agent→ Object→ Physical→ Entity

The final step of the system is generating the ontology. Ontology generation is another straightforward step in the process. This step gets the list of translated terms and the class hierarchy of each term and issues a file in the owl format. The Fig. 2 displays a small segment of a generated ontology for the electrical energy sector.

---

[5] https://books.google.com/ngrams.

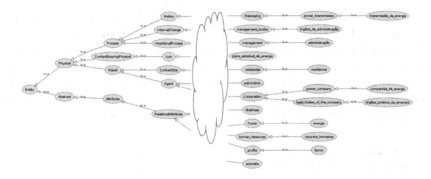

Fig. 2. A generated ontology sample.

4 Results

As stated previously, the system was applied to a list of terms extracted from a corpus related to the electricity sector. The list was generated automatically by the Exterm tool. Exterm extracted 4114 terms from which we randomly selected 100 terms for testing. The output of the system was analyzed to verify the system performance. The results were: 64 % of all terms were considered correctly classified; 16 % of all items were framed in a wrong sense, but somehow related; 17 % of all items were framed in a totally wrong sense (spurius); and 3 % of all items were not found in the WordNet database. A *related* framing is, for instance, to state that *Relationship agent* is a type of *Relation* when the correct would be to state that it is a type of *Agent*. A *spurius* framing is, for instance, to state that *electricity* is a type of *EmotionalState* when the correct would be to state that it is a type of *Energy*. All the terms have been translated correctly.

Examining the results presented certain conclusions can be drawn. Only three items in a hundred were not found in WordNet, which attests to its wide lexical coverage. Other lexical bases were attempted, such as DBpedia and FrameNet, but failed to obtain the same performance. The reasons for the wrong framing fall into two categories: (1) the term has multiple senses and the system chose the incorrect sense to the domain; or (2) there is no appropriate option to frame the term.

5 Conclusions

A system for automatic generation of ontologies from a list of terms over a domain and from a top level ontology was presented. The ontology generated can benefit from all the definitions and relationships designed for the SUMO ontology, but the system does not generate relationships between the terms of the domain. Thus, it's not captured relationships as *power company has shareholder*. This problem must be addressed in future versions.

The system was applied to a list of terms extracted from the electrical power domain. The generated ontology had 64 percent of its terms correctly classified and the ontology expressed in OWL language could be readily edited by various tools available. The database version of WordNet used by the system is 3.0.

The critical point of the process is the proper selection of the hypernym for the term translated. Is planned for inclusion in the next version of the system, a more suitable technique for selecting the best hypernym of a word among the options returned by WordNet. Probably, the use of a domain oriented corpus would produce better results, but that will be tested in a next version.

Acknowledgments. This research is supported in part by the funding agencies CEMIG, FAPEMIG, CNPq, and CAPES.

References

1. Baker, C.F., Fillmore, C.J., Lowe, J.B.: The berkeley framenet project. In: Proceedings of the 17th International Conference on Computational Linguistics, vol. 1. Association for Computational Linguistics (1998)
2. Dahab, M.Y., Hassan, H.A., Rafea, A.: TextOntoEx: automatic ontology construction from natural english text. Expert Syst. Appl. **34**(2), 1474–1480 (2008)
3. Dias-da-Silva, B.C.: The WordNet.Br: an exercise of human language technology research. In: Proceedings of the Third International WordNet Conference GWC, pp. 22–26 (2006)
4. Fellbaum, C.: A semantic network of English verbs. WordNet Electron. Lexical Database **3**, 153–178 (1998)
5. Wendt, I.d.S., et al.: Gerao automática de glossários de termos específicos de um corpus de Geologia. In: ONTOBRAS, 3. Anais, Florianópolis (2010)
6. Marchand, H.: The Categories and Types of Present-Day English Word-formation: A Synchronic-Diachronic Approach, 2nd edn. Munchen, Beck (1969)
7. Ruiz-Casado, M., Alfonseca, E., Castells, P.: Automatic extraction of semantic relationships for WordNet by means of pattern learning from wikipedia. In: Montoyo, A., Muñoz, R., Métais, E. (eds.) NLDB 2005. LNCS, vol. 3513, pp. 67–79. Springer, Heidelberg (2005)
8. Pease, A., Niles, I., Li, J.: The suggested upper merged ontology: a large ontology for the semantic web and its applications. In: Working Notes of the AAAI-2002 Workshop on Ontologies and the Semantic Web, Edmonton, Canada (2002)
9. Sanchez, D., Moreno, A.: Creating ontologies from web documents. In: Recent Advances in Artificial Intelligence Research and Development, vol. 113, pp. 11–18. IOS Press (2004)
10. Williams, E.: On the notions of lexically related and head of a word. Linguist. Inq. **12**, 245–274 (1981)

Computer-Generated Text Detection Using Machine Learning: A Systematic Review

Daria Beresneva[⊠]

Moscow Institute of Physics and Technology,
Russian Academy of National Economy and Public Administration,
Anti-Plagiat Research, Moscow, Russia
beresneva@phystech.edu

Abstract. Computer-generated text or artificial text nowadays is in abundance on the web, ranging from basic random word salads to web scraping. In this paper, we present a short version of systematic review of some existing automated methods aimed at distinguishing natural texts from artificially generated ones. The methods were chosen by certain criteria. We further provide a summary of the methods considered. Comparisons, whenever possible, use common evaluation measures, and control for differences in experimental set-up.

Keywords: Artificial content · Generated text · Fake content detection

1 Introduction

The biggest part of artificial content is generated for nourishing fake web sites designed to offset search engine indexes: at the scale of a search engine, usage of automatically generated texts render such sites harder to detect than using copies of existing pages. Artificial content can contain text (word salad) as well as data plots, flow charts, and citations. The examples of automatically generated content include text translated by an automated tool without human review or curation before publishing; text generated through automated processes, such as Markov chains; text generated using automated synonymizing or obfuscation techniques.

The aim of this paper is to review existing methods of artificial text detection. We survey these efforts, their results and their limitations. In spite of recent advances in evaluation methodology, many uncertainties remain as to the effectiveness of text-generating filtering techniques and as to the validity of artificial text discovering methods. This is a short version of review according to poster publication rules.

The rest of this paper is organized following systematic review guidelines. Section 2 presents the methods selected and their short description. A comparison of the methods is available in Sect. 3. Section 4 recaps our main findings and discusses various possible extensions of this work.

© Springer International Publishing Switzerland 2016
E. Métais et al. (Eds.): NLDB 2016, LNCS 9612, pp. 421–426, 2016.
DOI: 10.1007/978-3-319-41754-7_43

2 The Methods of Artificial Text Detection

As said before, the way of artificial content detection depends on the method by which it was generated. Let us consider various existing features, that authors use for classification. The information about results and datasets used for each method is given in the summary table.

2.1 Frequency Counting Method

To discover whether a text is automatically generated by machine translation (MT) system or is written/translated by human, the paper [1] uses the correlations of neighboring words in the text.

According to the author's hypothesis, the artificial text, the word's pair distribution (means the number of rare for language pairs) should be broken with function of "compatibility" of words with numbers i and j on the functions (means the number of rare for language pairs are longer than the standard and the number of frequent pairs) is understated.

2.2 Linguistic Features Method

One more way of machine translation detection [2] uses not only a statistical, but also linguistic characteristics of the text.

For each sentence, 46 linguistic features were automatically extracted by performing a syntactic parse. The features fall into two broad categories:

(1) Perplexity features extracted using the CMU-Cambridge Statistical Language Modeling Toolkit [8]
(2) Linguistic features felt into several subcategories: branching properties of the parse, function word density, constituent length, and other miscellaneous features.

2.3 The Method of Phrase Analysis

The method [6] involves the use of a set of computationally inexpensive features to automatically detect low-quality Web-text translated by statistical machine translation systems. The method uses only monolingual text as input; therefore, it is applicable for refining data produced by a variety of Web-mining activities.

The authors define features to capture a MT phrase salad by examining local and distant phrases. These features evaluate fluency, grammaticality, and completeness of non-contiguous phrases in a sentence. Features extracted from human-generated text represent the similarity to human-generated text; features extracted from machine-translated text depict the similarity to machine-translated text. By contrasting these feature weights, one can effectively capture phrase salads in the sentence.

2.4   Artificial Content Detection Using Lexicographic Features

By and large, the lexicographic characteristics can be used not only in machine translation detection, but in general case (patchwork, word stuffing or Markovian generators, etc.) as well. The method described in [6] uses feature set (a thorough presentation of these indices is given in [10]) to train the algorithm [5]. Here are some of them:

- the ratio of words that are found in an English dictionary;
- the ratio between number of tokens (i.e. the number of running words) and number of types (size of vocabulary), which measures the richness or diversity of the vocabulary;
- the χ^2 score between the observed word frequency distribution and the distribution predicted by the Zipf law;
- Honore's, Sichel's, and Simpson's scores;

2.5   Perplexity-Based Filtering

This method is based on conventional n-gram language models.

A language model is entirely defined by the set of conditional probabilities $\{p(w|h), h \in H\}$, where h denotes the $n - 1$ words long history (sequence of n-grams) of w, and H is the set of all sequences of length $n - 1$ over a fixed vocabulary.

A standard way to estimate how well a language model p predicts a text $T = w_1 \ldots w_N$ is to compute its perplexity over T, where the perplexity is defined as:

$$PP(pT) = 2^{H(T,p)} = 2^{-\frac{1}{N}\sum_{i=1}^{N} \log_2 p(w_i|h_i)} \tag{1}$$

Here, the perplexities are computed with the SRILM Toolkit [18].

2.6   A Fake Content Detector Based on Relative Entropy

This method seeks to detect Marcovian generators, patchworks and word stuffing, and it uses a short-range information between words [14]. Language model pruning can be performed using conditional probability estimates [15] or relative entropy between n-gram distributions [16].

The given entropy-based detector uses a similar strategy to score n-grams according to the semantic relation between their first and last words. This is done by finding useful n-grams, means n-grams that can help detect artificial text.

Let $\{p(\cdot|h)\}$ denote an n-gram language model, h' - the truncated history, that is the suffix of length $n - 2$ of h. For each history h, the Kullback-Leibler (KL) divergence between the conditional distributions $p(\cdot|h)$ and $p(\cdot|h')$ ([17]) is calculated. Then the penalty score assigned to an n-gram (h, w), which represents a progressive penalty for not respecting the strongest relationship between the first word of the history h and a possible successor: *argmax PKL(h, v)* is computed. The total score *S(T)* of a text *T* is computed by averaging the scores of all its n-grams with known histories.

2.7   Method of Hidden Style Similarity

The method [3] identifies automatically generated texts on the Internet using a (hidden) style similarity measure based on extra-textual features in HTML source code.

The first step is to extract the content by removing any alphanumeric character from the HTML documents and keeping into account the remaining characters using n-grams.

The second step is to model the "style" of the documents by converting the content into a model suitable for comparison.

The similarity measure of the texts is computed using Jaccard index.

For similarity clustering is used the technique called *fingerprinting:* the fingerprint of a document D is stored as a sorted list of m integers.

2.8   SciDetect Method

The next approach [21] was invented to detect documents automatically generated using the software SCIGen, MathGen, PropGen and PhysGen. For this, the distances between a text and others (inter-textual distances) are computed. Then these distances are used to determine which texts, within a large set, are closer to each other and may thus be grouped together. Inter-textual distance depends on four factors: genre, author, subject and epoch.

As the authors don't provide any numerical results of the method's work, we have implemented the method using the source Java code provided by SciDetect developers.

3   Choosing a Method

Every artificial content is generated for a specific purpose. Here we focus only on fake scientific paper for academic publishing or increase in the percentage of originality of the article.

It is important to recognize the aim of fake content for its subsequent detection. Various generation strategies require different approaches to find it. For example, algorithms for detecting word salad are clearly possible and are not particularly difficult to implement. A statistical approach based on Zipf's law of word frequency has potential in detecting simple word salad, as do grammar checking and the use of natural language processing. Statistical Markovian analysis, where short phrases are used to determine if they can occur in normal English sentences, is another statistical approach that would be effective against completely random phrasing but might be fooled by dissociated press methods. Combining linguistic and statistical features can improve the result of experiment. By contrast, texts generated with stochastic language models appear much harder to detect.

One also needs to estimate the data capacity. Text corpuses are taken depending on the aim of the experiment and capabilities of getting them. Like a generation strategy, every data capacity needs different approach. For instance, small trainings samples permit to use such indexes as Jaccard or Dice [5] to count the similarity measure or distance between documents. For big datasets, one can use some linguistics features

Table 1. Summing up the methods

The method	Dataset/Language	The best result
Frequency counting method [1]	2000 original texts, 250 artificial; Russian	90.61 % accuracy
The method of linguistic features [2]	2 k, 5 k, 10 k of generated words; Spanish-English	100 % F-measure for world stuffing
Phrase analysis method [6]	2 k, 5 k, 10 k of generated words	100 % F-measure for word stuffing
Lexicographic features method [14]	2 k, 5 k, 10 k of generated words	99 % F-measure for patchwork
Perplexity-based filtering [14]	English-Japanese parallel documents	97 % F-measure for 2nd Markov model
A fake content detector based on relative entropy [14]	English and French parallel corpora	99 % F-measure for 2nd Markov model
Hidden Style Similarity method [3]	A corpus of 5 million html pages	100 % accuracy at certain threshold
SciDetect method [19]	1600 artificial documents + 8200 original; English	100 % accuracy

and variations of Support Vector Machine and Decision Trees algorithms. Table 1 summarizes results of described methods. The numerical results are provided by the authors of the articles, except the last one.

4 Conclusion

This work presents the results of a systematic review of artificial content detection methods. About a hundred articles were considered for this review; perhaps one-sixth of them met our selection criteria. All the presented methods give good result in practice, but it makes no sense to choose the best one: every approach works under with different conditions like various text generation strategies, different dataset capacity, quality of data and other. Thus, before choosing which method to use, one needs to determine the features of artificial content generating method. Anyway, each approach involves trade off that requires further evaluation.

In future, we plan to compare all the presented approaches on standardized datasets and to do a robustness analysis across different datasets.

References

1. Grechnikov, E.A., Gusev, G.G., Kustarev, A.A., Raigorodsky, A.M.: Detection of artificial texts, digital libraries: advanced methods and technologies, digital collections. In: Proceedings of XI All-Russian Research Conference RCDL 2009, KRC RAS, Petrozavodsk, pp. 306–308 (2009)

2. Corston-Oliver, S., Gamon, M., Brockett, C.: A machine learning approach to the automatic evaluation of machine translation. In: Proceeding of 39th Annual Meeting on Association for Computational Linguistics, ACL 2001, pp. 148–155 (2001)
3. Urvoy, T., Lavergne, T., Filoche, P.: Tracking web spam with hidden style similarity. In: AIRWEB 2006, Seattle, Washington, USA, 10 August 2006
4. Witten, I.H., Frank, E.: Data Mining: Practical Machine Learning Tools and Techniques with Java Implementations. Morgan Kaufmann Publishers, Burlington (2011)
5. Arase, Y., Zhou, M.: Machine translation detection from monolingual web-text. In: Proceedings of 51st Annual Meeting of the Association for Computational Linguistics, Sofia, Bulgaria, pp. 1597–1607, 4–9 August 2013
6. Baayen, R.H.: Word Frequency Distributions. Kluwer Academic Publishers, Amsterdam (2001)
7. Clarkson, P., Rosenfeld, R.: Statistical language modeling using the CMU-Cambridge toolkit. In: Proceedings of Eurospeech 1997, pp. 2707–2710 (1997)
8. Chickering, D.M., Heckerman, D., Meek, C.: A Bayesian approach to learning Bayesian networks with local structure. In: Geiger, D., Shenoy, P.P. (eds.) Proceedings of 13th Conference on Uncertainty in Artificial Intelligence, pp. 80–89 (1997)
9. Vapnik, V.N.: The Nature of Statistical Learning Theory. Springer, New York (1995)
10. Chen, S.F., Goodman, J.T.: An empirical study of smoothing techniques for language modeling. In: Proceedings of 34th Annual Meeting of the Association for Computational Linguistics (ACL), Santa Cruz, pp. 310–318 (1996)
11. Honore, A.: Some simple measures of richness of vocabulary. Assoc. Lit. Linguist. Comput. Bull. 7(2), 172–177 (1979)
12. Sichel, H.: On a distribution law for word frequencies. J. Am. Stat. Assoc. 70, 542–547 (1975)
13. Lavergne, T., Urvoy, T., Yvon, F.: Detecting fake content with relative entropy scoring. In: PAN 2008 (2008)
14. Seymore, K., Rosenfeld, R.: Scalable backoff language models. In: ICSLP 1996, Philadelphia, PA, vol. 1, pp. 232–235 (1996)
15. Stolcke, A.: Entropy-based pruning of backoff language models (1998)
16. Manning, C.D., Schutze, H.: Foundations of Statistical Natural Language Processing. The MIT Press, Cambridge (1999)
17. Gyongyi, Z., Garcia-Molina, H.: Web spam taxonomy. In: 1st International Workshop on Adversarial Information Retrieval on the Web (AIRWeb 2005) (2005)
18. Heymann, P., Koutrika, G., Garcia-Molina, H.: Fighting spam on social web sites: a survey of approaches and future challenges. IEEE Mag. Internet Comput. 11(6), 36–45 (2007)
19. Labbé, C., Labbé, D.: Duplicate and fake publications in the scientific literature: how many SCIgen papers in computer science? Scientometrics, Akadémiai Kiadó, p. 10 (2012)

Sentiment Analysis for German Facebook Pages

Florian Steinbauer[1] and Mark Kröll[2](✉)

[1] University of Technology, Graz, Austria
`florian.steinbauer@student.tugraz.at`
[2] Know-Center GmbH Graz, Graz, Austria
`mkroell@know-center.at`

Abstract. Social media monitoring has become an important means for business analytics and trend detection, for instance, analyzing the sentiment towards a certain product or decision. While a lot of work has been dedicated to analyze sentiment for English texts, much less effort has been put into providing accurate sentiment classification for the German language. In this paper, we analyze three established classifiers for the German language with respect to Facebook posts. We then present our own hierarchical approach to classify sentiment and evaluate it using a data set of ∼640 Facebook posts from corporate as well as governmental Facebook pages. We compare our approach to three sentiment classifiers for German, i.e. AlchemyAPI, Semantria and SentiStrength. With an accuracy of 70%, our approach performs better than the other classifiers. In an application scenario, we demonstrate our classifier's ability to monitor changes in sentiment with respect to the refugee crisis.

Keywords: Sentiment analysis · German · Facebook posts

1 Introduction

Monitoring social media represents one means for companies to gain access to knowledge about, for instance, competitors, products as well as markets. Facebook, for instance, has more than 1.39 billion active users, 4.5 billion *likes* are created and on average 4.75 billion items are shared daily[1]; thus containing large amounts of potentially valuable information. As a consequence, social media monitoring tools have been gaining attention to handle this kind of (personal) information including the analysis of general opinions and sentiments, for example, towards a certain product or decision. While a lot of work has been dedicated to analyze sentiment for English texts, much less effort has been put into providing accurate sentiment classification for the German language.

In this paper, we implement a hierarchical classifier to analyze sentiment in German Facebook posts and compare its performance to three sentiment classifiers for German, i.e. AlchemyAPI, Semantria and SentiStrength. We evaluate

---

[1] http://blog.wishpond.com/post/115675435109/40-up-to-date-facebook-facts-and
-stats.

© Springer International Publishing Switzerland 2016
E. Métais et al. (Eds.): NLDB 2016, LNCS 9612, pp. 427–432, 2016.
DOI: 10.1007/978-3-319-41754-7_44

our approach using a data set of ~640 Facebook posts from Facebook pages. Our algorithm achieves an accuracy of 70 % which is higher compared to the other three libraries. In an application scenario we classify posts from Austrian and German Facebook pages and try to find notable examples in the timeline of *shitstorms*, events that trigger a negative outcry on social media platforms on the internet; in our case concerning the refugee crisis.

2 Related Work

While a lot of work has been dedicated to analyze sentiment for English texts, much less effort has been put into providing accurate sentiment classification for the German language. The interest group on German sentiment analysis IGGSA, formed by academic and industrial members, has published some 16 papers on the topic. Momtazi [4], for instance, compared different methods of sentiment detection on a corpus of German lifestyle documents pulled from Facebook, Blogs, Amazon and Youtube comments and received results of 69 % for positive and 71 % for negative documents using a rule-based method, outperforming Naive Bayes, SVM and Decision Trees. Kasper and Vela [2] analyze sentiment in German hotel reviews to facilitate decision making processes for customers as well as to serve as a quality control tool for hotel managers.

There is work which particularly addresses sentiment analysis/detection with respect to Facebook. Ortigosa et al. [6] presented *SentBuk* which classifies Spanish Facebook messages according to their polarity. SentBuk uses a hybrid approach combining lexical-based and machine-learning techniques and achieves accuracy values up to ~83.3 %. Neri et al. [5] studied ~1000 Facebook posts to compare the public sentiment towards Rai, the Italian public broadcasting service, vs. La7, a private competitor. Troussas et al. [8] experimented with several classification models on English posts; in their experiments the Naive Bayes classifier achieved the highest accuracy with 77 %. In [1], He et al. demonstrated the usefulness of social media monitoring by applying natural language processing to Twitter and Facebook content to compare pizza chains.

3 German Sentiment Analysis

3.1 Existing Libraries

In this section, we provide an overview of three existing libraries for German sentiment analysis, i.e. AlchemyAPI, Semantria and SentiStrength.

AlchemyAPI provides a palette of text processing and computer vision services since mid 2009 and added German sentiment analysis in 2012. AlchemyAPI combines linguistic and statistical analysis techniques, using the first on more structured documents as news articles and press releases and the latter on more unstructured dirty documents as Twitter tweets and Facebook posts. An evaluation by Meehan et al. [3] showed a 86 % accuracy level, based on a corpus of

5370 English tweets. Although these results appear promising, we were unable to reproduce these accuracy levels using our data set (38 % accuracy).

Semantria-a company owned by Lexalytics-is a web service similar to AlchemyAPI. Applying Semantria to our Facebook data set gives us an overall accuracy score of ~50 %. Semantria thus achieves a higher accuracy on our data set than AlchemyAPI. We observe that Semantria has some difficulties with short pieces of text. To give some negative examples, *"Da hilft wohl nur Boykott"* *(then boycott it is going to be)* and *"Eure Mozarella Sticks könnten auch nicht mehr winziger sein :-("* *(your mozarella sticks couldn't be tinier :-( )* are both rated positively although they convey negative sentiment.

The **SentiStrength** algorithm is based on a lexical approach created by Thelwall et al. in 2010 [7]. The algorithm achieves up to 69.6 % accuracy for positive and 71 % accuracy for negative German documents [4]. It is designed to extract sentiment out of short passages of text, like twitter status updates, forum entries or Facebook posts. One of the specific features of the SentiStrength classifier is its lack of any linguistic processing, i.e. it does neither use part-of-speech information nor any kind of stemming or morphological analysis. SentiStrength has problems with German negation words. It only checks for negation words right before opinionated verbs, while all other positions are ignored, for instance, "Ich liebe[+3] dieses Lied nicht" (I don't love this song) results in the positive rating $(3,-1)$ because *nicht (not)* is not used for rating.

3.2 A Hierarchical Approach to German Sentiment Analysis

Experimental Setup. Facebook provides a REST API to fetch posts written by users and posted publicly to a specific page. Collecting a one year span, roughly from the 10th of September 2014 until the 11th of September 2015, results in varying numbers of posts per page. Following companies and their brands were imported and used in our German Facebook data set:

Facebook page	#Posts	Facebook page	#Posts
McDonalds Deutschland	8942	SPAR Österreich	792
ÖBB	4092	derStandard.at	685
BurgerKing Deutschland	3210	Peek & Cloppenburg Deutschland	456
McDonalds Österreich	2193	Bipa	450
Billa	1620	Peek & Cloppenburg Österreich	283
DM Drogeriemarkt Österreich	1407	OEAMTC	212
Kronen Zeitung	1227	BurgerKing Österreich	194

Data Annotation. Four human annotators rated a set of ~100 posts (7 documents per crawled Facebook page) while the remaining ~600 posts were rated by one of the authors. All raters annotated each post positive (0 to 10) as well as negative (0 to 10) allowing a post to contain both polarities. 62.7 % of all ratings had complete agreement, 34.4 % had partial agreement and only 2.9 % of the posts had complete disagreement. A Fleiss' Kappa of $\kappa = 0.58$ ($\bar{P} = 0.74$ and $\bar{P}_e = 0.38$) shows an acceptable inter-rater agreement amongst our annotators. To improve the quality of the training data, we removed disputed annotations, i.e. the 2.9 % complete disagreement.

Hierarchical Classifier. The idea of our approach is to link two linear SVM classifiers and forward the result from the first classifier to the second. The first classifier determines whether a Facebook post is subjective or objective in nature (cf. [9]). If it is objective, neutral sentiment is assigned. If it is subjective, it is forwarded to the second classifier which determines its polarity.

We used the programming language Python to implement our algorithm, in particular the open-source machine learning library scikit-learn[2]. For tokenization, stop-word removal and stemming we used the open-source library Natural Language Toolkit (NLTK)[3]. In the tokenization step the posts are split into words and cast to all lower characters. After stop word removal, the tokens are converted to their word stem using the Snowball stemmer[4] with the German preset. While the first subjectivity/objectivity classifier yields best results with unigrams, the second polarity classifier performed marginally better combining unigrams and bigrams. Each post was represented by a tf-idf vector. As classification model we used a linear support vector machine[5], a state-of-the-art technique to classify textual content. Sentiment was distributed the following way in our training data set: negative posts: 231, neutral posts: 341 and positive posts: 65. We used 10-fold cross validation to evaluate the performance of the hierarchical classifier and achieved rather high F_1-scores for detecting negative (0.65) and neutral (0.74) sentiment and a low score for the positive sentiment (0.30). It appears that people tend to express negative sentiment more often than positive sentiment leading to a certain skewness in our training data and thus to fewer representatives to model positive sentiment within Facebook posts.

Table 1. Achieved accuracy values on the German Facebook data set; comparison of our hierarchical classifier to three German sentiment analysis libraries.

Sentiment classifier	Accuracy
Hierarchical classifier	0.70
Semantria	0.50
SentiStrength	0.49
AlchemyAPI	0.38

---

[2] http://scikit-learn.org/.
[3] http://www.nltk.org/.
[4] http://snowball.tartarus.org/.
[5] Classifiers such as Naive Bayes and Logistic Regression did not yield better results.

We compared our hierarchical classifier[6] to the three other sentiment analysis libraries with respect to their accuracy values. As illustrated in Table 1, our hierarchical approach to German sentiment analysis achieved the highest accuracy values. We used the publicly available version of the compared libries, i.e. we did not train them with respect to the Facebook domain - the most likely explanation for the lower accuracy values. In addition, we believe that separating opinionated from neutral posts also contributes to the higher accuracy.

4 Application Scenario

Social media monitoring represents a prominent proving ground for natural language processing techniques including sentiment analysis. In this section we show our algorithm's potential to analyze German Facebook entries posted by governmental or economic sources. In the following, we will examine changes in sentiment for the impact of the refugee crisis on Austrian railway operators. In mid summer of 2015 the influx of refugees via the western Balkan route hit a peek of up to 8.500 daily transiting refugees. One of the stops on this route was Vienna to change trains. In this timespan the atmosphere toward refugees was very positive, which is also reflected in the posts to the federal Austrian railway company ÖBB shown at the top of Fig. 1. The private railway operator Westbahn did not get involved in the affair and received close to none feedback

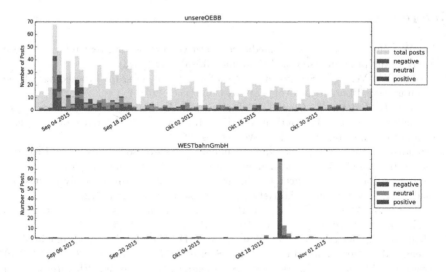

Fig. 1. Sentiment chart displaying classified posts containing the keywords *Flüchtling (refugee), Krise (crisis), Westbahnhof (west station)* for the Facebook pages unsereÖBB **(top)** and WESTbahn GmbH **(bottom)** from Aug. 27 2015 until Nov. 10 2015. (Color figure online)

---

[6] Experiments with a single layer classifier (same settings) led to an accuracy of 0.68 %.

on their Facebook page until they demanded compensation on 21th of October for the losses due to the refugee crisis. This motion had a very bad resonance in press and sparked a public outcry yielding 48 negative posts that day. Many of these posts accuse the company that they tried to make profit from the crisis.

5 Conclusion

In this paper, we implemented a hierarchical sentiment classifier which checks whether a post contains sentiment and if so, a subsequent classifier predicts its polarity. We evaluated our classifier using a set of $\sim$ 640 Facebook posts resulting in 70.0 % accuracy which achieves better results than three established libraries for German such as AlchemyAPI, Semantria and SentiStrength. To make the algorithm accessible, we implemented a CLI[7] where users can enter a Facebook page id and a sentiment chart for a desired time span will be plotted.

Acknowledgements. This work is funded by the KIRAS program of the Austrian Research Promotion Agency (FFG) (project nr. 840824). The Know-Center is funded within the Austrian COMET Program under the auspices of the Austrian Ministry of Transport, Innovation and Technology, the Austrian Ministry of Economics and Labour and by the State of Styria. COMET is managed by the Austrian Research Promotion Agency FFG.

References

1. He, W., Zha, S., Li, L.: Social media competitive analysis and text mining: a case study in the pizza industry. J. Inf. Manag. **33**(3), 464 (2013)
2. Kasper, W., Vela, M.: Sentiment analysis for hotel reviews. In: Proceedings of the Computational Linguistics-Applications Conference (2011)
3. Meehan, K., Lunney, T., Curran, K., McCaughey, A.: Context-aware intelligent recommendation system for tourism. In: IEEE International Conference on Pervasive Computing and Communications Workshops (2013)
4. Momtazi, S.: Fine-grained German sentiment analysis on social media. In: Proceedings of the 8th Conference on Language Resources and Evaluation (2012)
5. Neri, F., Aliprandi, C., Capeci, F., Cuadros, M., By, T.: Sentiment analysis on social media. In: Advances in Social Networks Analysis and Mining (2012)
6. Ortigosa, A., Martín, J.M., Carro, R.M.: Sentiment analysis in facebook and its application to e-learning. Comput. Human Behav. **31**, 527 (2014)
7. Thelwall, M., Buckley, K., Paltoglou, G., Cai, D., Kappas, A.: Sentiment in short strength detection informal text. J. Am. Soc. Inf. Sci. Technol. **61**(12), 2544 (2010)
8. Troussas, C., Virvou, M., Espinosa, K., Llaguno, K., Caro, J.: 4th International Conference on Information, Intelligence, Systems and Applications (2013)
9. Wiebe, J., Wilson, T.: Learning to disambiguate potentially subjective expressions. In: Proceedings of the 6th Conference on Natural Language Learning (2002)

---

[7] The application can be downloaded from the project GitHub repostory https://github.com/fsteinbauer/nldb16-sentiment-analysis.

An Information Retrieval Based Approach for Multilingual Ontology Matching

Andi Rexha[1(✉)], Mauro Dragoni[2], Roman Kern[1], and Mark Kröll[1]

[1] Know-Center GmbH Graz, Graz, Austria
{arexha,rkern,mkroell}@know-center.at
[2] FBK-IRST, Trento, Italy
dragoni@fbk.eu

Abstract. Ontology matching in a multilingual environment consists of finding alignments between ontologies modeled by using more than one language. Such a research topic combines traditional ontology matching algorithms with the use of multilingual resources, services, and capabilities for easing multilingual matching. In this paper, we present a multilingual ontology matching approach based on Information Retrieval (IR) techniques: ontologies are indexed through an inverted index algorithm and candidate matches are found by querying such indexes. We also exploit the hierarchical structure of the ontologies by adopting the PageRank algorithm for our system. The approaches have been evaluated using a set of domain-specific ontologies belonging to the agricultural and medical domain. We compare our results with existing systems following an evaluation strategy closely resembling a recommendation scenario. The version of our system using PageRank showed an increase in performance in our evaluations.

Keywords: Ontology matching · Multilingual · Information Retrieval · PageRank

1 Introduction

As a result of the continuous growth of available artefacts, especially in the Linked Open Data realm, the task of matching ontologies by exploiting their multilinguality has attracted the attention of researchers. If manually executed, it requires experts who have to deeply analyse the artefacts. Automatic systems, on the other hand, need to take into account a lot of aspects in order to suggest suitable matchings between the elements (i.e. "concepts" or "entities") of the artefacts.

The multilingual representation of the entities increases the probability of uniqueness concerning the labels used in a particular language. Such a probability further increases, if we take into account other directly connected entities (parents, children, etc.) as well. When label translations are done manually by experts, the choice of such translations have already been adapted to the

© Springer International Publishing Switzerland 2016
E. Métais et al. (Eds.): NLDB 2016, LNCS 9612, pp. 433–439, 2016.
DOI: 10.1007/978-3-319-41754-7_45

modelled domain. In this paper, we present an approach, inspired by the afore-mentioned principles, for finding matchings between ontologies in a multilingual setting. Such an approach has been implemented within a system based on the use of Information Retrieval techniques adopting a structured representation of each entity for finding candidate matchings.

We evaluate our approach on a set of domain-specific ontologies belonging to the medical and agricultural domains. Furthermore, we investigate the role of hierarchical information in the ontology matching task by applying the PageRank algorithm [9].

2   Related Work

Literature on ontology matching & alignment is very large and many systems and algorithms have been proposed (cf. [1,4]). However, research on ontology matching taking into account multilinguality is quite recent. In [15], the authors quantify the effects on different matching algorithms by discussing several learning matching functions. They show the effectiveness of the approach using a small set of manually aligned concepts from two ontologies. WordNet [6] was one of the first artefacts used for research in multilingual matching (cf. [2]) together with several projects focusing on the creation of WordNet versions for different languages. The two most important ones have probably been EuroWordNet [16] and MultiWordNet [7] whose tasks consist of building language-specific WordNets while keeping them aligned.

In [3], the authors constructed a Chinese-English lexicon and evaluated it in multilingual applications such as machine translation and cross-language information retrieval (CLIR). Another application of multilingual ontology matching for CLIR was described in [17]. Approaches based on the use of information retrieval techniques have not been widely explored even if they demonstrated their suitability for the ontology matching task. IROM [14] and LOM [11] implement a search & retrieve approach for defining mappings between ontologies. With respect to our approach, the IROM system does not take into account multilinguality or disambiguation capabilities for trying to reduce errors during the indexing operation. In conrast, LOM defines mappings by observing the ontologies in only one direction, from a source ontology O_s to a target ontology O_t, without refining these results by considering the contrary workflow.

Recently, ontology matching platforms have been developed and made accessible from external services. In [13], the authors describe an API for multilingual matching and implement two different matching strategies: (i) a direct matching between two ontologies, and (ii) a strategy based on composition of alignments. In [5], the authors discuss several approaches for ontology matching by examining similarities, differences, and identification of weaknesses. In addition to that, they propose a new architecture for a multilingual ontology matching service.

3   An IR-Based Approach for Multilingual Ontology Matching

We define a *matching* as a set of *correspondences or mappings* between entities asserting that a certain relation holds between these two entities. Formally, given two artifacts A_1 and A_2, a matching M between A_1 and A_2 is a set of 5-tuple: $\langle id, e_1, e_2, R, c \rangle$ such that id is a unique identifier of a single matching, e_1 and e_2 are entities of A_1 and A_2 respectively, R is the matching relation between e_1 and e_2 (for example, equivalence ($\equiv$), more general ($\supseteq$), or disjointness ($\perp$)), and c is a confidence measure, typically in the $[0, 1]$ range. In this work, we focus on the equivalence relation.

Our proposed approach takes inspiration by techniques used in the field of Information Retrieval and the process is split into three phases: *in the first one*, we create an index storing structured representation of all entities for each ontology involved in the matching operation. *In the second phase*, the indices built in the previous phase are exploited by performing queries. The structured representation of each entity stored in an index is extracted and converted into a query. Such a query is then invoked on the index containing the ontology to be matched and the top ranked result is then used for defining the mapping. *The third phase* uses the previously provided matches to build a weighted connected graph between the two considered ontologies. We try to overcome the locality of the second step by using the well known voting algorithm PageRank.

Availability of Multilingualism. As introduced in Sect. 1, our approach exploits ontology multilinguality for defining new mappings. Although it makes use of the presence of multiple languages within an ontology, it also works for monolingual or bilingual ontologies. This is an important property, as many of the existing ontologies are currently limited to just one or two languages.

Representation of Information. The main source of information for the matching is the textual information: the "label" associated with each concept described in an ontology, where with the term "label" we mean a string identifying the concept associated with its language tag (i.e. "concept_label@lang_code"). Based on this information we are able to make string-based operations, which are both effective and semantically sensible.

Index Construction. For each entity defined in the ontology we have a set of pairs "label-language", or, in presence of synonyms, we may have more pairs for the same language. Such labels are tokenized, stemmed[1], and normalized by using Natural Language Processing libraries[2]. Each token then is used as an index term with reference to the corresponding ontology entity [12].

---

[1] The list of supported languages is available in the Lucene analyser documentation.

[2] We used the text processors included in the Lucene (http://lucene.apache.org) library. In case of unavailability of libraries for a particular language, the original label is indexed as it is without being processed.

Matches Definition. Once all indices have been created, the matching operation between two ontologies is done by querying them. Given two ontologies, O_1 and O_2, the matching operation is done by performing the following seven steps:

1. For each entity stored in index of O_1, its structured representation is transformed into a query that is then invoked on the index containing the structured representation of O_2. The top ranked candidate matching (i.e. the first query result) is temporarily stored associated with its confidence score.
2. The same procedure is performed for entities contained in the index of O_2, where their structured representation is used for querying the index containing the structured representation of O_1.
3. For each matching $A_{O_1} \rightarrow B_{O_2}$ built by querying O_2 with information coming from O_1, we also verify if $B_{O_2} \rightarrow A_{O_1}$ holds.
4. If a matching $C_{O_2} \rightarrow A_{O_1}$ exists, we save the matching between A_{O_1} and the entity having the higher confidence score. If further entities from O_2 are candidate matchings for A_{O_1}, the highest confidence score is considered.
5. Otherwise if, by querying O_1 using structured representation stored in O_2, A_{O_1} is not mapped with any concept, the matching $A_{O_1} \rightarrow B_{O_2}$ is maintained and saved in the final output.
6. The same process is repeated using the results obtained by querying O_1 as well as using structured representations of O_2.
7. Each found candidate with lower confidence score than a threshold (0.5), is ignored.

As mentioned above, each query is created from the structured representation of entities stored in the indices. Once the query is built, a search operation on the related index is performed which returns for each match a score calculated from the following formula:

$$score(R_{c_1}, R_{c_2}) = coord(R_{c_1}, R_{c_2}) \cdot \sum_{x \in R_{c_1}} tf(x \in R_{c_2}) \cdot idf(x)^2, \qquad (1)$$

where R_{c_1} and R_{c_2} are respectively the representations of concepts defined in the source and in the target ontologies, $tf()$ and $idf()$ are the standard "term frequency" and "inverse document frequency" functions used in IR [12], and $coord(R_{c_1}, R_{c_2})$ represents the number of labels defined in the representation of c_1 occurring in the representation of c_2.

PageRank. To improve the matching concepts, we exploit the hierarchical structure of the ontologies/artifacts by using PageRank. In doing so, we aim to favor the correct matches and to penalize the mismatches. Given two ontologies O_1 and O_2, we conduct the following six steps to build the graph to be capable of applying the PageRank algorithm:

1. Each concept and individual of an ontology is represented as a vertex of the new graph.
2. Each hierarchical information (isA, broader, narrower, etc.) is represented as two directed edges in the graph. Each edge has a unitary weight.

3. Once the candidate matching is calculated by the Inverted Index, we create a new vertex and two edges with each of the matching candidates.
4. We weight each edge of the matching with the result given from the Inverted Index.
5. After creating the weighted graph, we apply the PageRank algorithm and extract the voting results from the voting vertices.
6. We rank the matching vertices in order to select the ones with the highest value (the correct candidate for our algorithm).

4   Evaluation

We evaluate our approach with two different setups and use the following ontologies that do provide multilingual labels: Agrovoc, Eurovoc, and Gemet that belong to the agriculture and environment domains, and MeSH, SNOMED, and MDR that belong to the medical one[3]. In the first setup, we evaluate our algorithm as a recommender system for matching concepts/individuals of ontologies. We compare the effectiveness of our indexing system with two state-of-the-art multilingual ontology matching algorithms only on the matching concepts/individuals. In particular, we consider the "WeSeE" [10] and the "YAM++ 2013" [8][4] systems, since these are the only ones implementing a multilingual ontology matching strategy. In the second setup, we evaluate the overall performance as an ontology matching system by using only the indexing approach. Finally, we estimate the impact of the PageRank algorithm applied to the hierarchy of the ontology.

Results. Table 1 shows the performance (expressed as F1-measure) of the solely indexing approach compared to the other systems. From the table we can see that our system outperforms "YAM++ 2013" and performs similar to "WeSeE". We believe that this is due to the use of external resources (search the web and using automatic translation of the labels) in contrast to our system which uses only the available labels.

In Table 2, on the left side we present the overall performance of the system evaluated on all the concepts/individuals. On the right side, we evaluate the system after the application of the PageRank algorithm and filter the matches.

The results show that the application of the PageRank algorithm has an impact of almost 10 % on the F1-measure of all the matchings. Another aspect to highlight is that in most of the cases the recall is much higher than the precision. We believe that this is due to the indexing approach. We filter the results of the query based on a threshold that could include a lot of noise and not existing matches, namely decreasing the precision of the system.

---

[3] The ontologies, their indexed version, and gold standard can be found at https://db.tt/p959AWhO.
[4] The 2012 version of the YAM++ system yields the same results as the 2013 version.

Table 1. Resulting F1-Measures of WeSeE, YAM++ and our system on the set of six multilingual ontologies. (Recommender setting)

Mapping set	Proposed	WeSeE	YAM++
	System		2013
Agrovoc ⇔ Eurovoc	**0.817**	0.785	0.615
Gemet ⇔ Agrovoc	**0.807**	0.726	0.579
MDR ⇔ MeSH	0.639	**0.749**	0.613
MDR ⇔ SNOMED	**0.677**	0.624	0.473
MeSH ⇔ SNOMED	0.507	**0.631**	0.458

Table 2. Resulting precision, recall and F1-Measure values of WeSeE, YAM++ and our system on the set of six multilingual ontologies. (Indexing and PageRank setting)

Mapping set	Indexing approach			PageRank approach		
	Precison	Recall	F-measure	Precison	Recall	F-measure
Agrovoc ⇔ Eurovoc	0.356	0.664	0.464	0.406	0.770	0.532
Gemet ⇔ Agrovoc	0.282	0.661	0.396	0.405	0.714	0.517
MDR ⇔ MeSH	0.173	0.405	0.242	0.264	0.420	0.324
MDR ⇔ SNOMED	0.307	0.314	0.311	0.316	0.396	0.351
MeSH ⇔ SNOMED	0.360	0.194	0.253	0.320	0.405	0.358

5   Conclusion

In this paper, we presented an approach for defining matches between ontologies based on PageRank, an IR-based technique, applied in either a cross-lingual and multilingual contexts. We implemented our approach in a real-world system and validated it on a set of domain-specific ontologies. The obtained results demonstrate the feasibility of the proposed approach from either the effectiveness and efficiency points of view.

Acknowledgements. The Know-Center is funded within the Austrian COMET Program under the auspices of the Austrian Ministry of Transport, Innovation and Technology, the Austrian Ministry of Economics and Labour and by the State of Styria. COMET is managed by the Austrian Research Promotion Agency FFG.

References

1. Bellahsene, Z., Bonifati, A., Rahm, E. (eds.): Schema Matching and Mapping. Springer, Heidelberg (2011)
2. Daudé, J., Padró, L., Rigau, G.: Mapping multilingual hierarchies using relaxation labeling. In: Joint SIGDAT Conference on Empirical Methods in Natural Language Processing and Very Large Corpora (1999)

3. Dorr, B.J., Levow, G.-A., Lin, D.: Building a Chinese-English mapping between verb concepts for multilingual applications. In: White, J.S. (ed.) AMTA 2000. LNCS (LNAI), vol. 1934, pp. 1–12. Springer, Heidelberg (2000)
4. Euzenat, J., Shvaiko, P.: Ontology Matching. Springer, Heidelberg (2007)
5. al Feel, H.T., Schäfermeier, R., Paschke, A.: An inter-lingual reference approach for multi-lingual ontology matching. Int. J. Comput. Sci. (2013)
6. Fellbaum, C.: WordNet: An Electronic Lexical Database. Bradford Books, Cambridge (1998)
7. Magnini, B., Strapparava, C., Ciravegna, F., Pianta, E.: Multilingual lexical knowledge bases: applied wordnet prospects. In: Proceedings of the International Workshop on the Future of the Dictionary (1994)
8. Ngo, D., Bellahsene, Z.: YAM++ results for OAEI 2013. In: Workshop on Ontology Matching (OM) (2013)
9. Page, L., Brin, S., Motwani, R., Winograd, T.: The pagerank citation ranking: Bringing order to the web. Technical report, Stanford InfoLab (1999)
10. Paulheim, H.: Wesee-match results for OEAI 2012. In: Workshop on Ontology Matching (OM) (2012)
11. Pirrò, G., Talia, D.: LOM: a linguistic ontology matcher based on information retrieval. J. Inf. Sci. **34**(6), 845 (2008)
12. van Rijsbergen, C.J.: Information Retrieval. Butterworth, Oxford (1979)
13. dos Santos, C.T., Quaresma, P., Vieira, R.: An API for multi-lingual ontology matching. In: Proceedings of the International Conference on Language Resources and Evaluation (2010)
14. Mousselly-Sergieh, H., Unland, R.: IROM: information retrieval-based ontology matching. In: Declerck, T., Granitzer, M., Grzegorzek, M., Romanelli, M., Rüger, S., Sintek, M. (eds.) SAMT 2010. LNCS, vol. 6725, pp. 127–142. Springer, Heidelberg (2011)
15. Spohr, D., Hollink, L., Cimiano, P.: A machine learning approach to multilingual and cross-lingual ontology matching. In: Aroyo, L., Welty, C., Alani, H., Taylor, J., Bernstein, A., Kagal, L., Noy, N., Blomqvist, E. (eds.) ISWC 2011, Part I. LNCS, vol. 7031, pp. 665–680. Springer, Heidelberg (2011)
16. Vossen, P.: Introduction to EuroWordNet. Comput. Humanit. **32**(2–3), 73 (1998)
17. Zhang, L., Wu, G., Xu, Y., Li, W., Zhong, Y.: Multilingual collection retrieving via ontology alignment. In: Proceedings of International Conference on Asian Digital Libraries (2004)

Automatic Text Summarization with a Reduced Vocabulary Using Continuous Space Vectors

Elvys Linhares Pontes[1](✉), Stéphane Huet[1], Juan-Manuel Torres-Moreno[1,2], and Andréa Carneiro Linhares[3]

[1] LIA, Université d'Avignon et des Pays de Vaucluse, Avignon, France
elvys.linhares-pontes@alumni.univ-avignon.fr
[2] École Polytechnique de Montréal, Montréal, Canada
[3] Universidade Federal do Ceará, Sobral, CE, Brazil

Abstract. In this paper, we propose a new method that uses continuous vectors to map words to a reduced vocabulary, in the context of Automatic Text Summarization (ATS). This method is evaluated on the MultiLing corpus by the ROUGE evaluation measures with four ATS systems. Our experiments show that the reduced vocabulary improves the performance of state-of-the-art systems.

Keywords: Word embedding · Text summarization · Vocabulary reduction

1 Introduction

Nowadays, the amount of daily generated information is so large that it cannot be manually analyzed. Automatic Text Summarization (ATS) aims at producing a condensed text document retaining the most important information from one or more documents; it can facilitate the search for reference texts and accelerate their understanding.

Different methodologies based on graphs, optimization, word frequency or word co-occurrence have been used to automatically create summaries [12]. In the last years, Continuous Space Vectors (CSVs) have been employed in several studies to evaluate the similarity between sentences and to improve the summary quality [1,5,10]. In this paper, we introduce a novel use of CSVs for summarization. The method searches for similar words in the continuous space and regards them as identical, which reduces the size of vocabulary. Then, it computes metrics in this vocabulary space seen as a discrete space, in order to select the most relevance sentences.

After a brief reminder on the implementation of neural networks to build CSVs in Sect. 2, Sects. 3 and 4 describe the previous works that used CSVs for summarization and our method respectively. In Sect. 5, we present the evaluation

E.L. Pontes and J.-M. Torres-Moreno — This work was partially financed by the European Project CHISTERA-AMIS ANR-15-CHR2-0001.

E. Métais et al. (Eds.): NLDB 2016, LNCS 9612, pp. 440–446, 2016.
DOI: 10.1007/978-3-319-41754-7_46

of the proposed approach with four ATS systems. Our results show that the reduced vocabulary in a discrete space improves the performance of the state-of-the-art. Section 6 concludes with a summary and future work.

2   Neural Networks and Continuous Space Vectors

Artificial Neural Networks (ANNs) have been successfully applied in diverse Natural Language Processing applications, such as language modeling [3,8], speech recognition or automatic translation. An ANN is a system that interconnects neurons and organizes them in input, hidden and output layers. The interest of these models has recently been renewed with the use of deep learning which updates the weights of the hidden layers to build complex representations.

Mikolov et al. [8] developed a successful approach with the so-called Skip-gram model to build continuous word representations, i.e., word embeddings. Their model aims at predicting a word, basing its decision on other words in the same sentence. It uses a window to limit the number of words used; e.g. for a window of 5, the system classifies the word w from the 5 words before and the 5 words after it. Given a sequence of training words $w_1, w_2, w_3, ..., w_N$, the objective of the Skip-gram model is to maximize the average log probability:

$$\frac{1}{N} \sum_{t=1}^{N} \sum_{-c \leq j \leq c, j \neq 0} \log p(w_{t+j} \mid w_t) \tag{1}$$

where c is the window size and N is the number of words in the training set.

3   Related Work

Several works have used word embeddings to measure sentence similarity, which is central to select sentences for text summarization. Kågebäck et al. [5] used continuous vectors as semantically aware representations of sentences (phrase embeddings) to calculate the similarity, and compared the performance of different types of CSVs for summarization. Balikas and Amini [1] also analyzed various word embedding representations to summarize texts in English, French and Greek languages. They proposed an autoencoder model to learn a language independent representation for multilingual sentences and determine the similarity between sentences. Phung and De Vine [10] used word embeddings to calculate the sentence similarity measures for the PageRank system to select the most significant sentences. They compare this PageRank version with other systems based on TF-IDF and different variants of Maximum Marginal Relevance (MMR).

Our methodology differs from the previously described methods because we reduce the vocabulary using a CSV-based similarity between words. This first step allows us to calculate more accurately the similarity and the relevance of sentences in a discrete space.

4   Reduced Vocabulary

Words can be represented by two main kinds of vectors: Discrete Space Vectors (DSVs) and CSVs. In DSVs, words are independent and the vector dimension varies with the used vocabulary. Thus similar words (i.e., "home" and "house", "beautiful" and "pretty") have different representations. For statistical techniques, this independence between similar words complicates the analysis of sentences with synonyms.

CSVs are a more compelling approach since similar vectors have similar characteristics and the vector dimension is fixed. For CSVs (word embeddings), it is possible to identify similar characteristics between words. For example, the words "home", "house" and "apartment" have the same context as "home" and have therefore similar vectors. However, the existing methods to calculate the sentence relevance are based on DSVs. We use CSVs to identify and replace the similar words to create a new vocabulary with a limited semantic repetition. From this reduced vocabulary, statistical techniques can identify with DSVs the similar content between two sentences and improve the results.

A general and large corpus is used to build the word embedding space. Our method calculates the nearest words in this space for each word of the texts to create groups of similar words, using a cosine distance. Then it replaces each group of similar words by the most frequent word in the group. For example, the nearest word of "home" is "house" and the word "home" is more frequent than "house" in the text, so we replace the word "house" by "home". Let us note that these substitutions are only used to compute sentence similarities but that the original words are kept in the produced summary. We devised the greedy Algorithm 1 to find the similar words of w in the texts among a pre-compiled list lcs of CSVs generated on the large corpus.

Algorithm 1. Reduce vocabulary of *text*

Input: n (neighborhood size), lcs (list of words inside continuous space), *text*
for each word w_t in *text* **do**
   if w_t is in lcs **then**
      $nset \leftarrow \{w_t\}$
      $nlist \leftarrow [w_t]$
      while $nlist$ is not empty **do**
         $w_l \leftarrow nlist.pop(0)$
         $nw \leftarrow$ the n nearest words of w_l in lcs
         $nlist.add((nw \cap$ vocabulary of $text) \setminus nset)$
         $nset \leftarrow nset \cup (nw \cap$ vocabulary of $text)$
      end while
      Replace in *text* each word of $nset$ by the most frequent of $nset$
   end if
end for
Return *text*

---

5   Experiments and Results

The reduced vocabulary approach was evaluated with four different systems. The first simple system (named "base") generates an extract with the sentences that are the most similar to the document. The second system (MMR) produces a summary based on the relevance and the redundancy of the sentences [2]. With the objective of analyzing different methodologies to calculate the relevance and the similarity of sentences (e.g. word co-occurrence, TF-ISF[1]...), we use two other systems: Sasi [11] and TextRank [7].

Pontes et al. [11] use Graph theory to create multi-document summaries by extraction. Their so-called Sasi system models a text as a graph whose vertices represent sentences and edges connect two similar sentences. Their approach employs TF-ISF to rank sentences and creates a stable set of the graph. The summary is made of the sentences belonging to this stable set.

TextRank [7] is an algorithm based on graphs to measure the sentence relevance. The system creates a weighted graph associated with the text. Two sentences can be seen as a process of recommendation to refer to other sentences in the text based on a shared common content. The system uses the Pagerank system to stabilize the graph. After the ranking algorithm is run on the graph, the top ranked sentences are selected for inclusion in the summary.

We used for our experiments the 2011 MultiLing corpus [4] to analyze the summary quality in the English and French languages. Each corpus has 10 topics, each containing 10 texts. We concatenated the 10 texts of each topic to convert multiple documents into a single text. There are between 2 and 3 summaries created by human (reference summaries) for each topic. We took the LDC Gigaword corpus (5th edition for English, 3rd edition for French) and the word2vec package[2] to create the word embedding representation, the vector dimension parameter having been set to 300. We varied the window size between 1 and 8 words to create a dictionary of word embeddings. A neighborhood of between 1 and 3 words in the continuous space was considered to reduce the vocabulary (parameter n of Algorithm 1). Finally, the summaries produced by each system have up to 100 words.

The compression rate using the Algorithm 1 depends on the number n of the nearest words used. Table 1 reports the average compression ratio for each corpus in the word embedding space for three values of n. For the English language, a good compression happens using 1 or 2 nearest words, while the vocabulary compression for the French language is not so high because the French Gigaword corpus is smaller (925M words) than the English Gigaword corpus (more than 4.2G words). Consequently, a higher number of words of the text vocabulary are not in the dictionary of French word embeddings.

In order to evaluate the quality of the summaries, we use the ROUGE system[3], which is based on the intersection of the n-grams of a candidate summary

---

[1] Term Frequency - Inverse Sentence Frequency.

[2] Site: https://code.google.com/archive/p/word2vec/.

[3] The options for running ROUGE 1.5.5 are -a -n 2 -x -m -2 4 -u -c 95 -r 1000 -f A -p 0.5 -t 0.

Table 1. Compression ratio of vocabulary for different numbers of nearest words (n) considered with CSVs.

Language	Compression ratio		
	$n = 1$	$n = 2$	$n = 3$
English	11.7 %	20.1 %	25.3 %
French	7.1 %	12.3 %	16.1 %

and the n-grams of a set of reference summaries. More specifically, we used ROUGE-1 (RG-1) and ROUGE-2 (RG-2). These metrics are F-score measures whose values belong to $[0, 1]$, 1 for the best result [6].

We evaluate the quality systems using DSVs, CSVs and our approach, which results in 3 versions for each system. The default version uses the cosine similarity as similarity measure for the base, MMR and Sasi systems with DSVs; the TextRank system calculates the similarity between two sentences based on the content overlap of DSVs. In the "cs" version, all systems use the phrase embedding representation for the sentences as described in [5] and employ the cosine similarity as similarity measure. Finally, the "rv" version (our method) uses a reduced vocabulary and the same metrics as the default version with DSVs. After selecting the best sentences, all system versions create a summary with the original sentences.

Despite the good compression rate with $n = 2$ or 3, the best summaries with a reduced vocabulary were obtained when taking into account only one nearest word and a window size of 6 for word2vec. Table 2 shows the results for the English and French corpora. Almost all the "cs" systems using the continuous space and the reduced vocabulary are better than the default systems.

For the English corpus, the "rv" versions obtain the best values, which indicates that the reduced vocabulary improves the quality of the similarity calculus and the statistical metrics. The difference in the results between English and French is related to the size of the corpus to create word embeddings. Since

Table 2. ROUGE F-scores for English and French summaries. The bold numbers are the best values for each group of systems in each metric. A star indicates the best system for each metric.

Systems	English		French		Systems	English		French	
	RG-1	RG-2	RG-1	RG-2		RG-1	RG-2	RG-1	RG-2
base	0.254	0.053	0.262	**0.059**	Sasi	0.251	0.053	0.248	0.047
base_cs	**0.262**	**0.054**	0.261	0.057	Sasi_cs	0.247	**0.058**	**0.251**	0.047
base_rv	**0.262**	**0.054**	**0.264**	0.054	Sasi_rv	**0.253**	0.053	0.244	**0.050**
MMR	0.262	**0.058**	0.270	0.059	TextRank	0.251	0.056	0.267	0.063
MMR_cs	0.260	0.053	**0.277***	**0.072***	TextRank_cs	**0.261**	0.056	**0.276**	**0.065**
MMR_rv	**0.265***	**0.058**	0.270	0.059	TextRank_rv	0.260	**0.062***	0.268	0.058

the French training corpus is not as big, the precision of the semantic word relationships is not accurate enough and the closest word may not be similar. Furthermore, the French word embedding dictionary is smaller than for English. Consequently, the "rv" version sometimes does not find the true similar words in the continuous space and the reduced vocabulary may be incorrect. The "cs" version mitigates the problem with the small vocabulary because this version only analyzes the words of the text that exist in the continuous space. Thus the "cs" version produces better summaries for almost all systems.

6   Conclusion

We analyzed the summary quality of different systems using Discrete Space Vectors and Continuous Space Vectors. Reducing the text vocabulary with a CSV-based similarity using a big training corpus produced better results than the methods described in [5] for English, but lower for French.

As future work, we will increase the French training corpus to extend the dictionary of word embeddings, and use other methodologies to create continuous space vectors, such as [3,9]. Furthermore, new methods have still to be devised to fully exploit CSVs to calculate the sentence relevance.

References

1. Balikas, G., Amini, M.R.: Learning language-independent sentence representations for multi-lingual, multi-document summarization. In: 17ème Conférence Francophone sur l'Apprentissage Automatique (CAp) (2015)
2. Carbonell, J., Goldstein, J.: The use of MMR, diversity-based reranking for reordering documents and producing summaries. In: SIGIR, pp. 335–336 (1998)
3. Collobert, R., Weston, J.: A unified architecture for natural language processing: deep neural networks with multitask learning. In: ICML, pp. 160–167 (2008)
4. Giannakopoulos, G., El-Haj, M., Favre, B., Litvak, M., Steinberger, J., Varma, V.: TAC2011 multiling pilot overview. In: TAC (2011)
5. Kågebäck, M., Mogren, O., Tahmasebi, N., Dubhashi, D.: Extractive summarization using continuous vector space models. In: 2nd EACL Workshop on Continuous Vector Space Models and their Compositionality (CVSC), pp. 31–39 (2014)
6. Lin, C.Y.: ROUGE: a package for automatic evaluation of summaries. In: ACL Workshop on Text Summarization Branches Out (2004)
7. Mihalcea, R., Tarau, P.: Textrank: bringing order into texts. In: EMNLP, pp. 404–411 (2004)
8. Mikolov, T., Sutskever, I., Chen, K., Corrado, G.S., Dean, J.: Distributed representations of words and phrases and their compositionality. In: NIPS, pp. 3111–3119 (2013)
9. Pennington, J., Socher, R., Manning, C.: Glove: Global vectors for word representation. In: Proceedings of the 2014 Conference on Empirical Methods in Natural Language Processing (EMNLP), pp. 1532–1543. Association for Computational Linguistics, Doha, October 2014
10. Phung, V., De Vine, L.: A study on the use of word embeddings and pagerank for vietnamese text summarization. In: 20th Australasian Document Computing Symposium, pp. 7:1–7:8 (2015)

11. Pontes, E.L., Linhares, A.C., Torres-Moreno, J.M.: Sasi: sumarizador automático de documentos baseado no problema do subconjunto independente de vértices. In: XLVI Simpósio Brasileiro de Pesquisa Operacional (2014)
12. Torres-Moreno, J.M.: Automatic Text Summarization. Wiley, Hoboken (2014)

A Platform for the Conceptualization of Arabic Texts Dedicated to the Design of the UML Class Diagram

Kheira Zineb Bousmaha[1](✉), Mustapha Kamel Rahmouni[1],
Belkacem Kouninef[2], and Lamia Hadrich Belguith[3]

[1] LRIIR, Department of Computer Science,
University of Oran 1, Ahmed Ben Bella, Oran, Algeria
kzbousmaha@yahoo.fr, kamelrahmouni1946@gmail.com
[2] National Institute of Telecommunication and Information and Communication
Technology of Oran (INTTIC), Oran, Algeria
bkouninef@ito.dz
[3] MIRACL, Faculty of Economics and Management of Sfax (FSEGS),
Department of Computer Science, University of Sfax, Sfax, Tunisia
l.belguith@fsegs.rnu.tn

Abstract. In many fields using information systems (IS), knowledge is often represented by UML models, in particular, by including class diagrams. This formalism has the advantage of being controlled by a large community and therefore the perfect means of exchange. The desire to use an automated tool that formalizes the intellectual process of the expert from an IS specification texts seems interesting. Our problem is devoted to the presentation of a new strategy that allows us to move from an informal to a semi-formal representation model, which is the UML class diagram. This issue is not new. It has aroused great interest for a long time. The originality of our work is that these texts are in Arabic.

1 Introduction

The specification of an IS is a mutual work done between the client, who is the only one to really know the problem and the designers who need to be helped to express it clearly. These needs have to be simple and specified in a language that can be understood by different sides of the project. The development of semi-formal models like UML and its derivatives from requirements specification can be long and tedious. Its automation leads to many challenges in different scientific fields such as requirements engineering, knowledge representation, the automatic processing of language, information extraction and knowledge engineering.

Several works are been interested to theses researches in the past [1, 2] and recently [3–5]. Also, many studies have focused on automating and semi automating this process to extract a UML model from natural language text [4–7]. The approaches proposed use most of time NLP (Natural Language Processing) techniques. The lack of formal semantics which hampers the UML language can lead to serious modeling

© Springer International Publishing Switzerland 2016
E. Métais et al. (Eds.): NLDB 2016, LNCS 9612, pp. 447–452, 2016.
DOI: 10.1007/978-3-319-41754-7_47

problems that generate contradictions in the developed models [3]. This motivated more work on the transition from UML to formal languages such as B [8], VDM [1], VDM++ [9], Z [10], Maude [11].

We propose in this paper a platform for the conceptualization of the texts of specification of an information system, dedicated to the design of the UML class diagram based on hybrid approaches that encompass both linguistic and statistical approaches.

The original contribution of our work is the fact that the specification texts are in Arabic. The Arabic language is especially challenging because of its complex linguistic structures. It has rich and complex morphological, grammatical, and semantic aspects since it is a highly inflectional and derivational language that includes root, prefixes, suffixes, and clitics. Particles are used to complete the meaning of verbs and nouns. Different words, with different meanings and pronunciations, often differ only by their diacritics. Besides the classical phenomena that exist in Latin languages, there are specific difficulties in the Arabic language that generate problems in its different process tasks, such as the agglutination, the free order language, the absence of short vowels which cause artificial syntactic and semantic ambiguities and complicate the grammar construction and other issues [12].

The rest of this paper is organized as follows; we present our platform approach and its different models in Sect. 2. Then we show the experiments we conducted and the first version results obtained and finally some perspectives for this work are given in Sect. 3.

2   The AL2UML Platform

The strategy we propose can be illustrated by the Fig. 1. The texts constitute the foundation of modeling. From them, it is possible to extract a linguistic model that contains the elements that are expressed in the texts. The conceptual plan corresponds to a result of the modeling whose automation is done through the linguistic model; but moving from the latter to the conceptual model can only be done by making significant modeling choices, which pertain to both the granularity of the desired description and the objectives of the modeling. The passage from the conceptual model represented by UML class diagram to the event_B model is in progress. Our ambition is to develop an IDM platform (model-driven engineering) dedicated to this purpose that will allow to link conduct the modeling activities and the formal validation of the functional model.

2.1   The Linguistic Model

To establish the linguistic model and proceed to the pretreatment of the texts of specification, we have designed a tool Alkhalil+ [12], it does a considerable amount of tasks on MSA[1], Segmentation[2], tokenization, morphological analysis[3], lemmatization,

---

[1] Modern Standard Arabic.

[2] STAR split the text into sentences we use 147 rules established by linguistics.

[3] Alkhalil Morpho Sys, Version 1.3, Un Open Source http://sourceforge.net/projects/alkhalil/.

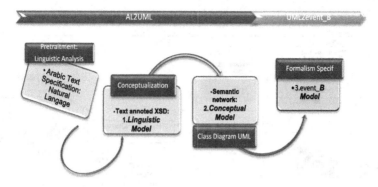

Fig. 1. The models of the platform AL2UML

part of-speech (POS) tagging[4], disambiguation and diacritisation[5]. The experiments have carried a disambiguation rate that is above 85 %. The output is an XML file containing the information that is necessary for the conceptual model. We have established 6 typology of verbs, for example *Class of verb of State*: includes the verbs that means an **hyponym**y. Example: تنقسم, تتجزأ, تتفرع.

2.2 The Conceptual Model: Architecture and Approach

As shown in Fig. 2, the proposed approach starts by extracting terms with the formula tf.idf and compound terms with a hybrid method which combine linguistic patterns with a statistical method based on the mutual information (MI) [13].

The second step is the design of the chunker. We have established a list of categories of chunks that were necessary for the classification of the sentences in order to extract the meaning. Only the chunks pertinent, supposed describe the structural description of the future IS are selected and which roles are attributed to them. We eliminate the treatment of temporal and behavioral syntagms types. Then we proceed to the classification of these sentences according to sentences patterns that we have already determined. The extracted information is then represented by a semantic network and a set of design patterns as well as heuristics containing the know-how and experience of the design experts are applied, which leads to generating the corresponding class diagram.

2.2.1 The Semantics Analysis

Our approach begins with the assignment of roles to the various components used in the sentence. Then, we proceed to a classification of its sentences into patterns of sentences. A semantic network is generated as Output in order to represent the whole of the extracted information.

---

[4] We implemented a set of grammar rules as a set of ATN (augmented transition network).

[5] We apply a method based on decision theory.

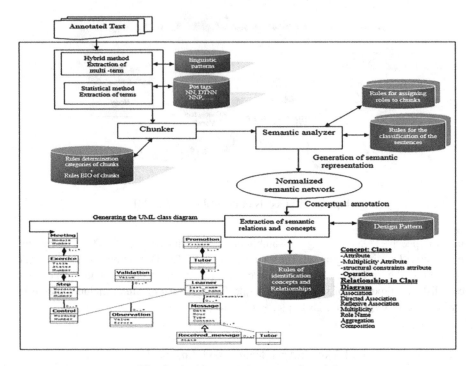

Fig. 2. Architecture of conceptual model

The Verbal sentence processing: it is recognized by the presence of a verbal chunk (VP). The VP is recognized by the identification of a verb, verbal noun, active participle, passive participle or the particle 'ل'. We look for the verb class and the recognition of the semantic relations that link the different chunks to the verb, in order to assign suitable roles (Ci) to the different words and chunks composing this sentence. We have identified 12 roles, these roles are inspired and adapted from the semantic of the casual theory of Fillmore. We have defined for each class of verb a role assignment algorithm.

The Nominal Sentence processing: The relationship is deduced by a prepositional chunk. It may even be implicit, in this case, we look for a pivot element (the subject of the action) in the nominal chunk, identified by its pos tag and by its position in this chunk. We have established algorithms that assign roles for each chunk of the sentence. Usually, the nominal sentence describes either an association relationship or an inheritance relationship.

The sentence pattern: The Sentences patterns allow us to stereotype the sentences; we have classified them according to 9 schemas (plans). This classification allows us to determine a first interpretation of the specification text.

Example of Action Schema: All sentences result from this combination of roles are classified as Action schema: *Verb from the Action class + actor Role + target Role + multiplicity Role,* The sentence below is ranked among the Action schema.

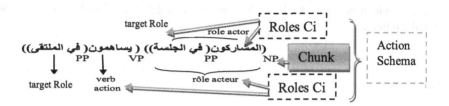

The role of multiplicity is automatically deduced from the multiplicity tag of the word "المشاركون".

2.2.2   The Semantic Network and UML Class Diagram Construction

We have represented this knowledge by semantic network. For each sentence schema, we have applied specific algorithms for processing nodes and arcs, for example, for the structural schema, the words having as Role possessor subject and the one of attribute will all be transformed into entity nodes linked together by a 'poss' arc, the entities of attribute roles will be the target nodes. The words or the chunks with a constraint Role in the former sentence, will be transformed into constraint nodes, then connected by arcs 'ct' with the entity target nodes. Once the network is generated, we proceed to its normalization. **For extracting the concepts of the class diagram,** we have defined rules data base encompassing a set of 60 rules and heuristics that we have applied to the semantic network in order to extract the concepts and relations for the class diagram Once the diagram is generated, we apply to it a set of design pattern in order to verify its conformity, its consistency and its completeness. We have considered two outputs, one in the form of an XSD file for better interoperability and another under graphical form.

3   Conclusion and Future Work

We have presented a platform (AL2UML) for the semi-formalization of specification requirements from NL to an UML class diagram. With this approach, we have achieved encouraging results. For the evaluation of our design, we have taken texts from the practical exercises of the 'software engineering' course taught to then 2nd year students in computer science of our university. There were 14 exercises in the chapter entitled 'modeling with UML'. We first, solved them manually then by using the tool. The first results we had, for a text comprising only simple sentences an f-measure greater than 93 % in the generation of the class diagram. The f-measure of each concept were in the order of 95 % for the extraction of class, and more than 92 % for the extraction of attribute, operation, and relation. This f-measure would decrease as the sentence became more complicated, containing negative forms (f-measure = 63,3825 %) or anaphora and ellipsis (f-measure = 41,3825 %) or a complex formulation.

As an extension, first we plan to take into account the treatment of the complex sentence, the anaphora and ellipsis, the synonymy by using Ontology AWN for the later. Second, we intent to complete the UML model by the OCL constraint expression language for processing constraints. Finally, we have to address the lack of formal semantics which penalizes UML in the developed models for a transition to Event_B formal language.

References

1. Meziane, F.: From english to formal specifications. Thesis, University of Salford, UK (1994)
2. Fougères, A., Trigano, P.: Rédaction de spécifications formelles: Élaboration à partir des spécifications écrites en langage naturel. Cahiers Romans de Sciences Cognitives 1(8), 29–36 (1997)
3. Sadoun, D.: Des spécifications en langage naturel aux spécifications formelles via une ontologie comme modéle pivot. Thesis, Université de Paris Sud (2014)
4. Landhäußer, M., Körner, S.J., Tichy, W.F.: From requirements to UML models and back how automatic processing of text can support requirements engineering. Softw. Qual. J. 22(1), 1–29 (2013)
5. Bajwa, I.S., Bordbar, B., Lee, M., Anastasakis, K.: Nl2alloy: a tool to generate alloy from nl constraints. J. Digital Inf. Manag. 10(6), 365–372 (2012)
6. Ilieva, M., Boley, H.: Representing textual requirements as graphical natural language for UML diagram generation. In: SEKE 2008, pp. 478–483 (2008)
7. Meziane, F., Vadera, S.: Obtaining E-R diagrams semi-automatically from natural language specifications (2010)
8. Mammar A., Laleau, R.: An automatic generation of B specifications from well-defined UML notations for database applications. In: International Symposium on Programming Systems, Alger, Algérie (2001)
9. Mit, E.: Developing VDM++ operations from UML diagrams. Thesis, School of Computing, Science and Engineering, University of Salford, UK (2007)
10. Shroff, M., France, R.: Towards a formalization of UML class structures in Z. In: Proceedings of the 21th International Computer Software and Applications Conference–COMPSAC 1997. IEEE (1997)
11. Mokhati, F., Badri, M.: Generating Maude specifications from UML use case diagrams. J. Object Technol. 8(2), 119–136 (2009)
12. Bousmaha, K.Z., Rahmouni, M.K., Kouninef, B., Belguith-Hadrich, L.: A hybrid approach for the morpho-lexical disambiguation of Arabic: Alkhalil+. J. Inf. Process. Syst. Jips (2016) (à paraitre)
13. Church, K., Hanks, P.: Word association norms, mutual information, and lexicography. Comput. Linguist. 16(1), 22–29 (1989). mars

Reversing the Polarity with Emoticons

Phoey Lee Teh[1(✉)], Paul Rayson[2], Irina Pak[1], Scott Piao[2], and Seow Mei Yeng[1]

[1] Department of Computing and Information Systems, Sunway University, Bandar Sunway, Malaysia
phoeyleet@sunway.edu.my, ipak1992@gmail.com, stephaniee29s@gmail.com
[2] School of Computing and Communications, Lancaster University, Lancaster, UK
{p.rayson, s.piao}@lancaster.ac.uk

Abstract. Technology advancement in social media software allows users to include elements of visual communication in textual settings. Emoticons are widely used as visual representations of emotion and body expressions. However, the assignment of values to the "emoticons" in current sentiment analysis tools is still at a very early stage. This paper presents our experiments in which we study the impact of positive and negative emoticons on the classifications by fifteen different sentiment tools. The "smiley" :) and the "sad" emoticon :(and raw-text are compared to verify the degrees of sentiment polarity levels. Questionnaires were used to collect human ratings of the positive and negative values of a set of sample comments that end with these emoticons. Our results show that emoticons used in sentences are able to reverse the polarity of their true sentiment values.

Keywords: Sentiment · Emoticons · Polarity · Emotion · Social media

1 Introduction

An emoticon is a symbol which includes "emotion" and "icon" [1]. Mobile and online text messaging platforms often include emoticons. It also known as "Emoji" or "Facemarks". In this paper, we study the effect of positive and negative emoticons by testing fifteen different sentiment tools in order to investigate the impact of the emoticons on the sentiment classification of the short messages by different sentiment analysis tools. In addition, we simulated 30 comments, similar to the comparison of sentiment tools, with each simulated comment containing "smiley" ☺, "sad" emoticon ☹, and without emoticon. We collected the opinion of "like" and "dislike" values for different variations of text in order to examine the impact of emoticon on the polarity of the text. Overall, our evaluation results show that emoticons are able to reverse polarity of the sentiment values.

2 Related Work

There have been some publications studying the behavioral usage of emoticons in text messaging, social media and the art of communication from social workplace. For example, Zhang et al. [2] stated that emoticons are strongly associated with subjectivity

© Springer International Publishing Switzerland 2016
E. Métais et al. (Eds.): NLDB 2016, LNCS 9612, pp. 453–458, 2016.
DOI: 10.1007/978-3-319-41754-7_48

and sentiment, and they are increasingly used to directly express a user's feelings, emotions and moods in microblog platforms. Yamamoto [3] verified the multidimensional sentiment of tweet messages by examining the role of the emoticon. Derks et al. [4] observed 150 secondary school students and studied the frequency of emoticons used to represent non-verbal face-to-face expressions. Ted et al. [5] studied the effect of emotions on emotional interpretation of messages in the workplace. Garrison et al. [6] studied the emoticons used in instant messaging. Tossell et al. [7] studied the emoticons used in text messaging sent from smart phones, particularly focusing on the gender difference in using emoticons. Another study by Reyes and Rosso [8] stated that emoticons are one of the features which are related to identifying irony within text.

3   Data and Experiment

In our earlier study [9] we showed that sentiment tools should consider the number of exclamation marks when detecting sentiment. In that study, we tested different numbers of exclamation marks, and our experiment showed that different numbers of exclamation marks have an impact on sentiment value of the text. We further argued that emoticons have a significant value for identifying the sentiment of comments on products based on testing of fifteen existing sentiment analysis tools. In this paper, we employ a similar method to explore the fifteen sentiment tools (Table 2) for positive and negative emoticons to investigate if the scores of polarity show any difference. We recorded the scores of: (1) unformatted text (e.g.: I like it), (2) comments that end with positive emoticon (e.g.: I like it :-)), and (3) comments that end with negative emoticon (e.g.: I like it :-(). We focus our experiment on these three types of comments, which can help to reveal the impact of the positive and negative emoticons. The impact is observed by comparing sentiment scores for three different types of comment.

Furthermore, we explore whether or not the emoticons affect the sentiment values of the comments assigned by human raters by conducting a survey. The data of comments on products used for the questionnaire survey were collected from online retails sites such as Ebay, Amazon and Lazada. Altogether, our corpus consisted of 1,041 raw comments were collected covering 10 different types of products, including (1) Beauty and Health, (2) Camera, (3) Computer, (4) Consumer Electronics, (5) Fashion, (6) Home appliance, (7) Jewellery and Watch, (8) Mobiles and Tables, (9) Sport goods, and (10) Toys and Kids. All of the collected comments were analyzed with Lancaster University UCREL's Wmatrix system to identify the emotional words that have the high frequencies of occurrence. Based on this, we manually categorized and extracted emotional words. Next, we designed the questionnaire with these 30 comments, using a 7- point Likert-type scale, where we repeated the questions to each of the listed comments. We collected the opinions of "like" and "dislike" values for different variations of text, employing similar approach to that which we used in comparing the sentiment tools. For example, three variants of the comment "I love it" were generated: (a) "I love it", (b) "I love it☺", and (c) "I love it☻", then respondents were asked to express their opinions at the levels of "like" and "dislike" about them. As a result, we collected rating data from 500 respondents. Figure 1 presents the fifteen strongest positive and negative comments with emotional words.

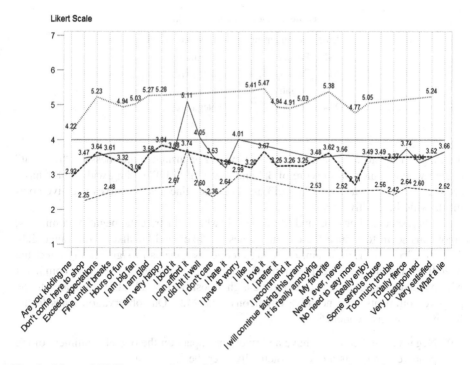

Fig. 1. Like and Dislike scores for positive and negative comments that end with emoticons (Color figure online)

4 Evaluation Results

The human rating data of the sentiment of comments of the 500 respondents were computed and analysed. Seven-point scale is used for both positive and negative comments, in which the mid-point of 4 represents neutral sentiment.

Figure 1 illustrates the mean scores with a chart. The four grey lines in the chart, denoted by M1, M2, M3 and M4, represent respectively the positive texts with positive emoticon, positive texts with negative emoticon, negative texts with positive emoticon, and negative texts with negative emoticon (see Table 1). Those comments on the X-axis are those simulated with both positive and negative emotions at the end of the messages that were given to the 500 respondents to assign sentiment values on the "like" and "dislike" scale.

With the neutral line N (see the red line) of the sentiment scale point 4 as the benchmark, the M1 line indicates that the mean sentiment rating scores of all the comments fall under the range of "slightly like" to "like" scales (corresponding to the numerical range of 4.22–5.24 points). In contrast, the whole M2 line falls below the neutral line N towards the "dislike" scale. This is an interesting and unexpected result, because M2 consists of positive texts but some of them have manual sentiment ratings that are even lower than the line M3 which consists of negative texts. On the other hand, line M3 appears as the second top level (although below line N) for many comments in

Table 1. Description of the lines

Line	Comment polarity	Emoticon used
M1	Positive	Positive
M2	Positive	Negative
M3	Negative	Positive
M4	Negative	Negative

the range of 3.2 to 5.11 sentiment scale points. For instance, the message "I hate it" ending with a smiley has a score of 3.2 and "I can afford it" ending with smiley has a score of 5.11. We categorize line M3 as "sarcastic" line since it mixes negative comments with the positive emoticon.

With respect to lines M3 and M4 in Fig. 1, which illustrate the negative comments, emoticons seem to reduce the overall values of "like" and "dislike" from range 2.3–5.11 to 2.36–3.74. Here again, changing emoticons appears to significantly affect the overall sentiment value of expressions. Overall, line M3 shows that negative comments with positive emoticons appear to have a higher positive overall rating compared to the positive comments with negative emoticons (line M2). Our observation leads to the following tentative conclusions:

(a) Negative emoticons can have an important impact on the overall sentiment of the positive comments and can potentially alter the sentiment.
(b) Various types of emoticons form an important factor that should be considered by the sentiment analysis tools.

We extracted scores for an example phrase "I like it" from fifteen freely available online sentiment analysis tools, as shown by Table 2. Our checking of the results shows that 9 out of the 15 tools include the value of expression of emoticons in their sentiment polarity classification process, which affects the positive or negative sentiments of the comments compared to the original text-only messages without smileys. For example, in our experiment Twitter Sentiment Analyzer, Lexalytics, Sentiment Analysis Engine, Sentiment Search Engine etc. produced negative sentiment scores for this message. These results also support our claim regarding the importance of emoticons for sentiment analysis. It is interesting that both Sentiment Analysis Engine and Sentiment Search Engine classified the message "I like it :-(" as totally negative when a negative emoticon was used. It shows that most of the sentiment tools included in our study consider emoticons as a relevant factor when determining the sentiment polarity of the text. However, they appear not to take account of the positive/negative interactions between text and emoticons, which we have highlighted in this study.

Table 2. Comparison of results of fifteen sentiment analysis tools.

Tools	Website	Score for "I like it :-)"	Score for "I like it :-("	Score for "I like it"
Twitter Sentiment Analyzer	http://textsentiment.com/twitter-sentiment-analyzer	0.705762	0.648023	0.565229
Lexalytics	https://www.lexalytics.com/demo	+0.5	−0.75	0
Sentiment Analysis with Python NLTK Text Classification	http://text-processing.com/demo/sentiment/	Pos: 0.6 Neg: 0.4 Overall: Positive	Pos: 0.4 Neg: 0.6 Overall: Negative	Pos: 0.5 Neg: 0.5 Overall: Positive
Sentiment Analysis Engine	http://www.sentimentanalysisonline.com/	Very Good 1	Very Bad −1	Neutral 0
Sentiment Analysis Opinion mining	http://text2data.org/Demo	Positive 0.178	Positive 0.099	Positive 0.167
Sentiment Search Engine	http://werfamous.com/	0.39	−0.42	0.08
Text sentiment analyzer	http://werfamous.com/sentimentanalyzer/	+50 %	−75 %	0 %
Meaning cloud	http://www.meaningcloud.com/	Positive 100 %	Positive 90 %	Positive 100 %
Tweenator Sentiment Detection	http://tweenator.com/index.php?page_id=2	Positive 88.3 %	Positive 33.57 %	Positive 75.65 %
LIWC	http://liwc.wpengine.com/	Positive 33.3	Positive 33.3	Positive 33.3
SentiStrenght	http://sentistrength.wlv.ac.uk/	Positive 2 Negative −1	Positive 2 Negative −1	Positive 2 Negative −1
TheySay	http://apidemo.theysay.io/	Positive 72.70 %	Positive 72.70 %	Positive 72.70 %
Selasdia Intelligent Sales Assistant	http://get.sendible.com/sendiblemonitoring/?gclid=CK6wm-27gcsCFZEK0wodDyoM5w	Positive	Positive	Positive
Sentiment Analyzer	http://sentimentanalyzer.appspot.com/	+100 %	+100 %	+100 %
EmoLib	http://dtminredis.housing.salle.url.edu:8080/EmoLib/en/	Positive	Positive	Positive

5   Conclusion

This study investigates the issue of the inconsistency of the sentiment value of messages and comments with respect to the positive and negative emotions. Using human raters, we have verified that the emoticons play a major role that can affect the overall sentiment value of a social media message or online review comment. Furthermore, our experiment reveals that most of the current sentiment tools that consider emoticons in their sentiment polarity classifications do not consider the interaction between conflicting positive/negative sentiment values of textual contents and emoticons when assessing the overall sentiment. In light of our experimental results, we propose that current sentiment tools should be improved by considering all these factors in their classification algorithms.

References

1. Kim, J., Lee, M.: Pictogram generator from Korean sentences using emoticon and saliency map. In: HAI 2015 Proceedings of the 3rd International Conference on Human-Agent Interaction, pp. 259–262. ACM, New York (2015). ©2015
2. Zhang, L., Pei, S., Deng, L., Han, Y., Zhao, J., Hong, F.: Microblog sentiment analysis based on emoticon networks model. In: Proceedings of the Fifth International Conference on Internet Multimedia Computing and Service - ICIMCS 2013, pp. 134–138. ACM, New York (2013). ©2013
3. Yamamoto, Y.: Role of emoticons for multidimensional sentiment analysis of Twitter. In: The 14th IIWAS, pp. 107–115. ACM, New York (2013). ©2014
4. Derks, D., Bos, A.E.R., Von Grumbkow, J.: Emoticons and social interaction on the internet: the importance of social context. Comput. Hum. Behav. 23(1), 842–849 (2007)
5. Ted, T., Wu, L., Lu, H., Tao, Y.: Computers in human behavior the effect of emoticons in simplex and complex task-oriented communication: an empirical study of instant messaging. Comput. Hum. Behav. 26(5), 889–895 (2010)
6. Garrison, A., Remley, D., Thomas, P., Wierszewski, E.: Conventional faces: emoticons in instant messaging discourse. Comput. Compos. 28(2), 112–125 (2011)
7. Tossell, C.C., Kortum, P., Shepard, C., Barg-Walkow, L.H., Rahmati, A., Zhong, L.: A longitudinal study of emoticon use in text messaging from smartphones. Comput. Hum. Behav. 28(2), 659–663 (2012)
8. Reyes, A., Rosso, P.: On the difficulty of automatically detecting irony: beyond a simple case of negation. Knowl. Inf. Syst. 40, 595–614 (2014)
9. Teh, P.L., Rayson, P., Piao, S., Pak, I.: Sentiment analysis tools should take account of the number of exclamation marks!!! In: The 17th IIWAS, pp. 4–9. ACM, New York (2015). ©2015

Spatial-Temporal Representation and Reasoning About Objects and Regions

Aldjia Aider[1(✉)], Fatiha Mamache[1], and Fatma-Zohra Belkredim[2]

[1] Laboratoire RIIMA, Département d'Informatique,
Faculté d'Electronique et d'Informatique, USTHB,
BP 32 El Alia, Bab-Ezouar, Alger, Algeria
{laider,amamache}@usthb.dz
[2] LMA Laboratoire de Mathématiques et Applications,
Département de Mathématiques, Faculté de Sciences,
Université Hassiba Benbouali Chlef, Chlef, Algeria
f.belkredim@univ-chlef.dz

Abstract. There are many Artificial Intelligence application that require reasoning involving spatial and temporal concepts. We propose a spatio-temporal formalism to represent the relationship between objects and regions. The formalism is based on a first order language augmented with operators which main aim is to facilitate the representation of spatial-temporal objects positions. This temporal logic allows to study the evolution of relative positions between entities during time. The trajectory of an object in areas is represented by equivalence classes of objects positions in present, past and future.

Keywords: Artificial intelligence · Knowledge representation · Temporal reasoning · Spatial-temporal logic · Qualitative spatial reasoning

1 Introduction

Artificial Intelligence Applications increasingly need representation and reasoning about space and time. In this paper, we propose a formalism to enrich Galton's work [1]. The aim of this language is to represent objects position in future and past. In our proposal, the first order logic is extended to include operators to represent the position of an object in the past, present and future. These operators allow the identification of the object's trajectory and simplifies its spatial-temporal representation. The set of time-elements when an object is present in an area is represented by an equivalence class of a time-element when the object was in this area for the first time. The remaining of the paper is organized as follows. In Sect. 2, we review some related works and present our language by defining its notations and terminologies to represent positions of objects. Section 3 is devoted to the definition of the spatial-temporal relationships between objects and regions. In Sect. 4, we present a deductive system and the semantics of the $\mathcal{SPTL}$ logic. We conclude and identify future developments in Sect. 5.

© Springer International Publishing Switzerland 2016
E. Métais et al. (Eds.): NLDB 2016, LNCS 9612, pp. 459–464, 2016.
DOI: 10.1007/978-3-319-41754-7_49

2   Related Work

Several formalism were proposed including spatial qualitative reasoning. The formalisms suggested are characterized by spatial entities as well as relationships between them. Within the framework of space theory based on areas, Galton [1,2] was the first to consider the movement of object from a qualitative point of view on a more general theory of time and action, whiah is different from the way it is defined before [3]. Galton does not distinguish between future and past. We propose a language which main aim is to facilitate the representation of the space-time positions of the objects in the future and the past. Galton [1,2] combined spatial theory with temporal theory, inspired by actions modeling formalism, more precisely by actions theory and Allen's time [3]. Using Mamache's formalism [4] to deal with this kind of situations, we present a logical formalism based on a first order language extended with special operators in order to represent spatial-temporal positions of objects.

3   Spatial-Temporal Reasoning About Objects and Regions

Within the framework of the formalization of a symbolic approach for Spatial-Temporal Reasoning, and inspired by Allen [3], McDermott [6] and Aider and Mamache [4,5] works, we propose a spatial-temporal formalism to reason about objects, regions and time.

3.1   Language, Notation and Terminology

We introduce the following language which is a first order language with equality:

- Connectors: $\neg, \vee, \wedge$ *and* $\supset$.
- Two signs of quantification noted $\exists$ and $\forall$.
- A symbol of equality, which we will note $\equiv$ to distinguish it from the sign $=$.
- A countable infinite collection of propositional variable.
- A set of operational signs or symbols functional.
- Three unary temporal operators: P_k (past), F_k (future), and P_0 (present).
- The expressions are the symbol strings on this alphabet.
- The set of the formulas noted Φ is by definition the smallest set of expressions which checks the following conditions:
  - Φ contains the propositional variables.
  - A set of elements called symbols of individuals.
  - If A and B are elements of Φ it is the same for $\neg A$ and $A \supset B$.
  - If A is an element of Φit is the same for P_k A, $F_k A$ and $P_0 A$.

The language, equally contains:

- A set of elements called symbols of individuals.
- A set of operative signs or functional symbols.
- A set of relational signs or symbols of predicates.

As Galton [7], the predicate *pos* is used to express that an object o is in a region r.

3.2   Atemporal Representation

Definition 1. An object o is an atemporal object if its position does not depend on time.

Let o be an atemporal object and r a region. To express that the object o is in the region r, we use the predicate "*pos*". We can generalize for m objects:

Let $o_1, o_2, ..., o_m$ m atemporal expressions of object type and r a region.

To express that the objects $o_1, o_2, .., o_m$ are in the region r, we use the predicate *pos* defined by:

$$pos(o_1, o_2, ..., o_m; r) \equiv pos(o_1; r) \wedge ... \wedge pos(o_m; r).$$

Example 1. $pos(plane, HouariBoumedieneAirport)$
means that the plane is in Houari Boumediene Airport.

Example 2. $pos(Algeria, Tunisia, Morocco; Africa)$:
means Algeria, Tunisia and Morocco are in Africa.

3.3   Temporal Representation

Definition 2. We call time-element an interval or a point of time [8,9].

Let I be a set of intervals, P a set of points of time and T the union of I and P then every object o is in an area r during a time-element t.

If the position of o in the area is instantaneous then t is a point of time and if it is so t is an interval.

If an object o would be (in the future) in an area r at a time-element t, we give the following definition:

Definition 3. Let o be an object type, r a region and $(t_1, t_2, \ldots, t_m)$ elements of time.

1. The formula $pos(o.t; r)$ expresses that the object o will be in the region r at the element of time t.
2. The formula $pos(t.o; r)$ expresses that the object o was in the region r at the element of time t.
3. The formula $pos(o.t_1, o.t_2, \ldots, o.t_m; r)$ expresses that the object o will be in the region r at the different elements of time $t_1, t_2, \ldots, t_m$.
4. The formula $pos(t_1.o, t_2.o, \ldots, t_m.o; r)$ expresses that the object o was in the region r at the different elements of time $t_1, t_2, \ldots, t_m$.

Example 3. 1. pos (plane.15: 00; Algiers) means: The plane will arrive at Algiers at 15 : 00.
2. pos (15: 00.plane; Algiers) means: The plane was in Algiers at 15 : 00.

Example 4. 1. *pos* (plane.15: 00; plane.18 : 00; plane.23 : 00; Algiers) means The plane will be in Algiers at 15: 00, 18: 00 and 23: 00.
2. *pos* (15: 00.plane; 18 : 00.plane; 23 : 00.plane; Algiers), means, The plane was in Algiers at 15: 00, 18: 00 and 23: 00.

Definition 4. Let T be a nonempty set of time-elements, O a set of objects and R a set of regions. We define a set $Pos(O.T, R)$ as the set of elements $pos(o.t; r)$.

$$Pos(O.T, R) = \{pos(o.t, r), o \in O, r \in R, t \in T/ \text{ will be in } r \text{ at } t\}.$$

To represent the fact that $o_1, o_2, \ldots, o_m$ will be in an area r respectively at $t_1, t_2, \ldots, t_m$ we use the notation:

$$pos(o_1.t_1, o_2.t_2, \ldots, o_m.t_m; r).$$

Definition 5. Let $Pos(T.O, R)$ be the set of elements $pos(t.o; r)$.

$$Pos(T.o, R) = \{pos(t.o, r), o \in O, r \in R, t \in T/o \text{will be in } r \text{ at } t\}.$$

To represent the fact that $o_1, o_2, \ldots, o_m$ will be in an area r respectively at $t_1, t_2, \ldots, t_m$ we use the notation

$$pos(t_1.o_1, t_2.o_2, \ldots, t_m.o_m; r).$$

Definition 6. A Trajectory T_j, in line with Kayser and Mokhtari [10] and Mamache [4] is a succession of time-elements t_j representing an evolution of the universe defined by the position of an object in an area of the universe.

4 Spatial-Temporal Relationships Between Objects and Regions

In our approach, we establish a relation between objects, areas and time when these objects are in these areas. To express this, we enrich our language by new operators. The notion of time (past, present and future) is represented by an integer k such as:

1. $k > 0$ represents the future. $F_k(pos(o; r_k))$ expresses the position of the object o in the area r_k in the future and,
2. $k < 0$ represents the past. $P_k(pos(o; r_k))$ expresses the position of the object o in the area r_k in the past.

If the object o would be in the future in different areas $r_1, r_2, \ldots, r_m$, we note:

$$F_1(pos(o; r_1)), F_2(pos(o; r_2)), \ldots, F_m(pos(o; r_m))$$

m is the number of areas where the object o would be in the future. If the object o was in different areas $r'_1, r'_2, \ldots, r'_s$:

$$P_{-1}(pos(o; r'_1), P_{-2}(pos(o; r'_2)), \ldots, P_{-s}(pos(o; r'_s)),$$

s is the number of areas where the object o was in the past.

The operator F_k allows us to determine all positions of the object o in the future and the operator P_k in the past (ramification).

4.1 $SPTL$ Spatial-Temporal Logic to Reason About Objects and Regions

We propose a spatial-temporal logic $SPTL$ to reason on the objects and regions.

Deductive System. The axioms of the spatial-temporal logic $SPTL$ are axioms of propositional logic [11]. The rules of deduction are:

1. The modus ponens [11].
2. Temporal generalization: If $pos(o;r)$ is a theorem, $F_k(pos(o;r))$, $P_k(pos(o;r))$ and $P_0(pos(o;r))$ are also theorems.

The theorems of $SPTL$ are by definition all the formulas deductible from the axioms by using the rules of deduction. In particular all the theorems of propositional calculus are theorems.

Semantic of $SPTL$. In the semantic of propositional calculus [10], an assignment of values of truth V is an application that for each propositional variable associates a value of truth. An assignment of value of truth describes a state of the world. In the case of $SPTL$, we choose as propositional variables the positions of an object o in an area r at a time-element t.

Definition 7. Let P(T) be the set of the parts of T and V the valuation defined by:

$$V : Pos(O, R) \to P(T) \to \{0, 1\}$$

$$pos(o;r)) \to T' = \{t/o \text{ is in r at t}\}$$

the valuation of $pos(o, r) = 1$ if $T' \neq \emptyset$ and $pos(o, r) = 0$ if $T = \emptyset$

4.2 Representation of the Movement When an Object is in Different Regions at Different Time-Elements

We note $Pos(o, R)$ the set of positions of an object o.

Definition 8. We define the application $Vt : Pos(o, R) \to P(T)$ as follows:
If o is in r_i at t_i, then $V_t(pos(o; r_i)) = \{t_i\}$

Definition 9. Let R_t a binary relation defined on the set $Pos(o; R)$ as follows:
$(pos(o; r_i)R_t pos(o; r_j) \Leftrightarrow V_t(pos(o; r_i)) = V_t(pos(o; r_j))$

Proposition 1. R_t *is a relation of equivalence as defined in the following diagram:*

$$P(O, R) \quad \overrightarrow{V_t} \ P(T)$$
$$s \downarrow \qquad\qquad \uparrow i$$
$$P(O, R)/R_t \ \overrightarrow{\overrightarrow{V_t}} \ ImV_t$$

$Pos(o, R)/R_t = \{pos(o, r) \in \underline{Pos(o; R)}\}$ is the set of equivalence classes of elements of $Pos(o; R)$ and $\overline{pos(o, r)} = \{pos(o, r_i) \in Pos(o, R)/pos(o, r_i)$ $R_t pos(o, ; r)\}$ is the class of equivalence of $pos(o; r)$. It contains all positions $pos(o, r_j)$ of the object o in different areas r_j (movement). V_t associates to any element of $Pos(o, R)$ a time-element t_i and $V_t(pos(o, r_i)) = \{ti\}$. The set of the time-elements when an object o is in different areas r_i or the movement of the object o is represented by the equivalence class of a position $pos(o, r_i)$ when o is in these areas, it is the representative of the class.

5   Conclusion and Perspectives

In this paper, we introduced a formalism based on a first order language extended with operators to represent spatial-temporal positions of objects in the past, present and future that makes the possibility to determine the Trajectory of this object. Also, we proposed an arborescent spatial-temporal logic to describe a state of the world. Our formalism can be used in the management of air, sea and rail transport in real time. Pieces of information will be centralized and the formalism will be the reference allowing rapid access. Furthermore it can be applied to moving objects as Animals, Humans or Robots.

References

1. Galton, A.: A critical examination of Allen's theory of action and time. J. Artif. Intell. **42**(2–3), 159–188 (1990)
2. Galton, A.: On the notions of specification and implementation. R. Inst. Philos. Suppl. **34**, 111–136 (1993)
3. Allen, J.: Towards a general theory of actions and time. J. Artif. Intell. **23**, 123–154 (1984)
4. Mamache, F.: A temporal logic for reasoning about actions. Int. J. Open Probl. Comput. Sci. Math. **4**(1), 97–112 (2011)
5. Aider, A., Mamache, F.: Objects, regions and spatio-temporal logic. Asian J. Math. Comput. Res. **9**(1), 77–91 (2016)
6. McDermott, D.: A temporal logic for reasoning about processes and plans. Process Cogn. Sci. **6**, 101–155 (1989)
7. Galton, A.: Spatial and temporal knowledge representation. Earth Sci. Inf. **2**(3), 169–187 (2009)
8. Birstouge, H., Ligozat, G.: Outils logiques pour le traitement du temps de la linguistique l'intelligence Artificielle. Masson, Paris (1989)
9. Knight, B., My, J., Peng, T.: Representing temporal relationships between events and their ef-fects. In: Proceedings of the International Fourth Temporal Workshop one Representation and Reasoning, pp. 148–152. IEEE Computer Society Press (1997)
10. Kayser, D., Mokhtari, A.: Time in a causal theory. Ann. Math. Artif. Intell. **22**(1–2), 117–138 (1998)
11. Kleene, S.C.: Mathematical logic. Collection U (1971)

Bias Based Navigation for News Articles and Media

Anish Anil Patankar and Joy Bose[(✉)]

Samsung R&D Institute, Bangalore, India
{anish.p,joy.bose}@samsung.com

Abstract. In existing news related services, readers cannot decide if there is another side of a news story, unless they actually come across an article representing a different perspective. However, it is possible to determine the bias in a given article using NLP related tools. In this paper we determine the bias in media content such as news articles, and use this determined bias in two ways. First, we generate the topic/bias index for one or more news articles, positioning each article within the index for a given topic or attribute. We then provide a user interface to display how much bias is present in the currently read article, along with a slider to enable the reader to change the bias. Upon the user changing the bias value, the system loads a different news article on the same topic with a different bias. The system can be extended for a variety of media such as songs, provided the lyrics are known. We test our system on a few news articles on different topics, reconfirming the detected bias manually.

Keywords: News · Media consumption · Clustering · Natural language processing · Classification · Sentiment analysis · Bias · Topic modelling

1 Introduction

There are positive and negative aspects of any event, and news articles from different sources covering the same event are typically not free from bias. Users cannot possibly browse and read all articles on a given topic from all sources. A related problem is: while reading a news article, readers have no way of determining if, and how much bias is present in the article. A study on social media users [1] found that people of polarized political orientations, such as liberal and conservative, generally only interact with others of a similar orientation and not with the differing groups. This shows that it is difficult for a user to come across articles that do not correspond closely to their already existing point of view.

In this paper we provide a system comprising of a web service to determine the bias in a given news article in real time by identifying the topic of the article, analyzing the words in an article for positive, negative or neutral connotations and determining a bias score as a ratio of the number of positive or negative words with the total number of words. This calculated bias score is displayed in the web browser using a suitable user interface, giving the user a chance to be aware of the bias of the currently read article and change it if they wish. The cloud server stores a list of articles on the same or

© Springer International Publishing Switzerland 2016
E. Métais et al. (Eds.): NLDB 2016, LNCS 9612, pp. 465–470, 2016.
DOI: 10.1007/978-3-319-41754-7_50

similar topics sorted by bias, which is used by the web service to provide a slider based interface. The user is able to view a different article on a given topic by choosing a different bias value on the slider.

2 Related Work

Slider user interfaces which provide a user a means to choose and personalize how much content related to a given topic they wish to read, are already available. Examples include the interface used by Google news [2]. Yahoo's Social Sentiment Slider [3] is a widget that provides a means to ask users relevant questions regarding how they feel about a given content, by sliding a cursor across a scale. However, a slider user interface that lets a user control the bias or sentiment related to a given article is not available at present.

A number of Twitter related tools and other works [4–6] are available that enable one to analyze or visualize the sentiment of tweets on a given topic or brand, using text mining. Similarly, methods using text mining to detect bias in news articles have been surveyed by Balyaeva [7]. Gupta [8] and Chu et al. [9] among others have analyzed online news articles to evaluate the present bias.

None of the above related work mentions any solutions to enable the user to select news articles based on the bias of the article related to a given topic. This is what we are proposing in this paper.

3 System Overview

Our overall system comprises a server component and the client user device such as a mobile phone with a web browser. On the client device, a web browser extension calculates the bias and topic of the already loaded article and then queries the server to find related articles with a different bias. The server returns the URL of the related articles to the web browser on the client device, which then loads the URL like any other webpage.

The server maintains a list of news articles on given topics, maintaining a database of URLs of articles on similar topics, along with the bias score for each article. The web browser extension on the client loads the related article URLs obtained from the server, displays them on the user device, and runs the user interface which displays the bias score for each article.

The system can alternatively be implemented wholly on the user device to calculate and display the bias score of the currently loaded article. However, the computation load and memory requirement in this case would be larger and slow down the device significantly. For this reason, a server based solution is preferable.

We use two methods for detection of bias, which are listed below.

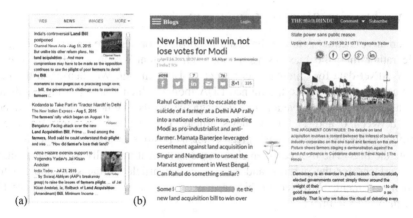

Fig. 1. (a) Illustration of the slider user interface for varying the bias in a given news article. The user can move the slider to get a similar news article with a different bias. (b) User action of moving the slider manually or swiping in right or left direction enables them to navigate to a different webpage on the same topic, with a different bias.

3.1   Detecting Bias When the Article Topic Is Known

We define a "topic" of an article as the keyword/s whose frequency of occurrence is highest. This may or may not coincide with the LDA definition (based on probability distributions) or knowledge base related definitions of topic.

If the topic of the news article is known, we first segment the article into sentences by using standard tools such as the OpenNLP sentence detector [12]. We then consider all the sentences which contain words corresponding to the known topic. For each of these sentences, we count the number of words with a positive or negative bias using the lexicon provided by Hu and Liu [15]. Then we calculate the ratio of positive to negative words, which would indicate whether a given article is positive or negative about the topic.

We also calculate the bias score for a given article in the following way: As before we consider sentences in the article containing the keywords related to the topic. Using the NPOV bias lexicon corpus built out of Wikipedia articles by Recasens et al. [10] (available from [13]) to identify biased language in articles, we count the total number of such biased words in sentences related to the topic and divide by the total number of words to calculate the bias score for the article. The news article is then placed on a scale based on the computed bias score for that topic.

3.2   Detecting Bias When the Article Topic Is not Known

If the topic of the article is not known, we first identify the topic of the article by using a method such as Rapid Keyword Extraction (RAKE) described by Rose et al. [11]. Then we use the previously described method to calculate the bias score.

Alternatively, we can simply calculate the ratio of biased words/total words to calculate the bias score.

3.3   Working with Sentiment Instead of Bias

An alternative method is to use the identified positive or negative sentiment in a given article rather than the bias, using a tool such as SentiWordNet [14]. We use the SentiWordNet data for training to construct an LDA model. We then input the news article as a bag of words to the trained LDA model, which then splits the words into two clusters, one for positive and one for negative words. We then calculate an overall score (between 0 and 1) for the news article. This method has the advantage of informing to the user how much a given article is positive or negative, rather than just indicate the bias. The user can then change the sentiment manually using the available user interfaces for the purpose and access another article on the same topic with a different sentiment. However, in this paper we have experimented with just the bias rather than the sentiment.

4   Experimental Setup with News Articles

For this experiment, we used 3 articles each, taken from popular news sources, on topics of contemporary relevance: Bernie Sanders and Donald Trump, prominent democrat and republican candidates for the US presidential nomination. We determined the bias score for each of these articles and tested the score using the previously described method, confirming the obtained relative bias score by asking 5 knowledgeable users aged 25–35 to rank the given articles for bias.

For the liberal democrat candidate (Sanders), we sourced a neutral article from Wikipedia, a positive article from a liberal website and a negative article from a conservative website (and reverse for the republican conservative candidate Trump).

The Tables 1 and 2 show the results of our test. In case of Trump, the bias score is lower for the neutral (Wikipedia) article than for the pro and anti articles, which is as expected. The same holds in case of pro-Bernie and anti-Bernie articles, although the difference is much less pronounced in this case. Since the biased and unbiased articles are correctly distinguished, it is feasible to separate them on the slider interface as we described earlier.

Table 1. Bias results for Donald Trump Related articles

Article	Bias score	Bias words	Total words	Ratio of positive/negative words
Pro Trump (a)	0.163	34	208	3.2
Anti-Trump (b)	0.142	82	579	0.81
Neutral (c)	0.107	303	2822	0.84

Sources (a) http://www.breitbart.com/big-government/2016/02/09/donald-trump-vows-to-be-the-greatest-jobs-president-god-ever-created (b) https://www.wsws.org/en/articles/2015/12/10/pers-d10.html (c) https://en.wikipedia.org/wiki/Donald_Trump

Table 2. Bias results for Bernie Sanders related articles

Article	Bias score	Bias words	Total words	Ratio of positive/negative words
Pro Bernie (a)	0.145	94	648	0.93
Anti-Bernie (b)	0.168	117	694	0.94
Neutral (c)	0.134	232	1733	1.08

Sources (a) http://www.alternet.org/election-2016/6-awesome-things-about-bernie-sanders-you-might-not-know (b) https://mises.org/library/economics-bernie-sanders (c) https://en.wikipedia.org/wiki/Bernie_Sanders

5   Use Cases and Applications

Our system, with a little modification, can also be used in a number of other use cases such as other media (songs, TV channels etc. with a given bias or sentiment related to a given topic) or product reviews (a method such as mentioned in Hu and Liu [15] can be used for the purpose). In each of these use cases, a similar method to the one described earlier in the case of news articles would be used.

6   Conclusion and Future Work

In this paper, we have described a slider based interface for navigating between news articles with a given bias. This would enable the user to be aware of and also control the amount of bias present in the news article they are currently reading. This new and interesting interface would lead to higher user engagement with news articles and other media. This can potentially change the way news content is accessed by users, and prevent users from being manipulated by bias in media.

In future we seek to improve the accuracy of the system, implement it for music and other use cases and use the (positive or negative) sentiment of the article rather than just bias. We also seek to extend the proposed algorithm to cover sub-topics within a news article.

References

1. Smith, M.A., Rainie, L., Shneiderman, B., Himelboim, I.: Mapping Twitter topic networks: from polarized crowds to community clusters. Pew Research Centre, vol. **20** (2014)
2. Google News Help: Personalizing Google News (2016). https://support.google.com/news/answer/1146405?hl=en
3. Yahoo Advertising Solutions: Introducing Yahoo!'s New Social Sentiment Slider. Yahoo News (2011). http://news.yahoo.com/blogs/advertising/introducing-yahoo-social-sentiment-slider-185414337.html
4. Tweet Sentiment Visualization App. http://www.csc.ncsu.edu/faculty/healey/tweet_viz/tweet_app/
5. Sentiment140. http://www.sentiment140.com

6. Agarwal, A., Xie, B., Vovsha, I., Rambow, O., Passonneau, R.: Sentiment analysis of Twitter data. In: Proceedings of Workshop on Language in Social Media (LSM 2011), Portland, Oregon, pp. 30–38. Association for Computational Linguistics, 23 June 2011

7. Balyaeva, J.: Learning how to detect news bias. Departmental Seminar, University of Ljubjana (2015)

8. Gupta, S.: Finding bias in political news and blog websites. Project report (2009). http://snap.stanford.edu/class/cs224w-2010/proj2009/report_Sonal_Gupta.pdf

9. Chu, A., Shi, K., Wong, C.: Prediction of average and perceived polarity in online journalism. Project report (2014). http://cs229.stanford.edu/proj2014/Albert%20Chu,%20Kensen%20Shi,%20Catherine%20Wong,%20Prediction%20of%20Average%20and%20Perceived%20Polarity%20in%20Online%20Journalism.pdf

10. Recasens, M., Danescu-Niculescu-Mizil, C., Jurafsky, D.: Linguistic models for analyzing and detecting biased language. In: Proceedings of ACL (2013)

11. Rose, S., Engel, D., Cramer, N., Cowley, W.: Automatic keyword extraction from individual documents. In: Text Mining: Applications and Theory, pp. 1–20. Wiley, New York (2010)

12. OpenNLP Sentence Detector. http://opennlp.apache.org/documentation/1.6.0/manual/opennlp.html#tools.sentdetect

13. Cristian Danescu-Niculescu-Mizil: Biased Language. http://www.cs.cornell.edu/~cristian/Biased_language.html

14. Esuli, A., Sebastiani, F.: SENTIWORDNET: a publicly available lexical resource for opinion mining. In: Proceedings of 5th International Conference on Language Resources and Evaluation (LREC-2006), Genoa, Italy. European Language Resources Association (ELRA) (2006). http://www.lrec-conf.org/proceedings/lrec2006/pdf/384_pdf.pdf

15. Hu, M., Liu, B.: Mining and summarizing customer reviews. In: Proceedings of ACM SIGKDD International Conference on Knowledge Discovery and Data Mining (KDD 2004), Seattle, Washington, USA, 22–25 August 2004

Learning to Rank Entity Relatedness Through Embedding-Based Features

Pierpaolo Basile, Annalina Caputo[(✉)], Gaetano Rossiello,
and Giovanni Semeraro

Department of Computer Science, University of Bari Aldo Moro, Bari, Italy
{pierpaolo.basile,annalina.caputo,gaetano.rossiello,
giovanni.semeraro}@uniba.it

Abstract. This paper describes the effect of introducing embedding-based features in a learning to rank approach to entity relatedness. We define several features that exploit word- and link-embedding approaches by relying on both links and the content that appear in Wikipedia articles. These features are combined with other state-of-the-art relatedness measures by using a learning to rank framework. In the evaluation, we report the performance of each feature individually. Moreover, we investigate the contribution of each feature to the ranking function by analysing the output of a feature selection algorithm. The results of this analysis prove that features based on word and link embeddings are able to increase the performance of the learning to rank algorithm.

1 Introduction

Entity relatedness tries to capture the strength of the relationship that ties together named entities or concepts. As a measure of semantic relatedness, it plays a key role in many computational tasks that require semantic processing of natural language, like question answering and information retrieval, text reuse, entity suggestion and recommendation, and it is at the core of several natural language processing algorithms for entity linking [2,4,8].

Over the past few years, many relatedness measures have been proposed which exploit Wikipedia in order to collect useful information about entities and their relationships. Each article in Wikipedia can be regarded as an entity or a concept of the real world. Then, the textual content of Wikipedia articles, their hyperlink graph, the hierarchical organization of content through categories all offer a rich amount of freely available information that has been harnessed to extract and weight entities and their relations.

Most of the methods that exploited Wikipedia for entity relatedness have focused on one single aspect at time, although every aspect of the Wikipedia content provides evidence of different kinds of relatedness. And in particular graph-theoretic methods have showed their capability in measuring relatedness both for computing semantic similarity [8] and for entity linking [7]. Moreover, the combination of such measures proved to be very effective in a learning to

© Springer International Publishing Switzerland 2016
E. Métais et al. (Eds.): NLDB 2016, LNCS 9612, pp. 471–477, 2016.
DOI: 10.1007/978-3-319-41754-7_51

rank framework [2]. Learning to rank for entity relatedness models the problem of ranking entities most related to a given target as a learning task [2,9]. This problem has some similarities with learning to rank in information retrieval where, trained over a set of features describing query-document pairs, the algorithm learns to rank documents according to their relevance to a given query. In the context of entity relatedness, the query is a target entity and the algorithm learns to rank entities on the basis of their relatedness to the target [2].

This paper introduces a new set of features based on the embedding of information about either the graph structure or the content of Wikipedia articles into a low-dimensional space. These embeddings are able to grasp the latent relationships that exist between words or links. For this reason, the resulting spaces contribute more to the learning of the ranking function than traditional vector spaces which exploit the same information.

2   Methodology

In this paper we investigate the problem of entity relatedness: given two entities e_i and e_j we want to define a function $r(e_i, e_j)$ able to predict their degree of relatedness. We model this problem taking into account several relatedness measures and combining them into a learning to rank framework following the approach proposed in [2]. The idea is that each feature is able to grasp a particular aspect of relatedness and we want to investigate the use of measures based on embeddings. All proposed measures rely on Wikipedia. More specifically, each page p of Wikipedia corresponds to an entity, which can be represented by three aspects: (1) $In(p)$, the set of links that point to p; (2) $Out(p)$, the set of pages pointed by p and 3) the content of p.

We expand the initial set of 27 features in [2] based only on $In(p)$ and $Out(p)$ by introducing three relatedness measures based on word/link embeddings. These embedding-based features exploit the word2vec [6] tool which implements a revised method of the Neural Network Language Model by using a log-linear approach. We build three separate corpora that take into account the Wikipedia page content:

Abstract (a). Each document in this corpus is the abstract of a Wikipedia page extracted from DBpedia. Since abstracts are the first paragraph of a Wikipedia page, which usually gives a definition of that concept, the idea behind this approach is to build a corpus of entities' definitions.

Entity (e). Each document is the list of entities that occur in the same Wikipedia page. The occurrence of an entity in a Wikipedia page is identified by the manual link that the editor of the page has associated with the entity anchor text. It is important to underline that the list of entities reflects their order of occurrence in the page.

Entity&Word (e&w). Each document is the list of both words and entities occurring in each Wikipedia page. Also, in this case the words/entities order is preserved. The idea behind this corpus is to learn embeddings taking into account both words and entities.

For each corpus we build a vector space using word2vec with the skip-gram model. We build a total of three vector spaces: $W2V_a$, $W2V_e$, and $W2V_{e\&w}$. In $W2V_e$ and $W2V_{e\&w}$ spaces we have a vector for each Wikipage page (entity). We can compute the relatedness between two entities as the cosine similarity between their corresponding vector representations. In the $W2V_a$ space, we adopt a different strategy to compute the entity vector since this space represents only words. We retrieve the abstract of each entity and build an abstract vector that is the sum of the word vectors occurring in the abstract. Then the relatedness between two entities is computed as the cosine similarity between their corresponding abstract vectors. In order to compare the proposed encoding-based features, we define two additional measures that compute the relatedness in a standard vector space of links. These measures are inspired by [1], where the semantics of an entity is represented by its distribution in the high dimensional concept space derived from Wikipedia. For each type of link ($In(p)$ and $Out(p)$) we build a vector space where each entity is represented through a vector whose dimensions are the link counts (ingoing and outgoing, respectively). Then, given two entities e_i and e_j, we define the following features:

1. $vsm_{in}(e_i, e_j)$: is defined as the cosine similarity between $In(e_i)$ and $In(e_j)$. This measure is equivalent to the one proposed in [1];
2. $vsm_{out}(e_i, e_j)$: is defined as the cosine similarity between $Out(e_i)$ and $Out(e_j)$.

Table 1 summarizes all the features exploited by our approach, where learning to rank and feature selection are performed using the same approach in [2].

Table 1. List of features exploited by our approach.

id	Description
1-27	Features proposed in [2] exploiting some statistical aspects: probability, joint probability, conditional probability, entropy, Kullback-Leibler (KL) divergence, and co-citation
28	Cosine similarity computed in the space built on the corpus of Wikipedia page abstracts ($W2V_a$)
29	Cosine similarity computed in the space built on the corpus of entities occurring in the Wikipedia pages ($W2V_e$)
30	Cosine similarity between the entity vectors computed in the space built on the corpus of both entities and words occurring in the Wikipedia pages ($W2V_{e\&w}$)
31	Cosine similarity between $In(e_i)$ and $In(e_j)$ ($vsm_{in}(e_i, e_j)$)
32	Cosine similarity between $Out(e_i)$ and $Out(e_j)$ ($vsm_{out}(e_i, e_j)$)

3 Evaluation

We evaluate the performance of the proposed features within a learning to rank approach to the problem of entity relatedness. The goal of the evaluation is twofold: (1) prove the effectiveness of the proposed relatedness measures based on embeddings; (2) provide a deep feature analysis by relying on the feature selection algorithm described in [2].

We adopt the same dataset proposed by [2], which was extracted from a subset of the CoNNL 2003 entity recognition task [5]. However, since we use a different version of Wikipedia, we remove from the dataset all Wikipedia pages that are no longer available. After this step the dataset consists of 957,622 pairs of entities for training, 361,984 pairs for validation, and 295,886 for testing. We set different parameters for each vector space as summarized in Table 2, where we report the vector dimension (Dim.), the number of vectors in the space and the *min-count* parameter[1]. All other word2vec parameters are set to default values. We rely on the Wikipedia dump released on 13th January 2016. The dump was processed using the Java Wikipedia Library (JWPL) of the DKPro project[2]. All the other components are developed in Java and Python by the authors.

Table 2. Settings about word embedding.

Space	Dim	#vectors	Min-count
$W2V_a$	200	227,239	5
$W2V_{e\&w}$	300	4,856,978	10
$W2V_e$	200	4,527,966	1

We use the RankLib library[3] to perform learning to rank by adopting LambdaMART algorithm and setting $nDCG$ at 10 ($n@10$) as the measure to optimize in the learning step. We use default values for all the other LambdaMART parameters. We adopt this setting in order to make our approach comparable with the one proposed by [2].

We first evaluate the performance of the learning to rank algorithm when the proposed features are combined to traditional link-based ones. We performed two runs: (1) we adopt only the state-of-the-art features introduced in [2] in order to compute a baseline (we refer to this group of features as SOA) and (2) we exploit all the features proposed in our methodology in addition to SOA (we refer to this second group as ALL). SOA achieves a $n@10$ value of 0.8050, while ALL obtains a $n@10$ value of 0.8187, which results in a 1.702 % of increment.

The second step of the evaluation is to assess the performance of each single feature. Results of this evaluation are reported in Table 3, in decreasing order of $n@10$ values. The table shows also the id and a short description for each feature, while in boldface are reported the measures introduced in this paper. Among the embedding-based features the most promising is the $W2V_{e\&w}$, while $W2V_a$, which is the only feature that exploits the page content, provides quite

---

[1] Words that appear less than *min-count* are discarded.

[2] Available on line: https://dkpro.github.io/dkpro-jwpl/.

[3] Available on-line: https://sourceforge.net/p/lemur/wiki/RankLib/.

Table 3. Performance of each feature evaluated in isolation.

Id	$n@10$	Description
21	0.7215	joint probability
14	0.2844	KL divergence between e_2 and e_1
2	0.4622	probability of e_2
29	0.5471	$W2V_e$
4	0.4622	entropy of e_2
26	0.6046	χ^2 using out links
30	0.5879	$W2V_{e\&w}$
24	0.6884	χ^2 using in links
28	0.4916	$W2V_a$
25	0.6668	χ^2 using both in and out links
16	0.5974	co-citation using both in and out links
17	0.5807	co-citation using out links
19	0.6327	Jaccard index using both in and out links
20	0.6071	Jaccard index using out links
23	0.6564	Avg. friend
27	0.4197	Point-wise mutual information (PMI)
32	0.5938	VSM_{out}
5	0.5333	conditional probability of e_1 given e_2
12	0.5315	Friend between e_2 and e_1
18	0.6651	Jaccard index using in links
6	0.7213	conditional probability of e_2 given e_1
15	0.6419	co-citation using in links
11	0.6450	Friend between e_1 and e_2
13	0.2343	KL divergence between e_1 and e_2
31	0.5028	VSM_{in}
10	0.3766	probability that e_2 is linked to e_1
8	0.4216	exists a link between e_2 and e_1
7	0.4906	exists a link between e_1 and e_2
9	0.4906	probability that e_1 is linked to e_2
22	0.4244	exists a bi-direct. link between e_1 and e_2
1	0.1120	probability of e_1
3 ˙	0.1120	entropy of e_1

Table 4. Results of the application of the feature selection strategy.

Id	$n@10$	$\%\Delta_{ALL}$	$\%\Delta_{SOA}$
21	0.6443	-21.30	-20.04
14	0.6657	-18.69	-17.39
2	0.6855	-16.27	-14.93
29	0.7595	-7.23	-5.75
4	0.7672	-6.29	-4.79
26	0.779	-4.85	-3.33
30	0.786	-3.99	-2.46
24	0.7913	-3.35	-1.80
28	0.7927	-3.18	-1.63
25	0.7929	-3.15	-1.60
16	0.8079	-1.32	0.26
17	0.8079	-1.32	0.26
19	0.8107	-0.98	0.61
20	0.81	-1.06	0.52
23	0.8147	-0.49	1.10
27	0.8138	-0.60	0.99
32	0.8158	-0.35	1.24
5	0.8172	-0.18	1.41
12	0.8169	-0.22	1.38
18	0.8155	-0.39	1.20
6	0.816	-0.33	1.27
15	0.8177	-0.12	1.48
11	0.8161	-0.32	1.28
13	0.8159	-0.34	1.25
31	0.8183	-0.05	1.55
10	0.8173	-0.17	1.43
8	0.8173	-0.17	1.43
7	0.8173	-0.17	1.43
9	0.8175	-0.15	1.45
22	0.8175	-0.15	1.45
1	0.8169	-0.22	1.38
3	0.8187	0.00	1.60

poor performance. Finally, in order to better understand the contribution of each feature to the learning to rank algorithm, we perform a feature selection approach as described in [2]. This algorithm introduces one feature at a time by starting from the first feature in the rank reported in Table 3. The remaining features are selected according to the greedy algorithm proposed in [3].

Table 4 shows the $n@10$ values obtained by the learning to rank algorithm applied to the features filtered with this selection strategy. The last two columns in the table report the difference in percentage with respect to the model that includes all the features (ALL) and the model based only on the features proposed in [2] (SOA). The analysis of the results highlights some important outcomes:

1. The use of only the top eleven features is able to outperform the model propose in [2] (the first horizontal line in Table 4 marks this first group of features).
2. All the embedding-based features belong to the top eleven features selected by the algorithm (reported in boldface in Table 4).
3. The most relevant improvement (about 9 %) is obtained by introducing the feature with id 29 based on link (i.e. entity) embeddings.

The $W2V_{e\&w}$ feature (id 30), which combines content (words) with entities, provides an interesting contribution to the performance of the algorithm.

4   Conclusions

This work proposed a new set of features based on word and link embeddings in a learning to rank framework for entity relatedness. We exploited the content of Wikipedia pages to build embeddings able to capture different kinds of relatedness. Measures based on these models have been evaluated in combination with other state-of-the-art features and vector space models built upon the Wikipedia link structure. We observed that all the embedding-based features boosted the performance of the learning to rank algorithm, and they appeared between the top ranked ones that were able to give the overall better performance.

Acknowledgments. This work is supported by the IBM Faculty Award "Deep Learning to boost Cognitive Question Answering" and the project "Multilingual Entity Liking" funded by the Apulia Region under the program FutureInResearch. The Titan X GPU used for this research was donated by the NVIDIA Corporation.

References

1. Aggarwal, N., Buitelaar, P.: Wikipedia-based distributional semantics for entity relatedness. In: 2014 AAAI Fall Symposium Series (2014)
2. Ceccarelli, D., Lucchese, C., Orlando, S., Perego, R., Trani, S.: Learning relatedness measures for entity linking. In: CIKM, pp. 139–148 (2013)
3. Geng, X., Liu, T.Y., Qin, T., Li, H.: Feature selection for ranking. In: SIGIR, pp. 407–414 (2007)
4. Hoffart, J., Seufert, S., Nguyen, D.B., Theobald, M., Weikum, G.: Kore: keyphrase overlap relatedness for entity disambiguation. In: CIKM, pp. 545–554 (2012)
5. Hoffart, J., Yosef, M.A., Bordino, I., Fürstenau, H., Pinkal, M., Spaniol, M., Taneva, B., Thater, S., Weikum, G.: Robust disambiguation of named entities in text. In: EMNLP, pp. 782–792 (2011)
6. Mikolov, T., Chen, K., Corrado, G., Dean, J.: Efficient estimation of word representations in vector space (2013). arXiv:1301.3781

7. Milne, D., Witten, I.H.: Learning to link with Wikipedia. In: CIKM, pp. 509–518 (2008)
8. Witten, I., Milne, D.: An effective, low-cost measure of semantic relatedness obtained from Wikipedia links. In: WIKIAI, pp. 25–30 (2008)
9. Zheng, Z., Li, F., Huang, M., Zhu, X.: Learning to link entities with knowledge base. In: Human Language Technologies: The 2010 Annual Conference of the North American Chapter of the Association for Computational Linguistics (2010)

Arabic Quranic Search Tool
Based on Ontology

Mohammad Alqahtani[(✉)] and Eric Atwell

School of Computing, University of Leeds, Leeds, UK
{scmmal,E.S.Atwell}@leeds.ac.uk

Abstract. This paper reviews and classifies most of the common types of search techniques that have been applied on the Holy Quran. Then, it addresses the limitations of these methods. Additionally, this paper surveys most existing Quranic ontologies and what are their deficiencies. Finally, it explains a new search tool called: a semantic search tool for Al-Quran based on Qur'anic ontologies. This tool will overcome all limitations in the existing Quranic search applications.

Keywords: Holy Quran · Information Retrieval (IR) · Natural Language Processing (NLP) · Ontology · Semantic search

1 Introduction

The Holy Quran (Al Quran) is sacred Arabic text [1]. Al Quran contains about 79,000 words forming 114 chapters [1]. The techniques used to retrieve information from Al Quran can be classified into two types: a semantic-based and a keyword-based technique. The semantic-based technique is a concept-based search tool that retrieves results based on word meaning, or concept match, whereas the keyword-based technique returns results based on letters matching word(s) queries [2]. The majority of Quranic search tools employ the keyword search technique.

The existing Quranic semantic search techniques are: an ontology-based [3], a synonyms-set [4] and a cross-language information retrieval (CLIR) technique [5]. The ontology-based approach searches for the concept(s) matching a user words query. Then, this technique returns verses related to these concept(s). The synonyms-set technique produces all synonyms of the query words using WordNet. After that, it finds all Quranic verses matching these words' synonyms. CLIR translates words of an input query to another language and then retrieves verses that contain words matching the translated words.

Several deficiencies exist with the Quranic verses (Ayat) retrieved for a query using the existing keyword search techniques. These problems are: irrelevant verses are retrieved, relevant verses are not retrieved or the order of retrieved verses is not ranked [4]. The keyword-based technique's limitations include misunderstanding the exact meaning of input words forming a query and neglecting some theories of information retrieval [6].

© Springer International Publishing Switzerland 2016
E. Métais et al. (Eds.): NLDB 2016, LNCS 9612, pp. 478–485, 2016.
DOI: 10.1007/978-3-319-41754-7_52

Moreover, current Quranic semantic search techniques have some limitations about finding requested information. This is because these semantic searches use uncompleted Holy Quran ontology. Additionally, these concepts have different scopes and formats [7].

This paper is organized as follows. Section 2 is Literature review containing a review of the structure of the Holy Quran, Quranic search applications, previous research on Quranic search tools and existing ontologies of Al Quran. Section 3 describes the methodology of Arabic Quranic Search Tool Based on Ontology. Finally, Sect. 4 concludes the critical points in this paper.

2 Literature Review

2.1 Structure of the Holy Quran

Challenging points regarding the natural structure of the Holy Quran exist when applying NLP technologies. First, a concept could be mentioned in different verses. For example, the concept of the Hell (النَّار) is discussed in various chapters and verses. Additionally, one verse may contain many themes. For example, verse 40 of Chapter 78 contains only seven words describing five different concepts such as Allah, Humans, chastisement, person and, the Judgment day [6]. Another unique style of the Quran is that one concept is mentioned using different words, depending on the context. For example, Muhammad (محمد) is the same as Ahmad (أحمد), and Mozzammil (مُزَّمَل). Additionally, a term may also refer to entirely different things, depending on the context: for example, L-jannat 'الجَنَّة' might refer to a heaven or a garden. Additionally, two different words may have the same letters but have different diacritics. For example, 'الجنة' represents two distinct words: 'الجَنَّة' means paradise, and 'الجِنَّة' means ghosts (see Table 1).

Table 1. Search results of الجنة based on concepts

Arabic pronunciation	English meaning	Arabic word	No. of verses
L-jannat	Paradise	الجَنة	109
aṣḥābu L-jannat	Companions of paradise	اصحاب الجنة	14
L-jannat	Garden	الجَنة	14
Jinnat	Ghost	الجِنّة	10
Junnat	cover	الجُنّة	2

Finally, the text of the Holy Quran is written in the classical Arabic language, which is slightly different from the modern Arabic language. This will cause a gap between the query and retrieved verses.

2.2   Quranic Search Applications

Desktop and Web applications have been developed to retrieve knowledge from Al Quran. The majority of these applications use keyword search techniques. However, some researchers have proposed frameworks for a Quranic semantic search tool based on concepts.

Khazain-ul-Hidayat [8], and Zakr [9] are free desktop applications that enable user to read, listen to and search the Quran in many different languages. These applications are mainly designed to be aid tools for teaching the Quran. A user can search Al Quran by querying a word or by entering a verse number. When the user queries a word, the results will include all verses containing any forms of this query word based on the stem of query words. For example, if the query word is 'ذكر', then the retrieved verses will contain other forms of 'ذكر', such as 'تذكرة', 'اذكر', 'تذكرة', 'الذاكرون', and 'ذكرى'.

Almonagib alqurany (المنقب القرآني) [10], Islam web [11], Tanzil [12], Quranic Arabic Corpus (QAC) [13], KSU Quran [15], The Quran [16] and the Noble Quran [14] are online Web applications that enable users to read, listen to and search Al Quran in different languages. Users can select a specific chapter, verse or word. In the case of searching by a word, these applications will return all verses that have words belonging to the same root of the query word.

Semantic Quran [17] is an online search tool application that allows a user to search verses based on concepts. In this application, each verse has a set of tags that are concepts. Additionally, not all verses are completely tagged. Therefore, the user can participate in tagging any verse. The idea behind this application is that many verses in the Holy Quran relate to certain concepts even though these verses do not have any common words.

2.3   Conducted Research on Quranic Search Tools

A lot of computational research has been carried out on Al Quran. The following review is about the IR and semantic search research.

[18] proposed a new Arabic question-answering system in the domain of Al Quran. The system prompts users to enter an Arabic question about Al Quran. Then, this system retrieves relevant Quranic verses with their Arabic descriptions from Ibn Kathir's book. This system uses 1,217 Quranic concepts integrated from the Quranic Arabic Corpus Ontology [13] and Quranic Topic Ontology [19]. It is claimed that retrieved results' accuracy can reach 65 % using the top result. This system has three phases for answering a question: question analysis using the 'Morphological Analysis and Disambiguation of Arabic' tool (MADA) [20], IR using 'explicit analysis approach' [21] and answer extraction. This proposed system does not recommend a solution for if the question terms do not match any concepts from the Quranic Ontology.

[22] suggested a Quranic semantic search tool by developing a simple domain of ontology for the animals mentioned in Al Quran. This paper concludes that the existing Arabic WordNet is not sufficient for finding synonyms for query words to increase

one's chances of retrieving information from a document, and based on this, it is suggested to develop Arabic WordNet for the Al Quran words.

[23] proposed a semantic search system for retrieving Quranic verses based on the enhanced ontology done by [13].

[24] recommended a semantic search for Al Quran based on CLIR. This research created a bilingual ontology: the English and Malay languages. This ontology is based on the ontology developed by [13]. They did this to experiment on this ontology for two translations of Al Quran. In the Malay translation, 5,999 verses are assigned to the concepts, and 237 verses do not relate to any concepts. In the English translation, they found 5,695 verses related to concepts in this ontology. On the other hand, 541 documents are not allocated to any concepts.

[25] has developed a tool called "Qurany" for searching the Quranic text in both Arabic and English. In this project, 6,236 HTML pages were created in which each HTML page contains one verse, its eight different English translations and the topic of this verse in both Arabic and English. The project's main idea is searching Al Quran's eight translations using the keyword search. This will enhance this tool's precision. Regarding [25], most of the available search tools on the Web use one English translation in the search process, with average recall and precision values of 54 % and 48 %, respectively.

In conclusion, all of the above research on Quranic semantic search are proposed frameworks for developing a semantic tool to search Al Quran. Moreover, no online tools for Quranic semantic search are currently up-to-date.

2.4 Research on Ontology of the Holy Quran

[26] developed a Semantic Quran dataset in an RDF format representing 42 different Al Quran translations. This dataset was built by merging data from two different semi-structured sources: the Tanzil project and the Quranic Arabic Corpus. This ontology has 7,718 links to DBpedia, 18,655 links to Wiktionary and 15,741,399 triples.

[22] developed an ontology for Al Quran in the scope of the animals found in Al Quran. The ontology provides 167 links to animals in Al Quran based on information found in "the Hewanat Al-Quran" book (حيوانات القران).

[27] rebuilt the existing ontology created by [15] using the Protégée tool and Manchester OWL. He increased the number of relationships from 350 to about 650 based on Al Quran, the Hadith and some online Islamic resource.

[25] has developed a tree of nearly 1,100 nodes representing Quranic abstract concepts. These concepts are linked to all verses of Al Quran. She used existing Quranic topics from the Islamic scholarly book called "Mushaf Al Tajweed" (مصحف التجويد). These concepts in the index have an aggregation relationship; the hierarchy of concepts is non-reflexive, non-symmetric and transitive.

[13] extracted 300 concepts and 350 relations from Al Quran using predicate logic. The relationship types connecting concepts are Part-of and IS-A. The ontology is based on a famous Al Quran discerption book called "Tafsir Ibn Kathir" (كثير ابن تفسير) [28].

[29] developed an ontology for Al Quran in the scope of pronoun antecedents. This ontology consists of 1,050 concepts and more than 2,700 relations. Additionally, the relationship types connecting concepts are has-antecedent, has-concept and has–a-segment. Additionally, he produced a dataset called "QurSim" containing 7,600 pairs of related verses that have similarity in the main topic. The scope of this dataset is the similarity of verses.

[30] unified Al Quran Arabic Corpus [13], Quran annotated with Pronominal Anaphora [28] and the Qurany project [25]. These datasets are merged in one XML file, and then the file is uploaded in the Sketch Engine tool as a unified Quranic corpus.

In conclusion, all developed Quranic ontologies have different scopes, such as animals, verses' similarity and Quranic topics. Additionally, these ontologies were built in different formats, such as XML, RDF and OWL. Moreover, these various datasets have some similarity in concepts. To overcome the limitations in existing AL-Quran Ontologies, we aligned the four main ontologies [13, 25, 26, 29]. Before the process of matching Ontologies, all ontologies is normalized to one file format. After unifying ontologies in one file format, the process of alignment follows the methodology of aligning ontology in [34, 35].

3   Arabic Quranic Semantic Search Tool

Figure 1 is a framework for a new semantic search tool called Arabic Quranic Semantic Search Tool based on ontology (AQSST). This search tool aims to employ both IR techniques and semantic search technologies. The design of this tool is constructed based on the theories in proposed previous research [23, 31–33]. AQSST is divided into six components: Quranic Ontology (QO), Quranic Database QDB, Natural Language Analyser (NLA), Semantic Search Model (SSM), Keyword Search Model (KSM) and Scoring and Ranking Model (SRM).

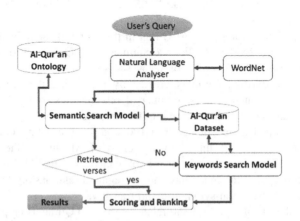

Fig. 1. Arabic Quranic semantic search tool structure (AQSST)

Quranic Ontology (QO) contains the new aligned Quranic ontology. QDB consists of Al Quran text in Arabic language and eight English translations of Al Quran, four different Tafsir (Description of AL-Quran), Al Quran words dictionary, Revelation reasons, concepts in Al Quran, and Named Entities (NE) based on Al Quran domain.

A user query undergoes different processes in NLA Model. This NLA parses a Natural Language query and then applies different NLP techniques on the tokenized query. These techniques are: spell correction, stop word removal, stemming and Part Of Speech (POS) tanging. After that, NLA uses Arabic WordNet to generate synonyms for the reformatted query words. Then, NLA adds semantic tags to these words using NE list, as shown in Table 2, and then sends the results to SSM.

Table 2. Example of analysing words in a query

Word	POS	Stem	Semantic tag	Synonym	Weight
الرحمن	NOUN	رحمن	Allah Name		2
الذئب	NOUN	ذئب	Animal	السرحان	2
يوسف	NOUN	يوسف	Prophet		2
البئر	NOUN	بئر	Well	الجُبّ	2
محمد	NOUN	محمد	Prophet	أحمد	2

SSM searches the Quranic ontology dataset by using SPARQL to find concepts related to the normalised query and then returns result to SRM. However, if no result is found KSM searches for verses contains words matching the analysed input words.

SRM filters the retrieved results from SSM and KSM; by eliminating the redundant verses (aya'at). Next, SRM ranks and scores the refined results based on the number of matching words in the results, the NE type of both the question and the answer, and the short distance between matched expressions in the retrieved results and question words. Finally, SRM provides the results to the user and then records the selected result. For instance, if a user search for l-jannat "الجنة" the retrieved result as shown in Fig. 2.

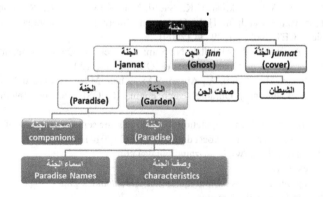

Fig. 2. Search result of (الجنة) in AQSST (dark colour is the most relevant result)

4 Conclusion

This paper summarises the search techniques used in existing search tools for AL Quran. Additionally, this paper studied the previous research have been conducted on Al Quran search methods and Al Ontologies. Depending on this study, many challenges have been found.

Firstly, Limitations of existing Al Quran search tools for retrieving all requested information. These search tools do not prompt users to search by concepts, phrases, sentences, questions or topics. Most search tools do not analyse the query texts by applying NLP techniques, such as parsing and spell check.

Secondly, absence of accurate and comprehensive resources for Islamic ontology. Existing Quranic datasets have different scopes and formats; and not follow ontology standards. Additionally, some Quranic ontologies are not available for usage.

The NER of the Arabic language is mostly focused on the modern Arabic language. Additionally, no well-formatted NER lists exist that are specialised for Quranic text, such as prophets' names, Allah's names, animals, times, religion, and etc. The Arabic language is an inflected language with complicated orthography.

Finally, all these limitations are highly considered in the design of the Arabic Semantic Quranic search tools.

References

1. Atwell, E., Brierley, C., Dukes, K., Sawalha, M., Sharaf, A.-B.: An artificial intelligence approach to Arabic and Islamic content on the internet. In: Proceedings of NITS 3rd National Information Technology Symposium (2011)
2. Sudeepthi, G., Anuradha, G., Babu, M.S.P.: A survey on semantic web search engine. Int. J. Comput. Sci. **9**(2), (2012)
3. Yauri, A.R., Kadir, R.A., Azman, A., Murad, M.A.A.: Quranic verse extraction base on concepts using OWL-DL ontology. Res. J. Appl. Sci. Eng. Technol. **6**(23), 4492–4498 (2013)
4. Shoaib, M., Yasin, M.N., Hikmat, U.K., Saeed, M.I., Khiyal, M.S.H.: Relational WordNet model for semantic search in Holy Quran. In: International Conference on Emerging Technologies, ICET 2009, pp. 29–34 (2009)
5. Yunus, M.A., Zainuddin, R., Abdullah, N.: Semantic query for Quran documents results. In: 2010 IEEE Conference on Open Systems (ICOS), pp. 1–5 (2010)
6. Raza, S.A., Rehan, M., Farooq, A., Ahsan, S.M., Khan, M.S.: An essential framework for concept based evolutionary Quranic search engine (CEQSE). Science International 26.1 (2014)
7. Alrehaili, S.M., Atwell, E.: Computational ontologies for semantic tagging of the Quran: a survey of past approaches. In: Proceedings of LREC (2014)
8. Khazain-ul-Hidayat. http://www.khazainulhidayat.com/
9. Zakr. http://zekr.org/
10. Almonagib alqurany. http://www.holyquran.net/search/sindex.php
11. Islam Web. http://quran.al-islam.com/
12. Zarrabi-Zadeh, H.: Tanzil (2007). http://tanzil.net/
13. Dukes, K.: Statistical Parsing by Machine Learning from a Classical Arabic Treebank. School of Computing, University of Leeds (2013)

14. Noble Quran. http://quran.com/
15. KSU Quran. http://quran.ksu.edu.sa/
16. The Quran. http://al-quran.info/
17. Semantic Quran. http://semquran.com/
18. Abdelnasser, H., Mohamed, R., Ragab, M., Mohamed, A., Farouk, B., El-Makky, N., Torki, M.: Al-Bayan: an Arabic question answering system for the Holy Quran. In: ANLP 2014, p. 57 (2014)
19. Abbas, N.H.: Quran 'search for a concept' tool and website. Diss. University of Leeds (School of Computing) (2009)
20. Habash, N., Rambow, O., Roth, R.: MADA+ TOKAN: a toolkit for Arabic tokenization, diacritization, morphological disambiguation, POS tagging, stemming and lemmatization. In: Proceedings of the 2nd International Conference on Arabic Language Resources and Tools (MEDAR), Cairo, Egypt, pp. 102–109 (2009)
21. Gabrilovich, E., Markovitch, S.: Computing semantic relatedness using Wikipedia-based explicit semantic analysis. IJCAI 7, 1606–1611 (2007)
22. Khan, H.U., Saqlain, S.M., Shoaib, M., Sher, M.: Ontology based semantic search in Holy Quran. Int. J. Future Comput. Commun. 2(6), 570–575 (2013)
23. Yauri, A.R.: Automated Semantic Query Formulation for Quranic Verse Retrieval. Putra Malaysia, Malaysia (2014)
24. Yahya, Z., Abdullah, M.T., Azman, A., Kadir, R.A.: Query translation using concepts similarity based on Quran ontology for cross-language information retrieval. J. Comput. Sci. 9(7), 889–897 (2013)
25. Abbas, N.: Quran 'search for a concept' tool and website. Unpublished dissertation, University of Leeds (2009)
26. Sherif, M.A., Ngonga Ngomo A.-C.: Semantic Quran a multilingual resource for natural-language processing. Semant. Web J. 6, 339–345 (2009)
27. Yauri, A.R., Kadir, R.A., Azman, A., Murad, M.A.A.: Ontology semantic approach to extraction of knowledge from Holy Quran. In: 2013 5th International Conference on Computer Science and Information Technology (CSIT), pp. 19–23 (2013)
28. Abdul-Rahman, M.S.: Tafsir Ibn Kathir Juz' 1 (Part 1): Al-Fatihah 1 to Al-Baqarah 141, 2nd edn. MSA Publication Limited, London (2009)
29. Sharaf, A.-B.M., Atwell, E.: QurAna: corpus of the Quran annotated with pronominal Anaphora. In: LREC, pp. 130–137 (2012)
30. Aldhubayi, L.: Unified Quranic Annotations and Ontologies. University of Leeds, Leeds, UK (2012)
31. Mony, M., Rao, J.M., Potey, M.M.: Semantic search based on ontology alignment for information retrieval. Int. J. Comput. Appl. 107(10), (2014)
32. Al-Yahya, M., Al-Khalifa, H., Bahanshal, A., Al-Odah, I., Al-Helwah, N.: An ontological model for representing semantic lexicons: an application on time nouns in the holy Quran. Arab. J. Sci. Eng. 35(2), 21 (2010)
33. Ou, S., Pekar, V., Orasan, C., Spurk, C., Negri, M.: Development and alignment of a domain-specific ontology for question answering. In: LREC (2008)
34. Doan, A., Madhavan, J., Dhamankar, R., Domingos, P., Halevy, A.: [GLUE] learning to match ontologies on the semantic web. VLDB J. Int. J. Very Large Data Bases 12(4), 303–319 (2003)
35. Taye, M., Alalwan, N.: Ontology alignment technique for improving semantic integration. In: Proceedings in the Fourth International Conference on Advances in Semantic Processing, Florence, Italy, pp. 13–18 (2010)

Author Index